Recent Advances in Aeroacoustics

Recent Advances in Aeroacoustics

Proceedings of an International Symposium held at Stanford University, August 22–26, 1983

Edited by
Anjaneyulu Krothapalli and Charles A. Smith

With Contributions by
W.K. Blake S.M. Candel D.G. Crighton
F. Farassat J.E. Ffowcs Williams A.R. George
M.S. Howe A.H. Nayfeh A. Powell N. Rott
F.H. Schmitz C.K.W. Tam Y.H. Yu

With 182 Illustrations

Springer-Verlag
New York Berlin Heidelberg Tokyo

Anjaneyulu Krothapalli
Department of Mechanical Engineering
FAMU/FSU College of Engineering
The Florida State University
Tallahassee, FL 32306
U.S.A.

Charles A. Smith
NASA Ames Research Center
Moffett Field, CA 94035
U.S.A.

Library of Congress Cataloging in Publication Data
Main entry under title:
Recent advances in aeroacoustics.
Proceedings of the International Symposium on Recent Advances in Aerodynamics and Acoustics, organized by the Joint Institute for Aeronautics and Acoustics.
1. Aerodynamic noise—Congresses. I. Krothapalli, A. (Anjaneyulu), 1950– . II. Smith, C. A. (Charles Arthur), 1945– . III. Agarwal, R. (Ramesh)
IV. Joint Institute for Aeronautics and Acoustics.
V. International Symposium on Recent Advances in Aerodynamics and Acoustics (1983 : Stanford University)
TL574.N6R43 1986 629.132′3 85-30256

Printed and bound by R.R. Donnelley & Sons, Harrisonburg, Virginia.
Printed in the United States of America.

9 8 7 6 5 4 3 2 1

ISBN 0-387-96088-0 Springer-Verlag New York Berlin Heidelberg Tokyo
ISBN 3-540-96088-0 Springer-Verlag Berlin Heidelberg New York Tokyo

Preface

The Joint Institute for Aeronautics and Acoustics at Stanford University was established in October 1973 to provide an academic environment for long-term cooperative research between Stanford and NASA Ames Research Center. Since its establishment, the Institute has conducted theoretical and experimental work in the areas of aerodynamics, acoustics, fluid mechanics, flight dynamics, guidance and control, and human factors. This research has involved Stanford faculty, research associates, graduate students, and many distinguished visitors in collaborative efforts with the research staff of NASA Ames Research Center.

The occasion of the Institute's tenth anniversary was used to reflect back on where that research has brought us, and to consider where our endeavors should be directed next. Thus, an International Symposium was held to review recent advances in the fields relevant to the activities of the Institute and to discuss the areas of research to be undertaken in the future. This anniversary was also chosen as an opportunity to honor one of the Institute's founders and its director, Professor Krishnamurty Karamcheti. It has been his creative inspiration that has provided the ideal research environment at the Joint Institute.

The International Symposium on Recent Advances in Aerodynamics and Acoustics was held at Stanford University, Stanford, California, U.S.A., August 22–26, 1983. Thirty-five distinguished scientists were invited to present a comprehensive review on the following subject areas: unsteady aerodynamics, jets and shear layers, V/STOL aircraft aerodynamics, rotor dynamics and aerodynamics, aeroacoustics, and rotor acoustics. The proceedings of the symposium are organized in two volumes. One volume contains the papers concerned with aerodynamics and the other volume those papers dealing with aeroacoustics and rotoracoustics.

Acknowledgments

During the preparation for the symposium a number of people, both at NASA Ames Research Center and at Stanford University, contributed in various ways. Here, we mention a few of those who represent the spirit and efforts of many. As honorary chairman of the Symposium, Mr. C. A. Syvertson served invaluably, not only in this capacity but also in several other essential roles before and during the symposium. Mr. C. T. Snyder and Mr. W. D. Deckert were instrumental in helping to organize the outstanding contribution and collaboration of NASA. Mr. David Hickey, through many discussions, provided suggestions and ideas which were incorporated in the running of the Symposium. As a co-chairman of the Symposium, Prof. Leonard Roberts coordinated many of the difficult organizing tasks.

We should like to note that the success of the technical sessions is a credit to the session chairmen, who effectively presided over the presentations and discussions. In this context we would like to extend our gratitude to Dr. Jack Nielsen, Prof. J. E. Ffowcs Williams, Prof. Brian Cantwell, Prof. Harold Levine, Prof. W. C. Reynolds, Dr. Gordon Banerian, Prof. Holt Ashley, and Dr. Craig Simcox.

The expert preparation of the proceedings was done by Ms. Doris Hsia and Ms. Dianne Budnik. It involved an immense effort on their part to get the proceedings in the present form. We give our sincere thanks and appreciation to both of them. We also wish to express our appreciation to Springer-Verlag for their understanding and patience.

An undertaking such as this would not be possible without the generous support of the NASA Ames Research Center. This support is gratefully acknowledged.

Finally, on a personal note we thank Prof. Krishnamurty Karamcheti. His continuous help, encouragement, and personal warmth has always been a source of stimulation for those of us associated with him. It is with a deep sense of appreciation that we dedicate these proceedings to him.

A. Krothapalli
C. A. Smith

Contents

List of Contributors

Blake, William
David Taylor Naval Ship
R&D Center
Bethesda, MD 20084

Candel, Sébastien, M.
Office National d'Études et de
Recherches Aérospatiales
BP72, 92322 Châtillon
France

Crighton, David
School of Mathematics
The Univ. of Leeds
Leeds, England LS2 9JT

Farassat, F.
NASA Langley Research Center
Hampton, VA 23665

Ffowcs Williams, J. E.
University Engineering Dept.
Trumpington St.
Cambridge, England CB2 1PZ

George, A. R.
Sibley School of Mechanical
and Aerospace Engineering
Upson Hall
Cornell University
Ithaca, NY 14853

Howe, M. S.
Faculty of Mathematical Studies
University of Southampton
Southampton, England S09 5NH

Nayfeh, Ali H.
Faculty of Engineering
Office of the Dean
Yarmouk University
Yarmouk, Jordan

Powell, Alan
David Taylor Naval Ship
R&D Center
Bethesda, MD 20084

Rott, Nicholas
Federal Institute of Technology
CH-8092 Zurich, Switzerland

Schmitz, F. H.
U. S. Army Research and
Technology Laboratories
NASA Ames Research Center
Mail Stop 215-1
Moffett Field, CA 94035

Tam, Christopher
Florida State Universityy
Tallahassee, FL 32306

Yu, Y. H.
U. S. Army Research and
Technology Laboratories
NASA Ames Research Center
Mail Stop 215-1
Moffett Field, CA 94035

Part I
Noise Due to Jets, Shear Layers, and Turbulence

Waves in Turbulent Mixing Layers

J. E. Ffowcs Williams
Cambridge University, Cambridge, England

Abstract

This paper discusses the reliability of theories which model waves on turbulent interfaces in terms of analogies. Those analogies often indicate the susceptibility of the wave generating elements of turbulence to external stimuli and imply some degree of control may be possible. But some instances in which the analogy is misleading have been suggested and as such they indicate that the theoretical modelling has lost touch with reality. Those cases are examined in depth with particular reference to wave generation on the turbulent interface of two fluids of different mean density. It is concluded that no inconsistency exists in that case and that there are no grounds for concern that the analogies are misleading.

1. Waves in Turbulent Mixing Layers

Turbulence induced vibration and noise is a process where random sources generate a relatively well ordered response. Vibrational modes are frequently weakly damped and have coherent length scales much greater than the size of the source eddies in the turbulence that excited them; large scale order in the response field seems to be created from the small scale chaos of turbulence. But definite models of the excitation process rarely maintain the view that response and wave scales are so disparate. Waves are coupled only to those source elements that precisely match their form; Fourier elements of the source and wave must have identical spatial and temporal structure. The useful fact that the bulk of the source is unlike the response field, and is therefore virtually independent of it, has been exploited in theories that treat the vibration as a passive by-product of turbulence. The response has, according to that view, a negligible influence on the dynamics of turbulence. Such theories have been very successful in establishing the scaling laws and determining the parameters most influencing turbulence induced vibration.

Those theories have been refined to model the facts that turbulence usually forms at unstable interfaces between fluids of different energy levels, and that a turbulent mixing layer separates the different states. The fact that the interface exists is probably significant to the wave field whose length scale is often large in comparison with the mixing layer thickness. Details of the layer are not in those cases as important as the fact that the interface separates material of different properties.

The success of those theories in explaining features of the response field that are influenced by the discontinuity in bulk properties, and the fact that Fourier elements of wave and source are matched, leads to the question of whether or not these theories can aid the understanding of turbulence itself—or at least those elements in the turbulence that are coupled to the response fields. That is the question underlying the material of this paper, which concentrates on areas in which the theoretical modelling appears to provide surprising results. The surprises are welcome if correct but probably spurious if definite examples where the theory misleads can be identified. What appeared to be one such example is treated in this paper, but the approach is then shown to be sufficiently robust to accommodate the issue.

The approach is based on Lighthill's analogy where the exact conservation equations are written in a form where *difficult* elements are gathered on one side of an equation and regarded as sources of waves that propagate according to the linear equation that forms the other side.

$$L\phi = S \tag{1.1}$$

Linear motions usually satisfy the homogeneous equation

$$L\phi = 0 \tag{1.2}$$

while all the non-linear elements are in S which is thought of as the wave source. Our interest lies in turbulence-generated waves, and the turbulence results from the fact that the linear disturbance field is unstable; the solutions of the homogeneous equation grow in either space or time or both without bound.

There are several different ways of expressing this kind of equation because in most models, the right hand side usually contains some linear term also. The left hand side is chosen so as to allow a tractable linear operator, leaving on the right hand side all other difficult-to-handle linear terms. The linear equation which applies over most of the field is usually selected for specifying L, leaving linear terms that persist over the remainder to be regarded as specified in S along with the genuine non-linear elements. Obviously the selection of L is a matter of individual choice and in that respect is non-unique and unphysical. This fact is embedded in the essential non-uniqueness of source fields. Though for a given source, the linear response might very well be unique, there is no unique prescription for the source of that response. What is or what is not a source is a matter of personal choice, and this is absolutely consistent with physics; the holographic reproduction of a wave field is indistinguishable from the real thing.

Our concern here is always with real systems governed exactly by the conservation equations incorporated into equation (1.1), so there is no doubt that if solutions of the homogeneous equation grow without bound, then the nonlinear terms must act to limit that growth. In that respect $L\phi = 0$ has no sensible global power of describing the field of turbulent flow, at least it cannot describe the field if subject to the usual constraints appropriate to a weak disturbance on a flow that is disturbed from an initially quiescent state. But that is often not our concern and constraints appropriate to an initially weak field are probably irrelevant to energetic motions. $L\phi = 0$ can have bounded solutions also and that property might give those solutions a greater relevance to continually driven finite fields. These theories are ones in which linear fields conforming to a chosen linear operator are superposed to mimick the complete flow. Individual elements of the linear fields do not necessarily have a realisable existence, but those that do have a more obvious relevance. Disturbances that grow quickly on the early laminar vortex layer emerging from a jet nozzle are more aptly modelled locally by growing solutions of the linear equation than by bounded solutions. On the other hand the large eddy fields of developed turbulence and their waves are better modelled by bounded solutions. The fact that they would pre-exist what for convenience is called their source is an added attraction because the turbulent *source* is actually the non-linear debris of the previously existing wave field.

These theories make a common prediction that those aspects of turbulence connected to the wavefield are inevitably sensitive to external influence. This prediction goes clear against the grain of the central foundation of the approach, that the turbulence is uninfluenced by the weak vibrational response it induces. At once this prediction is alarming and exciting, alarming because it might be an induction that the theoretical model has moved too far from the physics to provide a useful guide to reality, and exciting because it indicates that those (energetically weak) aspects of turbulence that are important to wave generation might be controllable by outside influence; turbulence might after all one day be silenced!

Suppose that in addition to the natural flow described by equation (1.1) there is an additional externally controlled source C, so that the complete field is described by

$$L\phi = S + C \tag{1.3}$$

S incorporates all the turbulent source elements, C is a supposedly independent control. Now select solutions of the elementary equation

$$LG = \delta \tag{1.4}$$

so that the Green function G conforms with the normal constraints on weak fields, i.e., waves are outgoing at infinity and do not anticipate their source. The operator L describes an unstable flow (otherwise there would be no turbulence and therefore no interest!) so that G must grow without bound.

The exact equation (1.3) has solution

$$\phi = S * G + C * G \tag{1.5}$$

and we know that ϕ is bounded. But $C * G$ is not bounded and from this it follows that the turbulent source S **cannot** be independent of the external C.

If on the other hand G is selected to be bounded, and this necessitates the selection of a non-causal Green function, the same kind of coupling of turbulence to external effects is evident. We know that the physical system described exactly by equation (1.5) cannot be one in which the field ϕ anticipates the action of the external source. But since $C * G$ does have non-causal elements, a contradiction (i.e., a mistake in what is after all an exact analysis) is avoided only if $S * G$ has an exactly cancelling non-causal element. Again S and C are essentially coupled—regardless of what the external control may be.

That the turbulence induced noise of jets is influenced by external sound fields has now become widely accepted (Bechert and

Pfizenmaier, 1975; Moore, 1977; Kibens, 1979) and it is perhaps only a question of time before that influence amounts to useful control. This exciting prospect is consistent with the deductions made through the kind of inhomogeneous *linear operator* equation described above and may be taken as evidence for the reasonableness of that modelling. Other large-scale aspects of turbulence may also be so modelled and subject to external control and the drag-producing large eddies in a turbulent boundary layer is one important candidate for immediate attention. But it may be that this kind of synthesis of a real field from *attractive* wavelets, exact though it may be, is in practice too precarious a process to yield reliable predictions. Slight errors can destroy the necessary cancelling of unreal effects that is built into the exact equation and if this were so these predictions would at best be unreliable.

Andrew Kempton has in a private communication put forward an exact calculation in which such a spurious conclusion seems apparent, and because of the importance such an example would have the rest of this paper is devoted to understanding his special case. We shall conclude that the theory is in fact robust enough to handle that case but in doing so we have to analyse a real problem in more penetrating depth than has hitherto been necessary. Kempton's example concerns sound generated by turbulent mixing flow at the interface of two fluids with different mean densities, a problem of the kind he had analysed in depth and described in his Ph.D. thesis (Kempton, 1977). We will start here by considering first some elementary aspects of sound generation by sources near density interfaces and then move onto the particular example that seemed to reveal an inconsistency.

2. Simple Sources in Density Inhomogeneous Fluids

This simple example concerns the ambiguities that can sometimes arise in descriptions of acoustic sources and particularly those that originate from inhomogenieties of density in the vicinity of the source region. Consider the source accompanying the fictitious creation of mass at a rate q per unit volume. At first sight the source is equal to $\frac{\partial q}{\partial t}$ but closer examination reveals that sound is generated only by the *volume* displaced by the newly created mass.

The linearised equation of *non-conservation*

$$\frac{\partial \rho}{\partial t} + \nabla \cdot \rho \mathbf{u} = q \tag{2.1}$$

combined with the linearised momentum equation in an inviscid fluid

$$\frac{\partial \rho \mathbf{u}}{\partial t} + \nabla p = 0 \tag{2.2}$$

combine to give

$$\frac{1}{c^2}\frac{\partial^2 p}{\partial t^2} - \nabla^2 p = \frac{\partial q}{\partial t} + \frac{1}{c^2}\frac{\partial^2}{\partial t^2}(p - c^2 \rho). \tag{2.3}$$

It is wrong to use the definition of the speed of sound, $c = (\frac{dp}{d\rho})^{1/2}$, assumed constant, to lead to the inhomogeneous wave equation in which the mass creation rate appears as the source of sound.

$$\frac{1}{c^2}\frac{\partial^2 p}{\partial t^2} - \nabla^2 p = \frac{\partial q}{\partial t} \tag{2.4}$$

i.e..

$$p = \frac{1}{4\pi}\frac{\partial}{\partial t}\int_v \left[\frac{q}{r}\right] dV$$

r is the distance between the source and observation points and the square brackets imply the time of reception retarded by the wave travel time r/c.

The failure of this argument is not in the neglect of the momentum inevitably injected into the moving fluid as mass is created but in the too casual use of the formula $\delta(p - c^2\rho) = 0$.

Sound production depends crucially on how the mass is created; whether it tends to be done at constant or variable density. When there is no mass creation, $q = 0$, then

$$\frac{\partial}{\partial t}(p - c^2\rho) = 0.$$

This balance might be disturbed by mass creation and so it might be reasonable to test its importance by making it proportional to the rate of mass creation. Let

$$\frac{\partial}{\partial t}(\rho - \frac{p}{c^2}) = \alpha q, \tag{2.5}$$

say, the value of α, which is probably confined to be within the range zero to unity, depending on the precise way in which mass is created. Then equation (2.3) becomes

$$\frac{1}{c^2}\frac{\partial^2 p}{\partial t^2} - \nabla^2 p = (1-\alpha)\frac{\partial q}{\partial t}$$

i.e.,

$$p = \frac{(1-\alpha)}{4\pi}\frac{\partial}{\partial t}\int_V \left[\frac{q}{r}\right] dv.$$

If $\alpha = 1$, the pressure is constant, the fluid motionless and $\frac{\partial \rho}{\partial t} = q$; these conditions are easily checked to be consistent with the defining equations (2.1), (2.2), and (2.5). That sort of mass creation is silent and might be thought of as brought about by the creation of extremely dense but small particles which are added to increase the mass density without significantly disturbing the fluid.

On the other hand if $\alpha = 0$, then equation (2.4) stands and sound is generated very effectively. In that case the newly created mass is fluid satisfying the same density/pressure relationship during its creation as does the ambient medium. The pressure, and hence density, change is in that event determined by the impedance offered by the exterior fluid which is very small for compact sources.

Where α is zero or unity, the pressure level within a very compact source is negligible, from which it follows that the real source strength

$$(1-\alpha)\frac{\partial q}{\partial t} = \frac{\partial}{\partial t}\nabla\cdot\rho\mathbf{u} + \frac{1}{c^2}\frac{\partial^2 p}{\partial t^2}$$

is effectively equal to $\frac{\partial}{\partial t}\nabla\cdot\rho\mathbf{u}$ distributed over the source volume, and the sound of a compact source is determined by how much volume is displaced during the mass creation phase

$$\begin{aligned} p &= \frac{1\partial}{4\pi\partial t}\int_{V(q\neq 0)}\left[\frac{\nabla\cdot\rho\mathbf{u}}{r}\right] dV \\ &= \frac{1\partial}{4\pi\partial t}\int_S\left[\frac{\rho_0 u_n}{r}\right] dV + \text{weak dipole term.} \end{aligned}$$

i.e.,

$$p = \frac{\rho_0}{4\pi}\frac{\partial}{\partial t}\left[\frac{Q}{r}\right]$$

Where S is the control surface surrounding the small volume V where mass is created and across which there is a volume outflow at a rate Q. It is this volume outflow that creates the sound, not the mass creation per se.

To be specific let q result from the creation of a *second immiscible phase* of fluid at mass density ρ_q occupying a (variable) fraction β of unit volume of two phase fluid. Let ρ_f be the density of the original fluid phase that occupies the fraction $(1-\beta)$ of unit volume. Then, by definition, $\delta(p - c^2\rho_f) = 0$ and since

$$q = \frac{\partial}{\partial t}(\beta\rho_q) \qquad \text{and } \rho = (1-\beta)\rho_f + \beta\rho_q$$

$$\delta(c^2\rho) = \delta p + \delta(c^2\beta(\rho_q - \rho_f))$$

and

$$\begin{aligned} \frac{1}{c^2}\frac{\partial^2}{\partial t^2}(p - c^2\rho) &= \frac{\partial^2}{\partial t^2}(\beta(\rho_f - \rho_q)) \\ &= \rho_0\frac{\partial^2\beta}{\partial t^2} - \frac{\partial q}{\partial t} \end{aligned} \tag{2.6}$$

ρ_0 being the mean value of the *fluid phase* density.

The linear source term in equation (2.3) is not simply $\frac{\partial q}{\partial t}$ because mass creation destroys the balance of p and $c^2\rho$, and the imbalance produces according to equation (2.6) the correct source which depends essentially on the rate of volume creation, β, not on the added mass.

$$\frac{1}{c^2}\frac{\partial^2 p}{\partial t^2} - \nabla^2 p = \rho_0\frac{\partial^2\beta}{\partial t^2}$$

$$p = \frac{\rho_0}{4\pi}\frac{\partial^2}{\partial t^2}\int_V [\beta]\frac{dV}{r}.$$

Similarly, if an external force of density f_i per unit volume were applied to linearly disturbed fluid, q in equation (2.1) would be zero but the left hand side of (2.2) would equal f_i and the linearised wave equation would then be

$$\frac{1}{c^2}\frac{\partial^2 p}{\partial t^2} - \nabla^2 p = \frac{1}{c^2}\frac{\partial^2}{\partial t^2}(p - c^2\rho) - \frac{\partial f_i}{\partial x_i}. \tag{2.7}$$

It would be wrong even though $\frac{D}{Dt}(p - c^2\rho) = 0$, to neglect $\frac{\partial^2}{\partial t^2}(p - c^2\rho)$ and write the solution as

$$p = -\frac{1}{4\pi}\frac{\partial}{\partial x_i}\int_V \left[\frac{f_i}{r}\right] dV$$

thereby giving the impression that the sound pressure is established regardless of the fluid density in the source region. In fact when second order quantities are neglected

$$\begin{aligned}\frac{\partial}{\partial t}(p - c^2\rho) &= \frac{D}{Dt}(p - c^2\rho) - u_i\frac{\partial}{\partial x_i}(p - c^2(\rho - \rho_0)) \\ &= c^2\frac{\partial}{\partial x_i}\{(\rho - \rho_0)u_i\} - c^2(\rho - \rho_0)\nabla\cdot\mathbf{u}\end{aligned} \tag{2.8}$$

and

$$\begin{aligned}\frac{\partial^2}{\partial t^2}(p - c^2\rho) &= c^2\frac{\partial}{\partial x_i}\left((\rho - \rho_0)\frac{\partial u_i}{\partial t}\right) - c^2(\rho - \rho_0)\frac{\partial}{\partial t}\nabla\cdot\mathbf{u} \\ &= c^2\frac{\partial}{\partial x_i}\left((\rho - \rho_0)\frac{\partial u_i}{\partial t}\right) + \frac{(\rho - \rho_0)}{\rho}\frac{\partial}{\partial t}\frac{Dp}{Dt} \\ &= c^2\frac{\partial}{\partial x_i}\left((\rho - \rho_0)\frac{\partial u_i}{\partial t}\right) + \frac{(\rho - \rho_0)}{\rho}\frac{\partial^2 p}{\partial t^2}.\end{aligned} \tag{2.9}$$

The second term is negligible for a compact linear source region so that equation (2.7) for that case is

$$\frac{1}{c^2}\frac{\partial^2 p}{\partial t^2} - \nabla^2 p = -\frac{\partial}{\partial x_i}\left(f_i - (\rho - \rho_0)\frac{\partial u_i}{\partial t}\right) \tag{2.10}$$

a form that makes it clear that when the mass density ρ in the source region is much higher than ρ_0 so that the total applied force balances the local momentum change $\int \rho \frac{\partial u_i}{\partial t}\, dV$ then the acoustic source is reduced in strength to $-\frac{\rho_0}{\rho}\frac{\partial f_i}{\partial x_i}$, or equivalently $-\rho_0 \frac{\partial}{\partial x_i}\{\frac{\partial u_i}{\partial t}\}$. The source of sound is then seen to be entirely due to kinematic and not to dynamic effects. The distributed source arising from the difference between p and $c^2\rho$ is not negligible whenever the mean mass density varies in or near the source region. Even in aerodynamically generated sound the linear source survives and in fact dominates the radiation field.

3. Sound Generated by Low Mach Number Turbulence in Inhomogeneous Flow

Consider now the particular case of a compact region of low Mach number turbulent flow in which the mean density of the fluid particles, $\overline{\rho}$, is inhomogeneous. Consider the speed of sound to be a constant c so that the pressure perturbation p is equal to the product $c^2(\rho - \overline{\rho})$, and the source terms of Lighthill's exact equation are then

$$\begin{aligned} \frac{1}{c^2}\frac{\partial^2 p}{\partial t^2} - \nabla^2 p &= \frac{\partial^2 \rho u_i u_j}{\partial x_i \partial x_j} + \frac{1}{c^2}\frac{\partial^2}{\partial t^2}\left(p - c^2(\rho - \rho_0)\right) \\ &= \frac{\partial^2 \rho u_i u_j}{\partial x_i \partial x_j} + \frac{\partial^2}{\partial t^2}(\rho_0 - \overline{\rho}). \end{aligned} \tag{3.1}$$

The mean density of a particle is of course constant so that

$$\begin{aligned} \frac{\partial(\overline{\rho} - \rho_0)}{\partial t} &= -u_i \frac{\partial(\overline{\rho} - \rho_0)}{\partial x_i} \\ &\doteq -\frac{\partial}{\partial x_i}((\overline{\rho} - \rho_0)u_i) \end{aligned} \tag{3.2}$$

at very low Mach number.

The second and dominant source term in (3.1) is therefore

$$\frac{\partial^2}{\partial t^2}(\rho_0 - \overline{\rho}) = \frac{\partial^2}{\partial x_i \partial t}((\overline{\rho} - \rho_0)u_i)$$

$$\frac{\partial^2}{\partial t^2}(\rho_0 - \rho) = \frac{\partial}{\partial x_i}\left((\overline{\rho} - \rho_0)\frac{Du_i}{Dt}\right) - \frac{\partial^2}{\partial x_i \partial x_j}(\overline{\rho} - \rho_0)\, u_i u_j \tag{3.3}$$

and so, in the low Mach number approximation,

$$\frac{1}{c^2}\frac{\partial^2 p}{\partial t^2} - \nabla^2 p = \rho_0 \frac{\partial^2}{\partial x_i \partial x_j} u_i u_j + \frac{\partial}{\partial x_i}\left((\bar{\rho} - \rho_0)\frac{Du_i}{Dt}\right). \tag{3.4}$$

It is clear from this that is is quite meaningless to enquire about the sound of the isolated Reynolds stress source term in (3.1) whenever variable mean density is involved. The sound of that term is in the *noise* of the calculation of the main sound source which is associated with the imbalance of pressure and density gradients. The Reynolds stress term cannot exist in isolation. In fact equation (3.3) displays explicitly the dependence of the dipole source strength on the difference between the force needed to accelerate unit volume of fluid whose density is $\bar{\rho}$ from that required to accelerate unit volume of ambient fluid. The Reynolds stress term, being quadrupole, is less efficient than this by a factor of the flow Mach number. It is probably not significant that the density in the Reynolds stress term in (3.1) is the local fluid density while that in equation (3.4) that displays the influence of density variations explicitly is ρ_0. Both terms are negligible in comparison with the leading dipole term at the low Mach number necessary for the straightforward interpretation of these fields by neglecting any subtle influence of retarded time variations.

Nonetheless, apparent inconsistencies can be revealed by the deliberate choice of an inappropriate form of the analogy and Kempton has found a particular case that emphasises these difficulties. Since the confidence one can have in the strength of the analogy in areas where it alone has been applied, and possibly predicted novel features, depends on experience in its application on more familiar ground, it is important to pinpoint the source of possible difficulty. For that reason we now discuss in detail the problem in which Kempton thought that the analogy would fail.

4. Sound Generation by Sources Near a Plane Layer of Density Gradient

The form of the acoustic analogy accounting specifically for the variations of bulk properties across a plane layer was derived by Ffowcs Williams (1974) and was later expanded and generalised by Dowling, Ffowcs Williams and Goldstein (1978). To illustrate the nature of the difficulty in interpreting this analogy consider the particular case of no mean motion and in which the speed of sound is a constant throughout all of space. The mean density varies however, across a turbulent interface, and the task is to determine the sound radiated from the turbulent sources in the vicinity of the interface, which is displaced a little from its undisturbed position at $x_3 = 0$ by the turbulence and acoustic motions. The interface remains close to its mean position but is diffused and distorted by the turbulence which is itself confined to the vicinity of the plane $x_3 = 0$; Figure 1 illustrates the geometry.

Ffowcs Williams (1974) shows that the sound can be described in one of two ways. The density interface is incorporated in an analogy with a plane sheet across which the mean density changes abruptly from ρ_0 to ρ_1. The interface of the analogy is placed in the linearly disturbed fluid, either above or below the real more diffuse interface, and the distant field at $(\mathbf{x}, t)$ is described exactly by the two alternate expressions depending on whether the interface is set at S_0 or S_1.

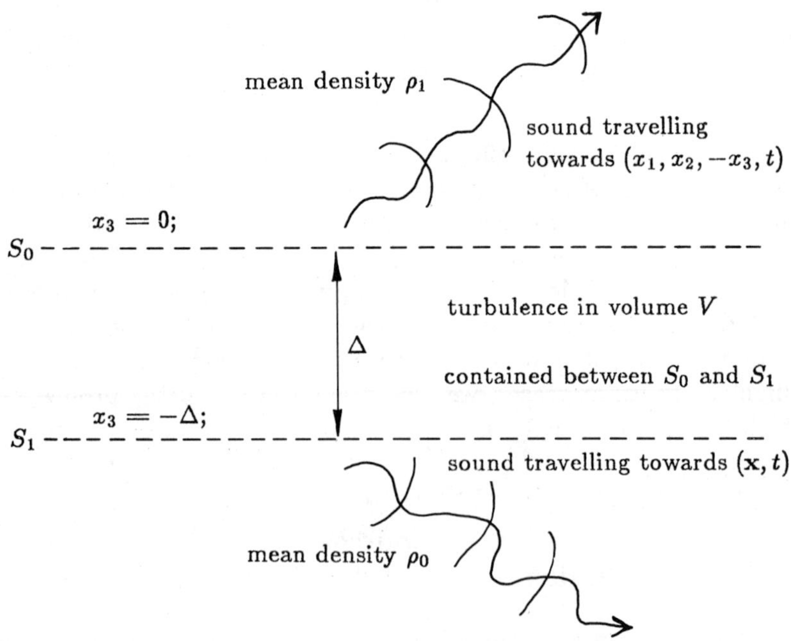

Figure 1. The geometry of the plane interface radiation problem.

$$(\rho - \rho_0)(\mathbf{x}, t) = \frac{1}{4\pi c^2 |\mathbf{x}|}(T_+^0 + R^0 T_-^0) \tag{4.1}$$

is equation (47) of reference 5, R^0 being the reflexion coefficient for plane waves leaving the interface in the direction of the distant observer at $\mathbf{x}$, but is in this particular case independent of that direction.

$$R^0 = \frac{\rho_1 - \rho_0}{\rho_1 + \rho_0} \tag{4.2}$$

$$T_+^0 = \frac{x_i x_j}{|\mathbf{x}|^2 c^2} \int_V \left[\rho \frac{D}{Dt}\left(\frac{1}{(1-M_r)} \frac{D}{Dt}\left(\frac{\rho u_i u_j + p_{ij} - c^2(\rho - \rho_0)\,\delta_{ij}}{\rho(1-M_r)}\right)\right)\right] d^3\mathbf{y} \tag{4.3}$$

where

$$M_r = \frac{u_r}{c} = \frac{x_i u_i}{|\mathbf{x}|c}.$$

T_-^0 is the image induced field at $(\mathbf{x}, t)$, i.e., it is the field that would be radiated to the point $(x_1, x_2, -x_3) = (\mathbf{x}', t)$ by the turbulent sources in V were they to exist in an unbounded homogeneous fluid at rest. The actual field radiated to the image point of $\mathbf{x}$ is, because of the continuity of pressure across the control surface S_0, equal to

$$(\rho - \rho_1)(\mathbf{x}, t) = \frac{1}{4\pi c^2 |\mathbf{x}|}(1 + R^0)\, T_-^0 \tag{4.4}$$

$$= \frac{1}{4\pi c^2 |\mathbf{x}'|}(T'_+ + R' T'_-) \tag{4.5}$$

where R' and T'_+ are the reflection coefficient and direct field for the alternative S_1 positioning of the control surface.

$$R' = \frac{\rho_0 - \rho_1}{\rho_0 + \rho_1} = -R^0 \tag{4.6}$$

$$T'_+ = \frac{x'_i x'_j}{|\mathbf{x}'|^2 c^2} \int_V \left[\rho \frac{D}{Dt}\left(\frac{1}{(1-M_r)} \frac{D}{Dt}\left(\frac{\rho u_i u_j + p_{ij} - c^2(\rho - \rho_1)\,\delta_{ij}}{\rho(1-M_r)}\right)\right)\right] d^3\mathbf{y} \tag{4.7}$$

The alternative expression to (4.1) for the field at $(\mathbf{x}, t)$ when the control interface is positioned at S_1 is

$$(\rho-\rho_0)(\mathbf{x},t)=\frac{1}{4\pi c^2|\mathbf{x}|}(1+R')\,T'. \tag{4.8}$$

When the turbulence is at low enough Mach number the dominant source term arises from the density element in the turbulence stress tensor; the Reynolds stress terms and the sources associated with pressure variations in compact sources are negligible by comparison. In that case

$$\begin{aligned} T_+^0 &\doteq \frac{x_i x_j}{|\mathbf{x}|^2 c^2}\int_V\left[\rho\frac{D}{Dt}\left(\frac{-c^2(\rho-\rho_0)\delta_{ij}}{\rho}\frac{DM_r}{Dt}\right)\right]d^3\mathbf{y} \\ &\doteq \frac{-\partial}{\partial t}\int_V\left[(\rho-\rho_0)\frac{DM_r}{Dt}\right]d^3\mathbf{y} \end{aligned} \tag{4.9}$$

i.e.,

$$\begin{aligned} T_+^0 &= \frac{-x_i}{c|\mathbf{x}|}\frac{\partial}{\partial t}\int_V\left[(\rho-\rho_0)\frac{Du_i}{Dt}\right]d^3\mathbf{y} \\ &= \frac{-x_\alpha}{c|\mathbf{x}|}\frac{\partial}{\partial t}\int_V\left[(\rho-\rho_0)\frac{Du_\alpha}{Dt}\right]d^3\mathbf{y}-\frac{x_3}{c|\mathbf{x}|}\frac{\partial}{\partial t}\int_V\left[(\rho-\rho_0)\frac{Du_3}{Dt}\right]d^3\mathbf{y} \end{aligned} \tag{4.10}$$

α being a summation variable over the indices (1, 2).

These two terms simplify further once it is recognized that the integral over the source volume of $\rho_0 u_3$ (and also $\rho_0 u_\alpha$, but for more subtle reasons) is negligible. This may be shown as follows.

Because $\frac{D\rho}{Dt}$ is negligible in the low Mach number compact source, it being of order $M^2\overline{\rho}$ times a typical source frequency and its effectiveness as a source a factor M^2 smaller than the retained terms, $\nabla\cdot\mathbf{u}$ is also negligible and

$$u_3=\frac{\partial}{\partial y_i}(y_3 u_i).$$

Therefore, with an error of order M times the retained term,

$$\frac{\partial}{\partial t}\int_V[u_3]\,d^3\mathbf{y}=\frac{\partial}{\partial t}\int_{S_0+S_1}[y_3 u_n]\,d^2\mathbf{y}.$$

On setting the surface S_0 at $y_3 = 0$ we find that

$$\begin{aligned} \frac{\rho_0}{c}\frac{\partial}{\partial t}\int_V \left[\frac{Du_3}{Dt}\right] d^3y &\doteq \frac{\rho_0}{c}\frac{\partial^2}{\partial t^2}\int_V [u_3]\, d^3\mathbf{y} \\ &= -\rho_0 \frac{\Delta}{c}\frac{\partial^2}{\partial t^2}\int_{S_1} [u_3]\, d^2\mathbf{y}. \end{aligned} \tag{4.11}$$

But Kirchhoff's theorem for a plane surface relates this integral directly to the distant sound field at $(\mathbf{x}, t)$.

$$\begin{aligned} \rho(\mathbf{x}, t) &= \frac{-\rho_0}{2\pi c^2|\mathbf{x}|}\frac{\partial}{\partial t}\int_{S_1} [u_3]\, d^2\mathbf{y} \\ &= \frac{-x_3}{2\pi c^3|\mathbf{x}|^2}\frac{\partial}{\partial t}\int_{S_1} [p]\, d^2\mathbf{y} \end{aligned} \tag{4.12}$$

so that equation (4.11) contains equalities between terms which are equal to

$$4\pi c^2|\mathbf{x}|\frac{2\Delta}{c}\frac{\partial\rho(x,t)}{\partial t}. \tag{4.13}$$

This is smaller than the other elements in T^0_+, which must amount collectively to a term of order $4\pi c^2|\mathbf{x}|\rho'(\mathbf{x}, t)$. or possibly $4\pi Rc^2|\mathbf{x}|\rho'(\mathbf{x}, t)$, by a factor equal to the compactness ratio based on the layer thickness Δ, $\frac{\Delta}{c}$ times the source frequency. This is negligible for compact low Mach number source layers.

The first term in equation (4.10) simplifies also because the integral of $\rho\frac{Du_\alpha}{Dt}$ is negligible over the planar source layer.

$$\begin{aligned} \int_V \left[\rho\frac{Du_\alpha}{Dt}\right] d^3\mathbf{y} &= -\int_V \left[\frac{\partial p}{\partial y_\alpha}\right] d^3\mathbf{y} \\ &\doteq \int_{-\Delta}^{0} [p]\begin{matrix} y_\alpha = +\infty \\ y_\alpha = -\infty \end{matrix}\, dy_3 \\ &\doteq 0; \end{aligned} \tag{4.14}$$

again this is because this term is smaller than the main field terms by the compactness ratio of the layer, so that equation (4.10) has the form:

$$T^0_+ = \frac{x_\alpha \rho_0}{c|\mathbf{x}|}\frac{\partial}{\partial t}\int_V \left[\frac{Du_\alpha}{Dt}\right] d^3\mathbf{y} - \frac{x_3}{c|\mathbf{x}|}\int_V \left[\rho\frac{Du_3}{Dt}\right] d^3\mathbf{y}. \tag{4.15}$$

T^0_- is the field that would be radiated by these sources to the image point of $(\mathbf{x}, t)$ in the surface S_0, were the fluid homogeneous in density i.e.,

$$T^0_- = +\frac{x_\alpha \rho_0}{c|\mathbf{x}|}\frac{\partial}{\partial t}\int_V \left[\frac{Du_\alpha}{Dt}\right] d^3\mathbf{y} + \frac{x_3}{c|\mathbf{x}|}\int_V \left[\rho\frac{Du_3}{Dt}\right] d^3\mathbf{y}. \tag{4.16}$$

Equations (4.1) and (4.2) then give the sound field at $(\mathbf{x}, t)$ as

$$\begin{aligned}(\rho - \rho_0)(\mathbf{x}, t) = {} & \frac{+1}{4\pi c^3}\frac{x_\alpha}{|\mathbf{x}|^2}\frac{2\rho_1\rho_0}{\rho_1+\rho_0}\frac{\partial}{\partial t}\int_V \left[\frac{Du_\alpha}{Dt}\right] d^3\mathbf{y} \\ & \frac{-1}{4\pi c^3}\frac{x_3}{|\mathbf{x}|^2}\frac{2\rho_0}{\rho_1+\rho_0}\frac{\partial}{\partial t}\int_V \left[\rho\frac{Du_3}{Dt}\right] d^3\mathbf{y}.\end{aligned} \tag{4.17}$$

This field is alternatively given by equation (4.8), in which, because T'_- is the field radiated by the sources to the point $(\mathbf{x}, t)$ as if the control surface was absent, T'_- takes the value of T^0_+ with ρ_1 replacing ρ_0, i.e., from equation (4.15)

$$T^0_- = \frac{+x_\alpha \rho_1}{c|\mathbf{x}|}\frac{\partial}{\partial t}\int_V \left[\frac{Du_\alpha}{Dt}\right] d^3\mathbf{y} - \frac{x_3}{c|\mathbf{x}|}\frac{\partial}{\partial t}\int_V \left[\rho\frac{Du_3}{Dt}\right] d^3\mathbf{y}. \tag{4.18}$$

and, because $(1 + R') = \frac{2\rho_0}{\rho_0+\rho_1}$, equation (4.8) becomes,

$$\begin{aligned}(\rho - \rho_0)(\mathbf{x}, t) = {} & \frac{1}{4\pi c^3}\frac{x_\alpha}{|\mathbf{x}|^2}\frac{2\rho_1\rho_0}{\rho_1+\rho_0}\frac{\partial}{\partial t}\int_V \left[\frac{Du_\alpha}{Dt}\right] d^3\mathbf{y} \\ & \frac{-1}{4\pi c^3}\frac{x_3}{|\mathbf{x}|^2}\frac{2\rho_0}{\rho_1+\rho_0}\frac{\partial}{\partial t}\int_V \left[\rho\frac{Du_3}{Dt}\right] d^3\mathbf{y}.\end{aligned} \tag{4.19}$$

in precise agreement with equation (4.17).

The second term in (4.19) is still linear even though its non-vanishing apparently depends on the variability of ρ because, by the

x_3 component of the momentum equation

$$\rho\frac{Du_3}{Dt} = -\frac{\partial p}{\partial y_3} \tag{4.20}$$

so that, with an error negligible for compact source layers, it equals

$$\frac{1}{4\pi c^3}\frac{x_3}{|\mathbf{x}|^2}\frac{2\rho_0}{\rho_1+\rho_0}\frac{\partial}{\partial t}\left(\int_{S_0}[p]\,d^2\mathbf{y} - \int_{S_1}[p]\,d^2\mathbf{y}\right). \tag{4.21}$$

The second of these terms, by Kirchhoff's theorem for a plane i.e., equation (4.12) equals $\frac{\rho_0}{\rho_0+\rho_1}(\rho-\rho_0)(\mathbf{x},t)$ while the first term, by the corresponding Kirchhoff expression for the field at $(\mathbf{x}',t)$ in terms of the pressure at S_0, equals $\frac{\rho_0}{\rho_0+\rho_1}(\rho-\rho_1)(\mathbf{x}',t)$.

Equations (4.17) and (4.19) are therefore equal to

$$(\rho-\rho_0)(\mathbf{x},t) = \frac{+1}{4\pi c^3}\frac{x_\alpha}{|\mathbf{x}|^2}\frac{2\rho_1\rho_0}{\rho_1+\rho_0}\frac{\partial}{\partial t}\int_V\left[\frac{Du_\alpha}{Dt}\right]d^3\mathbf{y}$$
$$+\frac{\rho_0}{\rho_0+\rho_1}((\rho-\rho_0)(\mathbf{x},t)+(\rho-\rho_1)(\mathbf{x}',t))$$

i.e.,

$$(\rho-\rho_0)(\mathbf{x},t) = \frac{x_\alpha}{4\pi c^3|\mathbf{x}|^2}2\rho_0\frac{\partial}{\partial t}\int_V\left[\frac{Du_\alpha}{Dt}\right]d^3\mathbf{y}+\frac{\rho_0}{\rho_1}(\rho-\rho_1)(\mathbf{x}',t) \tag{4.22}$$

Alternatively, by applying the same analogy to the field at the image point (x',t) equation (4.22) with ρ_0 and ρ_1 interchanged gives:

$$(\rho-\rho_1)(\mathbf{x}',t) = \frac{x_\alpha}{4\pi c^3|\mathbf{x}|^2}2\rho_1\frac{\partial}{\partial t}\int_V\left[\frac{Du_\alpha}{Dt}\right]d^3\mathbf{y}+\frac{\rho_1}{\rho_0}(\rho-\rho_0)(\mathbf{x}',t)$$

i.e.,

$$(\rho-\rho_0)(\mathbf{x},t) = \frac{-x_\alpha}{4\pi c^3|\mathbf{x}|^2}2\rho_0\frac{\partial}{\partial t}\int_V\left[\frac{Du_\alpha}{Dt}\right]d^3\mathbf{y}+\frac{\rho_0}{\rho_1}(\rho-\rho_1)(\mathbf{x}',t) \tag{4.23}$$

Together equations (4.22) and (4.23) then give, finally, a proof that $\int_V\left[\frac{Du_\alpha}{Dt}\right]d^3\mathbf{y}$ is negligible and that the leading term in the sound fields across the layer are:

$$(\rho - \rho_0)(\mathbf{x}, t) = \frac{\rho_0}{\rho_1}(\rho - \rho_1)(\mathbf{x}', t)$$
$$= \frac{-x_3}{4\pi c^3 |\mathbf{x}|^2} \frac{2\rho_0}{\rho_1 + \rho_0} \frac{\partial}{\partial t} \int_V \left[\rho \frac{Du_3}{Dt}\right] d^3\mathbf{y} \tag{4.24}$$

The sound due to these **nonlinear** density terms, non-linear in so much as they depend on the variability of ρ to prevent the vanishing of $\int_V \rho \frac{Du_3}{Dt}\, d^3\mathbf{y}$, is of the order M^{-1} bigger than that from the Reynolds stress source, and it is clear that the sound of such turbulence is radiated preferentially out into the heavier fluid in a direction normal to the mixing interface.

All that is perhaps not obvious but perfectly self consistent. The Kempton crisis comes when the two alternative expressions are used on a problem in which one is perfectly matched to the linear field it is mimicking and the other, though exactly equivalent, is not. That occurs where only **linear** motions are involved at the density interface and then the apparent source terms are much more subtle—they are not really sources at all but are some part of the various manifestations of the sound field.

5. Turbulence Near a Linearly Disturbed Plane Density Interface

Consider now the situation illustrated in Figure 2.

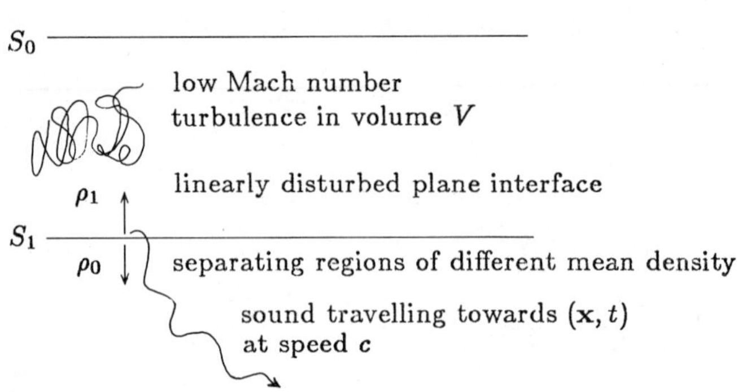

Figure 2. The geometry of the linear interface problem.

Equations (4.6), (4.7), and (4.8), where the control surface is naturally positioned give the field of this distribution of low Mach number turbulence as

$$(\rho - \rho_0)(\mathbf{x}, t) = \frac{1}{4\pi c^4} \frac{2\rho_0 \rho_1}{\rho_0 + \rho_1} \frac{x_i x_j}{|\mathbf{x}|^3} \frac{\partial^2}{\partial t^2} \int_V [u_i u_j] \, d^3\mathbf{y}, \tag{5.1}$$

whereas, the alternative if somewhat contrived positioning of the control surface at S_0 gives the integral equation for the field in the form determined by equations (4.1), (4.2), and (4.3).

$$(\rho - \rho_0)(\mathbf{x}, t) = \frac{1}{4\pi c^2 |\mathbf{x}|} \left(T_+^0 + \frac{\rho_1 - \rho_0}{\rho_1 + \rho_0} T_-^0 \right) \tag{5.2}$$

where

$$T_+^0 = \frac{x_i x_j}{|\mathbf{x}|^2 c^2} \int_V \left[\rho_1 \frac{D}{Dt} \left(\frac{1}{1 - M_r} \frac{D}{Dt} \left(u_i u_j - \frac{\delta_{ij} c^2 (\rho_1 - \rho_0)}{\rho (1 - M_r)} \right) \right) \right] d^3\mathbf{y} \tag{5.3}$$

V_1 being the volume between S_0 and S_1, and T_-^0 is simply T_+^0 with $\mathbf{x}' = (x_1, x_2, -x_3)$ replacing $\mathbf{x}$.

The sound at the image point of $\mathbf{x}$ in the surface S_0, i.e., at $x' = (x_1, x_2, -x_3)$ is

$$(\rho - \rho_0)(\mathbf{x}', t) = \frac{1}{4\pi c^2 |\mathbf{x}|} \frac{2\rho_1}{\rho_1 + \rho_0} T_-^0 \tag{5.4}$$

so that equation (5.2) is alternatively written as

$$(\rho - \rho_0)(\mathbf{x}, t) = \frac{T_+^0}{4\pi c^2 |\mathbf{x}|} + \frac{(\rho - \rho_1)}{2\rho_1}(x', t) \tag{5.5}$$

and retaining all terms possibly contributing to this low Mach number quadrupole radiation T_+^0 may be approximated as follows.

$$\frac{T_+^0}{4\pi c^2 |\mathbf{x}|} = \frac{x_i x_j}{4\pi c^4 |\mathbf{x}|^3} \rho_1 \frac{\partial^2}{\partial t^2} \int_V [u_i u_j] \, d^3\mathbf{y} + I \tag{5.6}$$

where

$$I = \frac{-(\rho_1 - \rho_0)}{4\pi^2 |\mathbf{x}|} \int_V \left[\rho \frac{D}{Dt} \left(\frac{1}{1 - M_r} \frac{D}{Dt} \left(\frac{1}{\rho (1 - M_r)} \right) \right) \right] d^3\mathbf{y}$$

which is not negligible. In fact

$$\begin{aligned}\frac{1}{1-M_r}\frac{D}{Dt}\left(\frac{1}{\rho(1-M_r)}\right) &= \frac{\frac{DM_r}{Dt}}{\rho(1-M_r)^3}+\frac{\nabla\cdot\mathbf{u}}{\rho(1-M_r)^2}\\ &\doteq \frac{1}{\rho}\frac{DM_r}{Dt}+\frac{3M_r}{\rho_1}\frac{DM_r}{Dt}+\frac{\nabla\cdot u}{\rho}\\ &= \frac{1}{\rho}\frac{\partial M_r}{\partial t}+\frac{1}{\rho_1}\frac{\partial}{\partial y_i}(u_iM_r)+\frac{3}{2}\frac{1}{\rho_1}\frac{\partial}{\partial t}M_r^2+\frac{\nabla\cdot u}{\rho}.\end{aligned} \tag{5.7}$$

Terms of order M_r^3, $(\rho_1-\rho)\nabla\cdot\mathbf{u}$ and $M_r\nabla\cdot\mathbf{u}$ are all smaller than the retained quadrupole terms and have been neglected in this approximation.

From these last two equations we can write I as

$$\begin{aligned}I &= \frac{-(\rho_1-\rho_0)}{4\pi c^2|\mathbf{x}|}\int_V\left[\frac{D}{Dt}\frac{\partial M_r}{\partial t}+\frac{\partial^2}{\partial t\partial y_i}(u_iM_r)+\frac{3}{2}\frac{\partial^2}{\partial t^2}M_r^2+\frac{\partial}{\partial t}\nabla\cdot\mathbf{u}\right]d^3\mathbf{y}\\ &= \frac{-(\rho_1-\rho_0)}{4\pi c^2|\mathbf{x}|}\int_V\left[\frac{\partial^2 M_r}{\partial t^2}+\frac{\partial}{\partial y_i}\left(u_i\frac{\partial M_r}{\partial t}\right)\right.\\ &\quad\left.+\frac{\partial^2}{\partial t\partial y_i}(u_iM_r)+\frac{3}{2}\frac{\partial^2 M_r^2}{\partial t^2}+\frac{\partial}{\partial t}\nabla\cdot\mathbf{u}\right]d^3\mathbf{y}\end{aligned} \tag{5.8}$$

Because conditions are linear on the two surfaces S_0 and S_1 that bound V, we can write

$$\int_V\left[\frac{\partial}{\partial y_i}u_iM_r\right]d^3\mathbf{y} \quad \text{as} \quad -\int_V\left[\frac{\partial}{\partial t}M_r^2\right]d^3\mathbf{y} \tag{5.9}$$

and this step actually eliminates all the non-linear terms in equation (5.8) because they then sum to zero leaving I as

$$\begin{aligned}I &= \frac{-(\rho_1-\rho_0)}{4\pi c^2|\mathbf{x}|}\frac{\partial}{\partial t}\int_V\left[\frac{\partial M_r}{\partial t}+\nabla\cdot\mathbf{u}\right]d^3\mathbf{y}\\ &\equiv \frac{-(\rho_1-\rho_0)}{4\pi c^2|\mathbf{x}|}\frac{\partial}{\partial t}\int_V\frac{\partial}{\partial y_i}[u_i]\,d^3\mathbf{y}\end{aligned} \tag{5.10}$$

i.e.,

$$I = \frac{-(\rho_1-\rho_0)}{4\pi c^2|\mathbf{x}|}\frac{\partial}{\partial t}\int_{S_0+S_1}[u_n]\,d^2\mathbf{y} \tag{5.11}$$

which, by Kirchhoff's theorem for a plane, equation (4.12), is in turn

$$I = \frac{-(\rho_1 - \rho_0)}{2}\left(\frac{\rho - \rho_1}{\rho_1}(\mathbf{x}', t) + \frac{\rho - \rho_0}{\rho_0}(\mathbf{x}, t)\right). \tag{5.12}$$

This result, inserted into equations (5.5) and (5.6) then give finally:

$$(\rho - \rho_0)(\mathbf{x}, t) = \frac{x_i x_j}{4\pi c^4 |\mathbf{x}|^3}\rho_1 \frac{\partial^2}{\partial t^2}\int_V [u_i u_j]\, d^3\mathbf{y} - \frac{\rho_1 - \rho_0}{2} \times \left(\frac{\rho - \rho_1}{\rho_1}(\mathbf{x}', t) + \frac{\rho - \rho_0}{\rho_0}(\mathbf{x}, t)\right) + \frac{(\rho - \rho_1)}{2\rho_1}(\mathbf{x}', t) \tag{5.13}$$

i.e.,

$$(\rho - \rho_0)(\mathbf{x}, t) = \frac{1}{4\pi c^4}\frac{2\rho_0\rho_1}{\rho_0 + \rho_1}\frac{x_i x_i}{|\mathbf{x}|^3}\frac{\partial^2}{\partial t^2}\int_V [u_i u_j]\, d^3\mathbf{y} \tag{5.14}$$

in agreement with equation (5.1), the direct form given by the more obvious positioning of the control surface at S_1. Obviously equation (5.1) which gives the result directly is more useful in this case—and is directly available. Only when there is doubt about the most relevant positioning of the surface is the matter apparently difficult to interpret, but then as we have already shown the variable density contrast scatters the turbulence with dipole efficiency and the field is given by the same formulae for both positionings of the control surface. It is however usually wrong to assume that the *source* structure is *known* to be a particular element of the Reynolds stress quadrupole, and to ignore the value of I in equation (5.6).

From all this we can conclude that the analogy is well posed but that care is needed in showing that the terms to be discarded are in fact negligible. The *source* is not always what it appears to be, and the most difficult terms to estimate casually are the linear elements which represent radiation rather than source terms whenever the interface of discontinuity of the analogy and reality fail to coincide.

Acknowledgements

The author is grateful to Andrew Kempton for having pointed out the care needed to properly interpret radiation problems near a linearly disturbed interface and for agreeing that that problem could be referenced prior to its publication.

References

[1] Bechert, D. and Pfizenmaier, E. "On the Amplification of Broadband Jet Noise by a Pure Tone Excitation," *J. Sound Vib. (43),* 1975, 581–587.

[2] Moore, C. J. "The Role of Shear Layer Instability Waves in Jet Exhaust Noise," *J. Fluid Mech. 80* (1977), 321–367.

[3] Kibens, V. *AIAA Paper 79-0592,* 1979.

[4] Kempton, A. J. "Efficient Sources of Aerodynamic Sound," *Ph.D. thesis,* Cambridge University, 1977.

[5] Ffowcs Williams, J. E. "Sound Production at the Edge of a Steady Flow," *J. Fluid Mech. 66* (1974), 791–816.

[6] Dowling, A. P., Ffowcs Williams, J. E., and Goldstein, M. E. "Sound Production in a Moving Stream," *Phil. Trans. Roy. Soc. A288, No. 1353* (1978), 321–349.

On Broadband Shock Associated Noise of Supersonic Jets

Christopher K. W. Tam
Department of Mathematics
Florida State University
Tallahassee, Florida 32306

1. Introduction

The noise of supersonic jets issued from choked or convergent-divergent nozzles is a subject of considerable current interest. When these jets are operated at off-design pressure ratios, measured noise spectra invariably show that they are made up of discrete and broadband components. The discrete tones or screech components are believed to be generated by an acoustic feedback phenomenon. One notable feature of these screech components is that the frequencies of the tones are the same in all directions. The broadband noise components have dominant frequencies which are usually higher than that of the fundamental feedback tone and range over several octave bands. At the present time, based on experimental and theoretical grounds, the broadband noise is generally further subdivided into two components each possessing distinctly different characteristics. These components are referred to as the turbulent mixing noise and the broadband shock associated noise or simply shock associated noise. The general consensus appears to be that the former is generated directly by the turbulence of the jet flow while the latter is produced by the unsteady interaction between the flow turbulence and the quasi-periodic shock cell structure of the jet. The primary objective of this paper is quite restricted. Our aim is to review briefly the observed characteristics of the broadband shock associated noise component, and its generation mechanism as proposed by various investigators.

The possibility that acoustic waves could be generated by shock-turbulence interaction in imperfectly expanded supersonic jets was recognized many years ago (see e.g., Lighthill (1953), Ribner (1955)). In these early works the turbulence was pictured as a random array of small eddies resembling what is now referred to as fine scale turbulence. To reduce the problem to a manageable form the quasi-periodic shock cell structure was all but discarded. It was usually replaced by a single shock wave of very large (infinite) spatial extent in the analysis. Basically the noise generation process as envisaged by these early investigators consists essentially of the random scattering of blobs of fine scale turbulence by plane shock waves. It should also be pointed out that no effort was ever made to correlate the predictions of this simple random scattering model to the measured spectral, directional and intensity characteristics of the broadband noise of supersonic jets. Thus these early theoretical results were largely academic and conceptual in nature. It turns out the observed shock associated noise are not generated by the random incoherent interaction between fine scale turbulence and the individual plane shock waves of a supersonic jet. Instead, the source of this noise component is spatially coherent over an extended length of the jet. Furthermore, the quasi-periodicity of the shock cell structure also plays a crucial role in the noise generation processes.

Current interest in broadband shock associated noise appears to begin with the work of Harper-Bourne and Fisher (1973) who carried out extensive experimental measurements of the noise of choked jets. By analyzing these data carefully they were able to identify a broadband noise component having spectral and directional characteristics which differ completely from those of the turbulent mixing noise. To find out how this noise component was generated they adopted a simple model comprising of an array of phased simple sources spaced at regular intervals apart. The distance between the point sources is equal to that of the shock cell spacing of the supersonic jet. By means of this model they were able to show that this new component of broadband noise was in all likelihood generated by the interaction between turbulence that are spatially quite coherent and the quasi-periodic shock cell structure of the jet. The crucial discovery that it requires a noise source which is spatial coherent and quasi-periodic over an extended length of the jet to account for the observed far field noise characteristics set their work completely apart from all the previous works. Since the pioneer study of Harper-Bourne and Fisher a good number of experimental and theoretical investigations on this subject have been carried out (e.g., Tanna (1977), Howe and Ffowcs Williams (1978), Seiner and Norum (1979, 1980), Pao and Seiner (1983), Seiner and Yu (1981), Tanna et al. (1981), Norum and Seiner (1982a, b), and Tam and Tanna (1982)). They provide the basis of our present day view of how broadband shock associated noise is generated. In this review, attention will be confined exclusively to the findings and theories of these more recent studies.

2. Characteristics of Broadband Shock Associated Noise

Broadband shock associated noise possesses many characteristics which are distinctly different from those of turbulent mixing noise. For instance, turbulent mixing noise is most intense in the downstream direction and peaks around $\theta = 30°$ measured from the jet axis. Broadband shock associated noise, on the other hand, is most prominent in the forward arc. Figure 1a shows the directivity of overall noise intensity for a fully expanded jet at a Mach number of 1.372 and a temperature ratio (jet to ambient temperature) of 0.73 measured by Tanna (1977). The jet is shock-free so that this is the directivity of turbulent mixing noise. Plotted in this figure also is the directivity of overall noise intensity of an underexpanded supersonic jet with the same fully expanded Mach number and jet operating conditions. The noise from this underexpanded jet contains both turbulent mixing noise as well as shock associated noise. The difference of these two directivities, therefore, would provide a good estimate of the distribution of shock associated noise. It is clear from this figure that shock associated noise is important mainly in the forward direction of the jet. Figure 1b shows a similar directivity plot for a hot jet with a temperature ratio of 2.27. Although quantitatively shock associated noise appears to be relatively less intense for hot jets yet it still predominates in the forward arc. Another way of seeing that the importance of shock associated noise is confined primarily to the forward directions of the jet is to compare the overall noise intensities between perfectly expanded and imperfectly expanded supersonic jets at different observation angles as the velocity of the jet increases from subsonic to supersonic. Data of this kind were measured by Tanna (1977) and are shown in Figure 2. Figure 2a shows that in the downstream direction the noise from a perfectly expanded and an underexpanded jet are nearly identical. In other words, shock associated noise makes no significant contribution to the overall noise intensity in this downstream direction. On the other hand, Figures 2b and 2c indicate clearly that at 90° and in the forward arc the underexpanded jet is noisier. Broadband shock associated noise is responsible for the difference in radiated sound pressure levels. The fact that shock associated noise can be easily identified in the forward arc does not mean that there is no shock

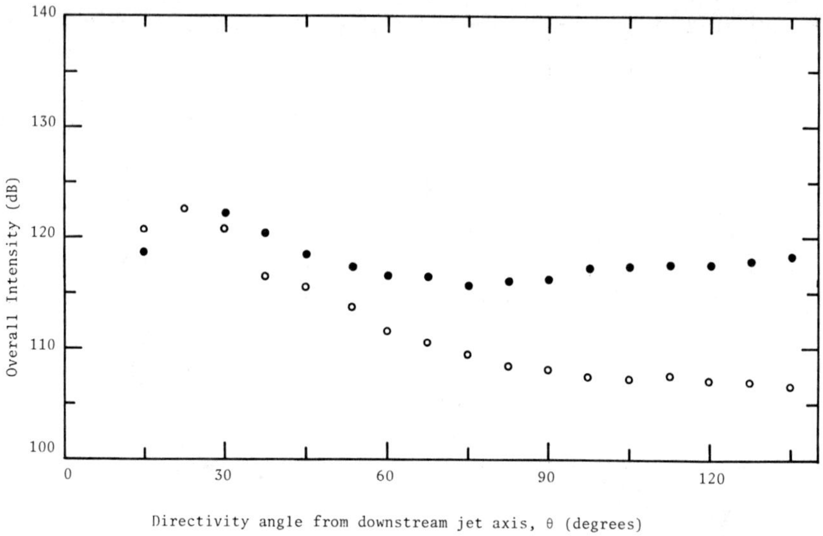

Figure 1a. Directivity of overall intensity, $M_j = 1.372$ (data from Tanna, 1977), $T_j/T_0 = 0.73$ (cold jet); o: fully expanded; •: under-expanded (convergent nozzle).

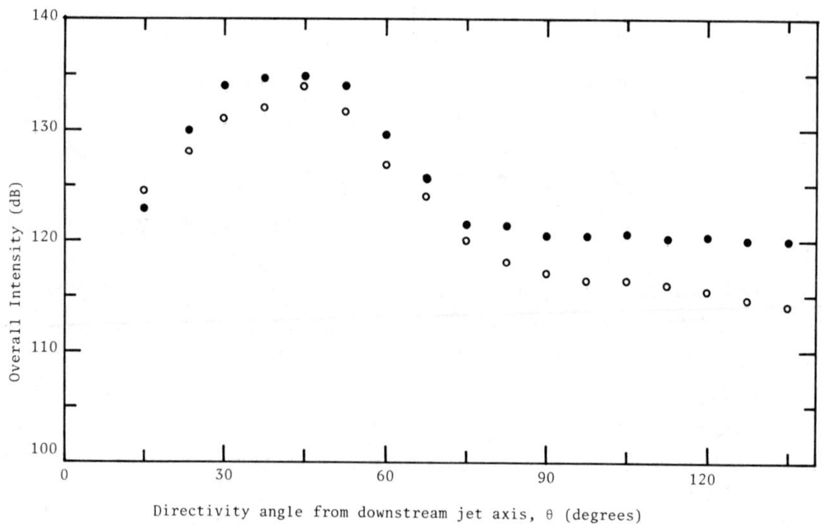

Figure 1b. Directivity of overall intensity, $M_j = 1.372$ (data from Tanna, 1977), $T_j/T_0 = 2.27$ (hot jet); o: fully expanded; •: under-expanded.

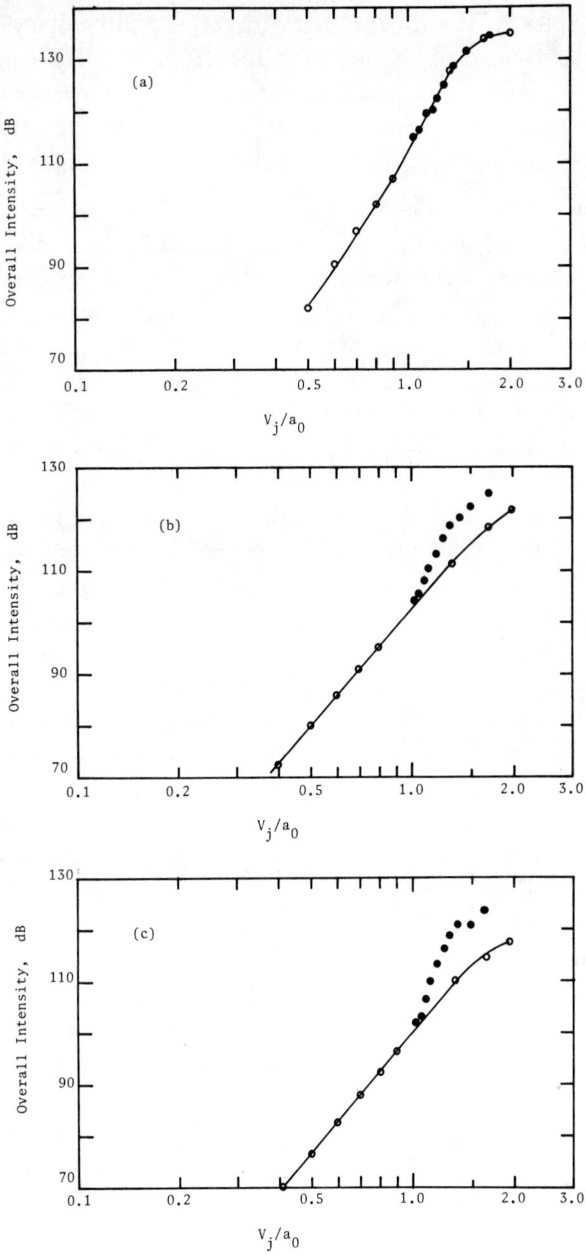

Figure 2. Overall intensity of shock associated noise (data from Tanna, 1977), (a) $T_j/T_0 = 1.0$, $\theta = 30°$; o: fully expanded; •: under-expanded; (b) $\theta = 90°$; (c) $\theta = 135°$.

associated noise in the downstream direction. Indeed, by analyzing the noise data carefully it is not difficult to identify the presence of this noise component in all directions. It is, however, a weaker contributor to the overall noise in the downstream direction relative to the turbulent mixing noise and hence is not as important.

For a given far field direction broadband shock associated noise are always observed to be concentrated in a relatively narrow frequency band as compared to say turbulent mixing noise. The noise spectrum consists essentially of a dominant characteristic peak. Figure 3 shows the noise spectra of three supersonic jets all operated at a fully expanded jet Mach number of 1.5. These spectra were measured at $\theta = 135°$ by Norum and Seiner (1982). The bottom curve is that of a perfectly expanded jet and hence represents the spectrum of turbulent mixing noise. It has a broad and smooth peak typical of noise generated by broadband turbulence as in the case of subsonic jets. The upper and the middle curves are the noise spectra of an underexpanded jet (using a convergent nozzle) and an overexpanded jet (using a convergent-divergent nozzle of design Mach number 2.0). As can be seen these spectra are dominated by a relatively narrow peak. The noise that form these peaks are not turbulent mixing noise since the peak frequencies are not the same as that of the bottom curve. Instead they are the broadband shock associated noise. The shape of the spectra shown here with a steep rise from the low frequency side to the peak followed by a gentle decrease on the high frequency side is typical of all shock associated noise spectra. Because of the range of frequency to be covered in a typical noise spectrum is fairly large, most of the measured spectra published in the literature are in the form of 1/3 octave intensity. These spectra give more weights to the high frequency components. They are, therefore, slightly distorted. However, qualitatively the above mentioned characteristics remain essentially true even if narrow band spectra are used. Figure 4 is a series of narrow band spectra at different θ of an overexpanded supersonic jet issued from a convergent-divergent nozzle of design Mach number 2.0 measured by Norum and Seiner (1982b). The operating fully expanded Mach number of the jet is 1.414. These narrow band spectra again confirm the observation that the broadband shock associated noise is made up of a dominant but slightly unsymmetrical peak with a relatively narrow half-width. In addition, it is seen from this figure that the peak frequency is a function of the observation angle. The peak frequency decreases as the angle of observation increases towards the

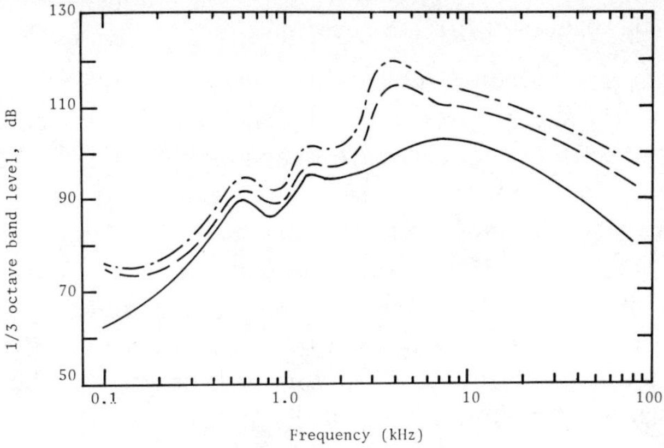

Figure 3. Noise spectra measured at $\theta = 135°$, $M_j = 1.5$ (data from Norum and Seiner, 1982b); —: perfectly expanded ($M_d = 1.5$); - - -: overexpanded ($M_d = 2.0$); – - – - – -: underexpanded ($M_d = 1.0$); (The nozzle exit diameters of the three nozzles are nearly the same.)

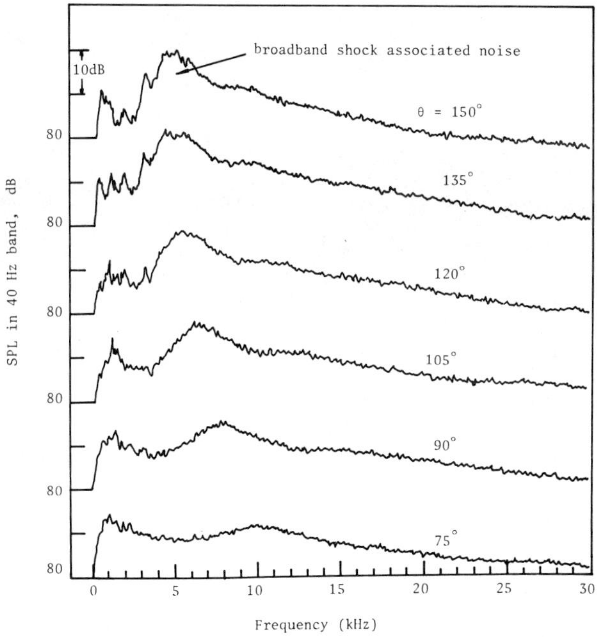

Figure 4. Narrow band noise spectra (data from Norum and Seiner, 1982b), $M_j = 1.414$, $M_d = 2.0$.

forward direction. This important characteristic of shock associated noise will be discussed further later on.

When a supersonic jet from a convergent-divergent nozzle is operated at the design pressure ratio the jet is shock free and hence the radiated sound is entirely turbulent mixing noise. If the pressure ratio of the jet is changed a quasi-periodic shock cell structure would form in the jet. The shock cell structure usually extends almost over the entire length of the supersonic part of the jet column. This is so regardless of whether the jet is overexpanded or underexpanded. In both cases shock associated noise would be produced thus increasing the noise intensity radiated to the forward directions. The intensity of shock associated noise is naturally a function of the operating fully expanded jet Mach number, M_j. The dependence of shock associated noise intensity on M_j is illustrated in Figure 5.

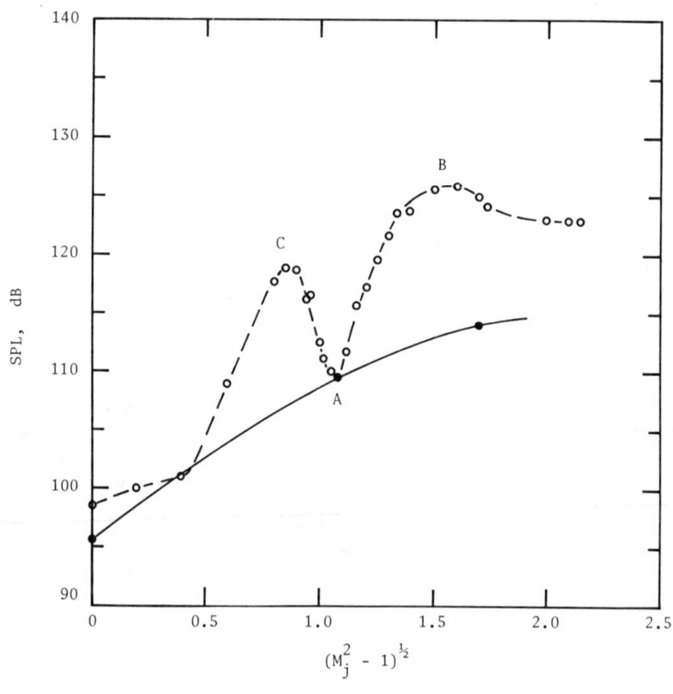

Figure 5. Overall noise intensity at 150° to the jet axis, $M_d = 1.5$ (data from Seiner and Yu, 1981); •: perfectly expanded jet. Copyright American Institute of Aeronautics and Astronautics, and reprinted with permission of the AIAA.

This figure shows the measured noise intensity at 150° to the jet axis as a function of jet Mach number obtained by Seiner and Yu (1981) using a convergent-divergent nozzle with design Mach number 1.5. At the design Mach number, there being no shock associated noise, the curve attains a local minimum. This is denoted in the figure by point A. With decreasing Mach number the total noise intensity increases following the curve AC. This is so even though the intensity of the turbulent mixing noise, given by the curve of the solid symbols, decreases. The reason for the increase in total noise intensity is that as the operating Mach number deviates more and more from the design Mach number the shock cell strength increases very rapidly giving rise to intense shock associated noise. At point C the mismatch between the pressure at the jet nozzle exit and the ambient pressure is so large that the oblique shocks of the jet can no longer form a quasi-periodic shock cell structure. A Mach disc would form near the end of the first shock cell. This in turn reduces the shock cell strength farther downstream. As a result the intensity of shock associated noise is reduced. If the jet is operated in the under-expanded mode the overall noise intensity follows the curve AB as the operating fully expanded Mach number increases. The point B at which the curve reaches the peak value corresponds approximately again to the condition for the appearance of a Mach disc in the jet. Beyond this point the shock structure is highly complicated with mixed subsonic and supersonic flows. Qualitatively Figure 5 is typical of all overall noise intensity curves for observation angle θ in the forward direction. The dip in the noise intensity curve at the nozzle design point A is, of course, a strong function of θ. This is illustrated in Figure 6. The data are taken from the measurements of Tam and Tanna (1982). The design Mach number of the convergent-divergent nozzle used in this series of experiments is 1.67. Figure 6c indicates that the characteristic dip at the nozzle design Mach number can be seen even at rear arc angles as small as $\theta = 60°$.

3. Sources of Broadband Shock Associated Noise

The work of Harper-Bourne and Fisher (1973) is undoubtedly one of the most important contributions to the theory of broadband shock associated noise in recent years. Not only did they identify the shock associated noise component experimentally but also they

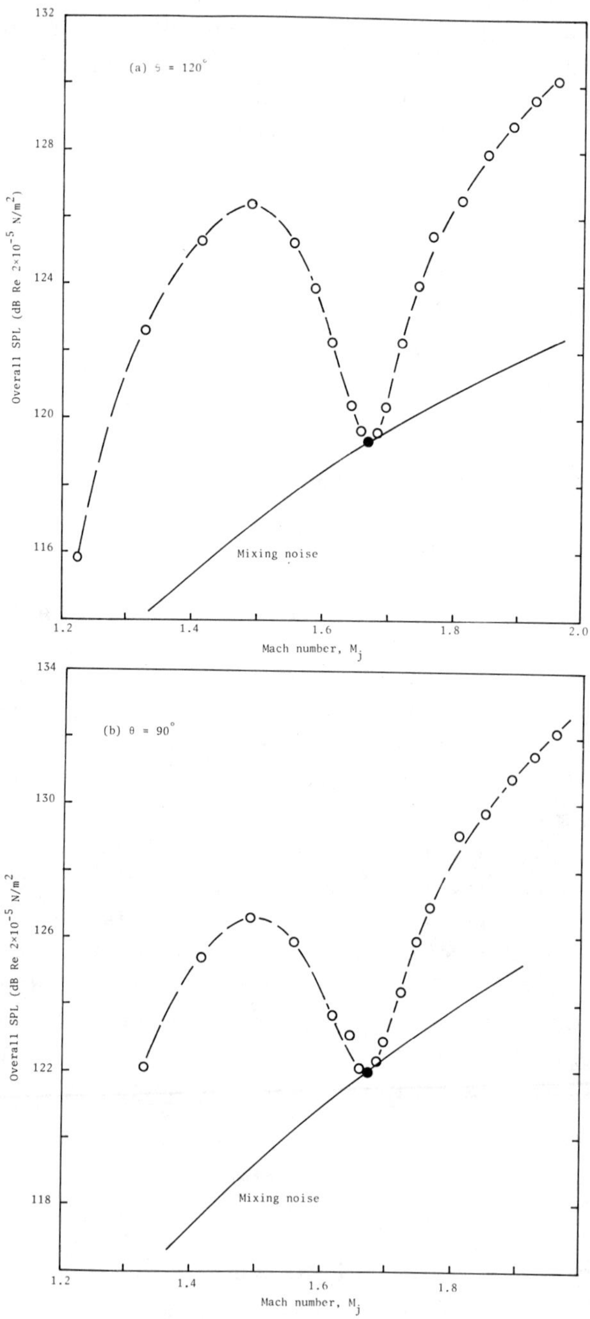

Figure 6. Shock associated and turbulent mixing noise components for a convergent-divergent nozzle operated with $T_j/T_0 = 1.0$. (a) $\theta = 120°$; (b) $\theta = 90°$.

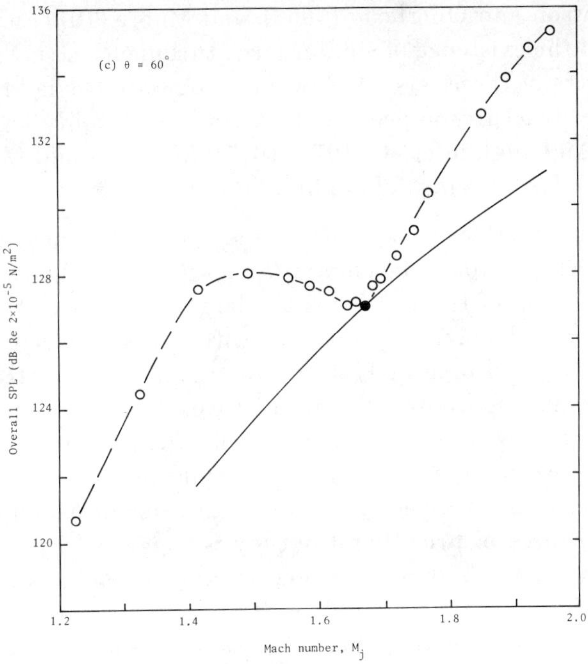

Figure 6. Shock associated and turbulent mixing noise components for a convergent-divergent nozzle operated with $T_j/T_0 = 1.0$. (c) $\theta = 60°$.

proposed a simple point-source array model to explain its generation process. In their model Harper-Bourne and Fisher introduced the concept that the noise source of shock associated noise is spatially coherent over an extended length of the jet. In other words, the sources are acoustically noncompact. This, of course, contradicts directly the older theories based on the notion of random scattering of point-like vortices of turbulence by the shock waves of the jet. The Harper-Bourne and Fisher model will be discussed in some detail below.

Since the work of Harper-Bourne and Fisher it was recognized by a number of investigators that the coherent sources of shock associated noise must ultimately be related to the large turbulence structures in the mixing layer of the jet. These large coherent structures were first observed in low speed shear flows by Brown and Roshko (1974), Roshko (1976) and Dimotakis and Brown (1976). In addition to being coherent these turbulence structures appear to be also spatially quasi-periodic rendering them in many ways wave-like characteristics. For high subsonic and supersonic jets, the visual observations of Dahan et al. (1978), Ozkul (1979), Salant (1971), Tam

(1972), Lowson and Ollerhead (1968), and Moore (1977) have clearly established the existence of similar large turbulence structures in the mixing layers of these jets. At low and moderate Reynolds numbers these large structures appear in the form of instability waves as observed by McLaughlin et al. (1975, 1977), Morrison and McLaughlin (1979), and Troutt and McLaughlin (1982).

In an imperfectly expanded supersonic jet a nearly periodic shock cell system would develop in the jet flow over the core and the transition region of the jet. Thus the large turbulence structures in such a jet would inevitably interact with the shock cell system as they are convected or propagate downstream. This unsteady interaction generates spatially coherent disturbances which could be the sources of broadband shock associated noise. The idea that the interaction between large scale structures and the quasi-periodic shock cell system of an imperfectly expanded supersonic jet might be an important source of broadband noise was suggested by Tam (1972) even before the work of Harper-Bourne and Fisher (1973). Recently the same idea was reconsidered by Tam and Tanna (1982). It became a key element of their proposed shock associated noise source model. In their work Tam and Tanna use simple but sufficiently realistic analytical models to represent the shock cell system and the large turbulence structures. By means of this model they were able to deduce a shock associated noise intensity scaling formula and a peak frequency formula. They tested these formulas against their experimental measurements and found very favorable agreements. At the present time this is the only experimentally verified shock associated noise intensity scaling formula which is applicable to supersonic jets issued from convergent-divergent nozzles. On following somewhat similar ideas Howe and Ffowcs-Williams (1978) modelled the shock associated noise source problem as one involving the interaction of a nearly periodic shock cell system and a moving thin cylindrical sheet of vortex rings. They used this model to provide a noise source for the determination of the far field sound through an acoustic analogy theory. Some very complicated formulas involving several parameters and adjustable constants were derived. Because of the complexity of these formulas their results cannot easily be interpreted in physical terms. Moreover, because of their complexity, as far as is known, these formulas have never been seriously tested against experimental data.

3.1 Phased Point-Source Array Model

Harper-Bourne and Fisher (1973) proposed that the source of broadband shock associated noise is in the form of a synchronized array of time periodic point monopoles. These monopoles are of equal strength and are spaced at a regular interval apart. Harper-Bourne and Fisher believed that the shock associated noise is emitted from the end-points of the shock cells of imperfectly expanded supersonic jets. The point monopoles are, therefore, used to simulate these very localized noise sources. Earlier a similar model was developed by Powell (1953) to describe the generation of screech tones in choked jets. In the Harper-Bourne and Fisher model the phases of adjacent monopoles are assumed to be correlated by the time taken for turbulence to be convected from one source point to the next as shown in Figure 7. Let L be the shock cell spacing and U_c be the turbulence convection velocity then the time taken for turbulence to be convected from source point A to source point B is $T_{AB} = L/U_c$. Thus the time origin of point source B is assumed to be delayed by an amount equal to T_{AB} relative to point source A. Similar time delay is applied to successive point sources of the array. Now consider sound radiation in the direction θ. The sound from source B will follow path BB' and that from source A, AA' etc. Clearly, path length AA' is longer than BB' by the length $AC = L\cos\theta$. The time needed for sound wave to travel the distance AC is $T_{AC} = L\cos\theta/a_\infty$ where a_∞ is the ambient speed of sound. Because of the difference in the propagation path length and in the origin of time the sound waves emitted from the two point sources A and B would be out of phase when they reach the observer far away. Thus there will be a tendency for them to partially cancel each other. It is, however, possible for the sound waves to arrive at the far field observer exactly in phase if certain conditions are met. When this happens there will be maximum constructive reinforcement of the sound intensity. For a given direction θ this condition is satisfied only for certain special frequencies, f_p. Hence one would expect the broadband shock associated noise spectra to peak at these frequencies of maximum reinforcement. The condition for maximum wave reinforcement occurs when the difference between the turbulence convection time T_{AB} and the sound propagation time T_{AC} is equal to an integral multiple of the period of oscillation. That is $T_{AB} - T_{AC} = n/f_p$ or

$$\left(\frac{L}{U_c} - \frac{L\cos\theta}{a_\infty}\right) = \frac{n}{f_p} \qquad n = 1, 2, 3, \cdots$$

The primary peak frequency corresponds to the case $n = 1$ so that from the above the peak frequency f_p is given by

$$f_p = \frac{U_c}{L(1 - M_c \cos\theta)} \tag{3.1}$$

In (3.1) $M_c = U_c/a_\infty$ is the convection Mach number based on the ambient speed of sound. By comparing equation (3.1) with their choked jet noise data Harper-Bourne and Fisher found reasonably good agreement. In this way they were able to offer an explanation to one of the most prominent characteristics of shock associated noise, namely, the shift of the peak frequency of the noise spectrum with observation angle as depicted in Figure 4.

In addition to deriving equation (3.1) Harper-Bourne and Fisher found semi-empirically, after extensive analysis of their choked jet noise data, that the shock associated noise intensity, I_s, varies as β^4 where $\beta = (M_j^2 - 1)^{1/2}$. The parameter β was introduced because Harper-Bourne and Fisher reasoned that it is the parameter which characterizes the pressure jump across a normal shock of upstream Mach number M_j. Figure 8 shows a comparison of the scaling formula $I_s \propto \beta^4$ and the measurements of Harper-Bourne and Fisher. The agreement is excellent. Subsequent to their work Tanna (1977) verified that this formula works not only for cold jets but for hot jets as well. The only restriction is that they are choked jets, i.e., jets issued from convergent nozzles.

3.2 Large Turbulence Structures—Shock Cells Interaction Model

Tam and Tanna (1982) proposed that broadband shock associated noise is generated by the weak interaction between downstream propagating large turbulence structures in the mixing layer of a jet and the quasi-periodic shock cells. To represent the large turbulence structures analytically the statistical model of Tam and Chen(1979) was used. Such a statistical model has also been employed successfully by Plaschko (1981). In this model the large turbulence structures are represented by a linear superposition of the normal hydrodynamic instability wave modes of the flow with random amplitude functions. The characteristics of each of the instability wave

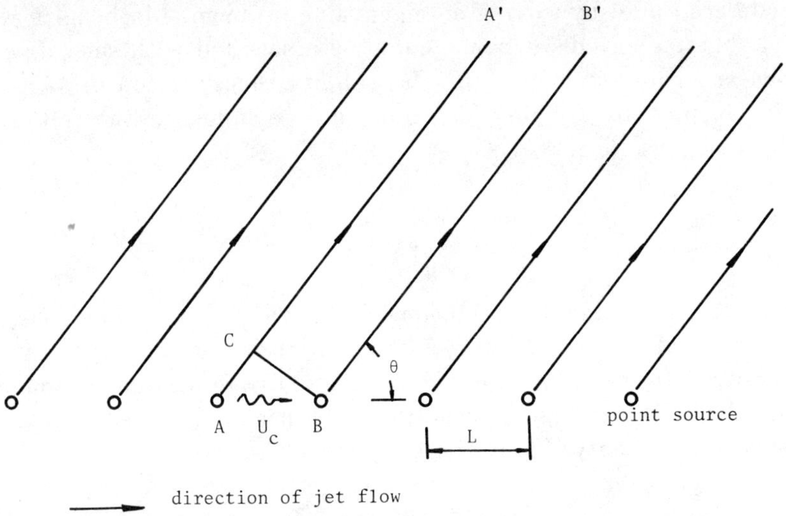

Figure 7. Schematic diagram of phased point-source array model.

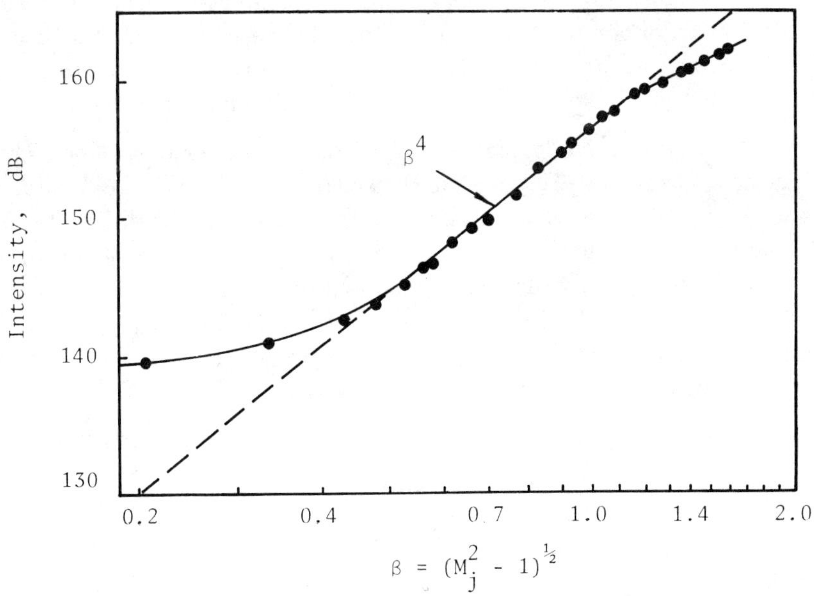

Figure 8. Variation of overall intensity at 90° to jet with β (data from Harper-Bourne and Fisher, 1973).

mode are found by solving an eigenvalue problem. Mathematically, if Φ denotes any disturbance quantity associated with such a wave then according to hydrodynamic stability theory it can be written, with respect to a cylindrical coordinate system, in the following dimensionless form

$$\Phi = \text{Re}\left[a(x)\psi(r,\theta)\, e^{i(k_r x - \omega t)}\right]; \qquad \text{Re}\,[\,] = \text{the real part of} \quad (3.2)$$

where ω is the angular frequency. k_r is the wave number. $\psi(r,\theta)$ is the eigenfunction and $a(x)$ is the spatial distribution of the wave amplitude (to be interpreted, in the context of representing the large turbulence structures, in a statistical sense). The wave speed or convection velocity, U_c, of the large structures is

$$U_c = \omega / k_r \tag{3.3}$$

The spatial amplitude function $a(x)$ describes the growth and decay of the instability wave. In general, at the nozzle exit $a(x)$ is small. It grows rapidly in the downstream direction and reaches a peak at some point and then decreases further on as illustrated in Figure 9. The location of the peak of $a(x)$ is a function of frequency. However, for the range of frequencies that are pertinent to shock associated noise it lies somewhere near the end of the potential core of the jet.

As a simple analytical model of the quasi-periodic shock cell structure, Tam and Tanna (1982) used the first order weak shock solution of Pack (1950) which has also been given by Tam (1972). In developing this solution the jet is modelled as a column of uniform flow bounded by a vortex sheet as shown in Figure 10. If p_s denotes the pressure perturbation of the shock cell structure, $p_s + p_a$ (p_a is the ambient pressure) being the actual pressure in the jet, the first order solution for the pressure ratio p_s/p_a is

$$p_s/p_a = \sum_{n=1}^{\infty} A_n \phi_n(r) \cos(\lambda_n x) \tag{3.4}$$

$$A_n = 2\Delta p / \mu_n p_a,$$
$$\lambda_n = 2\mu_n / D_j \, (M_j^2 - 1)^{1/2}; \quad n = 1, 2, 3, \cdots$$

In (3.4) $\phi_n(r)$ is a set of orthonormalized eigenfunctions. D_j is the effective jet diameter and μ_n are the zeros of the zeroth order Bessel

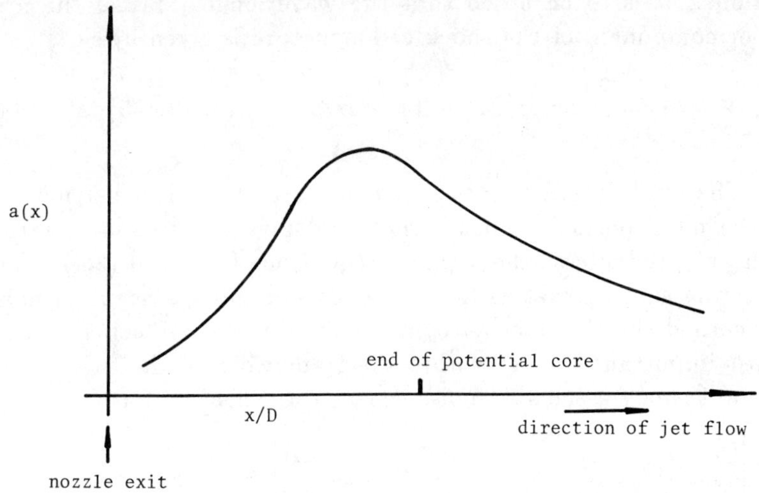

Figure 9. Typical spatial distribution of amplitude function $a(x)$.

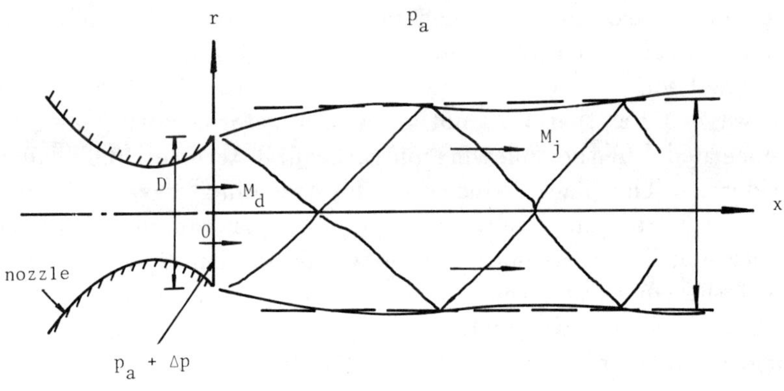

Figure 10. Schematic diagram of the shock cell system of a vortex sheet jet from a convergent-divergent nozzle.

functions. It is to be noted that the wave length, L_1, of the first Fourier component of the shock cell structure is given by

$$L_1 = \frac{2\pi}{\lambda_1} = \pi\left(M_j^2 - 1\right)^{1/2} D_j \mu_1 \quad (\mu_1 = 2.405) \tag{3.5}$$

The weak interaction between the large turbulence structures and the quasi-periodic shock cells would give rise to source terms which are products of the terms of equations (3.2) and (3.4). Tam and Tanna showed that as far as broadband shock associated noise is concerned the first Fourier component of the shock cell structure is most important. Hence within the framework of the theory, the dominant shock associated noise source term has the form

$$\begin{aligned} &\mathrm{Re}\left[a(x)A_1\psi(r,\theta)\phi_1(r)\cos(\lambda_1 x)\, e^{i(k_r x-\omega t)}\right] \\ &\quad = \mathrm{Re}\left[\frac{1}{2}a(x)A_1\psi(r,\theta)\phi_1(r)\left(e^{i(k_r-\lambda_1)x-i\omega t} + e^{i(k_r+\lambda_1)x-i\omega t}\right)\right] \end{aligned} \tag{3.6}$$

Now several important consequences can be deduced immediately from (3.6). Firstly, (3.6) represents propagating waves so that the noise sources are moving or nonstationary sources. This is in contrast to the Harper-Bourne and Fisher model which assumes spatially fixed sources. It turns out the second term of (3.6) gives subsonic wave and is, therefore, not important as far as noise generation is concerned. In the following only the first term of (3.6) will be considered. The phase velocity of this travelling wave is equal to $\omega/(k_r - \lambda_1)$ if the amplitude variation of $a(x)$ is ignored. This phase velocity can be supersonic relative to the ambient speed of sound even if the convection velocity, $U_c = \omega/k_r$, of the large structure is subsonic. By the wavy wall analogy if the travelling source component is supersonic it will lead immediately to Mach wave radiation as shown schematically in Figure 11. Suppose θ is the direction of Mach wave radiation and a_∞ is the ambient speed of sound. Then from the Mach angle relationship it is easy to obtain

$$a_\infty = [\omega/(k_r - \lambda_1)]\cos\theta$$

upon using equations (3.3) and (3.5) this relation may be rewritten in the form,

$$f = \frac{U_c}{L_1\,(1 - M_c\cos\theta)} \tag{3.7}$$

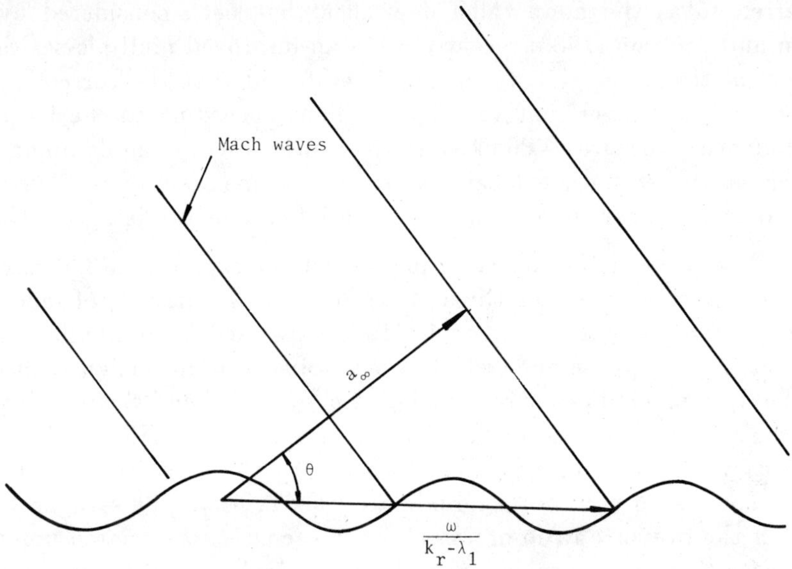

Figure 11. Mach wave radiation generated by a supersonic travelling wave source component.

where $f = \omega/2\pi$ is the frequency of oscillation and $M_c = U_c/a_\infty$ is the convection Mach number. According to the proposed Mach wave mechanism, one would, therefore, expect the shock associated noise radiated in the direction θ to have frequencies around that given by equation (3.7). In other words, the noise spectrum should peak at this frequency. A slight modification will, however, be needed. This arises from the weak dependence of $a(x)$ on x. This effect which also leads to a broadening of the wave number spectrum has been discussed at length by Tam and Morris (1980). Notice that (3.7) is identical to (3.1) if the shock spacing is taken to be equal to the wave length of the fundamental Fourier component of the quasi-periodic shock cells.

The second important implication of (3.6) is that the sources of shock associated noise is not located close to the nozzle exit. Instead, it is situated quite far downstream near where the amplitude of the large turbulence structures peaks. The axial distribution of the sources in this model is dictated primarily by that of $a(x)$. In reality the amplitude of the fundamental Fourier component of the shock cell structure, unlike that of expression (3.6), is not a true constant. It varies slowly in the downstream direction due to the spatial divergence of the jet flow. This effect which is commonly referred to as the nonparallel flow effect has been considered by Tam and Jackson (1983) recently using an improved multiple-scales model of the shock cell structure. They found that the correction required is, however, not very large. Thus according to the large turbulence structures—shock cells interaction theory the dominant shock associated noise is located where $a(x)$ is maximum (see Figure 9) somewhere near the end of the potential core of the jet.

The form of the source term as given by expression (3.6) also provides a means of estimating the variation of the intensity of shock associated noise with the operating fully expanded Mach number of the jet, M_j. Suppose one considers a jet operating at a slightly imperfectly expanded pressure ratio i.e., $\Delta p/p_a \ll 1$ then by expanding all the terms of expression (3.6) about the nozzle design condition it is easy to see that to first order in $\Delta p/p_a$ the magnitude of the noise source term varies approximately as $\Delta p/p_a$. This suggests that if the pressure ratio or M_j changes the shock associated noise intensity I_s would vary in proportion to $(\Delta p/p_a)^2$. By using gas dynamics relation for nozzle flows Tam and Tanna (1982) showed that for a given convergent-divergent nozzle $\Delta p/p_a$ is proportional to $(M_j^2 - M_d^2)$ provided $|M_j^2 - M_d^2| \lesssim 1$ where M_d is the nozzle design Mach number. It turns out $|M_j^2 - M_d^2| \lesssim 1$ is a sufficient condition for the jet flow to be free of Mach disc. Hence one obtains the following shock associated noise scaling formula

$$I_{(\text{shock associated noise})} \propto \left(M_j^2 - M_d^2\right)^2 \tag{3.8}$$

Expression (3.8) holds for any convergent-divergent nozzle of design Mach number M_d regardless of whether the jet is underexpanded or overexpanded. For choked jets M_d is unity so that in this case the semi-empirical scaling formula of Harper-Bourne and Fisher i.e., $I_s \propto \beta^4$ is recovered by putting $M_d = 1.0$ in expression (3.8).

The accuracy of the peak frequency formula (3.7) and the noise intensity scaling formula (3.8) has been tested by Tam and Tanna (1982) against their experimental measurements. Figure 12 shows a comparison of the measured and calculated peak frequencies using formula (3.7). In this figure M_j is 1.925 and M_d is 1.67 so that the jet is underexpanded. In the calculation the convection velocity U_c of the large turbulence structures has been taken to be equal to 0.7 times the velocity of the jet (see Harper-Bourne and Fisher (1973) and Seiner and Yu (1981) for experimental support of this assumption). As can be seen the agreement between the calculated and the measured values is good. Recently Tam and Jackson (1983) showed that the agreement can be improved if a more realistic shock cell model is used to estimate L_1. In the work of Tam and Tanna formula (3.7) was further tested by comparing it with isothermal and hot jet data at both underexpanded and overexpanded conditions. The overall agreement is very favorable. Figure 13 shows a comparison of the intensity scaling formula (3.8) and measured data for an underexpanded cold jet at $\theta = 120°$ and $90°$. Over the range in which the approximations used in deriving formula (3.8) are valid the agreement between the calculated results and data is very good. Similar good agreements were found by Tam and Tanna for isothermal and hot jets at underexpanded as well as overexpanded jet operating conditions.

3.3 Experimental Observations on the Sources of Broadband Shock Associated Noise

Recently Seiner and Norum (1980), Norum and Seiner (1982a) and Seiner and Yu (1981) carried out fairly detailed experimental investigations on the location and physical characteristics of the sources of broadband shock associated noise. In their experiments far field and near field microphones, a microphone array and hot-film probes were used. To deduce the location and size of the noise sources two-point correlations of the near and far field data were employed. Although some of their conclusions based on their measurements must be considered tentative and subjected to possibly other interpretations, yet by and large their efforts have provided a consistent and reasonably plausible physical picture of the actual broadband shock associated noise sources.

By using near field and far field microphone correlations with suitable frequency filtering, Seiner and Yu (1981) showed that the dominant source of shock associated noise is located quite far downstream of the nozzle exit more or less close to the end of the supersonic region or in the region of the last few quasi-periodic shock cells. This is consistent with the earlier measurements and conclusions of Seiner and Norum (1980) and Norum and Seiner (1982a). The effective size of the noise source region is quite large. The pressure-velocity correlation data of Seiner and Yu (1981) obtained

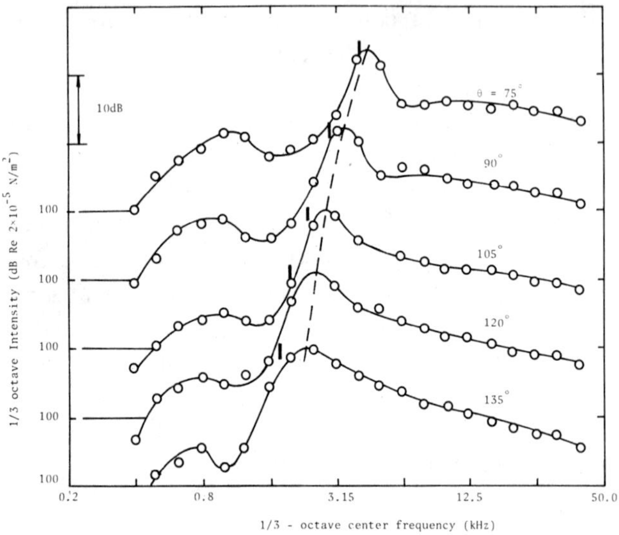

Figure 12. Variation of peak frequency with angle, $M_j = 1.925$, $M_d = 1.67$, $T_R/T_0 = 1.0$; | f_p calculated; - - - f_p experiment (from Tam and Tanna, 1982).

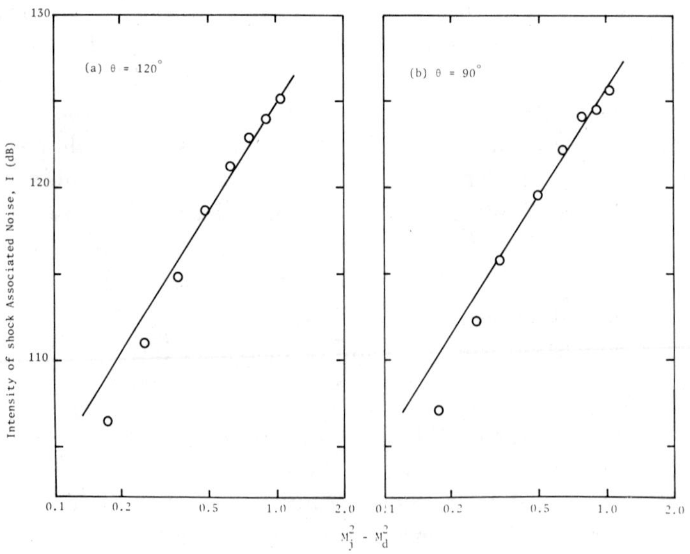

Figure 13. Shock associated noise intensity scaling. $M_d = 1.67$, $T_R/T_0 = 1.0$; o: experiment (underexpanded), —: $I \;\alpha\; (M_j^2 - M_d^2)^2$ (from Tam and Tanna, 1982).

by near field microphone and hot-film probe in the flow indicates that it extends over three to four shock cell spacings. Based on these measurements Seiner and Yu suggested that some large scale fluid motion existed that in some way collectively interacted and phased the motion of adjacent shocks. Thus according to these experimental findings the shock associated noise source region is spatially coherent and overlaps that of the turbulent mixing noise. The validity of this conclusion is further supported by the near field sound pressure contour measurements of Yu (1971). In the work of Yu 1/3 octave band near field sound pressure level contour maps of an underexpanded supersonic jet are provided, see Figure 14. The lobed nature of the contours permits easy identification of the direction of radiation and the possible origin and location of the sound source. In Figure 14 the downstream lobe is clearly associated with the dominant direction of the turbulent mixing noise and the slightly upstream lobe is that of the shock associated noise. Both seem to come from the same region of the jet approximately situated at the end of the jet core collaborating the results of Seiner and Norum (1980), Seiner and Yu (1981) and Norum and Seiner (1982a). Preliminary numerical calculations indicate that this is the same region at which an instability wave of the same frequency attains its maximum amplitude. In other words the location of the shock associated noise source of the model of Tam and Tanna (1982) is in agreement with all the most recent experimental observations.

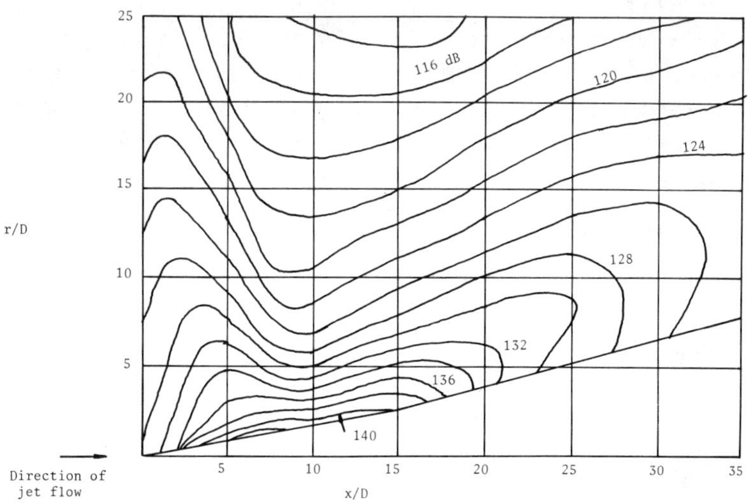

Figure 14. 1/3 octave near field sound pressure level contours. $f_0 = 20\ kHz$, $M_j = 1.67$, $M_d = 1.5$ (data from Yu, 1971).

To try to resolve the question of whether the source of shock associated noise are stationary or moving, Seiner and Yu (1981) measured the near field noise spectrum using an array of microphones positioned at 2.68 jet diameters away from the centerline of the jet. Their reasoning was that in the near field the sound would be dominated by the source closest to the microphone and that the cancellation effect which operates in the far field will not be as effective in the near field if the sources are fixed in space. In this case the near field spectra would not show the Doppler effect as given by equations (3.1) or (3.7). However, if the sources are moving then the same Doppler factor applies to the far field as well as to the near field sound. Their microphone measurements show unambiguous shift of the peak frequencies in the near field. Therefore, based on these results it would seem that there is a genuine possibility that shock associated noise are generated directly by moving sources. However, at this stage it is advisable to wait for the results of a more exhaustive investigation before making a definite conclusion.

4. Discussion

During the last several years some progress has been made in the identification and understanding of the characteristics and generation mechanism of broadband shock associated noise of supersonic jets from convergent-divergent nozzles. As yet no quantitative mathematical theory of this noise component other than scaling formula is available. At this time, in addition to the development of a mathematical theory, there are several aspects of broadband shock associated noise that obviously require further investigation. They include:

(a) The effect of jet temperature.

(b) The precise relationship between shock associated noise and turbulent mixing noise.

(c) The relationship, if any, between broadband shock associated noise and screech tones.

(d) Shock associated noise in other flow configurations, i.e., other than round jets.

Some preliminary works on several of the above areas have recently been carried out by Tanna et al. (1981), Nagel and Papathanasiou (1983), and Von Glahn (1980). Concerning items (b) and (c) it would be most interesting if it turns out that all the three noise components can be traced to one common noise source or link to the same flow mechanism. There are some experimental results which indicate that this is, indeed, a distinct possibility. Should this be the case, then the efforts needed to reduce supersonic jet noise may be greatly simplified for there will only be one noise source or flow mechanism to deal with.

Acknowledgements

This work was supported by NASA Lewis Research Center Grant NAG 3-182.

References

[1] Brown, G. L. and Roshko, A. "On Density Effects and Large Structure in Turbulent Mixing Layers," *J. Fluid Mech., 64* (1974), 775–816.

[2] Dahan, C., et al. "Coherent Structures in the Mixing Zone of a Subsonic Hot Free Jet," *J. Sound Vib., 59* (1978), 313–333.

[3] Das, I., Matambo, T. J., and Dosanjh, D. S. "Aeroacoustics of Supersonic Porous Plug-Nozzle Flows," *AIAA Paper 83-0775,* 1983.

[4] Dimotakis, P. E. and Brown, G. L. "The Mixing Layer at High Reynolds Number: Large Structure Dynamics and Entrainment," *J. Fluid Mech., 78* (1976), 535–560.

[5] Harper-Bourne, M. and Fisher, M. J. "The Noise From Shock Waves in Supersonic Jets," *AGARD CP-131,* 1973.

[6] Howe, M. S. and Ffowcs-Williams, J. E. "On the Noise Generated by an Imperfectly Expanded Supersonic Jet," *Phil. Trans. Roy. Soc. London, 289* (1978), 217–314.

[7] Lighthill, M. J. "On the Energy Scattered From the Interaction of Turbulence With Sound or Shock Waves," *Proc. Cambridge Phil. Soc., 49* (1953), 531–551.

[8] Lowson, M. V. and Ollerhead, J. B. "Visualization of Noise From Cold Supersonic Jets," *J. Acoustical Soc. Amer., 44* (1968), 624–630.

[9] McLaughlin, D. K., Morrison, G. L., and Troutt, T. R. "Experiments on the Instability Waves in a Supersonic Jet and Their Acoustic Radiation," *J. Fluid Mech., 69* (1975), 73–95.

[10] ————————————————. "Reynolds Number Dependence in Supersonic Jet Noise," *AIAA Journal, 15* (1977), 526–532.

[11] Moore, C. J. "The Role of Shear Layer Instability Waves in Jet Exhaust Noise," *J. Fluid Mech., 80* (1977), 321–367.

[12] Morrison, G. L. and McLaughlin, D. K. "The Noise Generation by Instabilities in Low Reynolds Number Supersonic Jets," *J. Sound Vib., 65* (1979), 177–191.

[13] Nagel, R. T. and Papathanasiou, A. G. "An Experimental Study of Supersonic Jet Shock Associated Noise," *AIAA Paper 83-0708,* 1983.

[14] Norum, T. D. and Seiner, J. M. "Broadband Shock Noise From Supersonic Jets," *AIAA Journal, 20* (1982), 68–73.

[15] ——————————. "Measurements of Mean Static Pressure and Far Field Acoustics of Shock—Containing Supersonic Jets," *NASA TM 84521,* 1982.

[16] Ozkul, A. "Investigation of Acoustic Radiation From Supersonic Jets by Double-Pulse Holographic Interferometry," *AIAA Journal, 17* (1979), 1068–75.

[17] Pack, D. C. "A Note on Prandtl's Formula for the Wavelength of a Supersonic Gas Jet," *Quart. J. Mech. Appl. Math., 3* (1950), 173–181.

[18] Pao, S. P. and Seiner, J. M. "Shock Associated Noise in Supersonic Jets," *AIAA Journal, 21* (1983), 687–693.

[19] Plaschko, P. "Stochastic Model Theory for Coherent Turbulent Structures in Circular Jets," *Phy. Fluids, 24* (1981), 187–193.

[20] Powell, A. "On the Mechanism of Choked Jet Noise," *Proc. Phy. Soc., B 66* (1953), 1039–56.

[21] Ribner, H. S. "Shock-Turbulence Interaction and the Generation of Noise," *NACA Report 1233,* 1955.

[22] Roshko, A. "Structure of Turbulent Shear Flows: A New Look," *AIAA Journal, 14* (1976), 1349–57.

[23] Salant, R. F., Zaic, G. F., and Kolezar, R. R. "Holographic Study of the Mach Wave Field Generated by a Supersonic Turbulent Jet," *KIT Proceedings of the Purdue Noise Control Conference,* Lafayette, Indiana, 1971.

[24] Seiner, J. M. and Norum, T. D. "Experiments on Shock Associated Noise of Supersonic Jets," *AIAA Paper No. 79-1526,* 1979.

[25] ——————————. "Aerodynamic Aspects of Shock Containing Jet Plumes," *AIAA Paper No. 80-0965,* 1980.

[26] Seiner, J. M. and Yu, J. E. "Acoustic Near Field and Local Flow Properties Associated With Broadband Shock Noise," *AIAA Paper No. 81-1975,* 1981.

[27] Tam, C.K.W. "On the Noise of a Nearly Ideally Expanded Supersonic Jet," *J. Fluid Mech., 51* (1972), 69–95.

[28] Tam, C.K.W. and Chen, K. C. "A Statistical Model of Turbulence in Two-Dimensional Mixing Layers," *J. Fluid Mech., 92* (1979), 303–326.

[29] Tam, C.K.W. and Jackson, J. A. "On the Shock Cell Structure and Noise of Supersonic Jets," *AIAA Paper No. 83-0703,* 1983.

[30] Tam, C.K.W. and Morris, P. J. "The Radiation of Sound by the Instability Waves of a Compressible Plane Turbulent Shear Layer," *J. Fluid Mech., 98* (1980), 349–381.

[31] Tam, C.K.W. and Tanna, H. K. "Shock Associated Noise of Supersonic Jets From Convergent-Divergent Nozzles," *J. Sound Vib., 81* (1982), 337–358.

[32] Tanna, H. K. "An Experimental Study of Jet Noise, Part II: Shock Associated Noise," *J. Sound Vib., 50* (1977), 429–444.

[33] Tanna, H. K., Tam, C.K.W., and Brown, W. H. "Shock Associated Noise Reduction From Inverted-Velocity-Profile Coannular Jets," *NASA CR-3454,* 1981.

[34] Troutt, T. R. and McLaughlin, D. K. "Experiments on the Flow and Acoustic Properties of a Moderate Reynolds Number Supersonic Jet," *J. Fluid Mech., 116* (1982), 123–156.

[35] Von Glahn, U. "New Interpretations of Shock Associated Noise With and Without Screech," *NASA TM 81590,* 1980.

[36] Yu, J. C. "Investigation of the Noise Fields of Supersonic Axisymmetric Jet Flows," *Ph.D. thesis,* Dept. of Mechanical Engineering, Syracuse Univ., 1971.

On the Absorbtion of Sound by Turbulence and Other Hydrodynamic Flows

M. S. Howe

University of Southampton, England S09 5NH

Summary

The dissipation of acoustic energy which occurs when sound generates vorticity at the boundaries of a fluid or propagates across a field of turbulence is examined. Vorticity generation occurs typically during scattering or diffraction by surfaces with corners or edges, and an understanding of the consequent attenuation is important in the design of mufflers and other devices used for suppressing acoustic and mechanical vibrations. The rate of dissipation is generally a nonlinear function of the acoustic intensity, but becomes linear and significantly greater in the presence of a mean flow.

When sound propagates through turbulence it is partly scattered and partly absorbed. Lighthill's (1953) theory of scattering is extended to include the possibility of acoustic absorbtion by a direct transfer of energy to the turbulent motions. A similar but greatly enhanced level of absorbtion takes place when sound propagates in turbulent pipe flow. The energy transfer occurs principally in the acoustic momentum and thermal boundary layers at the walls, whose properties are subject to the modifying influences of turbulence convection. This is discussed in relation to experimental data of Ronneberger and Ahrens (1977).

1. Introduction

The generation of noise by turbulence and other vortical flows has been studied extensively, especially in the period following the publication of Lighthill's (1952) pioneering paper on aerodynamic sound. Sound may frequently be regarded as a source of vorticity, however, in the sense that there can be a net transfer of energy from a compressible acoustic mode of fluid motion to one whose principal characteristics are those of an incompressible flow. This mechanism has important applications in the control of acoustic and mechanical vibrations in industrial systems. For example, large scale, coherent vortex shedding from the tube banks of a cross-flow heat exchanger often generates substantial levels of aerodynamic sound (Walker and Reising, 1968), which may be resonantly coupled to vibrations of tubes, plates and other structural members. The resulting oscillating stresses can correspond to sound pressure levels in excess of 150 dB (re 0.002μ bar) and be responsible for metal fatigue and ultimate structural failure. A particularly effective means of suppressing such resonances has been reported by Vér (1982), and involves the deployment of thin perforated plates aligned parallel to the flow in the heat exchanger cavity. Damping is obtained through the generation of vorticity in the perforations by the sound, the kinetic energy ceded to the vorticity being convected away by the mean flow, and ultimately dissipated into heat by the action of viscosity and thermal conduction. The utility of muffling devices of this kind, whose efficiency is explicitly dependent on the presence of mean flow, has been appreciated since the early part of the century (see Bechert, 1979a), and the governing mechanical principles have often been applied to dissipative processes in other branches of fluid mechanics (see, e.g., Lighthill 1979).

In the absence of surfaces or foreign bodies from which vorticity can be shed, an acoustic wave may still experience a net excess attenuation when traversing a region of vortical or turbulent flow. Part of the energy loss can be attributed to *scattering* by the turbulence, a phenomenon which was discussed in great detail by Lighthill (1953) and independently by Kraichnan (1953). This work, and much of that published more recently (see, e.g., Batchelor 1957, Lyon 1959, Crow 1969, Ffowcs Williams and Howe 1973, Brown and Clifford 1976, Korman and Beyer 1981, Kambe and Mya Oo 1981) implicitly assumes the turbulence to be *unchanged* by the passage of the sound, and accordingly that all of the energy diverted from the incident wave reappears as sound elsewhere in the medium. This approximation is acceptable only if the characteristic time scale of the turbulence is significantly different from that of the sound. When this is not the case a substantial amount of acoustic energy may be *absorbed* by the turbulence. In particular, sound of aerodynamic origin will be damped as it propagates through an extensive body of the turbulence which produced it. Although such *back-reaction* of the sound on the turbulence is unlikely to be significant in, say, estimating the order of magnitude of the noise produced by a turbulent jet (Noir and George 1978), it is a fundamental consideration in avoiding the prediction of an *acoustic catastrophe* in astrophysical applications (c.f. Crow 1967, Crighton 1969). Possible evidence of the importance of absorbtion is provided by the experiments Ross et al. (1983) and Guedel (1983) involving the propagation of sound through a turbulent shear layer.

The absorbtion is caused by the rotation and stretching of the turbulence vortex lines by the sound, i.e., by the strain imposed by the acoustic field. The evolution of the turbulence due to nonlinear self-interactions during the period of strain then provides the degree of irreversibility necessary to achieve a *net* transfer of energy from the sound to the turbulence. Now the attenuation is maximal when the turbulence time scale is of the same order as that of the sound. For turbulence of small characteristic Mach number M and eddy length scale l, it follows that sound of wavelength $\lambda \sim l/M \gg l$ will be preferentially absorbed, but only very weakly since the acoustic strain $\sim l/\lambda \ll 1$. The magnitude of the strain is significantly greater when the interaction takes place in the vicinity of a wall, for example, when sound interacts with a turbulent boundary layer or propagates in a pipe carrying a turbulent mean flow. The no-slip velocity and, say, constant temperature boundary conditions at the wall result in the formation of acoustic momentum and thermal boundary layers whose thicknesses are influenced by turbulent diffusion. For moderate frequencies these boundary layers extend beyond the viscous sublayer of the wall flow and produce turbulence strains having length scales which are typically an order of magnitude smaller than the acoustic wavelength. The corresponding enhanced rates of acoustic dissipation is confirmed by the experiments of Ahrens and Ronneberger (1971), Ingard and Singhal (1974), and Ronneberger and Ahrens (1977).

In this paper the theoretical bases of these mechanisms are discussed. The dissipation of sound due to vorticity production is examined in Section 2. A formula is given for the dissipation which is generally applicable in low Mach number flows wherein the length scale of the strain induced by the sound is much less than the acoustic wavelength. (Exceptionally, the formula turns out to be valid at arbitrary subsonic, uniform mean flow velocity in those situations which can be adequately treated by thin airfoil theory (Ashley and Landahl 1965).) Thus it provides an excellent description of the influence of mean flow on the dissipation of sound through vorticity production at corners, sharp edges or perforates where the flow can separate, but cannot be applied to estimate the absorbtion of sound propagating through an extensive region of turbulence. This problem is examined in Section 3 as an extension of the approach originally proposed by Lighthill (1953).

The very much more difficult problem of the interaction of sound with a turbulent wall layer is discussed in Section 4. The treatment here is strictly valid for infinitesimal mean flow Mach number, and is analytic only to the extent that the interaction is modelled by means of an eddy diffusivity (cf. Ronneberger and Ahrens 1977, Howe 1979c). The predicted absorbtion is in reasonable accord with experiment (except within the small interval of frequency noted by Ronneberger and Ahrens 1977), and application is made to the production and propagation of sound in fully developed turbulent pipe flow.

2. Attenuation of Sound by Vorticity Production at Edges and Corners

Vorticity is produced by the action of viscosity when sound impinges on a surface. In many situations, however, the relevant (mean flow or acoustic) Reynolds number is sufficiently large that viscosity can be ignored within the body of the flow, and its influence incorporated by means of a correction to the usual boundary conditions for an otherwise ideal fluid (e.g., by application of a Kutta condition). To fix ideas we shall frame the discussion in terms of the idealized configuration illustrated in Figure 1, involving the diffraction of sound by a rigid, circular cylindrical pipe in the presence of a low subsonic, isentropic flow produced by a fan within the pipe.

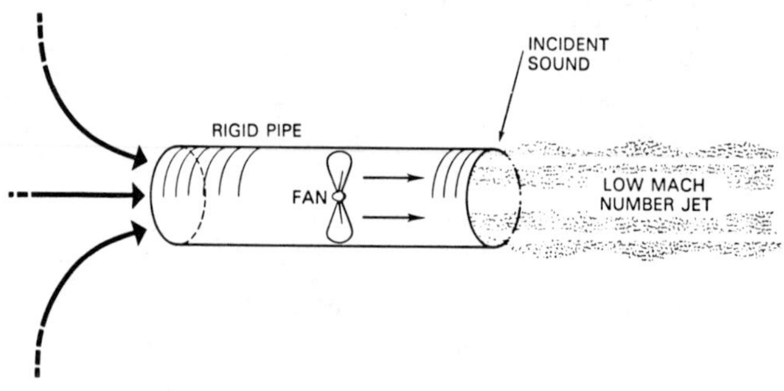

Figure 1. Diffraction of sound by a rigid, circular cylindrical pipe in the presence of a low, subsonic mean flow produced by a fan within the pipe.

Denote the vorticity by $\vec{\omega}(\mathbf{x}, t)$, and let v be the corresponding solenoidal velocity given in the absence of boundaries by (Batchelor 1967)

$$\mathbf{v}(\mathbf{x}, t) = \frac{1}{4\pi}\text{curl} \int \frac{\vec{\omega}(\mathbf{y}, t)}{|\mathbf{x} - \mathbf{y}|} d^3\mathbf{y}. \tag{2.1}$$

When boundaries are present this integral is assumed to be augmented by terms associated with the image system of the vorticity in the boundaries. This can always be done in a manner which ensures that div v = 0 and that the normal component of v vanishes on the instantaneous location of the boundaries. To account for prescribed motions of the boundaries, a velocity potential ψ is introduced which satisfies Laplace's equation $\nabla^2\psi = 0$ with normal component of velocity $\partial\psi/\partial x_n$, say, on a bounding surface equal to that of the surface. The remaining portion of the velocity field, accounting for the influence of fluid compressibility, is represented by a velocity potential ϕ which includes contributions from incident and diffracted sound waves, and also from compressible motions caused by the movement of the boundaries (the fan) and by the production of aerodynamic sound by the flow.

When the characteristic Mach number $M \ll 1$ the fluid density does not differ significantly from its undisturbed mean value ρ_o, and the total kinetic energy E is given, to within an error $\sim 0(M^2)$ relative to unity, by

$$\begin{aligned} E &= \frac{1}{2}\rho_o \int \{\mathbf{v} + \nabla(\psi + \phi)\}^2 \, d^3\mathbf{x} \\ &\equiv \frac{1}{2}\rho_o \int \mathbf{v}^2 d^3\mathbf{x} + \frac{1}{2}\rho_o \int [\nabla(\psi + \phi)]^2 \, d^3\mathbf{x}, \end{aligned} \tag{2.2}$$

where the second line is deduced by application of the divergence theorem and by noting that $\mathbf{v}(\psi + \phi) \sim 1/|\mathbf{x}|^3$ as $|\mathbf{x}| \to \infty$. The first term on the right of equation (2.2) is the leading approximation to the kinetic energy of the *vortical* motion. The second gives the contribution of the acoustic field and of the incompressible flow due to boundary motion. The latter is localized to the immediate vicinity of the moving boundaries (the fan in Figure 1) and characterises energy which is temporarily stored within the fluid, in the sense that the potential ψ is instantaneously reduced to zero when the boundaries are brought to rest (Batchelor 1967). Indeed, in the absence of dissipation and radiation damping, a continuous steady operation of the fan in Figure 1 would introduce vortical energy into the jet exhausting from the nozzle which is ultimately unbounded, whereas the energy associated with ψ remains bounded and localized.

Let

$$\Pi = \frac{1}{2}\rho_o \frac{\partial}{\partial t}\int \mathbf{v}^2 d^3\mathbf{x} \tag{2.3}$$

denote the rate of increase of the vortical kinetic energy. This can be expressed in terms of ψ and ϕ in high Reynolds number, isentropic flow by integrating over all space the scalar product of v and the momentum equation taken in the form

$$\frac{\partial \mathbf{v}}{\partial t} + \text{grad}\{\frac{\partial}{\partial t}(\psi+\phi) + \int \frac{dp}{\rho} + \frac{1}{2}(\mathbf{v}+\nabla\psi+\nabla\phi)^2\} = -\vec{\omega}\wedge(\mathbf{v}+\nabla\psi+\nabla\phi), \tag{2.4}$$

where p is the pressure and ρ the fluid density. Since div $\mathbf{v} = 0$ and $\mathbf{v} \sim 0(1/|\mathbf{x}|^3)$ as $|\mathbf{x}| \to \infty$, the gradient term in (2.4) makes no contribution to the integral, and

$$\Pi = \rho_o \int \vec{\omega}\wedge\mathbf{v}\cdot(\nabla\psi + \nabla\phi)\, d^3\mathbf{x}. \tag{2.5}$$

This expresses the production of kinetic energy in terms of the rate of working of the unsteady vortical lifting force $\rho_o\vec{\omega}\wedge\mathbf{v}$ due to boundary motion and the acoustic field.

In the absence of boundary motion (fan not operating in Figure 1), $\psi \equiv 0$ and vorticity production occurs solely as a result of the incident sound. It follows that Π is at most proportional to A^3 where A is a measure of the acoustic amplitude. When, however, boundary motions are sufficient to create a mean flow (the jet in Figure 1), both v and the vorticity $\vec{\omega}$ can have substantial mean components $\mathbf{U}$, $\vec{\Omega}$ independently of the incident sound. In this case Π is formally $0(A^2)$, which therefore suggests that mean flow considerably increases the rate of the production of vortical energy by the sound. In many instances the dominant interaction of the sound with the vorticity occurs in regions where $\nabla\psi$ is negligible (for example near the nozzle exit in Figure 1). The energy supplied to the vortical flow must then have been extracted directly from the incident sound, and Π determined by equation (2.5) is accordingly equal to the rate of dissipation of acoustic energy.

2.1 Influence of Mean Flow on Diffraction by a Semi-Infinite Plate

Figure 2(a) illustrates a two-dimensional diffraction problem which is useful for examining the implications of (2.5). A plane sound wave is incident on a semi-infinite rigid plate occupying the region $-\infty < x_1 < 0$, $x_2 = 0$, in the presence of a uniform subsonic flow at speed U in the x_1-direction. The combined action of viscosity and the sound is to cause flow separation (vortex shedding) at the edge of the plate. In the usual linearized, inviscid approximation of thin airfoil theory (Ashley and Landahl 1965) the vorticity is swept down-stream at speed U within a wake of infinitesimal thickness, and its strength is determined by application of the Kutta condition that the velocity and pressure are finite at the edge. This problem has been discussed by several authors (e.g., Jones 1972, Rienstra 1981), and it will be sufficient to quote the relevant formulae in the following.

Let the incident wave be specified by the velocity potential

$$\phi_i = Re\left[Ae^{i\{n_1x_1 - n_2x_2 - \omega t\}}\right], \tag{2.6}$$

where $\omega > 0$ is the radian frequency, and the wavenumber components n_1, n_2 are given by

$$n_1 = \frac{(\omega/c)\cos\theta}{1 + M\cos\theta}, \qquad n_2 = \frac{(\omega/c)\sin\theta}{1 + M\cos\theta}, \tag{2.7}$$

c being the speed of sound, $M = U/c$ the mean flow Mach number, and θ the angle between the normal to the wavefronts (surfaces of constant phase) and the negative x_1-direction. The coefficient A is defined in terms of the amplitude p_i of the incident acoustic pressure fluctuations by

$$A = \frac{p_i}{\rho_o\omega}(1 + M\cos\theta). \tag{2.8}$$

The vorticity in the wake is given by

$$\vec{\omega} = Re\left[\frac{-2i\beta\kappa A n_2 e\delta(x_2)\, e^{i\omega(\frac{x_1}{U} - t)}}{\sqrt{K_o} - N\sqrt{K_o + \kappa}}\right], \quad x_1 > 0 \tag{2.9}$$

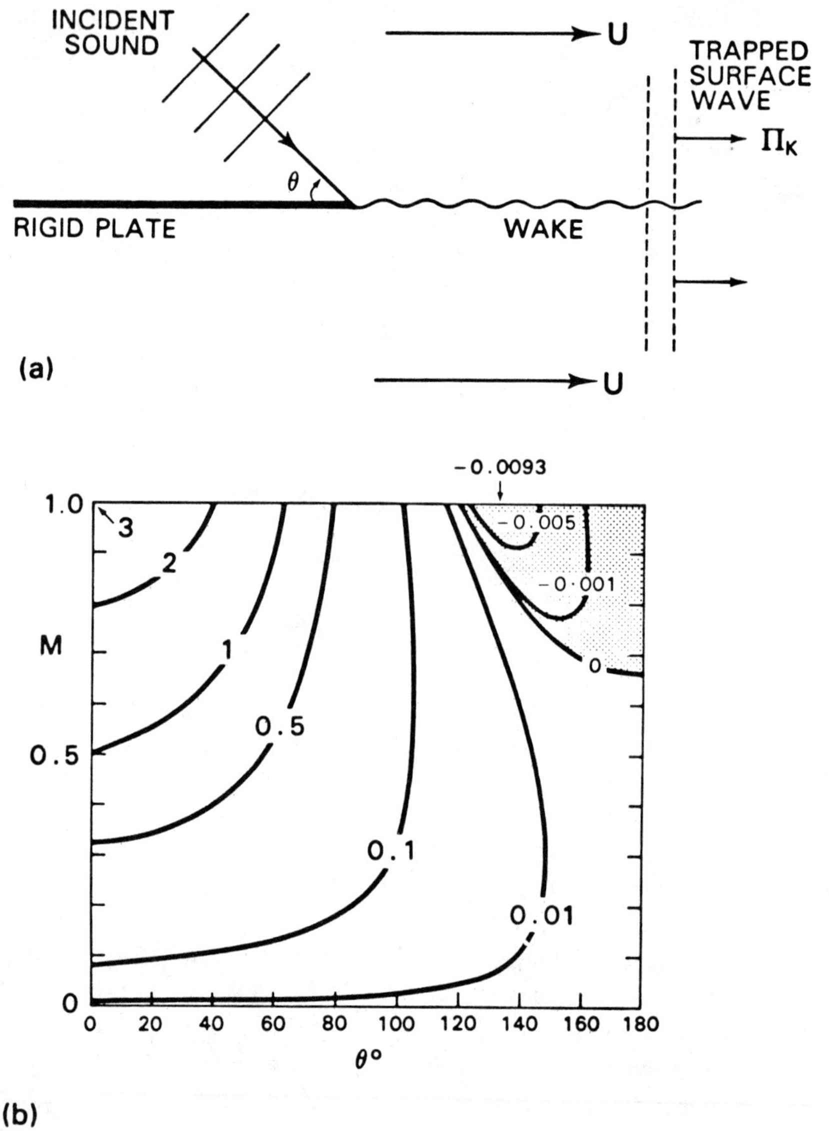

Figure 2. (a) Diffraction of a plane sound wave by a semiinfinite rigid plate in the presence of a uniform, subsonic mean flow. Π_K is the power transferred to the surface waves generated by the wake. (b) Contours of constant $\Pi\rho_o\omega/2|p_i|^2$, the acoustic power dissipated per unit spanwise length of the plate. The dissipation is negative in the shaded region where the power flux into the "trailing edge noise" field exceeds that dissipated due to vorticity production at the edge.

where **e** is a unit vector parallel to the edge of the plate and directed out of the plane of the paper in Figure 2(a), and $\delta(x_2)$ is the Dirac δ-function. The remaining terms in (2.9) are defined as follows:

$$\begin{gathered}\beta = \sqrt{1-M^2}, \quad \kappa = \omega/U, \quad K_o = \omega/\beta c, \\ N = K_o(M+\cos\theta)/(1+M\cos\theta).\end{gathered} \tag{2.10}$$

Since there are no moving surfaces in the vicinity of the edge, we can take $\psi = 0$ in (2.5). Of the remaining terms in the integrand, the vorticity $\vec{\omega}$ and the acoustic potential ϕ are each $0(A)$, whereas $\mathbf{v} \approx (U, 0)$ to leading order. Hence the principal contribution to the integral is $0(A^2)$, and the rate of dissipation of acoustic energy *per unit length parallel to the edge of the plate*, becomes

$$\Pi = \rho_o \int |\vec{\omega}| U \frac{\partial \phi}{\partial x_2}\, dx_1\, dx_2. \tag{2.11}$$

Now set

$$\phi = \phi_i + \phi_d + \phi_w, \tag{2.12}$$

where ϕ_d, ϕ_w respectively denote the contributions to the compressible motions due to diffraction in the absence of vortex shedding, and to the back-reaction of the shed vorticity on the plate ("trailing edge noise"). The integral in (2.11) is conveniently evaluated by employing the following Fourier integral representations of ϕ_d, ϕ_w:

$$\phi_d = \mp \frac{An_2}{2\pi i\sqrt{K_o - N}} \int_{-\infty}^{\infty} \frac{e^{i\{(k-K_oM)\frac{x_1}{\beta} \pm \gamma(k)x_2 - \omega t\}}}{\sqrt{K_o+k}\,(k-N+i0)}\, dk, \tag{2.13}$$

$$\begin{aligned}\phi_w = \pm \frac{An_2 e^{-i\omega t}}{2\pi i\sqrt{K_o-N}} \Bigg[&\int_{-\infty}^{\infty} \frac{e^{i\{(k-K_oM)\frac{x_1}{\beta} \pm \gamma(k)x_2\}}}{\sqrt{K_o+k}\,(k-\kappa-i0)}\, dk \\ &- \int_{\infty}^{\infty} \frac{\sqrt{\beta\kappa}\, e^{ikx_1 - |k|x_2}}{\sqrt{K_o+\kappa}\sqrt{k}(k-\beta\kappa-i0)}\, dk \Bigg],\end{aligned} \tag{2.14}$$

where real parts are to be taken on the right hand sides, $\gamma(k) = \sqrt{K_o - k}\cdot\sqrt{K_o + k}$, and all radicals involving k are either real and positive or positive imaginary for real k, and the upper/lower sign is taken according as $x_2 \gtrless 0$. The difference between the integrals in (2.14) represents the total perturbation potential due to the wake less that associated with the solenoidal velocity field $\mathbf{v}$.

Equations (2.9), (2.12) and the above integral representations are now used in (2.11). The various multiple integrals can be evaluated analytically and the result averaged with respect to time to yield the mean rate of acoustic dissipation

$$\Pi = \frac{2|p_i|^2}{\rho_o \omega} M \cos^2 \frac{\theta}{2}(1 + M\cos\theta)\left(1 + M\cos\theta - \frac{1}{2}M\right). \quad (2.15)$$

In the limit $M \ll 1$, $\Pi \approx 2|p_i|^2 M \cos^2 \frac{\theta}{2}/\rho_o\omega$, a result given by Howe (1981). But (2.15) is actually valid for arbitrary $M < 1$. To be sure, it has been derived on the basis of equations (2.3), (2.5) which neglect contributions to the vortical kinetic energy arising from fluid compressibility. However, when the vorticity convects at precisely the mean stream velocity U (as assumed in thin airfoil theory), there are neither pressure nor density fluctuations associated with the motion sufficiently far downstream, where the vortical energy must equal the kinetic energy of a fluid of density ρ_o. Hence the rate of increase of this component of the energy must coincide with the total acoustic dissipation rate.

The final term in the brace brackets of (2.15) makes a negative contribution to Π. It corresponds precisely to the contribution to the dissipation integral (2.11) made by ϕ_w, and actually represents the production of sound ("trailing edge noise") by the interaction of the wake with the plate. When M exceeds about 0.67 the acoustic energy so generated can exceed that dissipated during the formation of the vorticity and there is a net transfer of energy from the mean flow to the acoustic field. This possibility was noted by Rienstra (1981) who, however, gave an incorrect expression for Π, being less than (2.15) by an amount

$$\Pi_K = \frac{|p_i|^2}{\rho_o \omega} M(1 + M\cos\theta)\cos^2\frac{\theta}{2}. \quad (2.16)$$

The discrepancy arises because Rienstra mistakenly equated the dissipation to the net flux of energy *into the wake* through a control surface consisting of two semi-infinite planes, one just below and the other just above the wake, and connected at the edge of the plate. This procedure fails to account for the flux of energy into the trapped waves in the potential flow on either side of the wake generated by the vorticity. These waves propagate with the mean flow, exhibit no pressure or density fluctuations, and are maintained by an energy flux Π_K from the wake.

Figure 2(b) is a corrected form of Rienstra's (1981) Figure 3, and depicts contours of constant net sound power dissipation Π normalized by $2|p_i|^2/\rho_o\omega$ in the (M, θ)-plane. A net generation of acoustic energy is possible only for $M > 2/3$ and within the shaded region of the figure.

Situations involving a net conversion of mean flow energy into sound during the interaction of sound with a trailing edge occur often in practice. For example, transverse standing acoustic waves in a duct carrying mean flow can "lock-on" to the vortex shedding from heat exchanger tube banks, structural supports, etc., producing intense resonant tones. Welsh et al. (1983) have examined several such instances experimentally, and used (2.5) to estimate the contributions to the net mean flow/acoustic power transfer from individual vortices in the wake.

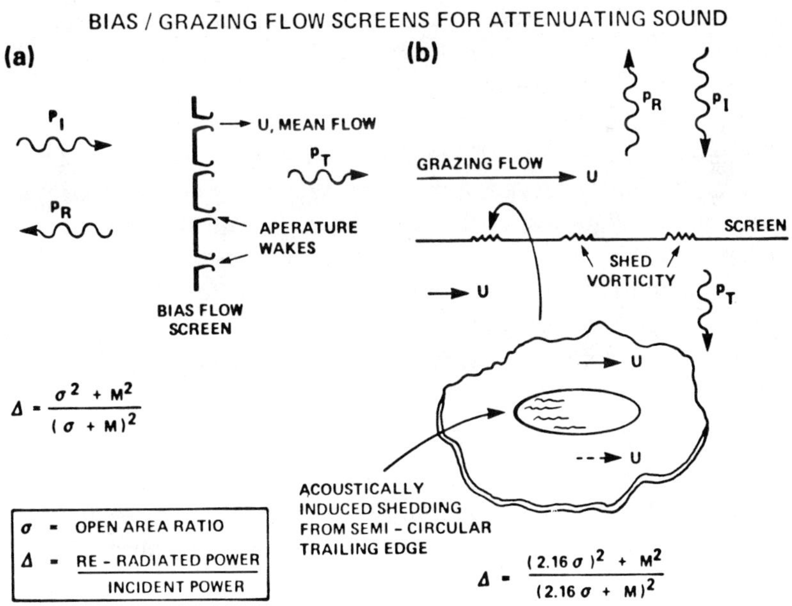

Figure 3. Influence of mean flow on the absorbtion of sound by vorticity production in perforates. (a) bias flow screen, (b) grazing flow screen. Δ is the absorbtion coefficient defined in the text.

2.2 Attenuation by Bias/Grazing Flow Perforated Screens

Detailed experimental confirmation of the influence of mean flow on the attenuation is furnished principally by studies of the emission of sound from the nozzle of a pipe flow. The loss of acoustic power due to "vortex ring" formation at the nozzle exit is determined by comparing the net power entering the nozzle from upstream within the pipe and the total radiated sound power, obtained by integration of the measured far field power flux (Bechert 1979a, b, Howe 1979a, Salikuddin et al. 1983).

Several other idealized configurations have been examined theoretically, two of which are illustrated in Figure 3. Figure 3(a) depicts a thin screen perforated with circular apertures through which is maintained a mean bias flow by a steady pressure differential across the screen. Vorticity produced in the perforates by, say, a normally incident pressure wave p_I is convected away from the screen in the bias flow jets. For screens of small fractional open area σ and at small bias flow Mach numbers, the absorbtion coefficient Δ (i.e., the ratio of the transmitted plus reflected sound powers to the incident sound power) is equal to $(|p_T|^2 + |p_R|^2)/|p_I|^2$, p_T, p_R being the transmitted and reflected waves respectively. In applications the Strouhal number $\omega R/U$ is usually small, R being the aperture radius and U the mean aperture velocity. In this case $\Delta = (\sigma^2 + M^2)/(\sigma + M)^2$, $M = U/c$ (Howe 1979b). Thus maximum attenuation occurs for $M = \sigma$, when 50% of the incident acoustic energy is absorbed. Evidently the deployment of a device of this kind could well provide considerable attenuation in a cavity, wherein multiple interactions of sound with the screen are possible. Furthermore, the attenuation rate can be controlled by varying the bias flow Mach number M, a property which might have applications in situations where the noise produced by fluid machinery depends on operating conditions (Dean and Tester 1978).

The case of a perforated screen in a uniform tangential mean flow (Figure 3(b)) has been examined by Howe (1980). Vorticity ejected from the apertures is convected along the screen, but at small Strouhal numbers this is accompanied by a net absorbtion of acoustic energy. Screens of this type are used to suppress resonant oscillations in mean flow ducts and cavities (Vér 1982).

3. The Absorbtion of Sound by Turbulence

Equation (2.5) cannot be used to determine the transfer of energy between a sound wave and a turbulent field through which it propagates. This is because the length scale of the straining of the turbulence is equal to the wavelength of the sound, and the rate of production of turbulence kinetic energy is then of the same order as the production of turbulence potential energy by compression of the flow. According to Lighthill (1953), propagation through high Reynolds number turbulence of infinitesimal Mach number is governed by the equation of aerodynamic sound which, to the desired order of approximation, may be taken in the form

$$\left(\frac{1}{c^2}\frac{\partial^2}{\partial t^2} - \nabla^2\right)p = \rho_o \frac{\partial^2 u_i u_j}{\partial x_i \partial x_j}, \tag{3.1}$$

where $\mathbf{u}$ is the total fluid velocity.

Set

$$u_i = U_i + V_i, \tag{3.2}$$

where $U_i(\mathbf{x}, t)$ is the turbulence velocity in the absence of an incident sound wave, and $V_i(\mathbf{x}, t)$ denotes the total perturbation caused by the wave. Substituting into the right hand side of (3.1) and discarding the terms U_iU_j, V_iV_j, which respectively describe the production of sound by the turbulence independently of the incident wave and the nonlinear self-interaction of motions produced by the sound, one finds

$$\left(\frac{1}{c^2}\frac{\partial^2}{\partial t^2} - \nabla^2\right)p = 2\rho_o \frac{\partial^2 U_i V_j}{\partial x_i \partial x_j}, \tag{3.3}$$

Henceforth we shall confine attention to incident sound whose wavelength greatly exceeds the correlation scale l, say, of the turbulence. This is the most important case in practice since, for infinitesimal Mach number M, the length scale of turbulence eddies having the same time scale (frequency) as the sound (and which therefore interact strongly with the sound) is $O(M)$ relative to the acoustic wavelength. In particular the theory will be applicable to the propagation of aerodynamically generated sound over large distances within the turbulence. This separation of length scales enables a *coherent* component of the sound to be introduced by averaging over regions of the turbulence which are large relative to l but small compared to the acoustic wavelength. Thus we shall write

$$p = \overline{p} + p', \quad V_i = \overline{V}_i + V'_i, \text{ etc.,} \tag{3.4}$$

where $\overline{p}$, $\overline{V}_i$, etc., are respectively the coherent acoustic pressure and perturbation velocity; p', V'_i are the corresponding fluctuating components (having zero means), and are statistically independent of the incident sound wave. When the acoustic wavelength $\gg l$ the intensity of the sound scattered out of the incident wave direction tends to be distributed evenly over all solid angles (Lighthill 1953). This permits the coherent wave to be interpreted as the propagating incident wave whose amplitude and phase are modified by scattering and *absorbtion* by the turbulence.

If the mean turbulence velocity vanishes, it follows from (3.3) that

$$\left(\frac{1}{c^2}\frac{\partial^2}{\partial t^2} - \nabla^2\right)\overline{p} = 2\rho_o \frac{\partial^2}{\partial x_i \partial x_j}\overline{U_i V'_j}, \tag{3.5}$$

and therefore that

$$\left(\frac{1}{c^2}\frac{\partial^2}{\partial t^2} - \nabla^2\right)p' = 2\rho_o \frac{\partial^2}{\partial x_i \partial x_j}U_i \overline{V}_j + 2\rho_o \frac{\partial^2}{\partial x_i \partial x_j}\left(U_i V'_j - \overline{U_i V'_j}\right). \tag{3.6}$$

The second of these equations will be used in conjunction with the perturbation momentum equation

$$\frac{\partial V_i}{\partial t} + \left(V_j \frac{\partial U_i}{\partial x_j} + U_j \frac{\partial V_i}{\partial x_j}\right) = -\frac{1}{\rho_o}\frac{\partial p}{\partial x_i} \tag{3.7}$$

to evaluate the interaction Reynolds stress $\overline{U_i V'_j}$ on the right of (3.5) in terms of $\overline{p}$. This will reduce (3.5) to an equation in $\overline{p}$ alone. To do this we use a *closure* approximation which is based on the observation that nontrivial contributions to $\overline{U_i(\mathbf{x},t)V'_j(\mathbf{x},t)}$ arise entirely from those components of $V'_j(\mathbf{x},t)$ which are correlated with the turbulent velocity $U_i(\mathbf{x},t)$ at $\mathbf{x}, t$. Such components of V'_j must have experienced an interaction with the turbulence within a distance $\leq O(l)$ from the point $\mathbf{x}$, and possible multiple interactions with the turbulence within a distance $\sim l/M$ from $\mathbf{x}$, i.e., within a distance travelled by a sound wave during the time scale of the turbulence. Hence, provided that the scattering of sound by the turbulence is sufficiently weak that multiple interactions over distances $\leq O(l/M)$

can be neglected (i.e., over distances $\sim$ wavelength of the sound), all terms describing such events may be discarded from (3.6), (3.7) when the solutions of these equations are to be used to calculate $\overline{U_i V'_j}$. In other words, in calculating p', V'_i in terms of $\overline{p}$, $\overline{V_i}$, it is sufficient to take (3.6), (3.7) in the form

$$\left(\frac{1}{c^2}\frac{\partial^2}{\partial t^2} - \nabla^2\right)p' = 2\rho_o \frac{\partial^2}{\partial x_i \partial x_j}(U_i \overline{V_j}) \tag{3.8}$$

$$\frac{\partial V'_i}{\partial t} = -\left(\overline{V_j}\frac{\partial U_i}{\partial x_j} + U_j \frac{\partial \overline{V_i}}{\partial x_j}\right) - \frac{1}{\rho_o}\frac{\partial p'}{\partial x_i}. \tag{3.9}$$

These equations constitute a *local Born approximation* for the random components p', V'_i, being the single order scattering product of the coherent wave and the turbulence.

When the solution of (3.8) is substituted into the final term on the right of equation (3.9), the latter determines $V'_i(\mathbf{x}, t)$ in terms of the turbulence velocity U_i and the coherent wave velocity $\overline{V_i}$. To the required order of approximation $\overline{V_i}$ may be expressed in terms of $\overline{p}$ by using the linearized coherent field momentum equation

$$\frac{\partial \overline{V_i}}{\partial t} = -\frac{1}{\rho_o}\frac{\partial \overline{p}}{\partial x_i}. \tag{3.10}$$

Having calculated $V'_i(\mathbf{x}, t)$ in this way it remains to insert the result into the right hand side of equation (3.5). In doing this velocity correlation products of the form $\overline{U_i(\mathbf{x}, t) U_j(\mathbf{X}, T)}$ are encountered. These may be simplified by assuming the turbulence to be *stationary random* in both position *and time*, in which case the correlation products are functions of $\mathbf{X}-\mathbf{x}$, $T-t$ alone, and a tensor correlation function $R_{ij}(\mathbf{x}, t)$ may be introduced with the property

$$R_{ij}(\mathbf{X}-\mathbf{x},\ T-t) = \overline{U_i(\mathbf{x}, t)\, U_j(\mathbf{X}, T)}. \tag{3.11}$$

Of course, the turbulence is not in general stationary in both time and space, and Crow (1967) overcomes this difficulty by introducing a system of random body forces (for example, buoyancy forces) which maintain the turbulence in equilibrium against dissipation. There are, however, several situations where the turbulence may be assumed to be stationary, or approximately so. For example, in the core region of fully developed turbulent pipe flow, and in the outer region of a turbulent boundary layer. Further, since the length scale of the turbulence responsible for absorbtion is much smaller than the acoustic wavelength, it is often true that, although the turbulence as a whole is decaying, the smaller scale motions may be assumed to be in equilibrium during the interval in which the interaction with the sound occurs.

Taking account of all these remarks, the coherent field equation (3.5) is reduced to an integro-differential equation of convolution-type (details of the calculation are similar to those described by Howe 1971). This equation has plane wave solutions $\bar{p} \propto \exp\{i(\mathbf{k}\cdot\mathbf{x}-\omega t)\}$, where $\mathbf{k}$ and ω are related by the *dispersion relation*

$$k^2 - \frac{\omega^2}{c^2} = \frac{2}{\omega}\int_{-\infty}^{\infty}\int_{-\infty}^{\infty} \frac{k_i k_j \psi_{ij}(\mathbf{k}-\mathbf{K}, \omega - \Omega)}{(\Omega + i0)} \times \left[\frac{2(\mathbf{K}\cdot\mathbf{k})^2}{[K^2 - (\Omega + i0)^2/c^2]} - k^2\right] d^3\mathbf{K}\, d\Omega. \tag{3.12}$$

In this result $\psi_{ij}(\vec{\kappa}, \Omega)$ is the wavenumber-frequency turbulence velocity spectrum, defined by

$$\psi_{ij}(\vec{\kappa}, \Omega) = \frac{1}{(2\pi)^4}\int_{\infty}^{\infty}\int_{\infty}^{\infty} R_{ij}(\mathbf{x}, t)\, e^{-i(\vec{\kappa}\cdot\mathbf{x} - \Omega t)}\, d^3\mathbf{x}\, dt, \tag{3.13}$$

and the integration with respect to Ω is assumed to pass above the simple poles at $\Omega = 0, \pm cK$. The latter condition derives from the requirement that the solutions of equations (3.8), (3.9) should be causally related to the interaction between the turbulence and the coherent wave field.

The integral on the right of (3.12) $\sim O(M^2)$ relative to each of the terms on the left hand side, so that the relevant solution $k \equiv k(\omega)$ of the dispersion relation lies in the neighborhood of $k_o = \omega/c$, appropriate for propagation in free space. To determine the correction consider a coherent wave propagating parallel to the unit vector $\mathbf{n}$, and set

$$\mathbf{k} = (k_o + i\mu)\,\mathbf{n}. \tag{3.14}$$

In general μ is a small complex quantity determined to leading order by inserting (3.14) into the left of (3.12) and setting $\mathbf{k} = k_o\mathbf{n}$ on the right hand side. This yields

$$\mu = \frac{ik_o^3}{\omega}\int_{-\infty}^{\infty}\int_{-\infty}^{\infty}\frac{n_i n_j \psi_{ij}(k_o\mathbf{n}-\mathbf{K},\,\omega-\Omega)}{(\Omega+i0)} \times\left[1-\frac{2(\mathbf{n}\cdot\mathbf{K})^2}{(k^2-(\Omega+i0)^2/c^2)}\right]d^3\,\mathbf{K}\,d\Omega. \tag{3.15}$$

The attenuation is determined by the real part of μ. Now $n_i n_j \psi_{ij}$ is real and non-negative for a stationary random medium (Stratonovich, 1963), and Re μ is therefore given by the contributions to the integral in (3.15) from the contour indentations around the poles $\Omega = \pm cK$, $\Omega = 0$. These correspond respectively to attenuation due to scattering and absorbtion by the turbulence, and we shall introduce the notation

$$\text{Re}\ \mu = \mu_S + \mu_A \tag{3.16}$$

to characterise their respective contributions to Re μ.

3.1 Attenuation Due to Scattering

Write $d^3\mathbf{K} = K^2 dK\, d\vec{\sigma}$, where $d\vec{\sigma}$ is the element of solid angle occupied by $(\mathbf{K}, d\mathbf{K})$. We then find from (3.15) that

$$\mu_S = \frac{\pi k_o^2}{c_o^4}\int\int \Omega^2(\vec{\sigma}\cdot\mathbf{n})^2 n_i n_j \psi_{ij}(k_o\mathbf{n} - \frac{|\Omega|}{c}\vec{\sigma}, \omega-\Omega)\, d\vec{\sigma}\, d\Omega, \tag{3.17}$$

the integration being over all directions of the unit vector $\vec{\sigma}$ and frequencies $-\infty < \Omega < \infty$.

This representation of μ_S associates the decay of the coherent wave with the scattering of energy into sound waves of frequencies Ω radiating into directions $\vec{\sigma}$. As noted previously by Lighthill (1953) and Kraichnan (1953), the scattered intensity vanishes at right angles to the incident wave direction ($\vec{\sigma}\cdot\mathbf{n} = 0$). Also, because $\kappa_i\kappa_j\psi_{ij}(\vec{\kappa},\omega) \equiv 0$ when the turbulence Mach number is sufficiently small for the motion to be regarded as incompressible, the scattered field vanishes in the back-scattering direction $\vec{\sigma} = -\mathbf{n}$, and in the forward direction $\vec{\sigma} = \mathbf{n}$ provided $\Omega \neq \omega$.

The detailed implications (3.17) will not be discussed here, since the attenuation defined by μ_S does not involve a transfer of acoustic energy to the turbulence, but merely its scattering into new directions of propagation. The interested reader is referred to the original papers of Lighthill and Kraichnan and to the additional references cited in the Introduction.

3.2 Attenuation Due to Absorbtion

This is given by the contribution to the Ω-integral in (3.15) from the indentation about the pole at $\Omega = 0$,

$$\mu_A = \frac{\pi k_o^2}{c} \int_{-\infty}^{\infty} n_i n_j \psi_{ij}(k_o \mathbf{n} - \mathbf{K}, \omega)\left[1 - \frac{2(\mathbf{K}\cdot\mathbf{n})^2}{K^2}\right] d^3\mathbf{K}, \tag{3.18}$$

an expression which confirms that the spectral components of the turbulence responsible for the attenuation have the same frequency as the incident sound. This is indeed necessary, since only those products of the wave-wave interactions of the sound with the turbulence which have zero (sum or difference) frequency can lead on average to a net transfer of energy. Note also that the pole at $\Omega = 0$ effectively localizes the contributions to (3.18) to non-propagating, incompressible perturbations of the turbulence caused by the passage of the sound wave, since it transforms the free space acoustic propagation operator $K^2 - \Omega^2/c^2$ to the Laplace operator K^2. In particular, of the two terms in the curly brackets in (3.18), the first corresponds to the effect of local straining of the vorticity by the sound (and arises from the first term on the right of (3.9)), whereas the second accounts for a longer range tendency of the incompressible components of the perturbation pressure to restore the turbulence to its original undisturbed state (second term on the right of (3.9)).

The integrand of (3.18) can be simplified by observing that turbulence velocity fluctuations at frequency ω are associated principally with wavenumbers $\vec{\kappa}$ satisfying $\kappa \sim \omega/U$, U being a typical turbulence velocity. Since $M \ll 1$, it follows that to a sufficient approximation we can write

$$\mu_A = \frac{\pi k_o^2}{c} \int_{-\infty}^{\infty} n_i n_j \psi_{ij}(\mathbf{K}, \omega)\left[1 - \frac{2(\mathbf{K}\cdot\mathbf{n})^2}{K^2}\right] d^3\mathbf{K}. \tag{3.19}$$

The order of magnitude of this expression is estimated by taking $\omega \sim U/l$, leading to

$$\mu_A \sim \frac{M^3}{l} \sim \frac{\epsilon}{c^3}, \tag{3.20}$$

where ϵ is the turbulence dissipation rate (see, e.g., Hinze 1975).

A more precise formula for μ_A can be given when the turbulence fluctuations (or rather the portion responsible for absorbtion) are assumed to be isotropic. In this case (Hinze 1975, §3.4)

$$\psi_{ij}(\mathbf{K},\omega) = \frac{E(K,\omega)}{4\pi K^4}(K^2\delta_{ij} - K_iK_j), \tag{3.21}$$

where the wavenumber-frequency energy spectrum $E(K,\omega)$ satisfies

$$\frac{1}{2}\overline{U_i^2} = 2\int_0^\infty \int_0^\infty E(K,\omega)\, dK\, d\omega, \tag{3.22}$$

and (3.19) becomes

$$\mu_A = \frac{2\pi k_o^2}{5c}\int_0^\infty E(K,\omega)\, dK. \tag{3.23}$$

This expression reduces to the attenuation coefficient obtained in the viscoelastic theory of Crow (1967) when $E(K,\omega)$ is assumed to have the separable form

$$E(K,\omega) = \frac{TE(K)}{\pi(1+\omega^2T^2)}, \tag{3.24}$$

where $E(K)$ is the usual three dimensional wavenumber energy-spectrum of the turbulence (Hinze 1975, §3.4), and T is a time scale governing the rate of return to isotropy of the strained turbulence. At sufficiently low frequencies this implies that $\mu_A \sim (k_o l)^2 M/l$, whereas as $\omega T \to \infty$, $\mu_A \sim M^3/l$.

An alternative prediction of the high frequency behaviour of μ_A can be made by noting that the main contribution to the integral in (3.23) would be expected to be from the region $K \sim \omega/u'$, where u' is a velocity characteristic of eddies of length scale $1/K$. If ω is large enough that K lies in the *inertial subrange*, dimensional arguments indicate that $u' \sim (\epsilon/K)^{\frac{1}{3}}$, and that the integral in (3.23) $\sim \sqrt{\omega/\epsilon}\, E(\sqrt{\omega^3/\epsilon})$. For such values of K, $E(K) \approx C\epsilon^{\frac{2}{3}}K^{-\frac{5}{3}}$, where $C \approx 1.7$ (Hinz 1975, §3.4), and therefore $\mu_A \sim \epsilon/c^3$, in agreement with the previous estimates.

Noir and George (1978) and Howe (1979c) have derived similar approximations for μ_A on the basis of a semi-empirical modelling of the perturbation turbulence Reynolds stress equation. There is, however, insufficient precise experimental data with which to compare these theoretical predictions, although order of magnitude comparisons made by Noir and George (1978) indicate that turbulence absorbtion cannot account for the whole of the observed excess attenuations (i.e., after the inclusion of attenuation by conventional relaxation mechanisms). Similar order of magnitude estimates suggest that the excess attenuation apparently evident in the shear layer experiments of Ross et al. (1983) are not satisfactorily explained by turbulence absorbtion.

In conclusion we note (following Crow 1967) that the predicted level of turbulence absorbtion is sufficient to ensure that the acoustic pressure fluctuations p_A, say, at a point x within an unbounded region of turbulence remain finite. These fluctuations are attributable to turbulence quadrupole sound sources (Lighthill 1952) located within a sphere of radius $\sim 1/\mu_A$ centered on x. The acoustic power emitted per unit volume of turbulence $\sim \rho_o U^3 M^5/l$, and (3.20) therefore implies that

$$\overline{p_A^2} \sim M p_t^2, \tag{3.25}$$

where p_t is the root mean square turbulence pressure ($\sim \rho_o U^2$).

4. Absorbtion of Sound by Wall Turbulent Shear Layers

There has been markedly less progress in analyzing the interaction of sound with turbulent boundary layers. Here we limit consideration to sound of wavelength $\gg \delta$, the boundary layer width. Introduce, as in Section 3, coherent components $\overline{p}$, $\overline{V}$, etc., of the sound, which is assumed to propagate along the x_1-axis, with the mean flow in the positive x_1-direction parallel to a smooth wall coinciding with $x_2 = 0$. At infinitesimal flow Mach numbers the principal component of the acoustic dissipation is given by

$$\Pi = \rho_o \int \left[\overline{\vec{\omega} \wedge \mathbf{u}} \cdot \overline{\mathbf{V}} + \nu \left(\frac{\partial \overline{V}}{\partial x_2} \right)^2 \right] d^3\mathbf{x}, \tag{4.1}$$

where $\overline{\mathbf{V}} = (\overline{V}, O, O)$, $\mathbf{u}$ is the turbulence velocity, ν the kinematic viscosity, and the integration is over the boundary layer turbulence (in $0 < x_2 < \delta$). The first term in the square brackets represents the rate of transfer of energy to the turbulence (Equation (2.5), and the second the dissipation due to viscous action. The latter is dominant in the viscous sublayer $x_2 v_*/\nu \lesssim 7$, v_* being the wall friction velocity. Note that $\overline{\mathbf{V}}$ is not irrotational in the neighborhood of the wall (as assumed in Section 2), but the derivation of (2.5) remains valid provided $\overline{\mathbf{V}}$ is statistically independent of the turbulence velocity fluctuations. $\overline{V}$ vanishes at $x_2 = 0$ and varies rapidly within an acoustic boundary layer of thickness $\delta_A \lesssim \delta$. Using the boundary layer approximation, the first term in the integrand of (4.1) may be replaced by $-\overline{u_1 u_2} \cdot \partial \overline{V}/\partial x_2$.

There is no deductive theory for the Reynolds stress $\overline{u_1 u_2}$ in a turbulent boundary layer, even when the flow is not modulated by incident sound. Recent analytical developments do not significantly advance the original phenomenological treatment due to Prandtl (Hinze 1975, Chapter 5), on which the following discussion is based.

Let $U(x_2)$ denote the mean flow velocity (whose dependence on x_1 is ignored) and suppose that the coherent velocity $\overline{V}$ represents a time dependent perturbation of the mean flow, such that

$$\overline{u_1 u_2} = -l_p^2 \left(\frac{\partial U}{\partial x_2} + \frac{\partial \overline{V}}{\partial x_2} \right)^2, \tag{4.2}$$

where the Prandtl *mixing length* l_p is assumed to be unaffected by the sound. The leading order approximation to (4.1) which accounts for the *net* (time averaged) dissipation of acoustic energy may then be expressed in the form

$$\Pi = \rho_o \int (2\epsilon_m(x_2) + \nu) \left(\frac{\partial \overline{V}}{\partial x_2} \right)^2 d^3\mathbf{x}, \tag{4.3}$$

where

$$\epsilon_m(x_2) = l_p^2 \frac{\partial U}{\partial x_2} \tag{4.4}$$

is the boundary layer *eddy viscosity* in the absence of sound.

The variation of $\overline{V}$ close to the wall is determined from the boundary layer approximation of the coherent part of the x_1-momentum equation (c.f. Howe 1979c). When $M \sim U/c \ll 1$, this reduces to

$$\frac{\partial \overline{V}}{\partial t} + \frac{1}{\rho_o}\frac{\partial \overline{p}}{\partial x_1} = -\frac{\partial}{\partial x_2}\left[\overline{u_1 u_2} - \overline{(u_1 u_2)_o}\right] + \rho_o \nu \frac{\partial^2 \overline{V}}{\partial x_2^2}, \tag{4.5}$$

where $-\overline{(u_1 u_2)_o}$ is the mean Reynolds stress in the absence of sound. When the acoustic wavelength $\gg \delta$ the variation of $\overline{p}$ across the boundary layer is negligible, so that if

$$\overline{V} = \hat{V}(x_2)\, e^{i(kx_1 - \omega t)}, \quad \overline{p} = \hat{p} e^{i(kx_1 - \omega t)}, \tag{4.6}$$

$\hat{p}$ may be taken to be constant, and just outside the boundary layer $\hat{V} = k\hat{p}/\rho_o\omega$. Substituting into (4.5) and using (4.2), (4.4), we find

$$\frac{\partial}{\partial x_2}\left[(\nu + 2\epsilon_m(x_2))\frac{\partial \hat{V}}{\partial x_2}\right] + i\omega\left(\hat{V} - \frac{k\hat{p}}{\rho_o\omega}\right) = 0, \tag{4.7}$$

which is to be solved subject to $\hat{V}(0) = 0$, $\hat{V}(\infty) = k\hat{p}/\rho_o\omega$.

In a fully developed turbulent boundary layer $\epsilon_m(x_2) \approx \kappa v_* x_2$ for $50 \lesssim x_2 v_*/\nu \lesssim L$ ($\sim 10^3$, typically) (see Hinze 1975, §7), where $\kappa \approx 0.41$ is the von Karman constant. For smaller values of x_2, $\epsilon_m \propto x_2^n$ ($n \geq 2$), but for the moment such departures from linearity are ignored except within the viscous sublayer where ϵ_m will be assumed to vanish. More precisely we shall set

$$\epsilon_m = \kappa v_*(x_2 - \delta_v), 0 \quad \text{for } x_2 \gtrless \delta_v, \tag{4.8}$$

where δ_v is the thickness of the viscous sublayer. Using this in (4.7) we find:

$$\hat{V} = \frac{k\hat{p}}{\rho_o\omega}\left(1 - \frac{H_0^{(1)}\left[a\left(1 + \frac{2\kappa v_*(x_2-\delta_v)}{\nu}\right)^{1/2}\right]}{[H_0^{(1)}(a)\cos b + H_1^{(1)}(a)\sin b]}\right), \quad x_2 \geq \delta_v$$

$$= \frac{k\hat{p}}{\rho_o\omega}\left[\frac{[H_1^{(1)}(a) + iH_0^{(1)}(a)]\sin\left(x_2\sqrt{\frac{i\omega}{\nu}}\right)e^{ib}}{[H_0^{(1)}(a)\cos b + H_1^{(1)}(a)\sin b]} + \left(1 - e^{ix_2\sqrt{\frac{i\omega}{\nu}}}\right)\right],$$

$$0 \leq x_2 < \delta_v \tag{4.9}$$

where $a = (i\omega\nu/\kappa^2 v_*^2)^{1/2}$, $b = \delta_v(i\omega/\nu)^{1/2}$, and $H_0^{(1)}$, $H_1^{(1)}$ are Hankel functions. In the absence of mean flow (4.9) becomes

$$\hat{V} = \frac{k\hat{p}}{\rho_o\omega}\left(1 - e^{ix_2\sqrt{\frac{i\omega}{\nu}}}\right), \quad x_2 \geq 0. \tag{4.10}$$

The general solution (4.9) may be used to determine the dissipation Π given by (4.3). This calculation is more easily performed by noting from (4.7) that the time averaged rate of dissipation can also be expressed in the form

$$E = \frac{k\nu}{4\omega}\left(\hat{p}^*\frac{\partial\hat{V}}{\partial x_2} + \hat{p}\frac{\partial\hat{V}^*}{\partial x_2}\right)_{x_2=0}, \tag{4.11}$$

an asterisk denoting the complex conjugate, where E is the dissipation *per unit area* of the wall. If E_ν is the corresponding dissipation in the absence of flow, due to viscous action alone, we find that

$$\frac{E}{E_\nu} = Re\left(\frac{\sqrt{2i}[H_1^{(1)}(a)\cos b - H_0^{(1)}(a)\sin b]}{H_0^{(1)}(a)\cos b + H_1^{(1)}(a)\sin b}\right). \tag{4.12}$$

This ratio gives a measure of the influence of the boundary layer turbulence on the sound, and its variation as a function of $\omega\nu/v_*^2$ is depicted by the solid curve in Figure 4(a) for the case $\delta_v v_*/\nu = 7$. According to this prediction the greatest departures from the viscous controlled dissipation occur for $\omega\nu/v_*^2 < 10^{-2}$. The dotted curve represents experimental results (discussed further below) reported by Ronneberger and Ahrens (1977), which reveal a discrete frequency range within which the simple theory is apparently inadequate. The dashed curve is the calculated fractional dissipation rate E_{δ_v}/E within the viscous sublayer (determined by restricting the integration in (4.3) to $0 < x_2 < \delta_v$). The dependence on frequency of the boundary layer thickness $\delta_A v_*/\nu$ (defined by the minimum value of x_2 at which $\hat{V} = k\hat{p}/\rho_o\omega$) is illustrated by the solid curve of Figure 4(b), the dashed curve is the ratio δ_A/δ_0, δ_0 being the acoustic boundary layer width in the absence of flow. According to these results δ_A is comparable with the width of the constant shear stress region of the turbulent boundary layer when $\omega\nu/v_*^2 \lesssim 10^{-3}$. At such frequencies, however, Figure 4(a) indicates that only about 50% of the dissipation actually occurs in the turbulent zone.

Ronneberger and Ahrens (1977) have estimated the dependence of E/E_ν on frequency by extrapolating to zero Mach number measurements made in an experimental study of sound propagation in turbulent pipe flow. Their results are reproduced in Figure 5, where E/E_ν is plotted against $\delta_+ = \sqrt{2v_*^2/\omega\nu}$ (Reynolds number based on friction velocity and the viscous length scale $\sqrt{2\nu/\omega}$.) The continuous curve gives the variation predicted by equation (4.12) for $\delta_v v_*/\nu = 7$. The latter exhibits a weak minimum at $\delta_+ \approx 6$ (where the predicted attenuation is less than in the absence of flow), but only qualitative agreement with experiment is evident in the range $4 \lesssim \delta_+ \lesssim 20$. In order to determine if this discrepancy is a consequence of the use of the crude approximation (4.8) to $\epsilon_m(x_2)$, the computations were repeated using the following empirical fit to turbulent pipe flow data assembled by Hinze (1975, p. 732):

$$\epsilon_m(x_2) = \kappa x_2 v_*\big[1 - \exp(-a_1 x_+ - a_2 x_+^2)\big], \quad x_+ = x_2 v_*/\nu. \tag{4.13}$$

where $a_1 = 0.0175$, $a_2 = 0.00015$. The boundary layer equation (4.7) must now be integrated numerically. Remarkably, the discrepancy between the corresponding prediction of E/E_ν (dashed curve in Figure 5) and experiment is greater than for the case of the simplified model (4.8)!

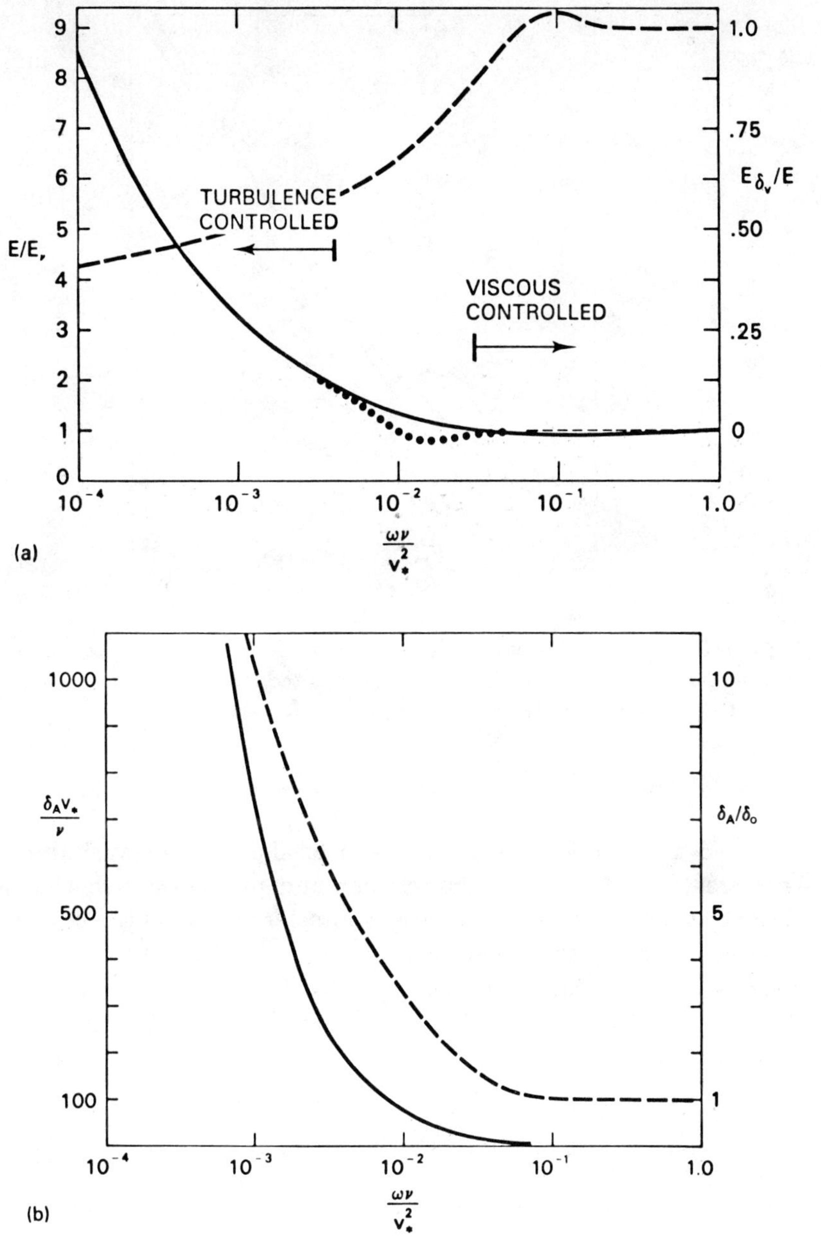

Figure 4. (a) Dependence of the attenuation on frequency: — E/E_ν, theory equation (4.12); ··· E/E_ν, experiment, Ronneberger and Ahrens (1977); - - - E_{δ_v}/E, theory. (b) Predicted dependence of acoustic momentum boundary layer thickness δ_A on frequency: — $\delta_A v_*/\nu$; - - - δ_A/δ_o.

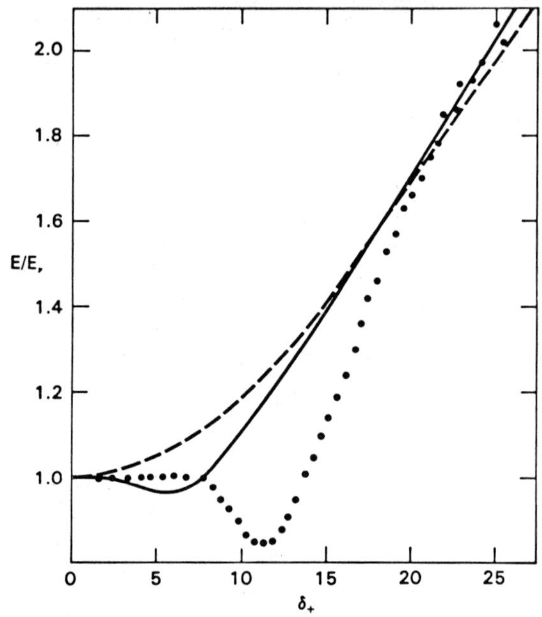

Figure 5. Comparison of theory and experiment E/E_ν: ··· experiment (Ronneberger and Ahrens 1977); — theory, equation (4.12); - - - theory, using the empirical eddy viscosity (4.13).

4.1 The Thermal Boundary Layer

No account has yet been taken of the periodic variations in temperature due to the compressions and rarefactions of the wall shear flow by the sound. The rate of transfer of heat in the boundary layer is increased by turbulent convection. We shall assume that the coherent temperature fluctuations $\overline{T} = \hat{T}(x_2)e^{i(kx_1-\omega t)}$ satisfy for $M \ll 1$ the following boundary layer approximation to the equation of heat transfer:

$$\frac{\partial}{\partial x_2}\left[(\chi + 2\epsilon_\theta(x_2))\frac{\partial \hat{T}}{\partial x_2}\right] + i\omega\left(\hat{T} - \frac{\hat{p}}{\rho_o c_p}\right) = 0, \qquad (4.14)$$

where χ is the thermometric conductivity of the fluid, $\epsilon_\theta(x_2)$ is the thermal eddy diffusivity in the absence of the incident sound, and c_p is the specific heat at constant pressure. We shall also set

$$\epsilon_\theta = \frac{\epsilon_m}{P}, \qquad (4.15)$$

where, according to Blom (1970), the turbulence Prandtl number P is approximately constant and lies between 0.7–0.8 for air. Bradshaw (1977) quotes a somewhat higher value of 0.91, however.

In equation (4.14) $\hat{p}/\rho_o c_p$ is the coherent temperature perturbation outside the boundary layer of the sound wave. When the dependence of ϵ_m on x_2 is modelled by (4.8), the solution $\hat{T}$ of (4.14) which satisfies $\hat{T}(\infty) = \hat{p}/\rho_o c_p$ and the usual wall condition $\hat{T}(0) = 0$, can be written down by making the obvious substitutions in equation (4.9). For the more general variation (4.13) of ϵ_m, the equation must be solved numerically.

4.2 Sound Propagation in Turbulent Pipe Flow

The above results are now applied to the propagation of sound in a smooth pipe in the presence of low Mach number, fully developed turbulent flow. The pipe has uniform cross-section $\mathcal{A}$ of perimeter $l_{\mathcal{A}}$, and the acoustic frequency is assumed to be sufficiently small that only plane waves can propagate (in the x_1-direction). An approximate equation for the coherent pressure $\overline{p}$ is derived by integrating the axial momentum equation and the continuity equation over the pipe cross-section. When account is taken of the momentum and thermal boundary layers at the pipe wall (but the very much weaker effect of the interaction of sound with the core turbulence, discussed in Section 3, is neglected), we find (c.f. Howe 1979c).

$$\begin{aligned}&\left[\frac{1}{c^2}\left(\frac{\partial}{\partial t}+U\frac{\partial}{\partial x_1}\right)^2-\frac{\partial^2}{\partial x_1^2}\right]\overline{p}\\ &\quad=\frac{\rho_o l_{\mathcal{A}}}{\mathcal{A}}\left[\nu\frac{\partial}{\partial x_1}\left(\frac{\partial\overline{V}}{\partial x_n}\right)-\frac{\chi}{T_o}\left(\frac{\partial}{\partial t}+U\frac{\partial}{\partial x_1}\right)\left(\frac{\partial\overline{T}}{\partial x_n}\right)\right],\end{aligned}\tag{4.16}$$

where the mean velocity U is in the positive x_1-direction and is regarded as constant outside the wall region, and T_o is the mean temperature. The derivatives on the righthand side are evaluated at the wall, x_n being a local wall-normal coordinate directed *into* the fluid. These can be calculated provided that solutions of the appropriate wall boundary layer equations are available at *the small but finite* Mach number $M = U/c$ of the mean flow. When this is so the dispersion relation for sound waves proportional to $e^{i\{kx_1-\omega t\}}$

becomes

$$k^2 - \left(\frac{\omega}{c} - Uk\right)^2 = \frac{il_{\mathcal{A}}k}{\mathcal{A}}F(k,\omega), \tag{4.17}$$

where F is a dimensionless function of k, ω.

The attenuation is determined by Re(F), and to leading order we have

$$k = \frac{\pm\omega/c}{(1 \pm M)} \pm i\alpha_{\pm}, \tag{4.18}$$

where the $\pm$ sign is taken according as propagation is in $\pm x_1$-direction, and the attenuation coefficient is given by

$$\alpha_{\pm} = Re\left[\frac{l_{\mathcal{A}}}{\mathcal{A}(1 \pm M)}F\left(\frac{\pm\omega/c}{1 \pm M}, \omega\right)\right]. \tag{4.19}$$

Now the earlier theoretical treatment of the acoustic boundary layers is strictly valid for $M \to 0$. The difficulty at finite values of M arises from the need to include in the boundary layer equations (4.7), (4.14) variable mean flow terms proportional to M. An approximate interpolation formula giving the dependence of $\alpha_{\pm}$ on M can be determined by means of the following argument. Suppose that $\omega\nu/v_*^2 \ll 1$ at the maximum relevant value of M. Asymptotically as $\omega\nu/v_*^2 \to 0$ the whole of the dissipation takes place in the turbulent flow where the mean velocity may be assumed constant and equal to U. In this case $\alpha_{\pm}$ is determined by replacing ω by $\omega - Uk$ in equations (4.7), (4.14). This provides a functional form of $F(k,\omega)$ in which k is subsequently replaced by $\pm(\omega/c)/(1 \pm M)$, as in (4.19). But when $M \to 0$, $\omega\nu/v_*^2 \to \infty$ for fixed ω, and the boundary layers are now entirely within the viscous sublayer. Since also $k \approx \pm\omega/c$, it is evident that the functional form of (4.19) obtained previously for *uniform mean flow* at finite M continues to give the correct attenuation as $M \to 0$, and will therefore be assumed to provide an interpolation formula for intermediate values of M.

The effectiveness of this procedure can be assessed by comparison with data published by Ronneberger and Ahrens (1977). The predicted and measured dependence of $\alpha_{\pm}/\alpha_o$ on M is depicted in Figure 6 for sound of frequency 630 Hz propagating in air in a circular pipe of internal diameter 2 cms, where α_o is the thermoviscous attenuation in the absence of flow, given by

$$\alpha_o = \frac{l_{\mathcal{A}}}{2\mathcal{A}c}\sqrt{\frac{\omega}{2}}(\sqrt{\nu} + (\gamma - 1)\sqrt{\chi}), \tag{4.20}$$

γ being the ratio of specific heats. The predictions have been made using the simple model (4.8) of ϵ_m and for a turbulence Prandtl number $P = 0.8$, since use of the more accurate representation (4.13) of ϵ_m is not expected to improve their accuracy. The moderate agreement between theory and experiment exhibits the greatest discrepancies in the range $0.1 \leq M \leq 0.2$, and corresponds to the discrepancy noted previously in Figure 5. This suggests that estimates of $\alpha_\pm$ which take explicit account of the variation of mean velocity within the acoustic boundary layers will not significantly improve the agreement with experiment.

Finally, let us estimate the magnitude of the mean square pressure fluctuations $\overline{p_A^2}$ of aerodynamic sound generated in fully developed, low Mach number turbulent pipe flow (c.f. Section 3). It is readily deduced (Howe 1979c, §4) that $\alpha_\pm \sim Ml_{\mathcal{A}}/\mathcal{A}$, and only turbulence quadrupoles within a distance $\sim \mathcal{A}/Ml_{\mathcal{A}}$ can influence the acoustic field at a given point. According to Davies and Ffowcs Williams (1968) the acoustic power generated per unit volume $\sim \rho_o U^3 M^3/l$, l being a turbulence length scale, and therefore $\overline{p_A^2} \sim (\mathcal{A}/l_{\mathcal{A}}l)Mp_t^2$, where p_t is the root mean square turbulence pressure averaged over a cross-section. This order of magnitude result is effectively identical with (3.25) for the case of an unbounded turbulent medium.

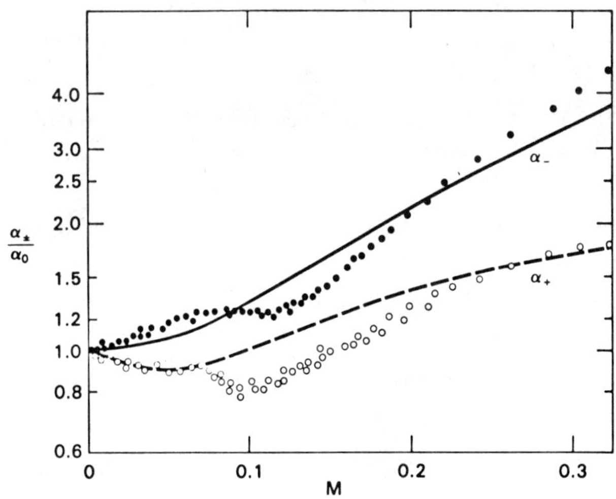

Figure 6. Attenuation of sound of frequency 630 Hz in turbulent pipe flow as a function of mean flow Mach number M. (i) Downstream propagation: ··· experiment (Ronneberger and Ahrens 1977), — theory; (ii) Upstream propagation: o o o o experiment, - - - theory.

References

[1] Ahrens, C. and Ronneberger, D. *Acustica 25* (1971), 150–157.

[2] Ashley, H. and Landahl, M. *Aerodynamics of Wings and Bodies,* Reading: Addison-Wesley, 1965.

[3] Batchelor, G. K. *Naval Hydrodynamics (Chapter 16),* Washington, D. C.: National Acad. Sci. Pub. 515 of Nat. Res. Counc., 1957.

[4] ————————. *An Introduction to Fluid Dynamics,* C.U.P., 1967.

[5] Bechert, D. W. *Sound Absorbtion Caused by Vorticity Shedding, Demonstrated With a Jet Flow,* AIAA Paper 79-0575, 1979.

[6] ———————. "Sound Sinks in Flows, a Real Possibility?" In *Mechanics of Sound Generation in Flows.* Ed. E. A. Müller. New York: Springer, 1979.

[7] Blom, J. *An Experimental Determination of the Turbulent Prandtl Number in a Temperature Boundary Layer.* Doctoral thesis, Eindhoven University of Technology, 1970.

[8] Bradshaw, P. *Ann. Rev. Fluid Mech. 9* (1977), 33–54.

[9] Brown, E. H. and Clifford, S. F. *J. Acoust. Soc. Am. 60* (1976), 788–794.

[10] Crighton, D. G. *Proc. Camb. Phil. Soc., 65* (1969), 557–566.

[11] Crow, S. C. *Phys. Fluids, 10* (1967), 1587–89.

[12] ——————. *J. Fluid Mech. 37* (1969), 529–563.

[13] Davies, H. G. and Ffowcs Williams, J. E. *J. Fluid Mech. 32* (1968), 765–778.

[14] Dean, P. D. and Tester, B. J. *Duct Wall Impedance Control as an Advanced Concept for Acoustic Suppression,* NASA CR-134998, 1975.

[15] Ffowcs Williams, J. E. and Howe, M. S. *J. Fluid Mech. 58* (1973), 461–480.

[16] Guedel, A. *Scattering of an Acoustic Field by a Free-Jet Shear Layer.* AIAA Paper 83-0698, 1983.

[17] Hinze, J. O. *Turbulence,* New York: McGraw-Hill, 1975.

[18] Howe, M. S. *J. Fluid Mech. 45* (1971), 785–804.

[19] ——————. *J. Fluid Mech. 91* (1979), 209–229.

[20] ——————. *Proc. R. Soc. A366* (1979), 205–223.

[21] ——————. *J. Fluid Mech. 94* (1979), 729–744.

[22] ——————. *Proc. R. Soc. A370* (1980), 523–544.

[23] ———. *Proc. R. Soc. A374* (1981), 543–568.

[24] Ingard, U. and Singhal, V. K. *J. Acoust. Soc. Am. 55* (1974), 535–538.

[25] Jones, D. S. *J. Inst. Math. Applics. 9* (1972), 114–122.

[26] Kambe, T. and Mya Oo, U. *J. Phys. Soc. Japan. 50* (1981), 3507–16.

[27] Korman, M. S. and Beyer, R. T. *J. Acoust. Soc. Am. 67* (1980), 1980–87.

[28] Kraichnan, R. H. *J. Acoust. Soc. Am. 25* (1953), 1096–1104.

[29] Lighthill, M. J. *Proc. R. Soc. A211* (1952), 564–587.

[30] ———. *Proc. Camb. Phil. Soc. 49* (1953), 531–551.

[31] ———. *J. Fluid Mech. 91* (1979), 253–317.

[32] Lyon, R. H. *J. Acoust. Soc. Am. 31* (1959), 1176–82.

[33] Noir, D. T. and George, A. R. *J. Fluid Mech. 86* (1978), 593–608.

[34] Rienstra, S. W. *J. Fluid Mech. 108* (1981), 443–460.

[35] Ronneberger, D. and Ahrens, C. D. *J. Fluid Mech. 83 (1977), 433-464.*

[36] Ross, R. et al. *Acoustic Wave Propagation Through the Shear Layer of the DNW Large Open Jet Wind Tunnel.* AIAA Paper 83-0699, 1983.

[37] Salikuddin, M. et al. *Refinement and Application of Acoustic Impulse Technique to Study Nozzle Transmission Characteristics.* NASA CR-3656, 1983.

[38] Stratonovich, R. L. *Topics in the Theory of Random Noise, Vol. 1.* New York: Gordon and Breach, 1963.

[39] Walker, W. M. and Reising, G.F.S. *Chem. Process. Engng. 49* (1968), 95–103.

[40] Welsh, M. C., Stokes, A. N., and Parker, R. *Flow-Resonant Sound Interaction in a Duct Containing a Plate: Part 1. Semicircular Leading Edge.* Submitted to *J. Sound Vib.*, 1983.

[41] Vér, I. L. "Perforated Baffles Prevent Flow-Induced Acoustic Resonances in Heat Exchangers." In *Fortschritte der Akustik (FASE/DAGA '82), Vol. 1,* Göttingen, 1982, pp. 531–534.

Part II
Rotor Acoustics

Analyses of Broadband Noise Mechanisms of Rotors

A. R. George
Cornell University
Ithaca, NY 14853

Abstract

The various source mechanisms which generate broadband noise on a range of rotors are reviewed. Analyses of these mechanisms are presented and compared to existing experimental data. The sources considered are load fluctuations due to inflow turbulence, due to turbulent blade boundary layers passing the trailing edge, and due to tip vortex formation turbulence. Vortex shedding noise due to laminar boundary layers and blunt trailing edges is not considered in detail as it can be avoided in most cases. Present analyses are adequate to predict the spectra from a wide variety of experiments on fans, helicopter rotors, and wind turbines to within about 5 to 10 dB. Better knowledge of the inflow turbulence improves the accuracy of the predictions. Inflow turbulence noise depends strongly on ambient conditions and dominates at low frequencies. Trailing edge and tip vortex noise are important at higher frequencies if inflow turbulence is weak. Boundary layer trailing edge noise increases slowly with angle of attack but not as rapidly as tip vortex formation noise. Tip noise can be important at high angles of attack for wide chord, square edge tips.

1. Introduction

Many aspects of rotor noise are still poorly understood, including the relative importance of various noise generation mechanisms. This is partly because there are a large number of noise mechanisms on rotors which can be important in different parts of the acoustic frequency spectrum, depending on the rotor parameters and operating environment. The wide variety of source mechanisms is due to various aeroacoustic effects including those due to: boundary layer, separated flow, and inflow turbulence; high Mach numbers, including nonlinear effects; blade-vortex interactions; nonuniform inflow;

etc. (George, 1978). In general, the mechanisms each affect different parts of the acoustic spectrum. Then, on craft with either tandem or with main and tail rotors, many of these mechanisms can interact with each other and between rotors. Thus in many cases it has not been clear what mechanisms are dominant in many operating conditions for full scale helicopters, propellers, etc.

This paper addresses a part of the problem: broadband noise. It will review broadband noise mechanisms, review various researchers' analyses of them and their assumptions, and compare calculations based on the available analyses to each other and to available experimental data. The aims of the work are to help understand which broadband noise mechanisms are important in which circumstances and to evaluate the various analytical approaches. It will be seen that several satisfactory analytical approaches are available. They are shown to predict absolute spectra to within about 5 dB of *clean* experiments. The paper also uses them to show which mechanism is important in which case and in what parts of the spectrum.

The frequencies of interest in rotor noise are usually determined by human annoyance (or by detection in some cases). The common measures of annoyance, such as the perceived noise level (PNdB) or A-weighted sound level (dBA) account for the fact that humans find low frequencies, say below a few hundred Hertz, much less annoying. On the other hand, if long distance propagation is a factor for the rotor in question, then high frequencies can be attenuated significantly by molecular absorption. For example, after propagation over 1 kilometer, frequencies above a few thousand Hz are attenuated drastically (George, 1978). Thus, frequencies of a few hundred to several thousand Hz are of primary interest.

The amount of power radiated by a rotor is generally extremely small compared to the aerodynamic power consumed by the rotor (by a factor of the order of 10^5 or more). Thus the acoustics do not affect the rotor performance to any extent. Noise is radiated by forces, volume displacements, or nonlinearities which are either unsteady or vary in their effects, when considered in terms of retarded time (George, 1978). In general, rotor noise can be divided into three categories:

1. **Discrete frequency noise** (sometimes called rotational or harmonic noise) is caused by steady or harmonically varying forces, volume displacements, or nonlinear flow effects. For

low to moderate blade tip Mach numbers these can be due to the basic blade rotation and forward flight of a helicopter or to steady inflow variations. These mechanisms were analyzed some time ago by Gutin, Deming, Lowson, Wright, etc. The steady loading noise is generally restricted to the first dozen or so harmonics of the blade passing frequency. For helicopter main rotors or large wind turbines these frequencies lie in the frequency range below 100 Hz where the ear is not very sensitive. These low order harmonics are, however, very important to the higher RPM propeller or tail rotor cases. Harmonically varying effects can, in principle, be important at higher frequencies for all rotors; but, in fact, the higher frequencies tend to be caused by loads, displacements, or nonlinearities which are impulsive, high Mach number, or randomly varying (caused by turbulence). In these other cases the phenomena are better analyzed specifically as impulsive or broadband, as discussed below.

2. **Impulsive noise** (sometimes called blade slap) consists of more or less distinct repeated pulses at blade-passing frequency. If these pulses are Fourier-analyzed they yield discrete or harmonic spectra, but their particular identity is due to their impulsive time histories and origins. In particular they are caused by events at certain blade azimuth angles such as blade-vortex interactions or local transonic blade motion toward the observer (say Mach numbers greater than approximately 0.75). These noise sources have been analyzed by Widnall (1971), Widnall and Wolf (1980), Ffowcs Williams and Hawkings (1969), Farassat (1975), Farassat and Succi (1980), and George and Chang (1983). Impulsive noise is unquestionably the most important noise source on helicopter or wind turbine rotors when it exists. However, a prime goal of aeroacoustic rotor design and operation is to avoid impulsive noise generation by control of blade-vortex interactions and avoiding high tip speeds. This leaves broadband noise as the important controlling factor in many situations where relative tip speeds are not transonic and blade-vortex interactions are avoided.

3. **Broadband noise**, which is the subject of this study, has a continuous (although sometimes humped or peaked) spectrum and is caused by disturbances which are not precisely repeated each blade revolution but which are basically random in nature. These random disturbances are generally due to some

sort of turbulence interacting with the rotor blades. The turbulence can either be incident on the blade from the ambient atmosphere or be generated by the blade motion itself. Recent reviews of broadband noise research may be found in Schlinker and Brooks (1982) and George and Chou (1983).

2. Broadband Noise Mechanisms and Analyses

In the present study emphasis is placed on *first principles* analyses, which make absolute predictions of noise spectra. These are not merely based on empirical correlation equations and thus do not require the determination of empirical constants for different families of rotors. A few comparisons are given for some of the scaling-law-based correlations available in the literature.

The first broadband noise prediction methods were based on empirical correlation of overall sound pressure level (OASPL) as, for example, by Widnall (1969). Previously, very early investigators had erroneously identified broadband noise with some sort of turbulent *vortex shedding* from the rear of the blades; hence the early name *vortex noise*. Actually, most broadband noise is due to force fluctuations on the blades which are in turn due to the influences of turbulent flows. Later it was found that, in the atypical case of laminar flow, the laminar boundary layers on blades can indeed shed nearly regular vortices at the trailing edge and thus radiate a narrow-peaked broadband sound, sometimes called *high frequency broadband noise*. (See e.g., Paterson et al. (1973), Aravamudan et al. (1979).) However, this source is not important on most full-scale rotors, except perhaps on helicopter tail rotors or small fans. Even then it can be eliminated easily by tripping the boundary layers (see Aravamudan et al., 1979). Blunt trailing edges can also radiate sound due to vortex shedding (Brooks and Hodgson, 1981) but, analogously, this can be avoided by using a sharp trailing edge.

Three types of analyses have been applied to date to calculate broadband radiation. The Lighthill formulation of aeroacoustics is:

$$\begin{aligned}\frac{1}{c_o^2}\frac{\partial^2 p}{\partial t^2}-\frac{\partial^2 p}{\partial x_i^2}&=\frac{\partial Q}{\partial t}-\frac{\partial F_i}{\partial x_i}+\frac{\partial^2 T_{ij}}{\partial x_i \partial x_j}\\&\equiv g\left(x_i,t\right)\end{aligned} \tag{2.1}$$

where

$p =$ far field perturbation pressure
$\rho =$ density
$c_o =$ undisturbed speed of sound
$Q =$ mass source strength: mass/volume/time
$F_i =$ force/volume $=$ momentum/volume/time
$T_{ij} =$ Lighthill stress $= \rho u_i u_j + (p - c_o^2 \rho)\, \delta_{ij} - \sigma_{ij}$
$\sigma_{ij} =$ viscous stress tensor.

and where the right hand side of equation (2.1), $g(x_i, t)$ is assumed to be known. Then the radiated sound can be found from the general solution

$$p = \frac{1}{4\pi} \int\!\!\int\!\!\int d^3 x_i \int dt'\, g(x_i', t) \frac{\delta\,(t - R/c_o - t')}{R} \tag{2.2}$$

of equation (2.1). For broadband sound the primary contribution to g is from the random forces applied to the fluid by the presence of the blades. (Hawkings and others have identified quadrupole terms which can be important in some high Mach number situations, but in these cases harmonic and impulsive radiation usually dominate broadband noise.)

The three approaches which have been applied so far to broadband noise are:

(1) Treat the general case of unsteady forces distributed in space following equation (2.1) with specialization to rotating blades.

(2) Approximate the distributed blade forces as rotating concentrated forces (dipoles) using a result of Ffowcs Williams and Hawkings (1969b).

(3) Find the radiation from an airfoil segment in rectilinear motion and then approximate a rotating blade by a sequence of straight line motions.

2.1 Inflow Turbulence Noise

The analysis of the sound generated by the unsteady forces due to turbulence on unducted rotors began with the quite general analysis of Homicz and George (1974) using the first method mentioned above. See Figure 1 for geometry. They devised an analysis for the sound radiated from arbitrarily-varying forces on a rotor disk, obtaining for the spectrum in the far field

$$\langle S(x_i, f)\rangle = \frac{f^2}{4\rho_o c_o^3}\frac{x_i x_j}{r^4}\int\int d^2x_i' \langle P_{L_iL_j}\left(x_{ij}' - \frac{f}{c_o}\hat{x}_i, f\right)\rangle \tag{2.3}$$

where $\langle P_{L_iL_j}(x_i, k_i, f)\rangle$ is the generalized time-averaged three-dimensional cross-spectral density of the rotor loads per unit area on the plane $x_3 = 0$. Then the analysis was applied to the varying loadings on a rotating blade in the disk. The loadings were obtained from an approximate compressible aerodynamic analysis for an inflow of isotropic turbulence defined by a Dryden spectrum. A good approximation to the rotor loads was shown to be chordwise compact but spanwise finitely correlated. Blade-to-blade correlations were treated rigorously. Due to the polar coordinates describing the blade motion and the Doppler shifts from the circular motions, the form of the final result includes a double sum over two series of Bessel functions as well as an integration over the inflow turbulence's wavenumber. The form of the result is:

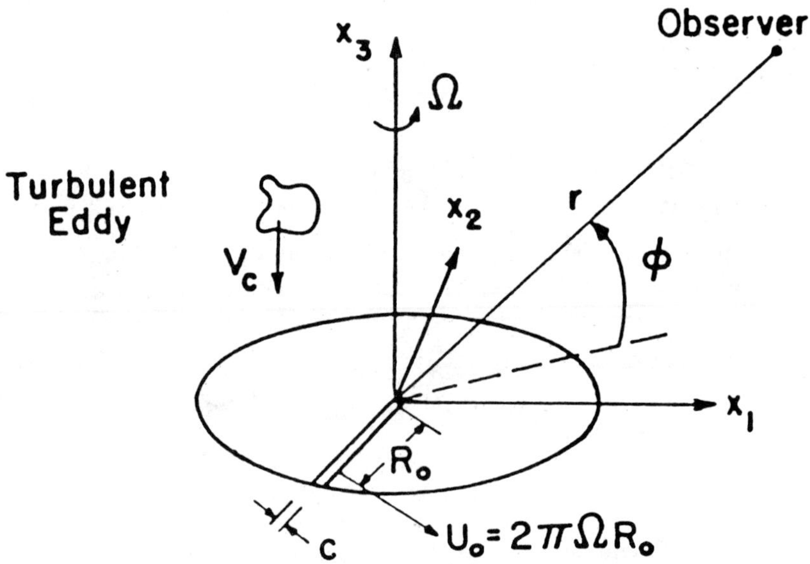

Figure 1. Rotor geometry showing definitions of symbols.

$$\langle S(x_i, f)\rangle \propto \frac{\Omega M_t^2 R_o^2}{\rho_o c_o^3 r^2} \int_{\xi_{min}}^{\infty} d\xi \left[\sum_n \sum_l G(\xi, n, l)\, J_{nB-l}^2\left(\frac{f}{\Omega}\alpha\right) \cdot J_l^2\left(\frac{M_o}{M_c}\left[\left(\frac{V_c}{\Lambda\Omega}\right)^2 \xi^2 - \left(\frac{f}{\Omega} - nB\right)^2\right]^{1/2}\right)\right] \tag{2.4}$$

where

$\langle S(x_i, f)\rangle$ = acoustic spectrum

ξ = dimensionless turbulent wave number

G = function including turbulence spectrum, unsteady aerodynamic, finite blade span effects, etc.

Λ = turbulence integral scale

Ω = shaft rotational frequency

B = number of blades

M_o = rotational Mach number at R_o

M_c = axial convection Mach number

M_t = r.m.s. turbulence Mach number

$\alpha = M_o \cos\phi$

Inflow turbulence was seen to be an important noise source over a range of frequencies. Also, the analysis explained the humped or peaked nature of the low frequency part of the spectrum as due to the larger scale components of the turbulent inflow. These large scale components give nearly periodic (i.e., modulated) disturbances as they are swept through the rotor plane and are intercepted by consecutive blade passages; this leads to nearly periodic but finite bandwidth radiated sound. Because of the complexities of evaluating the two infinite sums and the integration, the analysis is not well suited to computation at high frequencies. Computations using equation (2.4) become nearly prohibitively expensive for spectra past the range of about 500 to 1000 Hz for typical helicopter main rotor parameters. Thus high frequency analyses developed by George and Kim (1976) and by Amiet (1977), and variations on these two by Harris and co-workers (Aravamudan and Harris, 1979; Humbad and Harris, 1980).

The method of George and Kim (1976) is based on the second method mentioned above, assuming fluctuating forces rotating in a circle whose components in the observer's direction are statistically stationary. This effectively restricts the analysis to forces normal to the rotor plane and thus does not account for the smaller fluctuating torque forces which can be significant for observers near the plane of the rotor, nor does it account for the detailed radiation directionality of the blade elements. The assumption of circular motion also does not allow accurate treatment of forward-flight helicopter cases. However, as will be seen in this paper, these are not important restrictions except for within about 10 to 15 degrees of the rotor plane or for advance ratios greater than about 0.4.

The analysis starts from the result of Ffowcs Williams and Hawkings (1969) for the radiated spectral density of a single rotating point force:

$$\langle S(x_i, f)\rangle = \frac{f^2}{4\rho C_o^3 r^2} \sum_{n=-\infty}^{\infty} D_r(f - n\Omega)\, J_n^2\left(\frac{f}{\Omega}\alpha\right) \tag{2.5}$$

where D_r is the power spectral density of the force fluctuation in the observer direction. D_r is then derived for different mechanisms of blade loading including inflow turbulence (George and Chou, 1983; George and Kim, 1976), turbulent boundary layer-trailing edge noise (Kim and George, 1980; Chou and George, 1983), and tip vortex formation turbulence (George et al., 1980, 1983).

The analysis gives a reasonably simple equation for the radiated sound. It gives some analytical insight into the noise generation mechanism's dependence on rotor parameters, and it can also be numerically evaluated quite straightforwardly using various asymptotic methods.

The method of Amiet (1976a, 1977) is based on a different concept. Initially Amiet had analyzed the radiation of sound from a stationary, nonrotating airfoil in a uniform mean flow containing turbulence (Amiet, 1975; Paterson and Amiet, 1977). This analysis accounted for the full range of wavelength-to-chord ratios and accurately predicted the directionality of the radiation. (The directionality becomes multi-lobed at intermediate wavelength-to-chord ratios as can be seen from Figure 2, taken from Kaji (1975).) Then, later, as sketched schematically in Figure 3, Amiet used these results

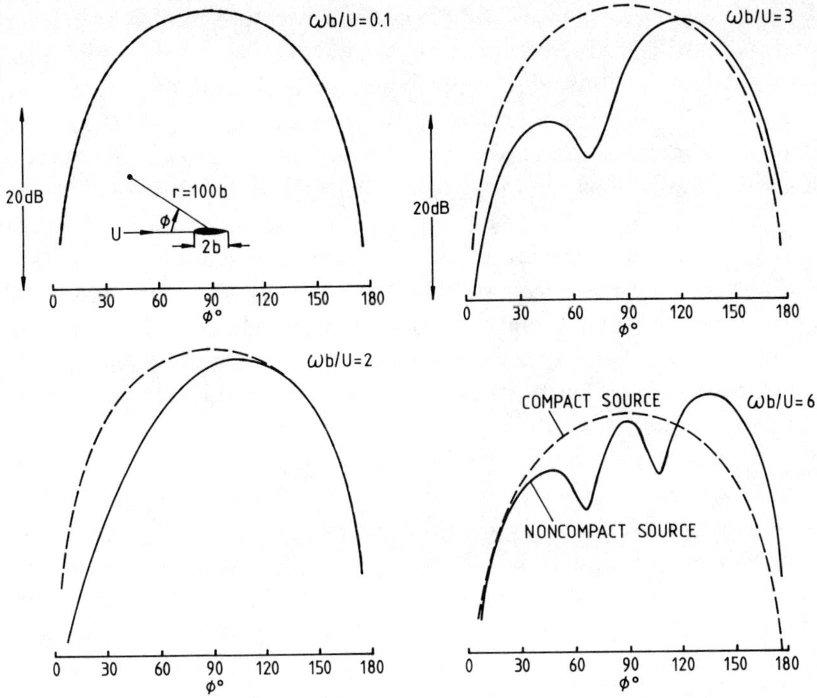

Figure 2. Comparison of directivity patterns of radiated sound predicted by compact and noncompact sources, $M = 0.5$.

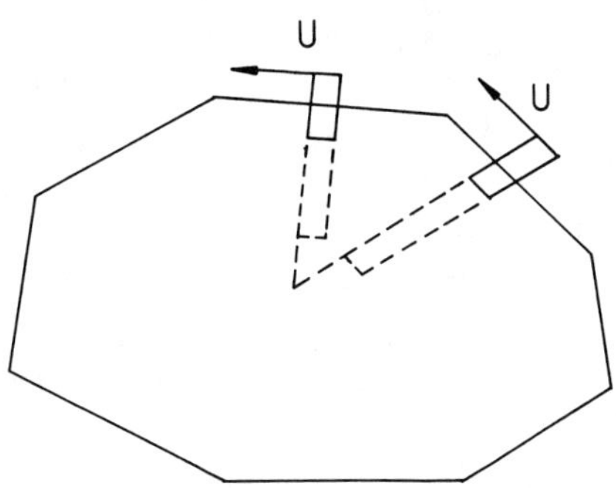

Figure 3. Schematic sketch of Amiet's method (actual increments used are 22.5°).

to synthesize the average radiation from rotating blades by numerically summing and averaging the radiation from a series of blade straight-line motions which approximate the circular (or epicycloidal) motion of a hovering (or translating) rotor (Amiet, 1976a, 1977). This approach can be shown to be accurate for high frequencies, in a similar manner as the analysis of George and Kim (1976). Amiet's approach also includes the effects of blade-to-blade correlations for turbulent scales which are small compared to the rotor size. Amiet's method has the advantage of being able to treat forward flight easily and being based on a more exact model of radiation directionality. However, at a fixed observer position, one sums and averages the multi-lobed radiation pattern over the range of directions from the rotating blade. Thus the pattern is smoothed out into a pattern which, except very near the rotor plane, is quite close to the simpler dipole pattern resulting from the approximations of George and Kim. This can be visualized by thinking about averaging Figure 2 over a range of angles and is borne out by the calculations presented in this paper and in George and Chou (1983) and Kim and George (1980). Other broadband noise analyses have been developed by Harris and co-workers (Aravamudan and Harris, 1978; Humbad and Harris, 1981) in conjunction with a series of experiments on broadband noise from model rotors. These analyses were based on variations of George and Kim's and Amiet's methods.

The methods of George and Kim and of Amiet were used in the present study for inflow turbulence calculations. The inflow turbulence itself can be due to the natural turbulence in the atmosphere or due to upstream disturbances as in the case of a helicopter tail rotor ingesting the wake of the main rotor. In making the calculations for this study the incident turbulence properties often had to be estimated based on the average measured properties of atmospheric turbulence (Lumley and Panofsky, 1964) or were rough estimates based on energy considerations (Tennekes and Lumley, 1972). In those cases alternative calculations were made to illustrate the sensitivity to the likely range of values. Another difficult problem in estimating the turbulent inflow properties is the anisotropy of the inflow due to the distortion of the turbulence by the contracting streamlines entering hovering rotors, stationary propellers, or fans (George, 1978; Hanson, 1974; Pegg et al., 1977). At present this effect also can only be estimated and alternative calculations made.

2.2 Boundary Layer-Trailing Edge Noise

Noise is also produced by the self-generated turbulence in a blade boundary layer passing over the trailing edge of the blade. This was recognized as far back as 1959 (Powell, 1959). Various investigators looked at very simplified models for this noise, but these early models were not complete and were primarily only useful as bases for correlations. Fink (1978), for example, used such a correlation to predict the on-axis noise of a rotor due to boundary layer trailing edge noise. Complete first-principles analyses of rotor trailing edge noise were developed more recently by Kim and George (1980), and by Schlinker and Amiet (1981). Also Hubbard et al. (1981) have recently proposed an OASPL and spectrum peak correlation for trailing edge noise on wind turbine rotors.

The analytical problem of sound radiating from the effects of turbulence being convected past a non-rotating trailing edge has been studied intensively since about 1970. A variety of model problems were studied (see the review of Howe, 1978) but these primarily resulted in scaling laws which needed empirical constants to be determined. There also remained a number of questions regarding the details of the modeling and the effects of the Kutta condition. On the other hand, Amiet developed a method which is based on solving the problem of a statistically stationary pressure field being convected past a trailing edge (Amiet, 1976b, 1978). This method has been compared to experimental findings by Brooks (1981) and found to be consistent. Amiet's results only depend on the pressure spectrum in the boundary layers being known from experiments.

In 1980 Kim and George (1980) constructed an analysis of boundary layer noise from rotors by using blade forces derived from Amiet's flow model to calculate D_r in the same manner as they had earlier for inflow turbulence noise. They integrated Amiet's pressure distributions to obtain the fluctuating force distributions needed for the dipole model used. Thus their analysis is again restricted to angles not too close to the rotor plane and to low advance ratios for helicopter forward flight. In the calculations given in this paper, an airfoil boundary layer thickness correlation was used (Kim and George, 1980; Chou and George, 1983) instead of the flat plate results used in the original publications. Later, Schlinker and Amiet (1981) used the same numerical summing and averaging method that Amiet (1977) had used for inflow turbulence to treat the trailing edge noise problem. Again we will see that the dipole method gives essentially the same results as Amiet's method except within about 15 degrees of the rotor plane where additional source terms should be included in both methods.

2.3 Tip Vortex Noise

Another source of broadband noise on airfoils or rotors is that of locally-separated flow from local stall or from tip vortex formation. Kendall (1978), Ahtye et al. (1979), and Fink and Bailey (1980) experimentally observed localized noise sources at wing and flap tips. Changes in noise from changes in rotor tip shape were experimentally observed some time ago by Lowson et al. (1972), although these effects may have been due to blade loading changes. Also earlier, Revell (1975), for the airframe noise case, had argued on an energy basis that vortex drag and associated turbulence in the trailing vortices must lead to additional noise in some manner. George et al. (1980) have identified this effect with the turbulence in the vortex formation and local separation region over the blade tip interacting with the trailing edge.

Their model starts from the experimental observations of separation on the suction side of blade or rotor tips, due to the boundary layer being swept around the tip by the pressure gradients at the tip. A separated vortex flow results which is very similar to the flow over the top surface of a sharp-edged delta wing in subsonic flow. Figures 4 and 5 are sketches of delta wing and wing/rotor tip flows.

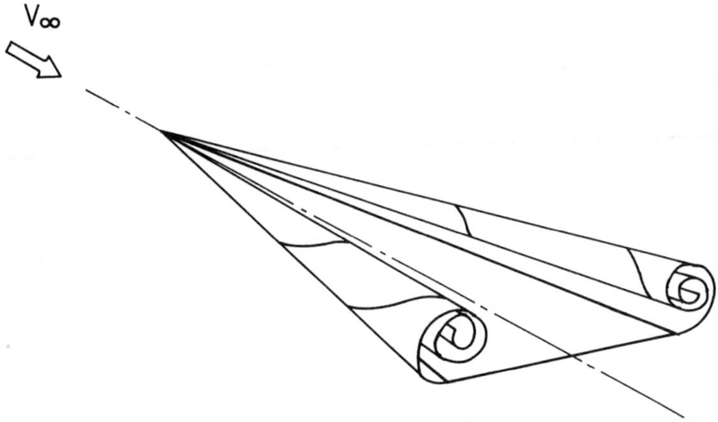

Figure 4. Vortex formation on a lifting delta wing.

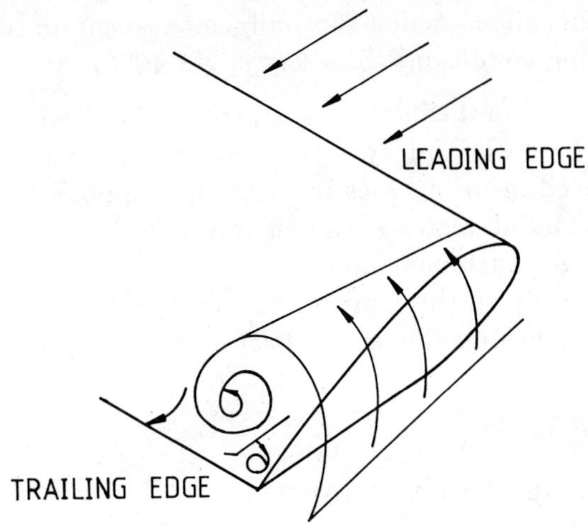

Figure 5. Vortex formation on the tip of a lifting wing or rotor blade.

It is known that these leading-edge vortices are quite turbulent. Large fluctuating pressures have been measured on the surfaces of delta wings under these vortices. George et al. (1980, 1983) used these data, pressure fluctuation data from two-dimensional flows, and data on the geometry and velocities associated with wing and rotor tip flows to estimate the separated turbulent pressure spectrum being convected past the trailing edge. Then this information was used to predict the load spectrum D_r and the resulting radiated sound in a manner similar to George and Kim's treatments of inflow turbulence and boundary layer turbulence noise. This tip vortex noise is shown to increase with blade loading as has been experimentally observed in many cases. The updated version of the analysis (George et al., 1983) uses turbulent pressure data measured under vortices on delta wings, which are then correlated with rotor tips using experimental length and velocity scales. This version is used to compute spectra for the various cases in the present study.

2.4 Other Mechanisms

Another local-separation turbulence-noise mechanism is that of local stall due to a high angle of attack from close proximity to a vortex from a previous blade passage. This phenomenon was studied experimentally by Paterson et al. (1975), but presently no analytical model of either the local separation or noise radiation is available. However, available detailed pressure measurements on rotor blades indicate that this phenomenon is usually not present on rotors under normal operating conditions (Shockey et al., 1976).

As mentioned in the introduction, trailing edge vortex-shedding from laminar blade boundary layers is a noise mechanism which can be eliminated in most cases by tripping the boundary layer. A similar mechanism of vortex shedding from blunt trailing edges has been identified for turbulent flows as well by Brooks and Hodgson (1981). Analogously to the laminar flow case, this noise source can be eliminated by using sufficiently sharp trailing edges.

3. Comparisons to Experiments

As discussed in more detail in George and Chou (1983) there are not very many experiments reported in the literature which present *clean* spectral data, unaffected or minimally affected by extraneous influences such as reverberation, engine or drive motor noise, etc. In order to find experiments suitable for comparisons, we also looked for the most complete measurements available of the inflow turbulence spectrum, and tried to choose representative data covering a range of rotors from wind turbines to helicopters to propellers. We did not consider data where only overall sound pressure level or octave band data were given, as this type of averaged data is inadequate to differentiate between source mechanisms and analyses.

In the comparisons, the data estimated as input to the analyses and to the correlations is given in the figure captions. Other input parameters were taken from the experimental papers. Inflow velocities were estimated using simple momentum theory with thrust determined by simple blade-element theory. As mentioned above, the inflow turbulence values were often estimated. In many cases we will show inflow turbulence noise calculations based on both the von Karman and the Dryden models for the inflow velocity spectrum. As will be discussed in Section 4, the von Karman spectrum is probably the better model for atmospheric turbulence.

In cases where separate calculations are shown for separate mechanisms, the results should be summed to compare to the experiments. However, in order not to clutter the figures, this was not done in most cases. It is comparatively easy to envision the sum on the decibel scale as the higher of the curves is essentially the level with only a 3 dB increase if two curves have the same level, and a 4.8 dB increase if three curves have the same level.

The cleanest full scale helicopter rotor data available are probably those of Leverton and Pollard (1973) and Leverton (1973). Leverton tested a full-scale rotor on a test rig, in an inverted position, in order to eliminate the effect of the recirculation which occurs when a rotor wake is directed toward the ground. He reported spectra measured using a tethered ballon at a range of angles to the rotor plane. His tests varied both load and RPM, although in the present paper we have chosen only a few cases for comparison. Unfortunately, neither the turbulent intensity nor scale were measured. As the inflow was drawn from near the ground, the integral scale could be quite reasonably estimated from the fairly well-established empirical relationship that $\Lambda = 0.9$ h, where Λ is the integral scale, and h is the height above the ground (Etkin, 1961). Similarly, values for turbulent intensity for various weather conditions can also be estimated from the extensive data and correlations in Lumley and Panofsky's (1964) monograph. Leverton also reported on some indoor model-scale tests, but we did not use those results, as the data did not include inflow turbulence information as did some other available model scale data.

Figure 6 compares a range of predictions to one set of data from Leverton (1973a). These data were taken at an angle of −75 degrees from the rotor plane where all of the analyses would be expected to be within the range of their assumptions. It is clear that the correlation of Pegg (1979) is too high for this case. It is also clear that at the lower frequencies, say below 1000 Hz, the boundary layer trailing edge noise and the tip vortex noise mechanisms are not important. However, at frequencies above about 1000 Hz they become important, with boundary layer trailing edge noise being the more important in this case. Fink's boundary layer noise correlation is seen to be a reasonable approximation to the more exact boundary layer noise calculations. Most of the noise below 1000 Hz is shown to be inflow turbulence noise based on the estimated turbulence properties. Both the analysis of George and Kim (1976) and that of Amiet (1977) agree within about 5 dB with each other, and with Leverton's data.

In Figure 7 a comparison is shown of calculations (based on the three mechanisms of George and co-workers, and on the two of Amiet) to some data of Leverton at −11.5 degrees from the rotor plane. As these analyses ignore in-plane force components, and as George and Kim use a blade dipole directivity, the agreement would not be expected to be quite as good; on the other hand, it is not clear how much of the differences are due to which effects.

To our knowledge, the spectra calculated by Johnson and Katz (1972) based on the measurements of noise from a UH-1 are still the only satisfactory full-scale complete helicopter broadband spectra available in the literature. The spectra, of course, contain both main and tail rotor noise as well as airframe, engine, and gearbox noise. In this case too, no inflow turbulence information was measured, so values of integral scale and intensity were again estimated using the helicopter altitude and assumed weather conditions. Figure 8 compares some calculations for the main rotor only with the full-scale complete spectrum. As mentioned previously, other sources besides the main rotor are present in this case, and it is difficult to estimate turbulence inflow properties. However, the results are reasonable, being within about 10 dB, and show the importance of all three mechanisms in various parts of the spectrum.

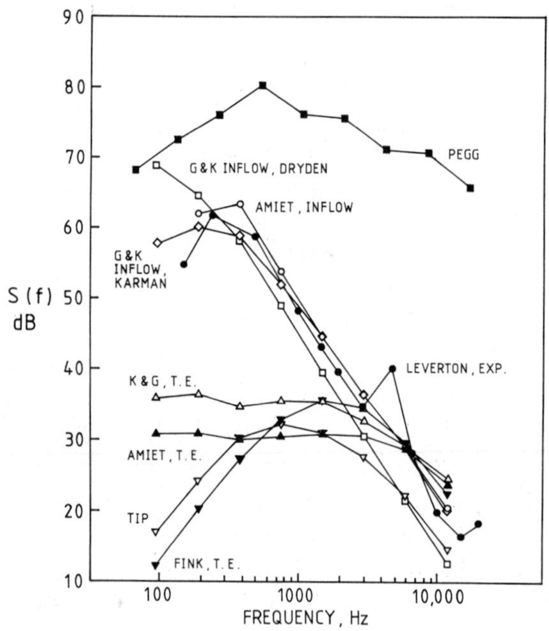

Figure 6. Comparison of the range of predictions for a full sized helicopter rotor with experiment of Leverton (1973a), $\phi = -75°$, $\Lambda = 0.57$ m, $\sqrt{\overline{w^2}} = 1$m/s.

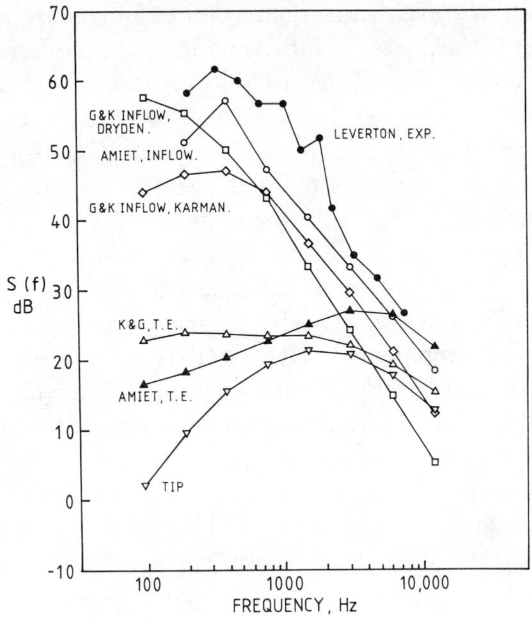

Figure 7. Comparison of predictions with the experiment of Leverton (1973a), $\phi = -11.5°$, same turbulence properties as in Figure 6.

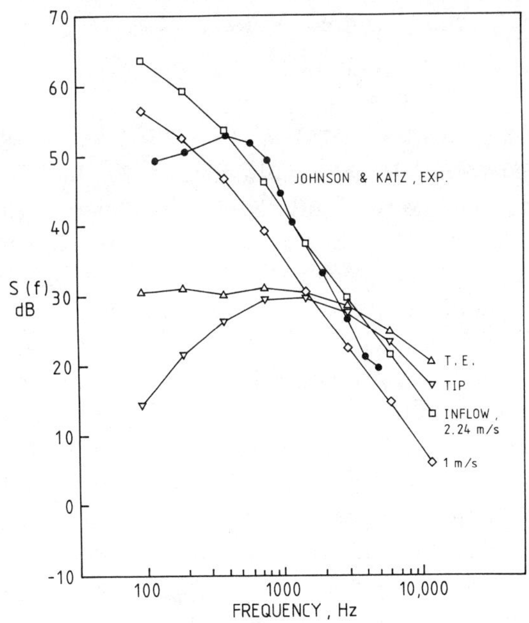

Figure 8. Comparison of main rotor calculations with experiment of Johnson and Katz (1972), for full-scale helicopter $\phi = -27°$, $\Lambda = 27$ m.

Two sets of suitable full-scale wind turbine noise measurements were found in the reports of Hubbard and his co-workers (Hubbard et al., 1981; Shepherd and Hubbard, 1982). Measurements were made for the MOD-OA turbine at ground level at a distance of 61 meters from the tower base in the rotor plane and at angles of 45 and 90 degress from it. Measurements for the MOD-2 turbine were made at ground level, both in the rotor plane and at several locations under the turbine axis. In both cases the background noise was measured and shown to be well below the measured spectra. However, no information on turbulence intensity or scale was measured. In these cases again the turbulence's characteristics were estimated, aided by the fact that at least the wind speed was known! (It should be noted that wind turbine noise annoyance may not be controlled by broadband noise, but by low frequency impulsive noise, due to interactions with specific atmospheric inhomogeneities or vortices shed by upwind towers.) Figures 9 and 10 show comparisons to this full-scale wind turbine data. Here again some extraneous sources may have been present and turbulence properties had to be estimated. Nevertheless, the agreement with the predicted absolute spectra is good, and it is seen that the primary source for frequencies above a few hundred Hertz is boundary layer trailing edge noise as was also suggested by Hubbard and co-workers.

Several sets of indoor tests of rotors in anechoic facilities are available for comparisons to analyses. One set was carried out by Paterson and Amiet (1979) in the UTRC anechoic wind tunnel facility on several model rotors. In these tests, both vertical ascent and forward flight were simulated, and different grids were used upstream to generate controlled inflow turbulence. Measurements were made of both the turbulent intensity and scale, so that in these cases none of the parameters needed for input to the analyses needed to be estimated. Figures 11 through 13 show comparisons of calculations to data presented by Paterson and Amiet (1979). In the no-grid (low inflow turbulence) case, it is clear that both tip and boundary layer noise are important at the higher frequencies. In all the cases the calculations and experiments agree to within about 5 dB. In their report, Paterson and Amiet have also shown good agreement with Amiet's inflow turbulence analysis whenever the primary noise source was inflow turbulence.

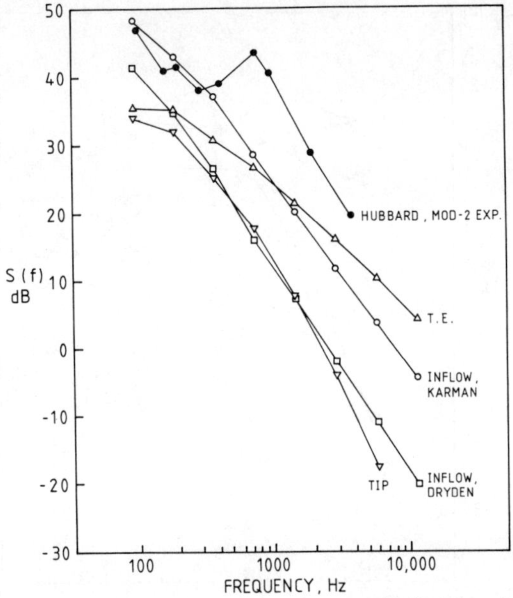

Figure 9. A comparison of predictions with Hubbard et al.'s (1981) experiment for a MOD-2 wind turbine, $\Lambda = 55$ m, $\sqrt{\overline{w^2}} = 1$m/s, ground distance = 69 m.

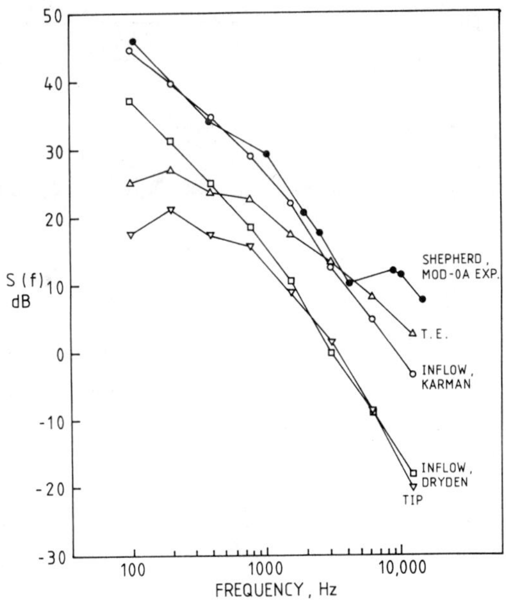

Figure 10. Comparison of predictions with Shepherd and Hubbard's experiment (1982) for MOD-OA wind turbine, $\Lambda = 27.5$ m, $\sqrt{\overline{w^2}} = 1$m/s, ground distance = 61 m.

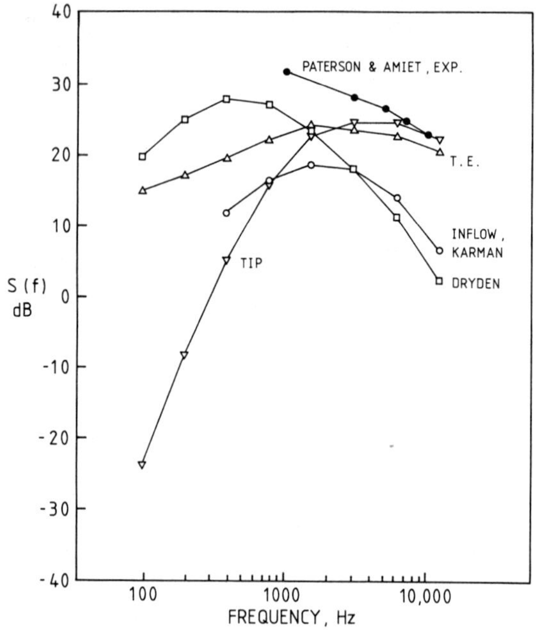

Figure 11. Comparison of predictions for model rotor, experiment of Paterson and Amiet (1979), no grid, Test VA-C-1.

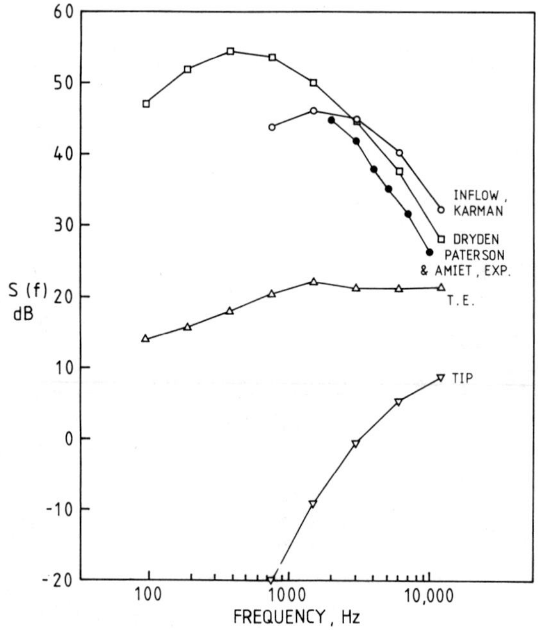

Figure 12. Comparison of predictions for model rotor, experiment of Paterson and Amiet (1979), medium grid, Test VA-M-4.

Similar sets of high quality model scale tests were carried out in the MIT anechoic wind tunnel and reported in a series of papers by Aravamudan and Harris (1978) and Humbad and Harris (1981). A number of the test results contained large amounts of laminar vortex shedding noise (high-frequency broadband noise) which partially covered the more interesting sources. However, in most cases, if this peaked part of the spectrum is ignored, significant parts of the measured spectra are still useful for comparisons. In these experiments, as in those of Paterson and Amiet, different grids were used upstream and the turbulent intensity and scale were measured. Figures 14–16 show similar satisfactory agreement to Humbad and Harris' (1981) model test data. The calculations satisfactorily account for varying turbulent intensity (from changing the upstream grid) and changing advance ratio (tunnel speed). Another similar set of experiments carried out earlier by Aravamudan and Harris (1978) also showed good agreement with calculations (George and Chou, 1983). The analyses in Harris and co-workers' original papers, which were based on those of George and Kim and of Amiet also showed good agreement with the data.

The final set of data we will use for comparison is Lowson et al.'s (1972) for a low speed fan. Tests were run in an anechoic room, with both ducted and unducted fans, and both before and after recirculation built up in the room. The non-recirculating conditions were better defined and were used for our comparisons. RPM, tip angle, and tip shape were varied as well. The turbulence was measured by a limited frequency range hot wire, giving reasonable estimates for the turbulent intensity in the room before recirculation set in. We estimated the turbulent integral scale to be 0.1 meter. The boundary layer was not tripped in these tests, thus leading to some laminar vortex shedding humps in the high frequency part of the spectrum which we again ignored. (It is interesting to note that when the tip was cut back in Lowson's tests, the laminar boundary layer hump was greatly reduced.) Figure 17 compares some of the data to correlations and calculations. Again the correlation of Pegg (1979) seems to be too high. Here the inflow turbulence noise is predicted to be important over the full range of frequencies, and the calculations agree very well. Figure 18 for a different RPM, and Figure 19 for a different blade pitch angle also show excellent agreement.

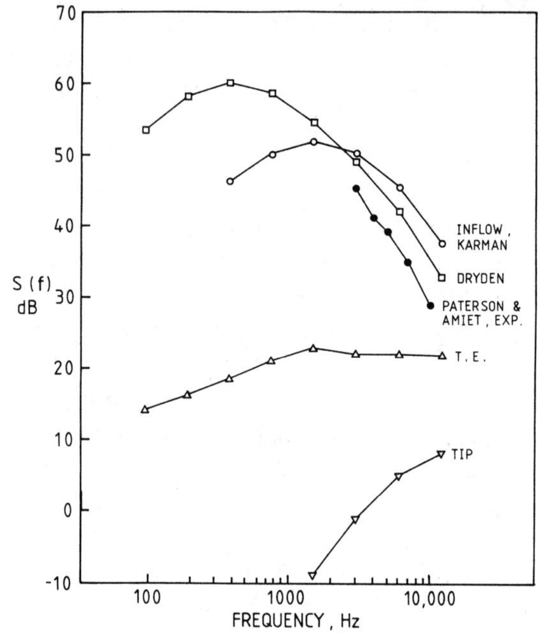

Figure 13. Comparison of predictions for model rotor, experiment of Paterson and Amiet (1979), large grid, Test VA-L-3.

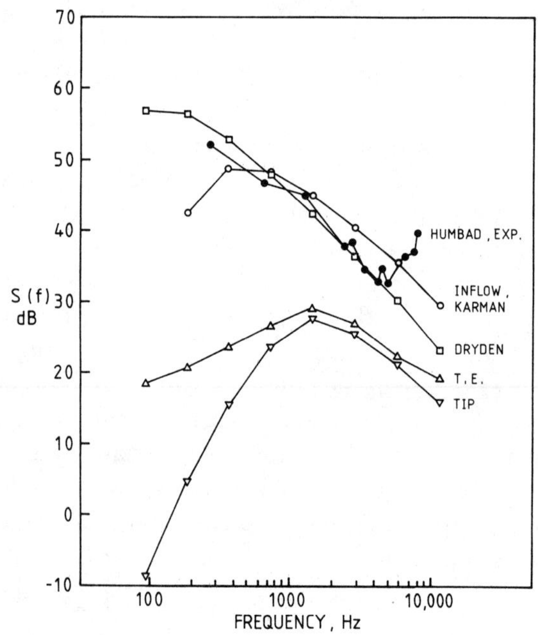

Figure 14. Comparison of predictions for model rotor, experiment of Humbad and Harris (1981), small grid, advance ratio = 0.125.

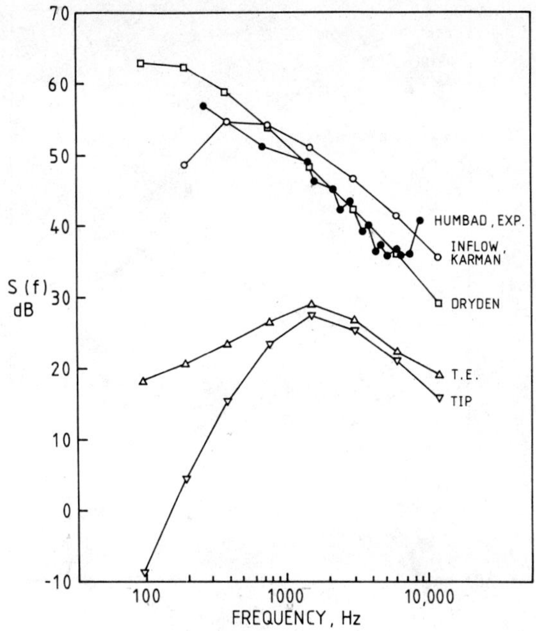

Figure 15. Comparison of predictions for model rotor, experiment of Humbad and Harris (1981), small grid, advance ratio = 0.25.

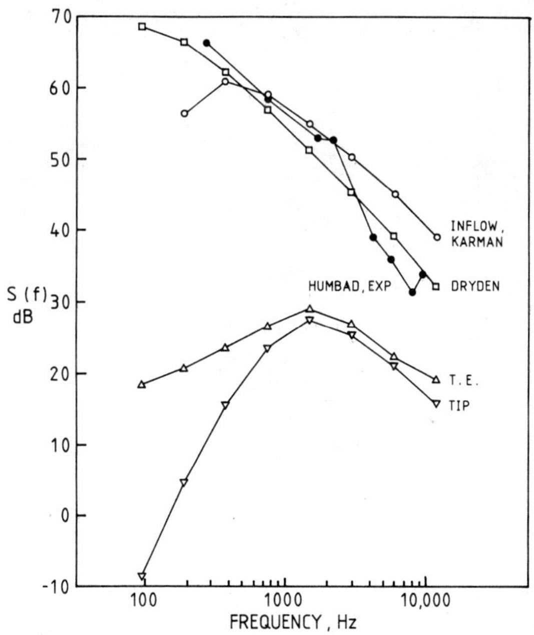

Figure 16. Comparison of predictions for model rotor, experiment of Humbad and Harris (1981), large grid, advance ratio = 0.25.

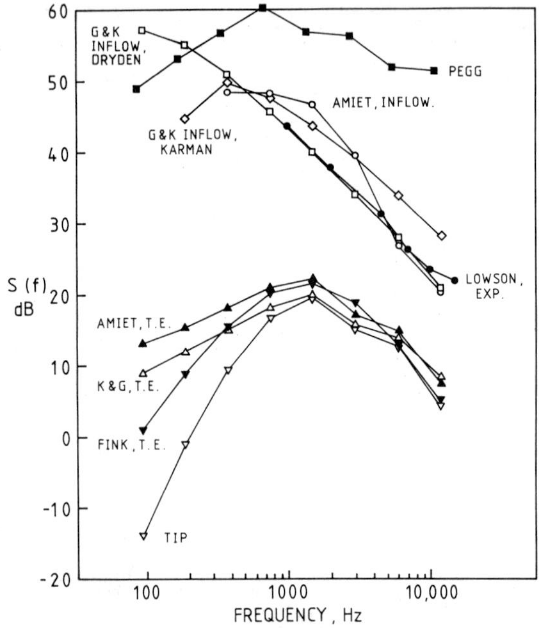

Figure 17. Comparison of predictions for low speed fan, experiment of Lowson et al. (1972), 1500 RPM, tip pitch = 15°.

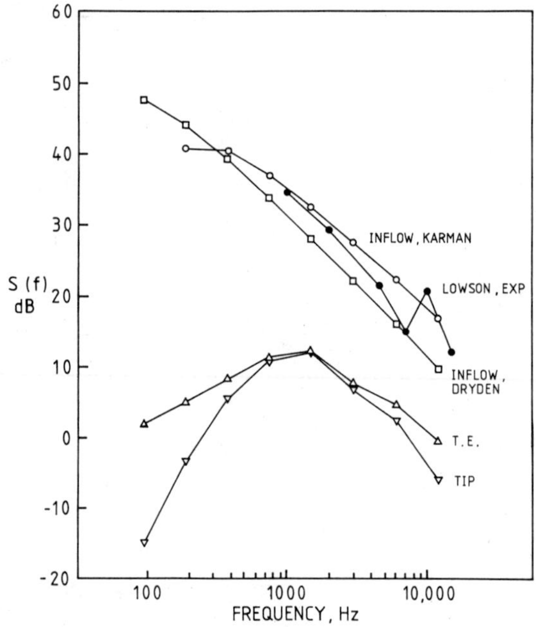

Figure 18. Comparison of predictions for fan noise, experiment of Lowson et al. (1972), 1000 RPM, tip pitch = 15°.

4. Comparisons of Analyses to Each Other

In this section the results calculated by the methods of George and co-workers and of Amiet and co-worker are compared to show the effects of the different assumptions in the analyses. The computational approach of Amiet allows treatment of forward flight (non-zero advance ratio) and of more accurate basic blade noise directionality. The George and Kim approach has been implemented for both the von Karman model and the Dryden model of the inflow turbulence spectrum. We will examine each of these effects by comparing the results of calculations made by the different methods.

Figure 20 gives plots of the Dryden spectrum, which is available in the George and Kim inflow-turbulence model, and of the Karman spectrum, which is available in both the George and Kim and in the Amiet models. It is clear that the Karman model contains more energy at high frequencies. Although the Doppler shifts make it more complicated, one can roughly identify a given frequency radiation with the inverse time for a blade to pass through a turbulence component of length $1/k$, where k is the wave number. Thus, for frequencies on the order of 10 KHz at a tip speed of 100 m/s, one is interested in turbulence components' wave number of order 100 m^{-1}. If, as for a full-scale helicopter, the integral scale is of order 100 m then the peak of the spectrum is at wave number k of order 0.01 m^{-1}, implying that the high frequencies come from wave numbers 10^4 higher than the inverse integral scale. Referring to Figure 20, we see that the difference between the two atmospheric turbulence models can be of order 10 dB at these wave numbers. For another way of looking at it, Figure 21 shows comparisons between inflow-turbulence noise calculations for a full-scale rotor for both 0.1 m and 67.0 m integral scales. It is apparent that for a small integral scale and low frequencies, the results for different assumed turbulence spectra are in close agreement, although even then the differences become more marked at high frequencies. One concludes that the von Karman spectra should be used for cases where the integral scale is large, such as in the atmosphere. For other cases, such as wind tunnel grid turbulence, the choice of spectrum models is not clear.

Figure 22 shows the effect of forward flight on inflow turbulence noise as calculated by Amiet's analysis, and compared to hover calculations based on George and Kim's analysis. It is notable that the advance ratio effect is not important for cases of interest

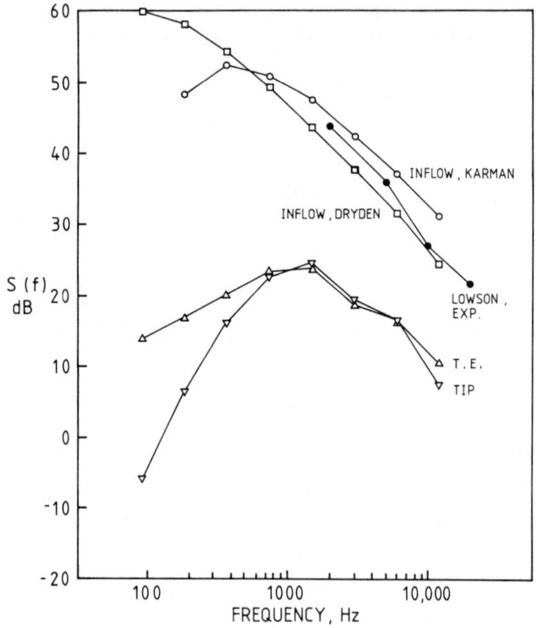

Figure 19. Comparison of predictions for fan noise, experiment of Lowson et al. (1972), 1600 RPM, tip pitch = 20°.

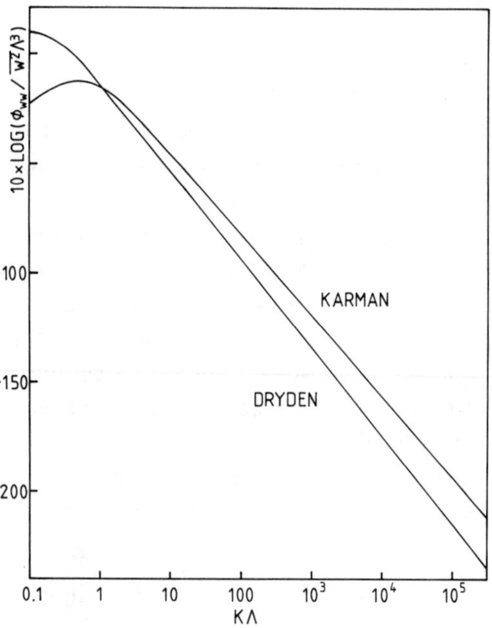

Figure 20. Comparison of von Karman spectrum and Dryden spectrum models of atmospheric turbulence.

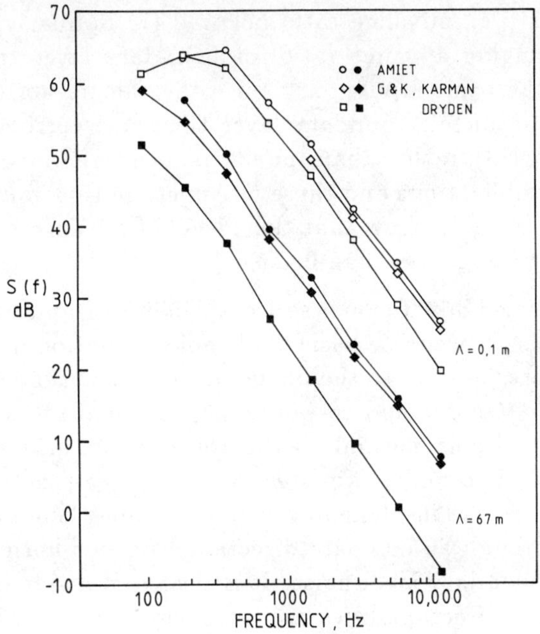

Figure 21. Effect of turbulent integral scale on rotor inflow turbulence noise calculations, UH-1, $\phi = 90°$, $\sqrt{\overline{w^2}} = \mathrm{lm}/s$.

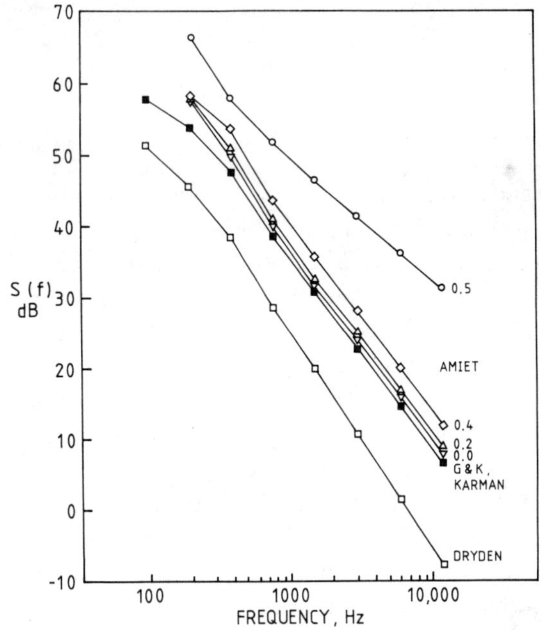

Figure 22. Effect of advance ratio on rotor inflow turbulence noise calculations, UH-1, $\phi = -60°$, $\Lambda = 67$ m, $\sqrt{\overline{w^2}} = \mathrm{lm}/s$.

to helicopters (i.e., advance ratio below 0.4). Similarly, Figure 23 compares changing advance ratio for boundary layer trailing edge noise. Here the basic inputs vary since the calculation of Kim and George uses an airfoil boundary layer thickness correlation (Chou and George, 1983), rather than the flat plate results incorporated in their earlier publications and those of Amiet. In this boundary layer case the results again show that the effects of advance ratio are not important for values less than 0.4.

As discussed in a previous section, Amiet's computational model incorporates an accurate basic blade noise radiation directionality for the pressures normal to the blade mean line. On the other hand, the methods of George and co-workers approximate the basic directionality by a dipole normal to the rotor plane, which would be expected to lead to underestimates for angles near the rotor plane. Both analyses ignore in-plane forces and other in-plane mechanisms. Figures 24 through 26 compare directionalities for both inflow turbulence and boundary layer noise. It is clear that, aside from overall differences, the directionalities are quite close except within about 10 to 15 degrees of the rotor plane.

5. Comparisons of Different Mechanisms in Different Situations

It is already apparent from the results presented so far that the various mechanisms can each be important in different situations. We have seen that inflow turbulence can dominate most other broadband noise sources when the inflow turbulence is strong. On the other hand, at high frequencies either boundary layer trailing edge noise or tip vortex noise can be important as can be seen from Figures 6–8 and 10–11. Both of these sources increase with blade angle of attack while tip vortex noise depends strongly on blade chord, and is more severe for square tip shapes. Some of these dependencies are shown in Figure 27. Calculated spectra are shown for pitch angles of 10 and 15 degrees for boundary layer noise and for tip vortex noise, based on both square and round tip shapes for a rotor similar to that of a UH-1 (compare with Figure 17). Clearly, tip vortex noise is favored by high angles of attack and wide tipe chords (low aspect ratio, untapered blades). On the other hand, wind turbines generally have tapered blades, and we have seen that their

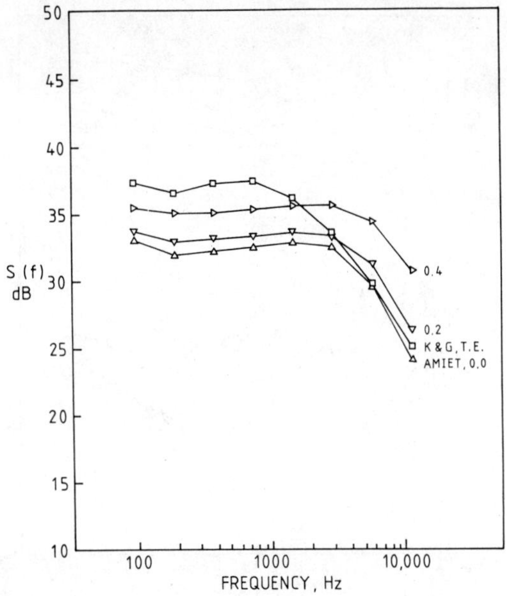

Figure 23. Effect of advance ratio on rotor trailing edge noise calculations, UH-1, $\phi = -90°$.

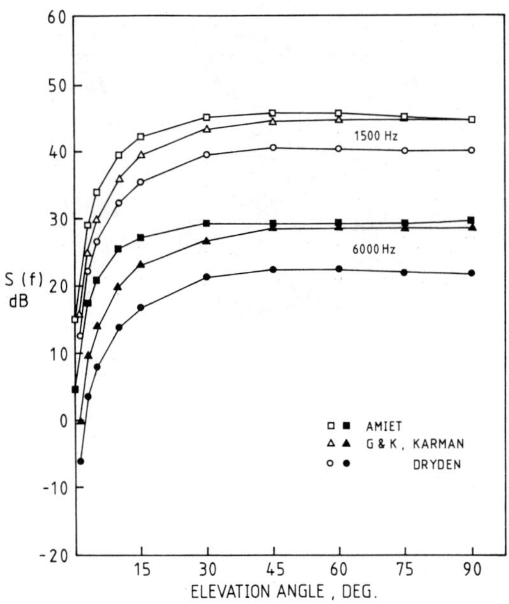

Figure 24. Directionality of inflow turbulence noise for both 1500 and 6000 Hz, S55 full scale rotor as in Leverton's (1973a) experiment, $\Lambda = 0.57$ m, $\sqrt{\overline{w^2}} = 1 \text{m}/s$.

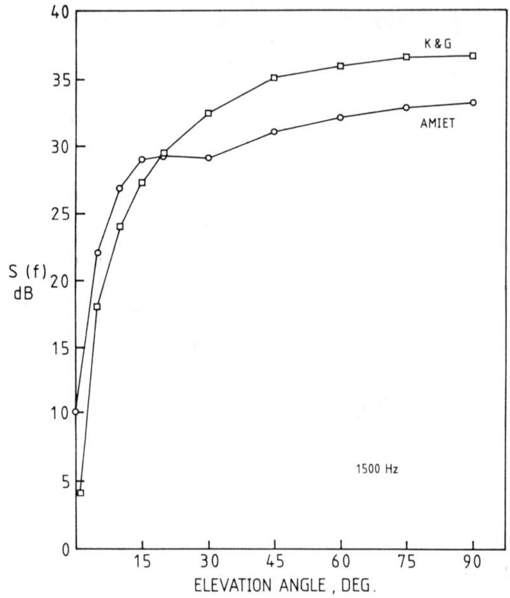

Figure 25. Directionality of rotor trailing edge noise, 1500 Hz, same rotor used as in Figure 24.

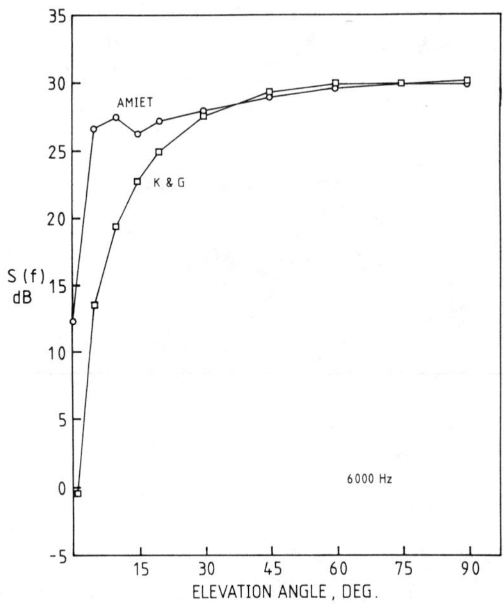

Figure 26. Directionality of rotor trailing edge noise, 6000 Hz, same rotor used as in Figure 24.

primary noise source in the frequency range of interest is boundary layer trailing edge noise (see Figures 18 and 19).

The relative importance of various mechanisms on a full helicopter configuration is an interesting but complex question, as various noise mechanisms exist on both the main and tail rotors. Some calculations showing the effects of varying the inflow turbulence scale are shown in Figure 28 for UH-1 helicopter main rotor parameters. The varying turbulent intensities give some idea of the variations expected between quiet nighttime conditions and typical daytime conditions. They could be considerably higher in windy conditions. Also, as first demonstrated experimentally by Hanson (1975) and later by Pegg et al. (1977), the contraction of the streamtubes into a hovering rotor or a stationary propeller or fan leads to an anisotropic and intensified inflow turbulence. This dynamic effect on the inflow turbulence has not yet been treated analytically or experimentally in any detail.

Next, we question what the relative importance of the tail rotor is compared to the main rotor. The primary added difficulty in dealing with tail rotor noise is that the tail rotor operates in the main rotor wake, which is itself not very well understood. The main rotor wake consists of a number of components: the overall downwash field; the tip vortices and other vortices shed from the main rotor blades; the wakes of the fuselage, engines, and rotor hub; and the turbulence present in these flows as well as that already existing in the atmosphere. The main rotor's mean wake can be approximately calculated by several methods but it has a fairly minor influence on the tail rotor noise. This is because it leads primarily to only low-order loading harmonics and noise. On the other hand, the tip vortices in the wake are quite localized, and lead to more annoying, higher frequency noise. The tip vortices' positions and structure are difficult to predict. In forward flight they follow initially epicycloidal paths which are perturbed by: (1) the roll-up of the main rotor wake into a horseshoe vortex system; (2) the self-induced instabilities of the vortex trajectories; and (3) the action of preexisting atmospheric turbulence. In addition, the actual structure of the vortices is not well known. They can contain axial velocities in their cores which are of the order of their maximum circumferential velocities. For the interaction geometry associated with forward flight, where the vortex core will be approximately perpendicular to the plane of the tail rotor, these axial velocities will be strong contributors to the loading fluctuations on the tail rotor blades. Also, the tip vortices are found

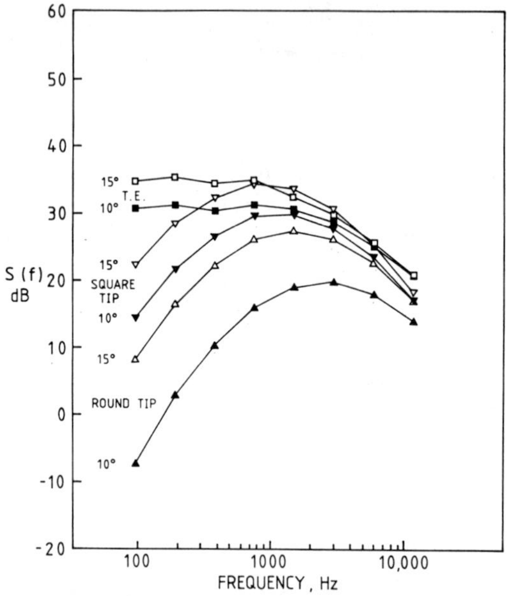

Figure 27. Comparison of trailing edge noise and tip vortex noise (with different tip shapes). Calculations for 10° and 15° rotor pitch, UH-1, $\phi = -27°$.

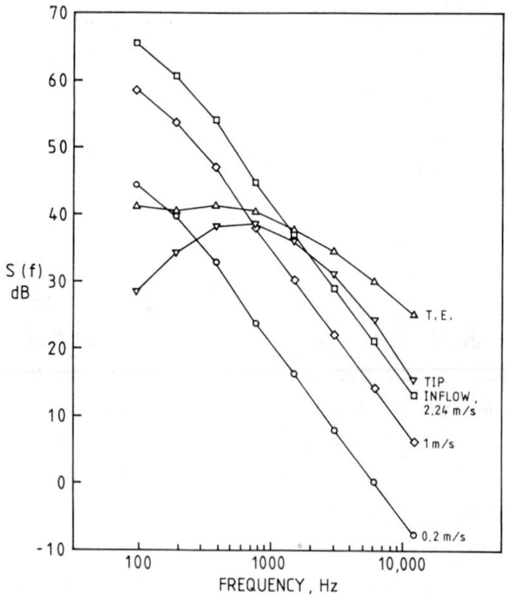

Figure 28. Noise calculations for UH-1 helicopter's main rotor, $r = 74.8$ m, $\phi = -78.5°$, $\Lambda = 67$ m.

to diffuse or *burst* under some conditions (the *vortex breakdown* phenomenon). After they break down, the vortices become more spread out and turbulent. This will strongly affect both the mean and the turbulent inflow seen by the tail rotor. The importance of wake ingestion on noise was also pointed out in some experiments of Levine (1976). He reported on some experiments where increases in 5 to 10 dB were found, in both narrowband and broadband noise, when a Sikorsky S-58T operated with the main rotor wake being blown into the tail rotor. Significant reductions in tail rotor noise were also reported by Barlow et al. (1972) for a redesigned OH-6A tail rotor. Another experimental study of tail rotor/main rotor wake interaction noise involved wind tunnel tests of a model with variable tail rotor position and direction of rotation (Balcerak, 1976; Pegg and Shidler, 1978). Balcerak (1976) made a parametric study varying tail rotor location, fin blockage area, tail rotor rotation direction, rotor speeds and thrusts, and tail rotor pusher/tractor configuration. Later Pegg and Shidler (1978) tested the same model, extending the work and emphasizing identifying the aeroacoustic mechanisms producing the noise. They found about a 12 dB increase in broadband noise when the main rotor flow was added to the tail rotor and significant increases in harmonics under a variety of conditions. These experiments are very important, and point out the complexity and the need for more analytical understanding of tail rotor broadband noise sources and how to reduce these sources.

In the present study, all we can do is present a few simple calculations to show the importance of the various mechanisms on tail rotors. Figure 29 presents UH-1 tail rotor noise for conditions corresponding to the main rotor calculations shown in Figure 28 (i.e., with no main rotor wake effects). Under those conditions tail rotor noise is clearly less important than main rotor noise. In contrast, Figure 30 presents calculations for inflow turbulence noise due to ingestion of assumed main rotor wake turbulence. (The tail rotor tip and boundary layer noise sources are unchanged from Figure 29.) The turbulence estimates range from ambient in the atmosphere to large, and the length scales are alternatively taken as the main rotor chord or radius. Although these estimates range widely, it is clear from comparing Figure 30 to Figure 28 that tail rotor ingestion of the main rotor wake is very important and deserves much more attention in the future.

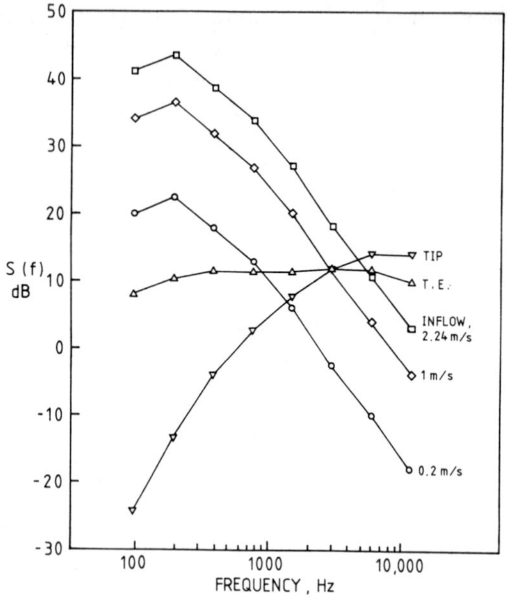

Figure 29. Noise calculations for UH-1 helicopter's tail rotor, same conditions as in Figure 28.

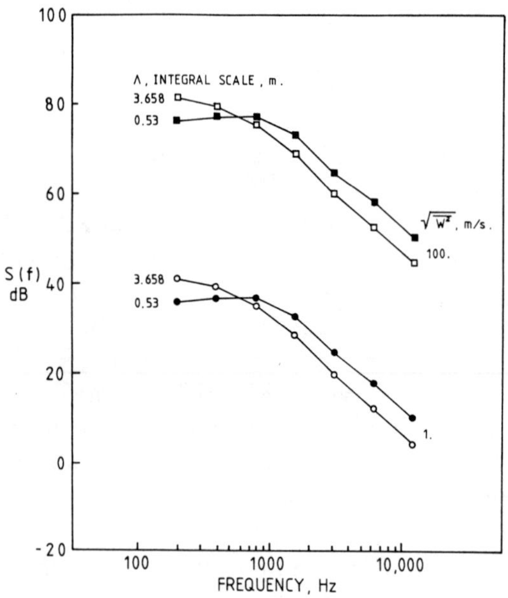

Figure 30. Effect of main rotor wake being ingesting by the tail rotor, UH-1; various assumptions for turbulent intensity and scale.

The final comparisons of mechanisms are for a typical light aircraft propeller. Figure 31 compares calculations for a static propeller on the ground. A number of different inflow turbulence intensity values are used, ranging from values representative of quiet nighttime values to typical daytime values. It is evident that under high atmospheric turbulence conditions, inflow turbulence is dominant over almost all of the spectrum. On the other hand, for low turbulence intensities, or for a flyover case as shown in Figure 32, we find that tip or trailing edge noise can be important, depending on blade angle of attack. Figure 32 shows calculations somewhat arbitrarily based on 4.0 and 8.5 degree angles of attack. The higher sensitivity of tip vortex noise to blade thrust (angle of attack) is quite evident.

6. Conclusions

The understanding of and the ability to predict broadband rotor noise are approaching a satisfactory state in many respects. Until about ten years ago, understanding was essentially qualitative, sometimes erroneous, and several mechanisms were not even recognized. As shown in this paper, the important broadband noise mechanisms are now understood well enough to be able to make predictions to within about 5 dB of experimental data. This understanding should enable designers to minimize broadband noise in cases where it is a controlling factor in a design.

The calculations and comparisons shown indicate that inflow-turbulence-induced lift fluctuations are the most important broadband noise sources at low frequencies. This radiation can be predicted down to the lowest blade-passing frequencies, including the smooth-peaked spectral structure, by the method of Homicz and George (1974). For the higher frequencies, which are of more practical interest, the methods of George and Kim (1976) and of Amiet (1977) are as satisfactory and much easier to compute. When the same inflow turbulence spectrum is used, both of these methods seem to agree well with each other, and with measurements over a full range of parameters except for angles within about 10 to 15 degrees of the rotor plane. Further research is needed on broadband rotor noise near the rotor plane.

Most experiments do not include enough inflow-turbulence data to define the inputs to the analyses. In particular, the effect of

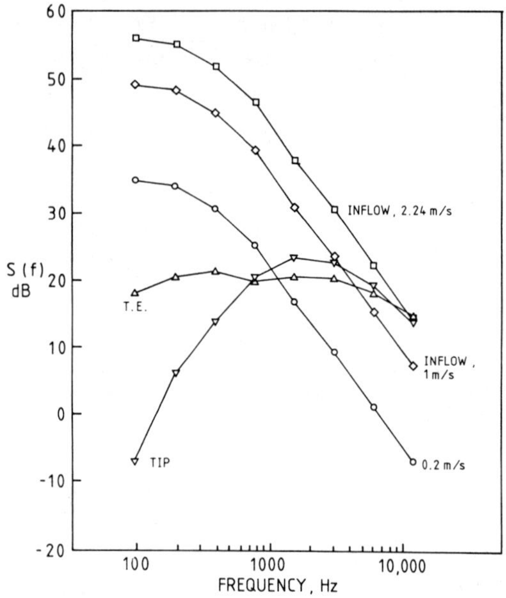

Figure 31. Noise calculations for a typical light aircraft propeller, static on ground, $\phi = -90°$, $\Lambda = 2$ m, 4° angle of attack.

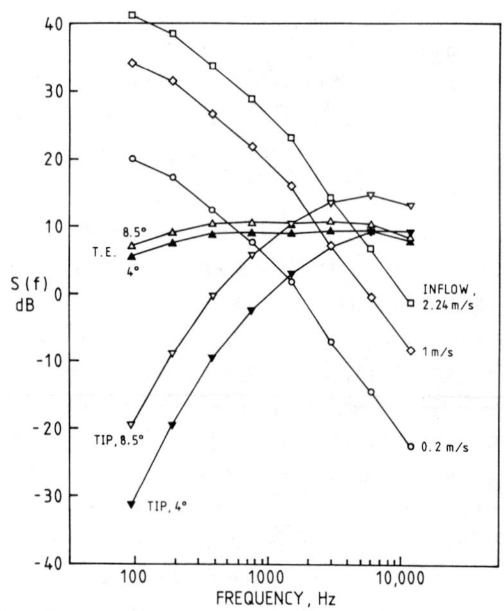

Figure 32. Noise calculations for a typical light aircraft propeller, fly-over, $r = 74.8$ m, $\phi = -78.5°$, $\Lambda = 67$ m, 4° and 8.5° angle of attack.

streamtube contraction on generating anisotropic and larger scale inflow turbulence needs more theoretical and experimental attention.

Boundary layer trailing edge noise is now well understood. The analyses of Kim and George (1980) and of Schlinker and Amiet (1981), and the correlation of Fink (1978) all appear to give results which agree reasonably well with experiments. This source was seen to be the primary broadband noise source for full scale wind turbines. It often is the most important noise source at high frequencies on large rotors when inflow turbulence is weak. It increases significantly with angle of attack due to the increase of boundary layer thickness.

Tip vortex formation noise seems to be satisfactorily predicted although it is not sufficiently dominant in any of the experiments to definitively establish the precise accuracy of the model of George et al. (1980, 1983), which uses delta wing leading edge vortex data to model the tip vortex. Much more experimental data are needed on flows and fluctuating pressures on different shapes of rotor and wing tips. Tip vortex noise is most important for square tip shapes and for wide tip chords. It probably can be reduced significantly by detailed tip shape changes, but this is presently unexplored.

The noise radiated from helicopter tail rotors remains poorly understood. The main rotor wake is very complex and itself poorly understood, although it is the primary input needed for tail rotor noise prediction.

Finally, comparisons of the mechanisms to better-defined experimental measurements are still needed to establish the analyses' accuracies and enable further improvements of the broadband noise analyses and noise minimization techniques.

Acknowledgement

The assistance of S. T. Chou with this work is gratefully acknowledged. This work was supported by NASA-Langley Research Center, Dr. F. Farassat, technical monitor.

References

[1] Ahtye, W. F., Miller, W. R., and Meecham, W. C. "Wing and Tip Noise Measured by Near- and Far-Field Cross-Correlation Techniques," *AIAA Paper 79-0667,* 1979.

[2] Amiet, R. K. "Acoustic Radiation From an Airfoil in a Turbulent Stream," *J. of Sound and Vibration, 41* (1975), 407–20.

[3] ————. "Noise Produced by Turbulent Flow Into a Propeller or Helicopter Rotor," *AIAA Paper No. 76-560,* July 1976a.

[4] ————. "Noise Due to Turbulent Flow Past a Trailing Edge," *J. of Sound and Vibration, 47* (1976b), 387–93.

[5] ————. "Noise Produced by Turbulent Flow Into a Propeller or Helicopter Rotor," *AIAA Journal, 15* (1977), 307–8.

[6] ————. "Effect of the Incident Surface Pressure Field on Noise Due to Turbulent Flow Past a Trailing Edge," *J. of Sound and Vibration, 57* (1978), 305–06.

[7] Aravamudan, K. S. and Harris, W. L. "Experimental and Theoretical Studies on Model Helicopter Rotor Noise," *MIT FDRL Report No. 78-1,* Jan. 1978.

[8] ————————. "Low-Frequency Broadband Noise Generated by a Model Rotor," *J. Acoustical Soc. America, 66* (1979), 522–33.

[9] Aravamudan, K. S., Lee, A., and Harris, W. L. "An Experimental Study of High Frequency Noise From Model Rotors: Prediction and Reduction," *Vertica, 3* (1979), 47–63.

[10] Balcerak, J. C. "Parametric Study of the Noise Produced by the Interaction of the Main Rotor Wake With the Tail Rotor," *NASA CR-145001,* 1976.

[11] Barlow, W. H., McCluskey, W. C., and Ferris, H. W. "OH-6A Phase II Quiet Helicopter Program," *U. S. Army USAAMRDL Tech. Report 72-29,* Sept. 1972.

[12] Brooks,T.F. "Trailing Edge Noise Prediction Using Amiet's Method," *J. of Sound and Vibration, 77* (1981), 437–39.

[13] Brooks, T. F. and Hodgson, T. H. "Trailing Edge Noise Prediction From Measured Surface Pressures," *Journal of Sound and Vibration, 78* (1981), 69–117.

[14] Chou, S. T. and George, A. R. "Effects of Boundary Layer Thickness on Rotor Trailing Edge Noise," *Cornell University Fluid Dynamics and Aerodynamics Program Report FDA-83-03,* April 1983.

[15] Etkin, B. "Theory of the Flight of Airplanes in Isotropic Turbulence—Review and Extension," *AGARD Report 372,* 1961.

[16] Farassat, F. "Theory of Noise Generation From Moving Bodies With an Application to Helicopter Rotors," *NASA TR R-451,* 1975.

[17] Farassat, F. and Succi, G. P. "A Review of Propeller Discrete Frequency Noise Prediction Technology With Emphasis on Two

Current Methods for Time Domain Calculations," *Journal of Sound and Vibration, 71, No. 3* (1980), 399–419.

[18] Ffowcs Williams, J. E. and Hawkings, D. L. "Sound Generated by Turbulence and Surfaces in Arbitrary Motion," *Philosophical Transactions of the Royal Society of London, A264* (1969a), 321–42.

[19] ——————————. "Theory Relating to the Noise of Rotating Machinery," *J. Sound and Vibration, 10* (1969b), 10–21.

[20] Fink, M. R. "Minimum On-Axis Noise for a Propeller or Helicopter Rotor," *J. Aircraft, 15* (1978), 700–02.

[21] Fink, M. R. and Bailey, D. A. "Airframe Noise Reduction Studies and Clean Airframe Noise Investigation," *NASA CR-159311,* April 1980.

[22] George, A. R. "Helicopter Noise: State-of-the-Art," *Journal of Aircraft, 15, No. 11* (1978), 707–15.

[23] George, A. R. and Chang, S. B. "Noise Due to Transonic Blade-Vortex Interactions," *American Helicopter Society, Paper A-83-39-50-D000,* May 1983.

[24] George, A. R. and Chou, S. T. "Broadband Rotor Noise Analyses," *NASA Contractor Report,* to be published (1984).

[25] George, A. R. and Kim, Y. N. "High-Frequency Broadband Rotor Noise," *AIAA Journal, 15* (1977), 538–45; also *AIAA Paper No. 76-561,* July 1976.

[26] George, A. R., Najjar, F. E., and Kim, Y. N. "Noise Due to Tip Vortex Formation on Lifting Rotors," *AIAA Paper AIAA-80-1010,* June 1980.

[27] ——————————. "Noise Due to Tip Vortex Formation on Lifting Rotors," *Cornell University Fluid Dynamics and Aerodynamics Program Report FDA-83-02,* April 1983.

[28] Hanson, D. B. "Spectrum of Rotor Noise Caused by Atmospheric Turbulence," *J. of Acoustical Society of America, 56* (1974), 110–26.

[29] ——————. "Measurements of Static Inlet Turbulence," *AIAA Paper 75-467,* March 1975.

[30] Homicz, G. F. and George, A. R. "Broadband and Discrete Frequency Radiation From Subsonic Rotors," *J. Sound and Vibration, 36, No. 2* (1974), 151–77.

[31] Howe, M. S. "A Review of the Theory of Trailing Edge Noise," *J. of Sound and Vibration, 61* (1978), 437–65.

[32] Hubbard, H. H., Shepherd, K. P., and Grosveld, F. W. "Sound Measurements of the MOD-2 Wind Turbine Generator," *NASA*

Contractor Report 165752, July 1981.

[33] Humbad, N. G. and Harris, W. L. "Effects of Tip Geometry on Model Helicopter Rotor Low Frequency Broadband Noise," *MIT FDRL Report No. 81-2,* May 1981.

[34] ——————————. "Model Rotor Low Frequency Broadband Noise at Moderate Tip Speeds," *AIAA Paper AIAA-80-1013,* June 1980.

[35] Johnson, H. K. and Katz, W. M. "Investigation of the Vortex Noise Produced by a Helicopter Rotor," *USAAMRDL TR-72-2,* Feb. 1972.

[36] Kaji, S. "Noncompact Source Effect on the Prediction of Tone Noise From a Fan Rotor," *AIAA Paper 75-446,* March 1975.

[37] Kendall, J. M. "Measurements of Noise Produced by Flow Past Lifting Surfaces," *AIAA Paper 78-239,* 1978.

[38] Kim, Y. N. and George, A. R. "Trailing-Edge Noise From Hovering Rotors," *AIAA Journal, 20* (1982), 1167–74; also *American Helicopter Society Preprint 80-60, 36th Annual Forum Proceedings,* 1980.

[39] Leverton, J. W. "The Noise Characteristics of a Large *Clean* Rotor," *J. of Sound and Vibration, 27* (1973a), 357–76.

[40] Leverton, J. W. and Pollard, J. S. "A Comparison on the Overall and Broadband Noise Characteristics of Full-Scale and Model Helicopter Rotors," *J. of Sound and Vibration, 30* (1973b), 135–52.

[41] Levine, L. S. "An Analytic Investigation of Techniques to Reduce Tail Rotor Noise," *NASA CR-145014,* July 1976.

[42] Lowson, M. V., Whatmore, A., and Whitfield, C. E. "Source Mechanisms for Rotor Noise Radiation," Loughborough Univ. of Technology, Dept. of Transport Technology *Report TT-7202,* Feb. 1972.

[43] Lumley, J. L. and Panofsky, H. A. *The Structure of Atmospheric Turbulence,* New York: Wiley, 1964, Chapters 4 and 5.

[44] Paterson, R. W. and Amiet, R. K. "Noise and Surface Pressure Response of an Airfoil to Incident Turbulence," *J. Aircraft, 14* (1977), 729–36.

[45] ——————————. "Noise of a Model Helicopter Rotor Due to Ingestion of Turbulence," *NASA CR-3213,* Nov. 1979.

[46] Paterson, R. W., Amiet, R. K., and Munch, C. L. "Isolated Airfoil-Tip Vortex Interaction Noise," *J. Aircraft, 12* (1975), 34-40.

[47] Paterson, R. W., et al. "Vortex Noise of Isolated Airfoils," *J. Aircraft, 10* (1973), 296–302.

[48] Pegg, R. J. "A Summary and Evaluation of Semi-Empirical Methods for the Prediction of Helicopter Rotor Noise," *NASA TM-80200,* Dec. 1979.

[49] Pegg, R. J. and Shidler, P. A. "Exploratory Wind Tunnel Investigation of the Effect of the Main Rotor Wake on Tail Rotor Noise," *Proceedings Inter. Specialists Symposium on Helicopter Acoustics*, NASA Langley Research Center, Hampton, VA, *NASA CP2052*, May 22–24, 1978.

[50] Pegg, R. J., Magliozzi, B., and Farassat, F. "Some Measured and Calculated Effects of Forward Velocity on Propeller Noise," *ASME Paper 77-GT-70*, 1977.

[51] Powell, A. O. "On the Aerodynamic Noise of a Rigid Flat Plate Moving at Zero Incidence," *J. Acoustical Soc. America, 31* (1959), 1649–53.

[52] Revell, J. D. "Induced Drag Effect on Airframe Noise," in *Progress in Astronautics and Aeronautics, 45, Aeroacoustics*. Ed. I. R. Schwartz, AIAA and MIT Press, 1976, 221–35; also *AIAA Paper 75-487*, 1975.

[53] Schlinker, R. H. and Amiet, R. K. "Helicopter Rotor Trailing Edge Noise," *AIAA Paper AIAA-81-2001*, October 1981.

[54] Schlinker, R. H. and Brooks, T. F. "Progress in Rotor Broadband Noise Research," presented at the 38th Annual Forum of the American Helicopter Society, Anaheim, CA, May 4–7, 1982.

[55] Shepherd, K. P. and Hubbard, H. H. "Sound Measurements and Observations of the MOD-OA Wind Turbine Generator," *NASA Contractor Report 165856*, Feb. 1982.

[56] Shockey, G. A., Williamson, J. W., and Cox, C. R. "Helicopter Aerodynamics and Structural Loads Survey," 32nd Annual National V/STOL Forum of the American Helicopter Society, Washington, DC, May 1976.

[57] Tennekes, H. and Lumley, J. L. *A First Course in Turbulence*, Cambridge, Mass.: MIT Press, 1972, Chapter 8.

[58] Widnall, S. E. "A Correlation of Vortex Noise Data From Helicopter Main Rotors," *J. Aircraft, 6* (1969), 279–81.

[59] ————. "Helicopter Noise Due to Blade-Vortex Interaction," *J. Acoust. Soc. Amer., 50, No. 1, Pt. 2* (1971), 354–65.

[60] Widnall, S. E. and Wolf, T. L. "Effect of Tip Vortex Structure on Helicopter Noise Due to Blade-Vortex Interaction," *J. Aircraft, 17, No. 10* (1980), 705–11.

The Evolution of Methods for Noise Prediction of High Speed Rotors and Propellers in the Time Domain

F. Farassat
NASA Langley Research Center
Hampton, VA 23665

Nomenclature

$\vec{b}$	$= \lambda \vec{M}_t + \lambda_1 \vec{t}_1$; $b = \lvert\vec{b}\rvert$.
b_ν	$= \vec{b} \cdot \vec{\nu}$.
c	speed of sound.
$F(\vec{y};\vec{x},t)$	$= f(\vec{y}, t - r/c) = [f(\vec{y},\tau)]_{ret}$.
$F_i(t)$	net force acting on the medium by a point source in motion; $F_r = F_i \hat{r}_i$.
$f(\vec{x},t)$; $f(\vec{y},\tau) = 0$	the equation of the blade surface in the frame fixed to the undisturbed medium.
$f_m(\vec{x},t) = 0$	the equation of the blade mean surface in the frame fixed to the undisturbed medium.
g	$= \tau - t + r/c$.
$H(f)$	Heaviside function.
H	local mean curvature of the blade surface.
h_n	$= \lambda M_n + \lambda_1 \cos\theta$.
I_ϵ	integrals in acoustic formulations over $f = 0$ with a hole of dimension ϵ removed at the observer position on the surface.
$L(\vec{x},t)$	maximum dimension of the surface produced by the intersection of $g = 0$ and a source distribution in motion.
l	length scale of a source distribution.
l_i	local force/unit area acting on the medium at the blade surface.
l_r	$= l_i \hat{r}_i$.
$\vec{M}, M_i$	local Mach number vector based on c, $M_n = \vec{M} \cdot \vec{n}$, $M_r = \vec{M} \cdot \vec{\hat{r}}$.

$\vec{M}_p$	the projection of the local Mach number vector on the local plane normal to the edges (e.g., TE) of the blade surface, $M_p = \lvert\vec{M}_p\rvert$.
$\vec{M}_t$	the projection of $\vec{M}$ on the local tangent plane of the blade surface for fixed source time τ, $M_t = \lvert\vec{M}_t\rvert = v_t/c$.
$\vec{n}, n_i$	unit outward normal to $f = 0$.
p'	acoustic pressure.
p'_L, p'_T	loading and thickness acoustic pressure.
p	surface pressure $p_a - p_0$ on $f = 0$.
$\tilde{p}(\vec{y}, \tau)$	$= p_a + (\gamma - 1)p_0$, p_a = absolute blade surface pressure, p_0 = ambient pressure in the undisturbed medium.
$\tilde{p}_B(\vec{\eta}, \tau)$	$\equiv p(\vec{y}(\vec{\eta}, \tau), \tau)$ blade surface pressure described in a frame moving with the blades.
Q_i $(i = 1\text{–}4)$	surface source distribution, $Q_r = Q_i\hat{r}_i (i = 1\text{–}3)$.
Q_F, Q_N	see equations 4.13a, 4.13b.
$\vec{r}, r_i$	$= \vec{x} - \vec{y}$, $r = \lvert\vec{x} - \vec{y}\rvert$.
$\vec{\hat{r}}, \hat{r}_i$	unit radiation vector $\vec{r}/r$.
$\vec{\hat{r}}_p$	unit vector in the direction of the projection of $\vec{\hat{r}}$ on the local plane normal to the edges (e.g., TE) of blade surface.
$r_m(\vec{x}, t)$	the radius of the sphere $g = 0$ as it leaves a source distribution in motion.
S	(in dS) element of blade surface area.
T_{ij}	the Lighthill stress tensor.
T	typical period of a fluctuating source.
t	observer time.
$\tilde{t}$	upper limit of a given time interval.
$\vec{t}_1$	the projection of the unit radiation vector $\vec{\hat{r}}$ on the local tangent plane to $f = 0$, τ-fixed. Not unit vector, $\lvert\vec{t}_1\rvert = \sin\theta$.
v_n, v_t	local normal and tangential velocity of $f = 0$.
$\vec{x}$	observer position.
$\vec{x}_s(t)$	the position vector of a point source in motion.
$\vec{y}$	source position.

Greek Symbols

β_n	$= (1 - M_n^2)^{1/2}$.
γ	ratio of specific heats; (in $d\gamma$) length variable along the intersection of an edge of $f = 0$ (e.g., TE) and the collapsing sphere $g = 0$.
Γ	(in $d\Gamma$) length variable of the arc of intersection of surfaces $f = 0$ and $g = 0$.
Δp	$p_l - p_u$ on the mean surface of the blade $f_m = 0$.
$\Delta\tau\,(\vec{x}, t)$	the time that $g = 0$ takes to cross a source distribution.
$\delta(f)$	the Dirac delta function.
ϵ	a small positive quantity.
θ	the angle between $\vec{n}$ and $\hat{\vec{r}}$.
κ_1, κ_2	principal curvatures.
λ	$= (\cos\theta - M_n)/\tilde{\Lambda}^2$.
λ_1	$= (\cos\theta + M_n)/\tilde{\Lambda}^2$.
Λ	$= (1 + M_n^2 - 2M_n\cos\theta)^{1/2}$.
$\tilde{\Lambda}$	$= (\Lambda^2 + \sin^2\theta)^{1/2}$.
Λ_0	$= \left[M_p^2\cos^2\psi + (1 - \vec{M}_p \cdot \hat{\vec{r}}_p \sin\psi)^2\right]^{1/2}$.
$\vec{\nu}, \nu_i$	unit inward surface vector perpendicular to an edge (e.g., TE) of the surface $f = 0$, τ-fixed (known also as the geodesic normal.)
ρ_0	density of undisturbed medium.
Σ	(in $d\Sigma$) surface area of $F = 0$.
$\sigma, \sigma_1, \sigma_b$	the length parameter on $f = 0$ along $\vec{M}_t$, $\vec{t}_1$, and $\vec{b}$, respectively.
σ_{11}, σ_{22}	two components of tensor $(\vec{t}_1\vec{t}_1 - \vec{M}_t\vec{M}_t + \vec{t}_1\vec{M}_t + \vec{M}_t\vec{t}_1)/\tilde{\Lambda}^2$.
ψ	the local angle between $\hat{\vec{r}}$ and an edge of $f = 0$.
$\vec{\Omega}, \Omega_i$	$= \vec{n} \times \vec{\omega}$.
$\vec{\omega}$	angular velocity.

Subscripts and Other Symbols

l	lower surface of $f = 0$.
m	mean surface of the blade.
r	component of a vector in radiation direction.
u	upper surface of $f = 0$.
ν	component of a vector in the direction of unit vector $\vec{\nu}$ tangent to $f = 0$.
$\Box^2$	D'Alembertian, $1/c^2\ \partial^2/\partial t^2 - \nabla^2$.

1. Introduction

This article is a survey of time domain methods available for prediction of discrete frequency noise of rotating blades. It describes mainly the evolution of time domain analysis at the NASA Langley Research Center in the past decade. The theoretical aspect of noise prediction is stressed in this article. The results presented here are all based on the linear wave equation. Although nonlinear effects are important for high speed rotating machinery with thick blades, the related mathematical theory, at least as used by practicing acousticians, is in its infancy. It is a major challenge of the coming decade to rigorously analyze nonlinear effects.

The problem of noise prediction which we will discuss can be described by a linear wave equation with sources which lie on a moving boundary (i.e., the blade surface). The governing equation can be written in several forms and solved by amazing varieties of methods. Our analysis will be based on the Ffowcs Williams-Hawkings (FW-H) equation originally published in 1969 (Ffowcs Williams and Hawkings, 1969). It is a fundamental equation which can be used for studying both the acoustics and aerodynamics of moving bodies. This equation is a linear wave equation with two surface and one volume source terms. In our analysis, only the surface source terms will be retained. The volume term, which accounts for nonlinearities, depends on the knowledge of the flow-field around the blades and is thus generally not easily available. Although methods exist to account for nonlinearities using this volume source, further theoretical work is needed for justification of its use.

Because of the availability of two recent survey articles by the author (Farassat and Succi, 1980; Farassat, 1981), most of the details of derivations of results in common with this paper will be left out. The mathematical tools needed to follow the arguments of the present paper are described in Farassat (1977) and the references cited there. In addition to acoustics, some linear aerodynamic results which are based directly on acoustic formulas are also presented. These are new applications of acoustic formulations in the time domain which appear very promising in supplying the aerodynamic load data needed for noise prediction.

2. Ffowcs Williams-Hawkings Equation and Some Variations

This equation is derived by extending the fluid inside a moving body, assuming that the fluid is at rest there, and applying the condition that the body is impenetrable. There are several methods of deriving this equation (Ffowcs Williams and Hawkings, 1969; Farassat, 1975, 1977; Goldstein, 1977). Perhaps an easy method is by converting ordinary derivatives in the mass continuity and momentum conservation equations into generalized derivatives realizing that there are jumps in flow parameters across the blade surface (Farassat, 1975). The resulting equations, known as the generalized mass continuity and momentum equations, are then manipulated, as in the derivation of Lighthill's equation to get the FW-H equation:

$$\Box^2 p' = \frac{\partial}{\partial t}[\rho_0 v_n |\nabla f| \delta(f)] - \frac{\partial}{\partial x_i}[l_i |\nabla f| \delta(f)] + \frac{\partial^2}{\partial x_i \partial x_j}[T_{ij} H(f)] \tag{2.1}$$

Here p' is the acoustic pressure, v_n and l_i are the local normal velocity and force intensity acting on the fluid medium, respectively. The surface of the moving body is described by $f(\vec{x}, t) = 0$ where $f > 0$ outside the body. The density in the undisturbed medium and the Lighthill stress tensor are denoted ρ_o and T_{ij}, respectively. In this equation $\delta(f)$ stands for the Dirac delta function and $H(f)$ is the Heaviside function.

The three inhomogeneous terms on the right of equation (2.1) are known as the thickness, loading and quadrupole sources, respectively. As mentioned earlier, we will neglect the quadrupole source term in the remainder of this paper. One variation of the FW-H equation is obtained by combining the thickness and loading terms into a single term using a result due to Isom (1975). A simple derivation of this result is by Ffowcs Williams (reported by Farassat (1979)). It states that thickness noise is equivalent to the loading noise produced by a uniform pressure distribution of magnitude $\rho_o c^2$. Using this result, the FW-H equation can be written as

$$\Box^2 p' = -\nabla \cdot [\tilde{p}\vec{n}|\nabla f|\delta(f)] \tag{2.2}$$

where $\tilde{p} = p_a + (\gamma - 1)p_o$, p_a, and p_o are the local absolute pressure on the blade and ambient pressure, respectively. The ratio of specific heats is denoted γ. Here we have assumed $l_i = pn_i$, where $p = p_a - p_o$. We have neglected the viscous stress term in l_i since numerous numerical experiments have shown that, in the region of peak directivity, the viscous stress does not generate significant noise as compared to the sources which are retained. The above equation has been used to develop an equation for noise prediction which will be discussed later.

We will now present an approximation of the FW-H equation for thin airfoils. As is common in linear aerodynamic theory, it is sometimes desirable to specify the boundary conditions on the mean surface of the airfoil section. We can do a similar approximation for the thickness and loading terms in equation (2.1) (Farassat, 1982). Let $f_m(\vec{x}, t) = 0$ be the equation for the mean surface of the blade. Then the FW-H equation can be approximated as

$$\begin{aligned} \Box^2 p' = {} & \frac{\partial}{\partial t}\{\rho_o[(v_n)_u + (v_n)_l]|\nabla f_m|\delta(f_m)\} \\ & - \frac{\partial}{\partial x_i}\{[(l_i)_u + (l_i)_l]|\nabla f_m|\delta(f_m)\} \end{aligned} \tag{2.3}$$

where the subscripts u and l stand for the upper and lower surfaces, respectively. The solution of this equation in various forms constitute thin airfoil acoustic theory such as that by Hanson (1976, 1979, 1980).

Note that in equation (2.3), we should take $l_i = pn_i$ where n_i is the unit normal on the *actual* blade surface. Except near a blunt leading edge, one may use $(l_i)_u + (l_i)_l = -\Delta p(n_i)_m$ where $\Delta p = (p)_l - (p)_u$ and $(n_i)_m$ is the unit normal to the mean surface pointing to the suction side of the blade.

3. Time Domain vs. Frequency Domain Analysis

Since the mid-seventies, many acoustic formulations for prediction of the noise of rotating blades have been published. A major impetus has been the publication of the paper by Ffowcs Williams and Hawkings on the sound from moving surfaces (Ffowcs Williams and Hawkings, 1969). Many of these formulations have been reviewed by Farassat (1981). There are two basically different approaches available for finding a solution to the FW-H equation. One is based on time domain analysis and the second on frequency domain analysis. Since this article emphasizes time domain analysis, it is appropriate to discuss the differences between the two approaches.

A comparative listing of the advantages of one approach over the other follows:

(1) The analytic expressions of time domain formulations are generally simpler and do not require special functions. They appear to be easier to code on a computer.

(2) A single expression is used for both the near and far-field in the time domain methods while one has to develop two separate expressions for near and far-fields in the frequency domain. The same expression in the time domain is also valid for a moving or stationary observer while different expressions must be derived in the frequency domain.

(3) In general, approximations or restrictions, such as the assumptions of thin airfoil sections or propeller-like motion, are not needed in time domain methods. One is almost always forced to introduce some approximations or restrictions in the frequency domain analysis.

(4) Because of the removal of the dimension of time in the frequency domain, some find it easier to work in this domain. It must be mentioned that by sufficient practice, one can also get used to thinking in time domain.

(5) One can get more qualitative information from frequency domain formulations. This is an important advantage of frequency domain analysis which is valuable in theoretical studies. One should remember that this advantage is achieved at the expense of some approximations which are not required in the time domain analysis.

In conclusion, it is perhaps fair to say that time domain formulations should be used in quantitative analysis of the noise of rotating blades. Frequency domain formulations are suitable for qualitative analysis of the noise field. In this respect, both approaches are complementary in the understanding, prediction, and reduction of the noise of rotating blades.

4. Acoustic Formulation

4.1 Compact Source Results

Although this article's main theme is the prediction of the noise of high speed rotating blades, it is useful to know when compact source formulations can be employed. These formulations are easier to code and execute on a computer. We give two formulas here by Lowson (loading noise) and Succi (thickness noise). Before we present these, we quickly review the conditions of compactness of acoustic sources (Farassat, 1975, 1981).

Let a source distribution of length scale l be in motion. For a given observer time t in $[o, \tilde{t}]$ and observer position, find the maximum dimension $L(\vec{x}, t)$ of the effective radiating volume or surface as follows: Construct the object in space which is generated by the intersection of the collapsing sphere $g = \tau - t + r/c = 0$ and the source distribution. The length $L(\vec{x}, t)$ is the maximum dimension of this object. In general, it is different from l. Let $\Delta\tau(\vec{x}, t)$ and $r_m(\vec{x}, t)$ be the time it takes for the sphere $g = o$ to pass through the source distribution and the radius of the sphere as it leaves the source region, respectively. Finally, let the typical period of fluctuation of the source distribution be T. Then the distribution is considered compact if both of the following conditions are satisfied.

(i)

$$\text{Min}_{t\epsilon[o,\tilde{t}]} r_m(\vec{x}, t) \gg \text{Max}_{t\epsilon[o,\tilde{t}]} L(\vec{x}, t) \tag{4.1a}$$

(ii)

$$T \gg \text{Max}_{t\epsilon[o,\tilde{t}]} \Delta\tau(\vec{x}, t) \tag{4.1b}$$

Lowson's equation is the solution of the wave equation

$$\Box^2 p' = -\frac{\partial}{\partial x_i}\{F_i(t)\,\delta[\vec{x} - \vec{x}_s(t)]\} \tag{4.2}$$

where $F_i(t)$ is the force acting on the medium and $\vec{x}_s(t)$ is the position of the force. The solution of equation 4.2 is Farassat (1981, 1982) and Lowson (1965).

$$4\pi p'(x,t) = -\frac{\partial}{\partial x_i}\left[\frac{F_{i(\tau)}}{r|1-M_r|}\right]_{ret} \tag{4.3a}$$

$$= \frac{1}{c}\frac{\partial}{\partial t}\left[\frac{F_r}{r|1-M_r|}\right]_{ret} + \left[\frac{F_r}{r^2|1-M_r|}\right]_{ret} \tag{4.3b}$$

where $F_r = F_i\hat{r}_i$. Equation 4.3a in expanded form was given by Lowson. Equation 4.3b by Farassat is useful in writing the far-field approximation to Lowson's equation using the result

$$\frac{\partial}{\partial t}[\cdot]_{ret} = \left[\frac{1}{1-M_r}\frac{\partial}{\partial \tau}(\cdot)\right]_{ret} \tag{4.4}$$

(see Farassat, 1981). The important conclusion from Lowson's equation is that source fluctuations as well as acceleration are significant contributors to the sound field of a moving point force.

Succi's thickness noise formula is based on the solution of

$$\Box^2 p' = \frac{\partial^2}{\partial t^2}\{\rho_o\psi\delta[\vec{x}-\vec{x}_s(t)]\} \tag{4.5}$$

where ψ is the volume of the source and $\vec{x}_s(t)$ describes the position of the source. The solution of this equation is [Farassat and Succi (1980), Farassat (1981), and Succi (1979)].

$$4\pi p'(\vec{x},t) = \frac{\partial^2}{\partial t^2}\left[\frac{\rho_o\psi}{r|1-M_r|}\right]_{ret} \tag{4.6b}$$

$$= \left\{\frac{1}{(1-M_r)}\frac{\partial}{\partial \tau}\left[\frac{1}{(1-M_r)}\frac{\partial}{\partial \tau}\left(\frac{\rho_o\psi}{r|1-M_r|}\right)\right]\right\}_{ret} \tag{4.6a}$$

Succi's equation has been used for propeller noise calculations (Farassat and Succi, 1980; Succi, 1979). As in Lowson's formula, the source acceleration is significant in noise generation.

In the application of Succi's equation, it is important to note that if ψ is allowed to fluctuate, then conditions (4.1b) must be satisfied. If not, then more complicated analysis such as Dowling's (1976), shows that equation 4.3a does not correctly describe the far-field dependence of the sound field on the Doppler factor. However, in discrete frequency noise calculations, Succi's equation has been found very useful when no blade vibrations are involved.

Both of the above compact source results are included in the NASA propeller noise prediction program.

4.2 Noncompact Source Results

We present four expressions of the solution of the FW-H equation which are designated formulations 1, 1-A, 2, and 3. The reason for deriving so many different expressions will become clear shortly. We briefly discuss the solution of the wave equation

$$\Box^2\phi = \frac{\partial}{\partial x_i}[Q_i|\nabla f|\delta(f)] \tag{4.7}$$

where $i = 1$ to 4, $x_4 =$ ct. The solution of this equation is Farassat (1975, 1981),

$$4\pi\phi(\vec{x},t) = \frac{\partial}{c\partial t}\int_{F=0}\frac{1}{r}\left[\frac{-Q_r + cQ_4}{\Lambda}\right]_{ret} d\Sigma - \int_{F=0}\frac{1}{r^2}\left[\frac{Q_r}{\Lambda}\right]_{ret} d\Sigma \tag{4.8}$$

where $Q_r = Q_i\hat{r}_i$, $i = 1$ to 3 and $F = [f(\vec{y},\tau)]_{ret}$. In the FW-H equation, without the quadrupole term we have $Q_i = (-\vec{l}, \rho_o v_n)$. It is also shown in the above references that

$$\frac{d\Sigma}{\Lambda} = \frac{dS}{|1 - M_r|} = \frac{c\,d\tau\,d\Gamma}{\sin\theta} \tag{4.9}$$

where dS is the element of the surface area of $f = 0$, M_r is Mach number in the radiation direction and $d\Gamma$ is the element of length of the curve of intersection of $g = 0$ and $f = 0$.

Formulation 1: This formulation was derived in Farassat (1975). It is based on equation 4.8 and is

$$4\pi p' = \frac{1}{c}\frac{\partial}{\partial t}\int_{f=0}\left[\frac{\rho_o c v_n + l_r}{r|1-M_r|}\right]_{ret} dS + \int_{f=0}\left[\frac{l_r}{r^2|1-M_r|}\right]_{ret} dS \quad (4.10a)$$

$$= \frac{\partial}{\partial t}\int_{-\infty}^{t}\int_{\substack{f=0\\g=0}} \frac{(\rho_o c v_n + l_r)}{r\sin\theta}\, d\Gamma d\tau + \int_{-\infty}^{t}\int_{\substack{f=0\\g=0}} \frac{c l_r}{r^2\sin\theta}\, d\Gamma d\tau \quad (4.10b)$$

Because of the generality of equation 4.10b, it was originally used exclusively for both subsonic and supersonic rotating blades (Farassat and Brown, 1977). However, it was found that the implementation of the collapsing sphere approach on a computer is time consuming. Also the code is sensitive to numerical error at low tip speeds. Later a new code was written that used equation 4.10a for both subsonic and supersonic blades when $|1 - M_r|$ was not near zero and used equation 4.10b when $|1 - M_r|$ was small (Nystrom and Farassat, 1980).

Equations 4.10a and 4.10b are the simplest noncompact source formulations available. The observer time derivatives in these equations are taken numerically. This causes two problems which led the author to search for alternate formulations. First, the execution time on a computer is long for high-speed blades. Second, the program is sensitive to numerical errors of time differentiation resulting in high frequency oscillation. The removal in time differentiation has a beneficial effect in subsonic cases also by speeding up the execution time. Therefore, for the low speed case, formulation 1-A, closely related to formulation 1 was derived.

Formulation 1-A: Taking the time derivative in equation 4.10a inside the first integral and using equation 4.4 results in the following equation.

$$\begin{aligned} 4\pi p'_L(\vec{x},t) = & \frac{1}{c}\int_{f=0}\left[\frac{\dot{l}_i \hat{r}_i}{r(1-M_r)^2}\right]_{ret} dS \\ & + \int_{f=0}\left[\frac{l_r - l_i M_i}{r^2(1-M_r)^2}\right]_{ret} dS \\ & + \frac{1}{c}\int_{f=0}\left[\frac{l_r(r\dot{M}_i\hat{r}_i + cM_r - cM^2)}{r^2(1-M_r)^3}\right]_{ret} dS, \end{aligned} \quad (4.11a)$$

$$4\pi p'_T(\vec{x},t) = \int_{f=0} \left[\frac{\rho_o v_n (r\dot{M}_i \hat{r}_i + cM_r - cM^2)}{r^2(1-M_r)^3} \right]_{ret} dS \qquad (4.11b)$$

$$p'(\vec{x},t) = p'_L(\vec{x},t) + p'_T(\vec{x},t), \qquad (4.11c)$$

Here p'_L and p'_T stand for the acoustic pressure due to loading and thickness, respectively. The dot on $\dot{l}_i$ denotes the rate of variation of this vector with respect to source time. Note that v_i is the local velocity of the blade surface with respect to a frame fixed in the undisturbed medium. A similar equation for the mean surface approximation of FW-H equation (equation 2.3) can also be written easily. The appearance of high order singularities in the denominators when $M_r = 1$ makes this equation unsuitable, as it is written, for supersonic rotating blades. Note also that the integral with r^{-2} dependence is singular as the observer is moved onto the surface for aerodynamic calculations. This problem will be addressed in the discussion on Long's aerodynamic integral equation. Equation 4.11, as well as the corresponding solution of equation 2.3 for the blade mean surface, are currently coded in the NASA propeller noise prediction program. These equations will be used for helicopter rotor noise prediction.

Formulation 2: This formulation was the first of its kind without observer time differentiation which was suitable for supersonic rotating blades. The idea behind the derivation was taken from linear aerodynamic theory. One can establish considerable similarity, and indeed equivalence, between acoustic and aerodynamic results (Farassat, 1982). If a thin wing is moving forward at uniform speed U along the x_1-axis, then it can be shown that $\partial/\partial t$ in our acoustic equations (e.g., equation 4.10) in the frame fixed to the wing is $-U\,\partial/\partial x_1$. The procedure in aerodynamics which effectively gets rid of time differentiation without increasing the order of the singularity is as follows:

(i) Bring $U\,\partial/\partial x_1$ inside the integral and replace $\partial/\partial x_1$ by $-\partial/\partial y_1$ where $\vec{x}$ and $\vec{y}$ are the observer and source positions in the moving frame. It can be shown that if U is constant, $r|1-M_r|$ is a function of components of $\vec{x}-\vec{y}$ in the moving frame.

(ii) Integrate by parts with respect to y_1 to get rid of the high order singularity in the denominator. In this step, line integrals along the leading and trailing edges of the thin wing will appear in the aerodynamic formulation.

The above idea together with the following three assumptions—(a) a thin airfoil section lying on the helicoidal surface generated by the blade motion; (b) uniform angular and forward speed (propeller-like motion), (c) steady blade surface pressure—led to the derivation of formulation 2 (Farassat, 1982). As expected, line integrals over the leading and trailing edges of the blade appeared in the acoustic result. A test of the derived expression on a computer proved its superiority to formulation 1 both in speed of execution and accuracy. This result will not be presented here because it is superseded by formulation 3 to be discussed below. It was a stepping stone toward a more sophisticated result without the above three restrictive assumptions (a), (b), and (c).

Formulation 3: To improve on formulation 2, one possible approach is the regularization of equation 4.10a retaining the Doppler factor. However, it can be shown that the resulting expression does not allow efficient coding for computers. The reason is that, for every observer time within a portion of every period of the noise, an elaborate integration scheme to remove the Doppler singularity must be used. This may slow the running time compared to formulation 1 but may improve precision. This approach was rejected in view of its inefficiency on a computer.

An ideal formulation should use information about p on the surface (i.e., should not include $\partial p/\partial n$), should allow an easy and simple procedure for handling singularities and should be amenable to efficient computer coding. A formulation satisfying these requirements was published recently (Farassat, 1983a). The derivation is based on equation 2.2 and uses some results from generalized function theory and differential geometry. It is a result which is applicable to the actual blade surface rather than its thin blade approximation. It is intended for use in the prediction of the noise of that portion of the blade which runs above sonic speed. The reason is that, for subsonic speed, formulation 1-A is very efficient and needs no improvements.

The step-by-step derivation of formulation 3 is described in Farassat (1983a). It will not be repeated here. Briefly, equation 2.2 is first solved for an open surface and then the result is applied to the blade which is formed of two surfaces patched along curves which have discontinuities in p or surface slope. The final result which consists of surface and line integrals is as follows:

$$
\begin{aligned}
4\pi p'(\vec{x},t) = & \int_{F=0} \frac{1}{r}\left[\frac{1}{\Lambda}\left(\tilde{p}Q_F + \frac{\lambda}{c}\frac{\partial \tilde{p}_B}{\partial \tau} - b\frac{\partial \tilde{p}_B}{\partial \sigma_b}\right)\right]_{ret} d\Sigma \\
& + \int_{F=0} \frac{1}{r^2}\left[\frac{\tilde{p}Q_N}{\Lambda}\right]_{ret} d\Sigma - \int_{TE} \frac{1}{r}\left[\frac{(\tilde{p}b_\nu)_u + (\tilde{p}b_\nu)_l}{\Lambda_0}\right]_{ret} d\gamma \\
& - \int_{\substack{shock\\ traces}} \frac{1}{r}\left[\frac{b_\nu \Delta\tilde{p}}{\Lambda_0}\right]_{ret} d\gamma - \int_{DEFD} \frac{1}{r}\left[\frac{b_\nu \tilde{p}}{\Lambda_0}\right]_{ret} d\gamma
\end{aligned}
\tag{4.12}
$$

Here the curve γ is generated by the intersection of the collapsing sphere $g = \tau - t + r/c = 0$ and the edges and shock traces on the blade, see Figure 1. Other symbols are defined in the *Nomenclature*. In addition, we define

$$
\begin{aligned}
Q_F = & \frac{1}{c}\left[(2\lambda^2 - \frac{1}{\tilde{\Lambda}^2})\dot{M}_n - 2\lambda\vec{b}\cdot\vec{\Omega} - \frac{1}{\tilde{\Lambda}^2}(\vec{t}_1 - \vec{M}_t)\cdot\vec{\Omega}\right] \\
& + 2b^2\kappa_b + \sigma_{11}\kappa_1 + \sigma_{22}\kappa_2 - 2Hh_n
\end{aligned}
\tag{4.13a}
$$

$$
Q_N = \lambda[2(\cos\theta - M_n)\lambda_1 + 1] \tag{4.13b}
$$

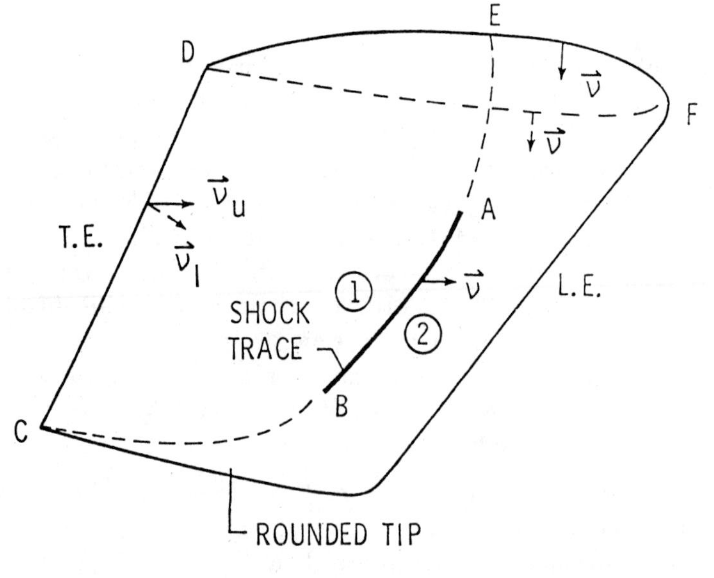

Figure 1. The geometry of the supersonic portion of the blade.

Note that we have corrected a misprint in Q_F and an error in Q_N in the original (Farassat, 1983a). Also the line integral over the trailing edge (TE) is written in a more general form which is valid when the Kutta condition does not hold. The subscripts u and l stand for the upper and lower surfaces, respectively.

By analogy with aerodynamic formulas, the appearance of the surface gradient of $\tilde{p}$ in this equation is expected. As seen from equation 4.12, no normal gradient of $\tilde{p}$ is required. The curvature terms appear because the actual blade surface is used in the derivation. It can be shown that the appearance of singularities in equation 4.12 is rare and can be avoided altogether if airfoils with sharp leading edges are used in computation. In the next section, the observer is moved onto the blade to obtain an aerodynamic integral equation.

5. Aerodynamic Formulations

Aerodynamic input data to the noise prediction program are required in all the acoustic formulations discussed earlier. This is because of the use of the acoustic analogy in the derivation of the FW-H equation. There is a close relationship between linear aerodynamics and acoustics. Indeed, it can be shown that many linear aerodynamic formulas can easily be derived from the FW-H equation. In particular, one can show that the thickness and lifting problems of wing theory are related to our thickness and loading source terms (Farassat, 1982).

Our direct way to derive aerodynamic formulations from our acoustic results is to move the observer onto the surface and obtain an integral equation. This is because $p'(\vec{x},t)$ becomes $p(\vec{x},t)$ when the observer reaches the blade surface. There is, however, a mathematical difficulty. Some of the surface integrals in the acoustic formulations become divergent when the observer lies on the blade surface. However, it can be shown that these surface integrals are semi-convergent, i.e., they can be regularized by a procedure which will be described below. This regularization is in fact taking the principal value of the divergent integrals.

The regularization procedure is as follows. Select a small hole of a shape which will make the singular integrals convergent. This is possible because of the semi-convergence property of the integrals. The shape of the hole is not unique and one can find several different shapes for regularization. The shape of the hole which is selected must be tailored to the numerical computation scheme. Assume this hole is of dimension ϵ. Now set the observer at a distance $\delta > 0$ above the cap of the hole, calculate the limit of the singular integrals as $\delta \to 0$, keeping ϵ fixed. Then let $\epsilon \to 0$. We will present two integral equations by L. N. Long and the present author based on formulations 1-A and 3 and using different regularization procedures.

5.1 Long's Subsonic Integral Equation

This integral equation is based on formulation 1-A taking $l_i = pn_i$. For regularizing the divergent integral (depending on r^{-2}, equation 4.11, Long (1983a, 1983b) selected a square hole of side length ϵ and the center at the observer position on the surface. Two opposite sides of this square are parallel to the local tangential Mach number. When the observer is moved onto the cap of this hole, some additional terms appear making the integral equation as follows:

$$4\pi\left(1 - \frac{1}{2\beta_n^2}\right)p = \frac{2\pi}{\beta_n^2}\rho_o v_n^2 + I_\epsilon \tag{5.1}$$

Here $\beta_n^2 = 1 - M_n^2$ and I_ϵ stands for the integrals in equation 4.11 over $f = 0$ with a square hole as described above at the observer position. The above integral equation has been solved numerically for lifting and nonlifting wings and rotors (Long, 1983a, 1983b). The results have been satisfactory and, thus equation 5.1 provides a new method for prediction of the load data on the blades.

5.2 Farassat's High Speed Integral Equation

Using formulation 3, the present author has derived another aerodynamic integral equation for high speed rotating blades (Farassat, 1984). The regularization procedure used involves a hole of the shape shown in Figure 2. This hole is produced by the collapsing sphere of radius $r \leq \epsilon$ for some small $\epsilon > 0$. The observer which is at the center of the sphere, reaches the blade surface as the radius of the sphere shrinks to zero. It can be shown that the singular integral (depending on r^{-2}) in equation 4.12 becomes convergent. The contribution of the cap of this hole is then added to the remaining

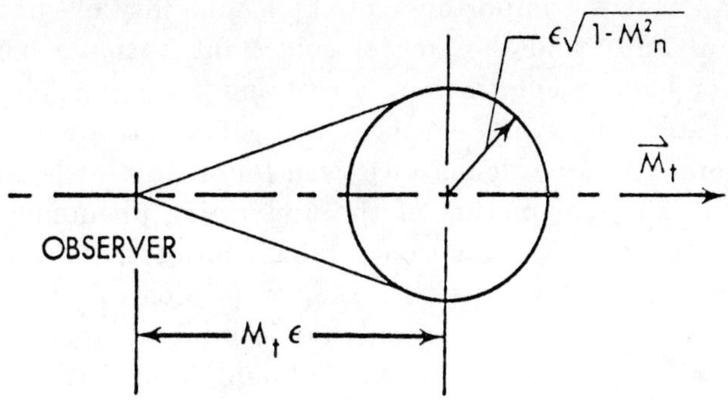

Figure 2. The shape of the hole used on the body surface to regularize the divergent integral of equation 5.2.

integral to give

$$4\pi\left[1-\frac{1}{2(1-M_n)}\right]p=\frac{2\pi\rho_o c^2}{1-M_n}+I_\epsilon \tag{5.2}$$

Here, as before, I_ϵ stands for the integrals in equation 4.12 over the blade $f = 0$ with a hole of the shape shown in Figure 2 at the observer position.

There are several methods of solving the above integral equation such as the collocation and Galerkin methods. Applications of equation 5.2 are at present under study by the author.

6. Concluding Remarks

This article summarizes the work at NASA Langley in deriving acoustic formulations in the time domain for the prediction of the discrete frequency noise of rotating blades. The major aim of this work has been development of results with as few approximations and restrictions as possible. Throughout this work, the efficient execution on a computer and the precision of the predicted noise were the motivating forces. It now appears that a combination of formulation 1-A, for the subsonic portion of the blade, and formulation 3, for the transonic-supersonic part of the blade, results in the most efficient codes. The selection of either of the two formulations must be based on the value of M_r on the panel on the blade at the time (or times) of emission.

Because of the importance of the availability of good aerodynamic input data, and the lack of consistent, rational, and simple methods for load calculations on a rotating blade, we have tried to adapt acoustic equations for such a task. The introduction of time into the aerodynamic calculations, even for steady loads, appears to enhance the physical picture of the underlying phenomenon. Also Long's experience with equation 5.1 has shown that this equation is relatively easy to work with and invert. Most of the numerical difficulties inherent in other aerodynamic integral equations do not appear here. Equations 5.1 and 5.2 should, therefore, allow us to calculate blade loads by a simple, direct and efficient method.

References

[1] Dowling, A. P. "Convective Amplification of Real Simple Sources," *Journal of Fluid Mechanics, 74* (1976), 529–46.

[2] Farassat, F. "Theory of Noise Generation From Moving Bodies With an Application to Helicopter Rotors," *NASA Technical Report R-451,* December 1975.

[3] ———. "Discontinuities in Aerodynamics and Aeroacoustics: The Concept and Applications of Generalized Derivatives," *Journal of Sound and Vibration, 55* (1977), 165–93.

[4] ———. "Extension of Isom's Thickness Noise Formula to the Near Field," *Journal of Sound and Vibration, 67* (1979), 280–81.

[5] ———. "Linear Acoustic Formulas for Calculation of Rotating Blade Noise," *AIAA Journal, 19, No. 9* (1981), 1122–30.

[6] ———. "Advanced Theoretical Treatment of Propeller Noise" in *Propeller Performance and Noise," Lecture Series 1982-08,* von Karman Institute for Fluid Dynamics, May 24–28, 1982.

[7] ———. "The Prediction of the Noise of Supersonic Propellers in Time Domain—New Theoretical Results," *AIAA Paper 83-0743,* April 1983a.

[8] ———. "A New Aerodynamic Integral Equation for High Speed Rotating Blades Based on an Acoustic Formula in the Time Domain," to appear as *Technical Note* in *AIAA Journal,* 1984.

[9] Farassat, F. and Brown, T. J. "A New Capability for Predicting Helicopter Rotor and Propeller Noise Including the Effect of Forward Motion," *NASA Technical Memorandum TMX-74037,* June 1977.

[10] Farassat, F. and Succi, G. P. "A Review of Propeller Discrete Frequency Noise Prediction Technology With Emphasis on Two

Current Methods for Time Domain Calculations," *Journal of Sound and Vibration, 71* (1980), 399–419.

[11] Ffowcs Williams, J. E. and Hawkings, D. L. "Sound Generated by Turbulence and Surfaces in Arbitrary Motion," *Philosophical Transactions of Royal Society of London, A264* (1969), 321–42.

[12] Goldstein, M. W. *Aeroacoustics,* New York: McGraw-Hill, 1977.

[13] Hanson, D. B. "Near Field Noise of High Tip Speed Propellers in Forward Flight," *AIAA Paper 76-565,* 1976.

[14] ————. "The Influence of Propeller Design Parameters on Far Field Harmonic Noise in Forward Flight," *AIAA Paper 79-06709,* 1979.

[15] ————. "Helicoidal Surface Theory for Harmonic Noise of Propellers in Forward Flight," *AIAA Journal, 18* (1980), 1213–20.

[16] Isom, M. P. "The Theory of Sound Radiated by a Hovering Transonic Helicopter Blade," Polytechnic Institute of New York, *Report Poly-AE/AM, No. 75-4,* 1975.

[17] Long, Lyle N. "The Compressible Aerodynamics of Rotating Blades Using an Acoustic Formulation," *Ph.D. dissertation,* the George Washington University, April 1983a.

[18] ————. "Aerodynamics of Propellers and Rotors Using an Acoustic Formulation in the Time Domain," *AIAA Paper 83-1821,* AIAA Applied Aerodynamics Conference, July 13–15, 1983b, Danvers, MA.

[19] Lowson, M. V. "The Sound Field for Singularities in Motion," *Proceedings of the Royal Society of London, A286* (1965), 559–72.

[20] Nystrom, P. A. and Farassat, F. "A Numerical Technique for Calculation of the Noise of High Speed Propellers With Advanced Blade Geometry," *NASA Technical Paper 1662,* July 1980.

[21] Succi, G. P. "Design of Quiet Efficient Propeller," *SAE Paper 790584,* 1979.

Helicopter Impulsive Noise: Theoretical and Experimental Status

F. H. Schmitz and Y. H. Yu
Aeromechanics Laboratory
U. S. Army Research and Technology Laboratories
AVSCOM
Ames Research Center
Moffett Field, CA 94035

Abstract

The theoretical and experimental status of helicopter impulsive noise is reviewed. The two major source mechanisms of helicopter impulsive noise are addressed: high-speed impulsive noise and blade-vortex interaction impulsive noise. A thorough physical explanation of both generating mechanisms is presented together with model and full-scale measurements of the phenomena. Current theoretical prediction methods are compared with experimental findings of isolated rotor tests. The noise generating mechanisms of high speed impulsive noise are fairly well understood—theory and experiment compare nicely over Mach number ranges typical of today's helicopters. For the case of blade-vortex interaction noise, understanding of noise generating mechanisms and theoretical comparison with experiment are less satisfactory. Several methods for improving theory/experiment are suggested.

1. Introduction

Over the past 15 years, many researchers have devoted a significant portion of their lives attempting to understand the basic generating mechanisms of sound produced by rotors. In many instances, these efforts have been quite successful, with each new advance bringing the goals of understanding, modeling, and acoustic prediction of rotor noise much closer. Obviously, this rosy outlook is not true in every instance, and there are areas in which much work is needed. However, as it is with the squeaky wheel, the noises that are loudest—the more important external noise sources—are getting the most attention. For helicopters, these loudest and often the most annoying sounds are impulsive in nature, originating on both main and tail rotors.

The treat (and follow-through) of noise rules for helicopters is having a beneficial effect on all acoustic aspects of rotors. In many cases, noise is now treated seriously early in the design process instead of as a problem to be fixed during production. However, there still remains the important trade-off of noise and performance. If reducing noise increases performance, the likelihood of an *acoustic design change* is enhanced. However, if the acoustic change decreases vehicle performance, the change must be weighed against other competing factors. If the acoustician is unsure of his design guidelines, his position in the design group is seriously weakened, and, in general, he will have less influence in the entire design process. Thus, it is imperative that the major sources of noise be addressed even at the expense of perhaps more interesting research phenomena.

Complicating this entire picture is the degree to which the vehicle itself enters into the noise-generation process. In the case of the helicopter, almost the entire aerodynamic noise-radiation process is tied to the aerodynamic state of the main and tail rotors. The disciplines of aerodynamics, rotor-body dynamics, and rotor-blade dynamics can all influence the exact state of the rotor. Theoretically, the problem is almost intractable. There are so many possible factors that can influence the radiated noise that some experimental guidance is necessary to sort out important controlling parameters.

Because there are so many sources of noise on a helicopter and because it would take a much larger document than this monograph to review adequately the status of all of them, we will not attempt it. Instead, we will concentrate on a status report on helicopter impulsive noise which includes high-speed (HS) impulsive as well as blade-vortex interaction (BVI) impulsive noise. The reader is asked to consult Lowson (1973), George (1977), and White (1980) for a more general review of the status of helicopter noise. In particular, George (1977) and White (1980), respectively, give excellent theoretical and experimental reviews. Limiting this effort to the impulsive noise of helicopters is not as restrictive as it first seems, for these noise sources have received and probably will continue to receive most of the research attention, probably because they are responsible for much of the radiated noise on today's helicopters.

Most of what we shall review can be mathematically represented by the following general, well-known integral equation governing the noise that is radiated from a body in arbitrary motion:

$$\begin{aligned} 4\pi a_0^2 \overline{\rho}'(\vec{x},t) = {} & \frac{\partial^2}{\partial x_i\,\partial x_j}\int\!\!\int\!\!\int_V \left[\frac{\overline{T}_{ij}}{r|1-M_r|}\right]_\tau dV(\vec{\eta}) \\ & - \frac{\partial}{\partial x_i}\int\!\!\int \left[\frac{P_{ij}n_j}{r|1-M_r|}\right]_\tau dS(\vec{\eta}) \\ & + \frac{\partial}{\partial t}\int\!\!\int \left[\frac{\rho_0 v_n}{r|1-M_r|}\right]_\tau dS(\vec{\eta}) \end{aligned} \tag{1.1}$$

where

$$T_{ij} = \rho u_i u_j + P_{ij} - a_0^2 \rho' \delta_{ij}$$

This equation was derived in Ffowcs Williams and Hawkings (1969) and expanded upon by many researchers (e.g., Hawkings and Lowson, 1974; Farassat, 1974; Isom, 1975). Far-field acoustic pressure is explicitly expressed in terms of integrals over the blade surface and the surrounding volume in a reference frame moving with the blade surface.

We shall begin by looking at some experimental results to see if we can correlate specific terms in the theory (equation 1.1) to specific features of the radiated noise. Before we delve into quantitative comparisons, it is helpful to look at helicopter deterministic noise sources qualitatively. Fortunately, the loudest and most annoying sounds are often the easiest to relate to aerodynamic events on the blade. Therefore, we shall begin with a physical explanation of impulsive noises of the helicopter. These arguments will then be used to explain many of the measured characteristics of helicopter impulsive noise. Finally, we shall compare theory with measured data and explain what additional effects are needed to model the resulting acoustic radiation.

Much of the work presented here has borrowed from our research and from that of colleagues at the U. S. Army Aeromechanics Laboratory, Research and Technology Laboratories (AVRADCOM), over the past ten years. We have tried to use the research of others to complement and enhance our interpretation of the present status of helicopter impulsive noise. Because our point of view is one of an involved researcher, there are undoubtedly some unintentional biases in the monograph for which we offer our apologies.

2. Qualitative Arguments of Helicopter Impulsive Noise

It is possible to analyze measured acoustic data in the frequency or time domains. However, in instances when we are dealing with helicopter impulsive noise, it is much easier to diagnose the cause of that radiated sound if a time-domain analysis is used. The sequence of events that cause the radiated noise are all related to events that happen as the rotor rotates (changes azimuth angle Ψ). Thus, it is natural to think of Ψ as a *source time* which tends to order the events an observer may hear.

For simplicity in these first arguments, we shall only consider terms on the right-hand side of equation (1.1) which can be derived by linearizing the governing mass and momentum equations. Therefore, only the last two terms of equation (1.1) are discussed. It turns out that the last integral governs *linear* high-speed impulsive noise, and the second terms is a large contributor to blade-vortex interaction impulsive noise.

2.1 Linear-Thickness Impulsive Noise

Linear-thickness noise can be represented by considering equation (1.1) with only the third term appearing on the right-hand side. The resulting simplified integral equation becomes

$$P'(\vec{x},t) = \frac{1}{4\pi}\frac{\partial}{\partial t}\int\int\left[\frac{\rho_0 v_n}{r|1-M_r|}\right]_\tau dS\,(\vec{\eta}) \qquad (2.1)$$

This equation states that the acoustic pressure is really the sum of a distribution of singularities over the blade surface. It is important to remember that sound generated by each singularity must travel a slightly different path to the obsever's location and, therefore, will arrive at a different observer time t. The simplest way of describing this integration is to divide the tip section of the blade into two chordwise panels. (We only treat the blade-tip region because we know that it is the source of most of the acoustic radiation.) The first panel is composed of *sources* and the second of *sinks*, as shown in Figure 1 for a single-bladed rotor.

One of the most interesting aspects of the evaluation of the thickness integral

$$\int\int\left[\frac{\rho_0 v_n}{r|1-M_r|}\right]_\tau dS\,(\vec{\eta})\,.$$

is that the integration is a function of $1/(r|1 - M_r|)$ which depends on the observer's location. The factor $1/(1 - M_r)$ represents the Doppler amplification of acoustic signals and is a strong function of the Mach number in the radiation direction, M_r. As shown in Figure 2, for an observer in the disk plane located directly ahead of the rotor's flight path, M_r becomes a maximum when $\psi \approx 90°$. Thus, we would expect the thickness noise peak to originate near $\psi = 90°$.

Now let us sketch a graphical outline of the integration of the integral for the in-plane observer located directly ahead of the rotor (Figure 3). First, consider the simple source ($\rho_o v_n = q$). Then

$$\int\int\left[\frac{q}{r|1 - M_r|}\right] dS\,(\vec{\eta})$$

becomes as in Figure 3. Similarly, the integral of the simple *sink* becomes the same curve shifted (delayed) in time $c/2(\Omega R)$ sec.

$$t_{\text{source}} \approx t_{\text{sink}} + \frac{c \sin \psi}{2(\Omega R)}$$

Although not explicitly shown, all of the above arguments depend on the correct evaluation of the retarded time equation, $\tau = t - r/a_o$. The simple shift in observer time causes the two sources to not cancel. This is the major mechanism of linear-thickness noise. It is characterized by a large negative-pressure pulse; a similar set of arguments was presented in Schmitz and Yu (1979) (see Figure 4).

Consider the same integral for an observer situated directly beneath the rotor's axis. (See Sketch 1.) In this location, the observer sees no difference in time or in attitude for the source and sink integration. Thus, when both contributions are summed and the time-derivative taken, little radiation is found. The contributions from the simple source and sink cancel, causing no acoustic radiation. These simple arguments explain why the linear-thickness noise only radiates near the plane of the rotor disk. This will be confirmed when we look at some experimental data.

2.2 Blade-Vortex Interaction Impulsive Noise

The qualitative characteristics of blade-vortex interaction (BVI) noise can be deduced in a similar manner. Again, we shall perform

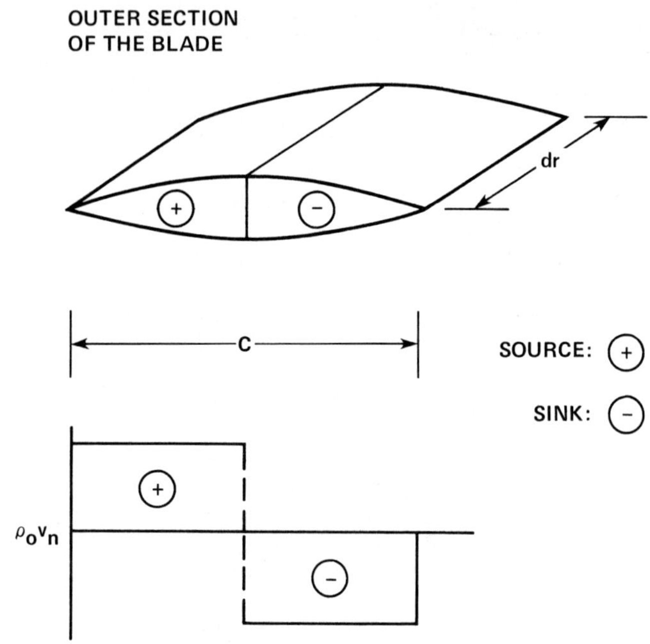

Figure 1. Acoustic sources and sinks.

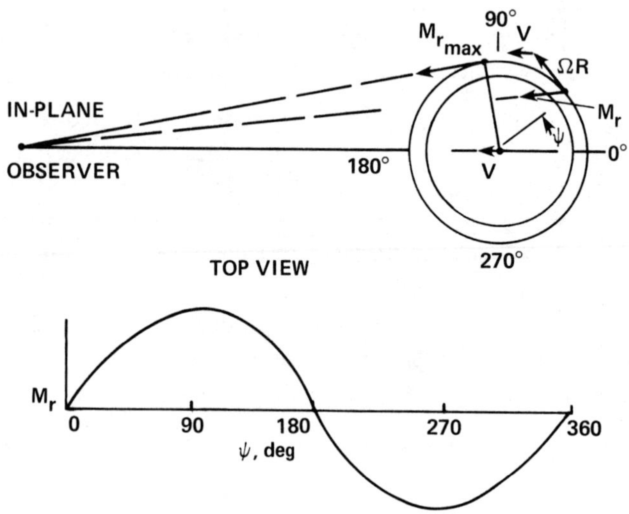

Figure 2. Variation of M_r over azimuthal angles.

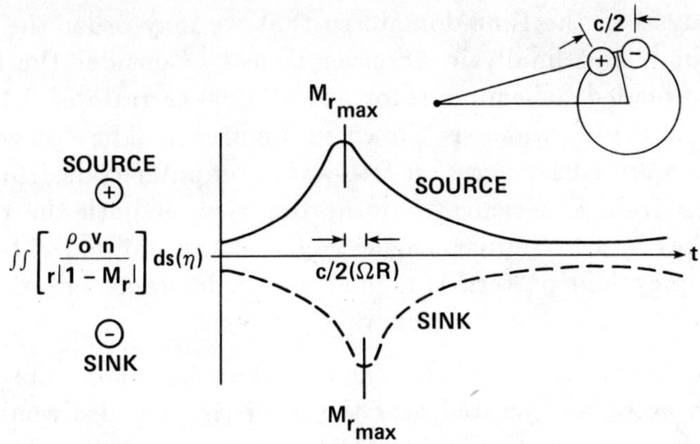

Figure 3. Variation of M_r over observer time.

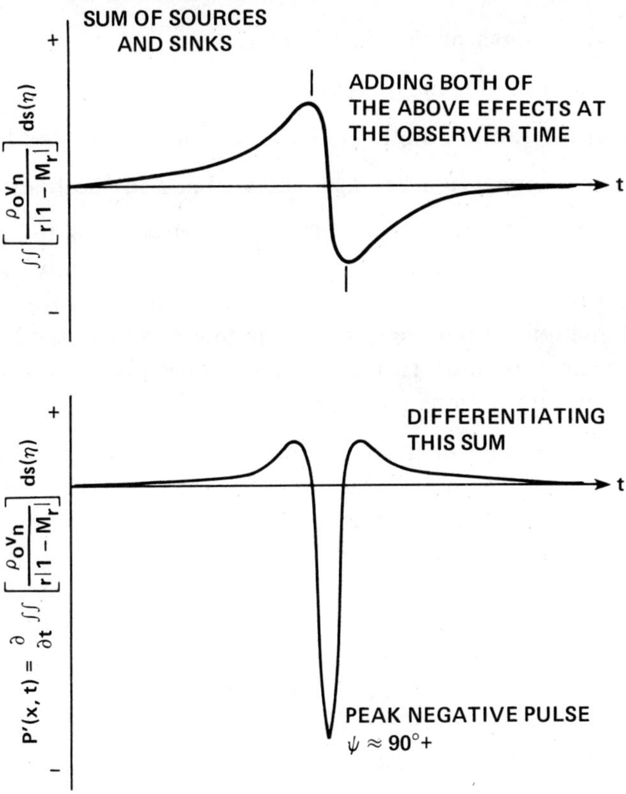

Figure 4. Summation of sources and sinks for given observer time and its derivative.

our analysis in the time-domain so that we may order the acoustic events in ψ and finally in observer time t. Consider the top view of a two-bladed helicopter rotor at an advance ratio of 0.145 ($\mu = V/\Omega R = 0.145$) which is shown in Figure 5. The epicycloid-like patterns were derived from a *free-wake* computer code. However, it is known from experience that the top view of both the rigid and free wakes is quite similar, suggesting that the simple rigid-wake tip vortex epicycloid pattern is a good first-order representation of the tip-vortex patterns of the free-wake analysis.

We know from theoretical considerations that most of the radiated noise is generated near the rotor tip. We also would expect blade-vortex interaction noise to occur when the rotor blade (outer 20–30%) passes close to the trailing-tip vortices. As seen from Figure 5, there are seven possible blade-vortex interactions (labeled 1–7).

The strength of each interaction is governed by

1. Local strength of the tip vortex.
2. Core size of the tip vortex.
3. Local interaction angle of the blade and the vortex line.
4. Vertical separation between the vortex and the blade.

In general, the induced velocity of the rotor disk tends to make all of the tip vortices pass under the rotor disk in level steady-state flight conditions for $\mu \approx 0.15$. However, if the rotor operates in steady descending flight, then the positive inflow (upflow) tends to force the epicycloid-type pattern into the rotor disk plane, causing strong blade-vortex interactions (see Figure 6).

The net result of such considerations is shown in Figure 7. Of these potential seven blade-vortex interaction (BVI) encounters, only a few are known to radiate strong impulsive noises. Consider encounters numbered 1–4, which are all on the advancing side of the rotor disk. Number 3 in particular is an encounter in which the blade and the vortex are almost parallel. In this case, we can use some simple two-dimensional arguments to estimate the correct shape of such a pulse. An advancing blade might see the following (see Figure 8). A conceptual view of the angle-of-attack time-history as the vortex passes near the airfoil is shown in Figure 9. For incompressible flow, this will result in a net lift versus time on the rotor, which is sketched in Figure 10. In this simple arguments, the entire blade is assumed to feel the presence of the changing angle of

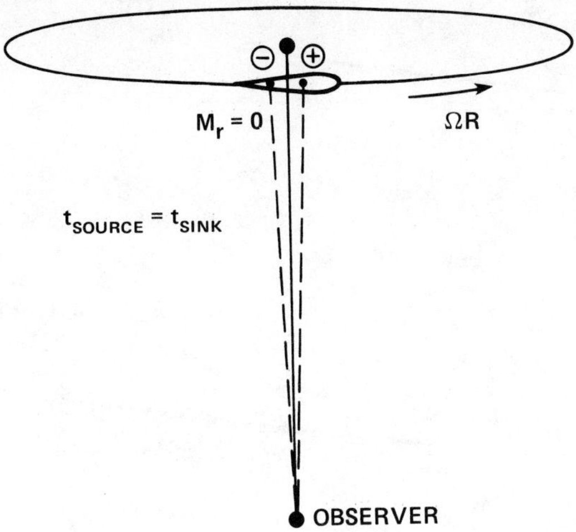

Sketch 1. Observer beneath rotor axis.

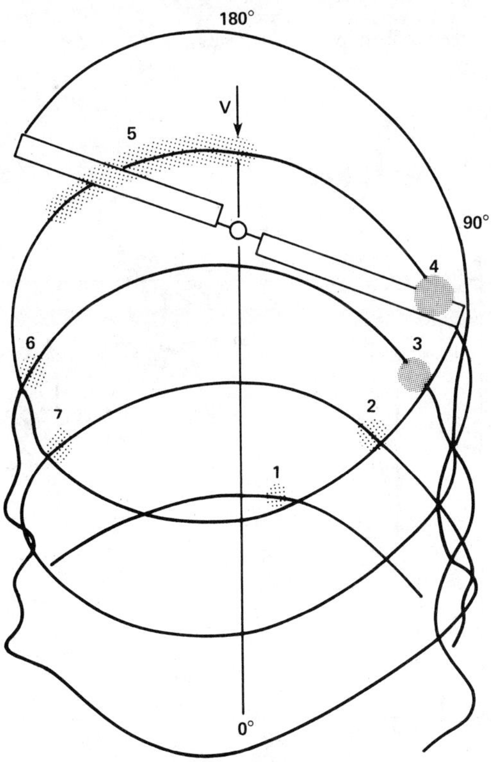

Figure 5. Blade-vortex intersections during partial-power descent (from Tangler, 1977).

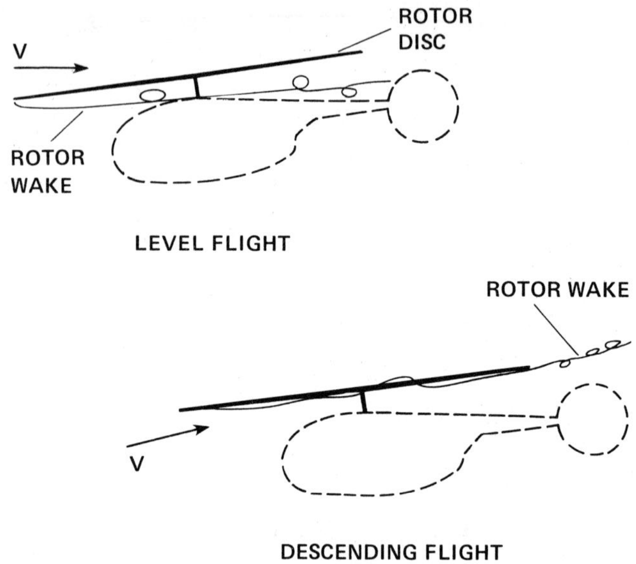

Figure 6. Interaction of tip vortices with a rotor disk for level and descending flights.

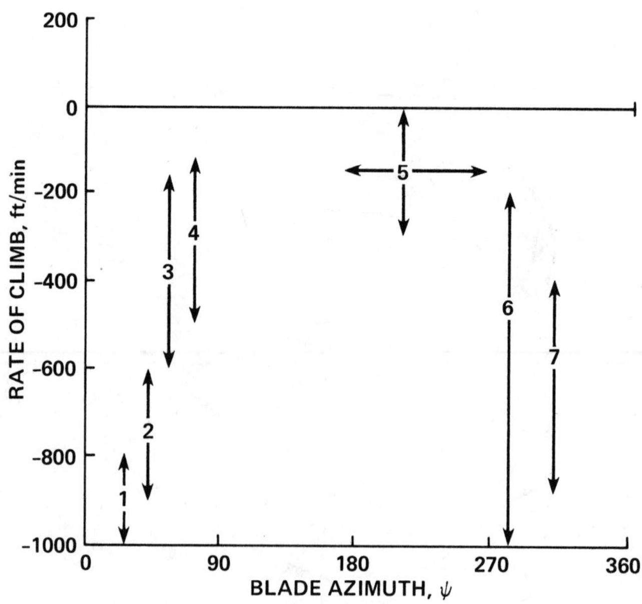

Figure 7. Flight conditions for potential blade-vortex interactions.

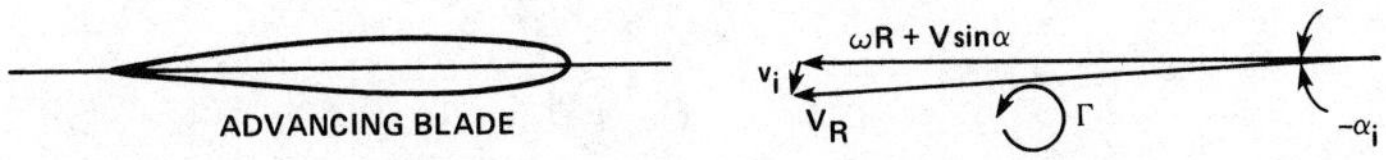

Figure 8. Blade-vortex interaction on an advancing side.

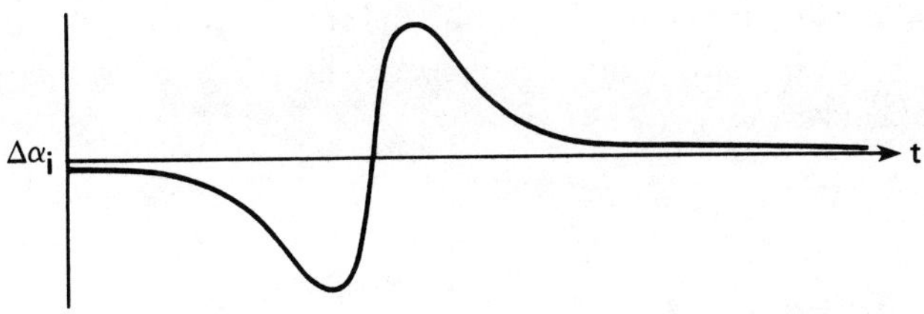

Figure 9. Angle-of-attack time-history for an advancing blade.

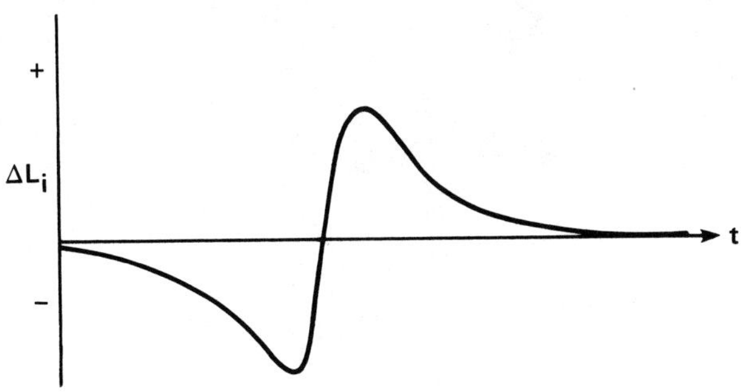

Figure 10. Lift history of an advancing blade.

attack. The resulting time-varying force field is impulsive in nature. The radiated noise is given by

$$P'(\vec{x},t) = -\frac{\partial}{\partial x_i} \int\int \left[\frac{P_{ij}n_j}{r|1-M_r|}\right]_\tau dS(\vec{\eta}) \qquad (2.2)$$

Treating the entire blade as a single radiating body (an acoustically compact body) and considering radiation to the far-field, this expression can be rewritten as

$$P'(\vec{x},t) \approx \frac{\partial}{\partial t}\frac{1}{a_o}\left[\frac{\Delta L_i \cos\theta_i}{r|1-M_r|}\right]$$

where θ_i is the angle between the surface normal in the direction of the force on the fluid and a line from the point of the applied force to the observer.

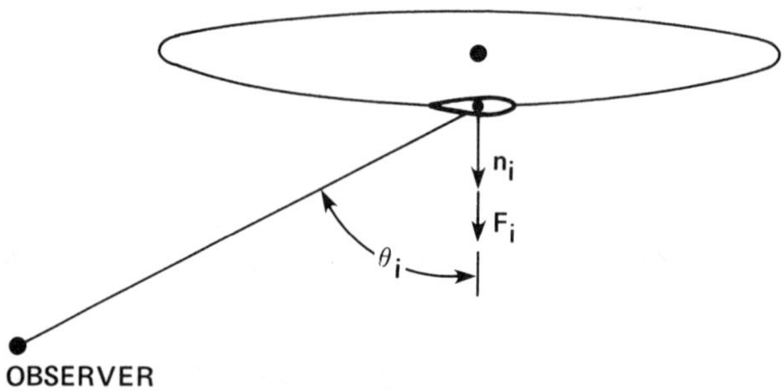

Figure 10a.

This expression governs the shape of the BVI noise. Similar to thickness noise, the *Doppler amplification* alters the magnitude of the radiation force field, but not the basic character. Thus, the shape of the radiated acoustic pressure becomes that shown in Figure 11.

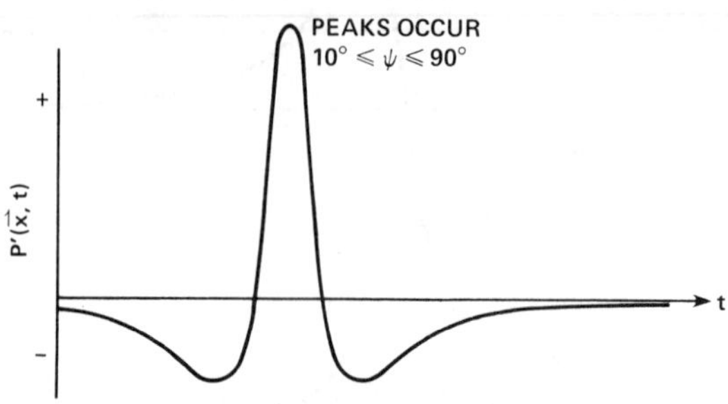

Figure 11. Acoustic pressure signature of blade-vortex interaction.

The net effect of blade-vortex interaction disturbances on the advancing side of the rotor disk is acoustic radiation in the form of a sequence of predominantly positive spikes similar to the above. These near discontinuities are of varying strengths and occur between $\psi = 0°$ and $\psi = 90°$. For the observer in the far-field, these positive pressure impulses arrive before the large negative thickness-noise pulse. Notice that the acoustic radiation decreases as the observer approaches the plane of the rotor disk (i.e., $\cos\theta_i \to 0$).

Now, consider the BVI on the retreating side of the rotor disk (numbers 5, 6, 7). Again, using our simplified two-dimensional qualitative model, we have the following geometry:

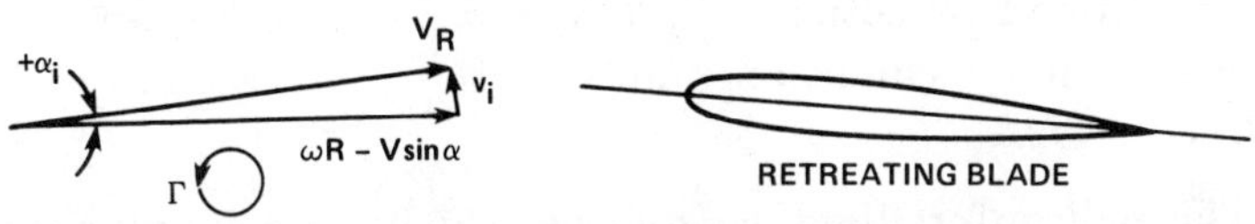

Figure 11a.

For retreating BVI, the sign of the approaching vortex is opposite to BVI on the advancing side. Therefore, by similar argument the net effect is a predominantly negative radiated acoustic signature for each BVI on the retreating side, as shown in Figure 12. All but the No. 5 BVI impulses occur later in time than the thickness-noise pulse. These two pieces of information help isolate the general origins of the radiated noise. Obviously, these simple qualitative arguments do not tell us many of the more interesting details. However, these arguments do help us interpret measured impulsive-noise acoustic data. (Similar arguments can be found in Nakamura (1981).)

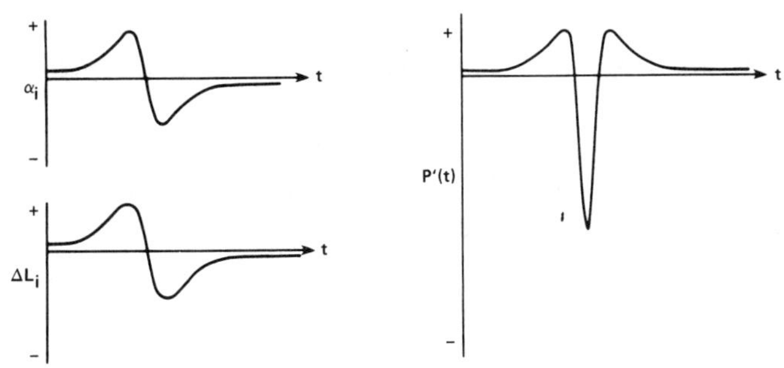

Figure 12. Angle of attack, lift, and acoustic pressure time history of a retreating blade.

3. Measured Impulsive Noise

Over the past ten years, a novel in-flight method of gathering helicopter impulsive-noise data was developed by the Aeromechanics Laboratory (U. S. Army Research and Technology Laboratories, AVRADCOM). In the early stages of the work, an OV-1C Mohawk aircraft was instrumented with a microphone on the top of its vertical tail and flown in formation with the subject helicopter, as shown in Figure 13. Because the Mohawk was fairly quiet and the impulsive signals quite intense, good-quality acoustic data were acquired. (Additional details can be found in Schmitz and Boxwell (1976).) The major advantages of gathering data in this manner are (1) no ground reflections; (2) long and steady data samples; and (3) helicopter flight conditions and directivity profiles are easily explored.

The helicopter flight conditions which were investigated for the UH-1H helicopter are shown in Figure 14. High-speed impulsive noise was measured in high-speed forward flight, and BVI impulsive noise was measured in moderate forward-speed but descending flight. Also illustrated in this figure are contours of BVI noise as heard in the helicopter cabin. In early experiments, it was thought that cabin noise was a good indicator of radiated impulsive noise. This new in-flight technique showed that cabin measurements were necessary but not sufficient conditions for noise radiation. There is radiated noise in other directions that is not heard in the cabin. Therefore, if a pilot were to fly so as to minimize the cabin impulsive noise, he might still be radiating BVI noise to ground observers.

It was generally observed from the measured data that the far-field acoustic waveform radiated by each blade was multipulse in nature. As many as three distinct pressure disturbances could be repetitively identified in the acoustic waveform. For identification of this waveform structure and familiarity with data presented in the following sections, an idealized composite drawing of the acoustic waveform showing this multipulse character is presented in Figure 15. In this figure, peak pressure amplitude of the acoustic signature is illustrated versus one-half revolution (one-blade passage) in time, with time increasing from left to right. The peak pressure amplitude scale used here is an absolute scale measured in dynes per square centimeter. On this scale, a sinusoidal-shaped waveform with a peak

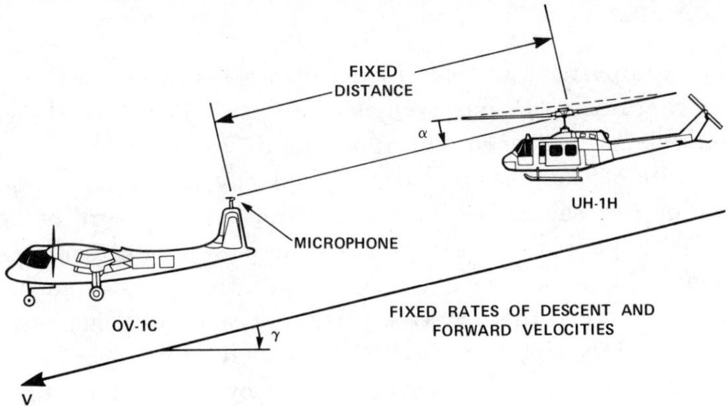

Figure 13. In-flight technique.

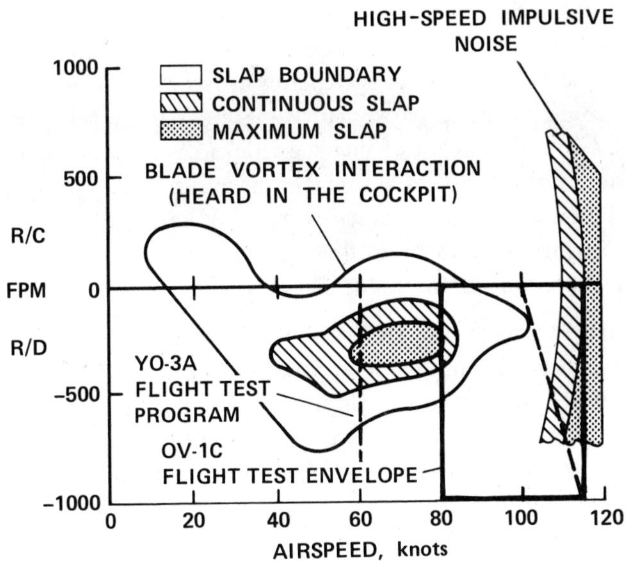

Figure 14. Impulsive-noise boundaries for UH-1 series helicopters.

pressure amplitude of 512 dynes/cm^2 would exhibit a root mean square (rms) sound pressure level of 124 dB (ref. 0.0002 dyne/cm^2 rms).

The composite waveform model illustrates three predominant pressure characteristics observed in the data. They are shown in the same relative sequence and approximate pulse width that were characteristic of the measured data. Typically, the sequence began with one or two successive increases in positive pressure of *triangular* pulse shape (No. 1 in Figure 15). These positive pressure peaks were followed by a large near-triangular negative pressure pulse. At high advance ratios, the negative pressure rise (No. 2) increased in amplitude slightly slower than its subsequent rapid decrease (No. 3), and the waveform is represented more by a saw-tooth or half-triangular pulse. Finally, when it was observed to occur, an extremely narrow positive-pressure spike followed immediately after or as a result of the extremely rapid increase in pressure.

Using the qualitative arguments presented at the beginning of this chapter, it is possible to trace the origins of the radiated noise. Referring to the composite illustration, the negative pulse (No. 2, Figure 15) is associated with high-speed thickness effects. As we have shown, it occurs in source coordinates at about $\psi \approx 90°$. The initial series of positive pulses (No. 1) is a direct result of blade-tip vortex interaction on the advancing side of the rotor disk. As we have shown, they occur before the thickness-noise impulse, occurring in source coordinates at $\psi \approx 10° \rightarrow 80°$. Characteristic No. 3 is really just a manifestation of intense thickness noise. It occurs when the thickness noise (and its associated aerodynamics) is so large that local shocks on the blade radiate to the far-field. In this latter case, nonlinear terms need to be added to the simple linear calculations to predict the acoustic far-field.

Figure 16 presents a performance matrix of measured acoustic data at flight conditions between 80 and 115 knots IAS and rates of descent of 0 to 800 ft/min (see Figure 14). These unaveraged acoustic waveforms, corresponding to two consecutive blade passages, were recorded at a nominal hub-to-microphone separation distance of 95 ft with the microphone positioned directly ahead of the helicopter ($\beta = 0$) and nearly within the plane of the rotor tips ($\alpha \approx 0$). Each of the acoustic time-histories has the same amplitude scale shown, for example, with the upper right waveform in the performance matrix.

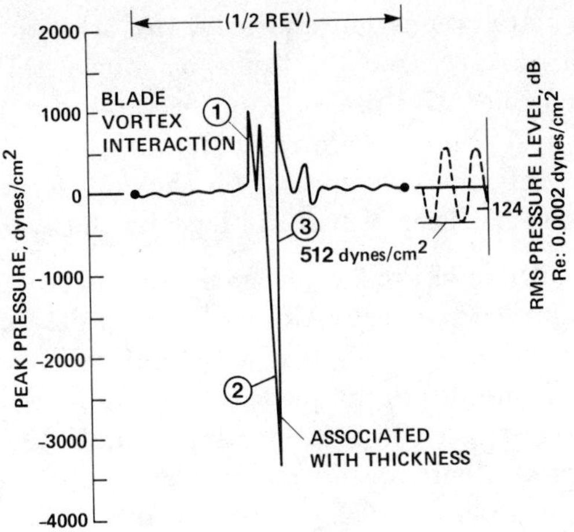

Figure 15. Composite illustration showing dominant UH-1H acoustic waveform features.

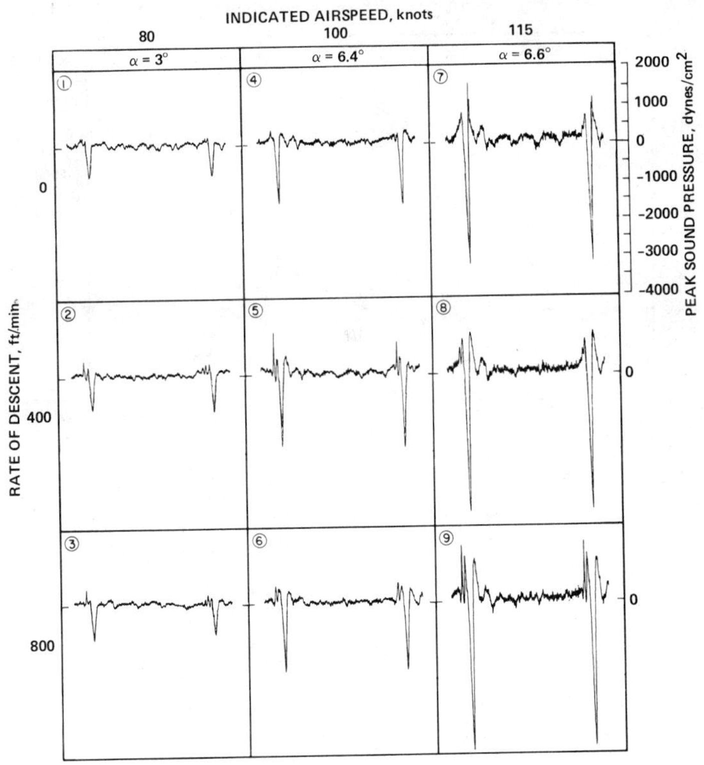

Figure 16. Unaveraged Acoustic signature of UH-1H impulsive noise versus forward airspeed and rate of descent.

To show data trends more clearly, the acoustic waveform for each condition was averaged 128 times, as shown in Figure 17. The large, consistent negative-pressure pulse was used to trigger the averaging process. Since blade-to-blade variability is removed by this technique, only one blade passage is shown. Notice that random background noise has been reduced in level by the averaging process.

Peak amplitude of the large negative-pressure pulse is strongly dependent on forward speed. Although the width of the negative pulse appears to decrease slightly with increasing speed, no consistent trends in amplitude or pulse width could be deduced with changes in descent rate. It is interesting to note that under level-flight conditions at all air-speeds, no impulsive noise was heard in the cabin, indicating that for all flight conditions tested, the pilot was unaware that the helicopter was radiating that part of the impulsive noise waveform associated with the negative pressure peak.

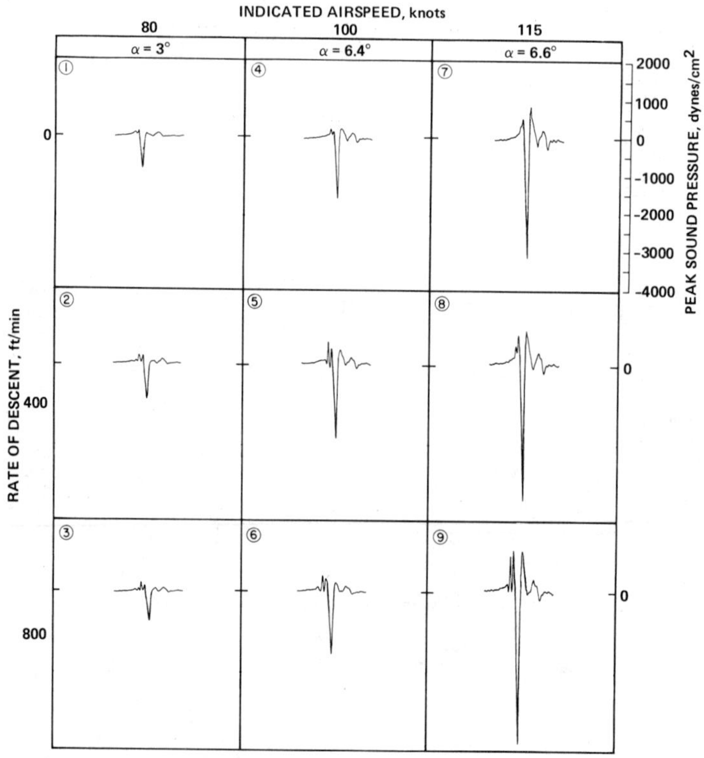

Figure 17. Averaged acoustic signature of UH-1H impulsive noise versus forward airspeed and rate of descent.

At the high forward speed conditions of 115 knots, the large negative-pressure peak, when measured nearly in-plane, was followed by a positive-pressure increase which exhibited some variability from blade to blade. This extremely rapid rise in pressure documented here was so intense that it was heard directly in the cockpit of the Mohawk, over and above the aircraft's own internal noise levels. However, no apparent slapping was heard in the cabin of the helicopter at any speed above 100 knots IAS regardless of rate of descent. To the pilot of the helicopter, a moderate increase in vibration level was the only noticeable effect, even though the UH-1H was radiating tremendous amounts of acoustic energy.

Blade slap was heard in the cabin under partial-power descents at forward speeds below 100 knots. Blade slap appeared to be most intense within the helicopter at about 80 knots IAS at a rate of descent of 400 ft/min. The occurrence of this cabin noise correlates with the positive impulsive pressures which precede the large negative-pressure pulse on the acoustic waveforms. The occurrence of these positive-pressure pulses is sensitive to rates of descent and resulting rotor-wake geometry, thus confirming that these pulses are a direct result of blade-tip vortex interaction.

Directivity profiles of the UH-1H impulsive noise were measured throughout a sweep of angular microphone positions for two operating conditions: 80 knots IAS, 400 ft/min rate of descent, and 115 knots IAS, 0 ft/min rate of descent. The longitudinal and lateral angles, α and β, respectively, were measured from a line in space drawn between the rotor hub and microphone to the rotor tip-path plane for longitudinal directivity and to the forward velocity vector for lateral directivity.

Figures 18 and 19 present the longitudinal and lateral directivity profiles of the UH-1H helicopter in an 80 knot IAS, 400 ft/min rate of descent. As discussed previously, this operating condition produced blade-slap noise that was audible to the helicopter pilot.

The longitudinal-directivity signatures contain both positve- and negative-pressure pulses, the former exhibiting considerable variability from blade to blade. These positive-pressure pulses, which result because of blade-tip vortex interaction, become large for longitudinal angles (α) between 10° and 52°. Even at the nearly in-plane or overhead position, some blade-tip vortex interaction radiation is evident, indicating that a wide angular distribution of acoustic energy is radiated to the far-field in the longitudinal plane. However,

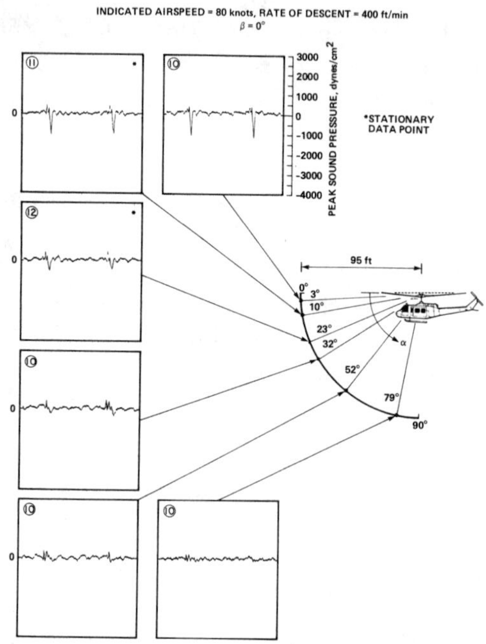

Figure 18. UH-1H longitudinal acoustic directivity at 80 knots IAS and 400 ft/min R/D ($\beta = 0°$).

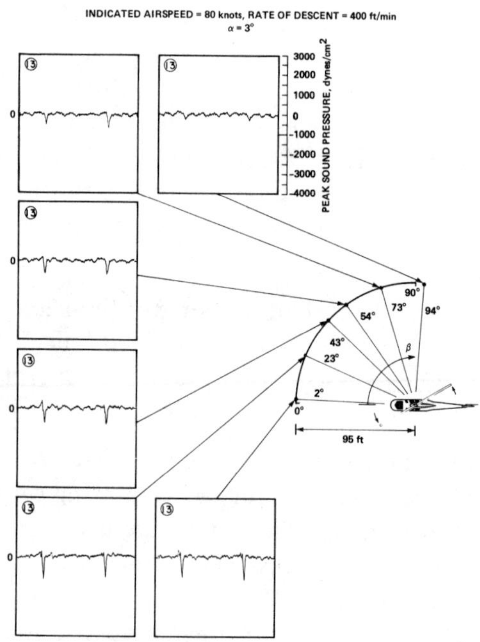

Figure 19. UH-1H lateral acoustic directivity at 80 knots IAS and 400 ft/min rate of descent ($\alpha = 3°$).

the negative pressure peak exhibits quite different radiation characteristics. It reaches its maximum level near the in-plane positions of the rotor disk but decreases rapidly to half amplitude by the 23° position and continues to decrease uniformly with increasing angle until it is hardly discernible above background noise levels at the 52° angular position. These experimental findings are in agreement with the qualitative arguments of the previous section. Thickness noise is a maximum in the plane of the rotor and BVI noise is small at in-plane positions and increase under the rotor.

Lateral-directivity signatures, shown in Figure 19, for the same operating conditions reflect a rapid decrease in the impulsive nature of the positive-pressure pulse for measurements to the advancing-blade of the rotor and a gradual disappearance into background noise by the 54° point. Again, variability between blade signatures is present. The negative-pressure pulse is shown to decay less rapidly in-plane than out-of-plane as the directivity angle is increased. It is approximately half amplitude at 73° and still discernible to the side of the helicopter ($\beta = 94°$). Although the helicopter pilot cannot hear any slapping noise associated with the negative-pressure pulse, it is clear that near the tip-path plane of the helicopter, large angular distributions of acoustic energy are being radiated.

The longitudinal- and lateral-directivity profiles for the high-speed level flight are shown in Figures 20 and 21. Although the positive-pressure pulses associated with blade-tip vortex-interaction impulsive noise are noticeably absent for this flight condition, there are rapid increases first in the large-amplitude negative pressure and then in the positive pressure. The negative-pressure peak is predominantly caused by transonic thickness effects and, although much larger in amplitude for this high-speed condition, varies in longitudinal and lateral directivity in much the same way as in the low-speed case. It is at a maximum near the tip-path plane of the rotor and falls off uniformly with increasing lateral directivity angles where, at $\beta = 84°$, it is barely noticeable. As indicated previously, the pilot cannot identify this large angular distribution of radiated impulsive noise from inside the helicopter.

The extremely sharp positive-pressure rise shown to exist in a narrow angular region near and above the rotor's tip-path plane and in the direction of forward flight resulted in very intense radiated noise levels. However, similar to the negative-pressure pulse, no impulsive-noise radiation could be detected by the helicopter pilot.

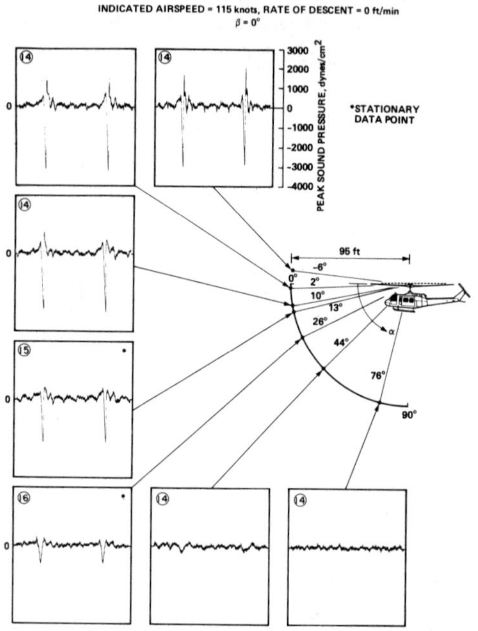

Figure 20. UH-1H longitudinal acoustic directivity, level flight, 115 knots IAS ($\beta = 0°$).

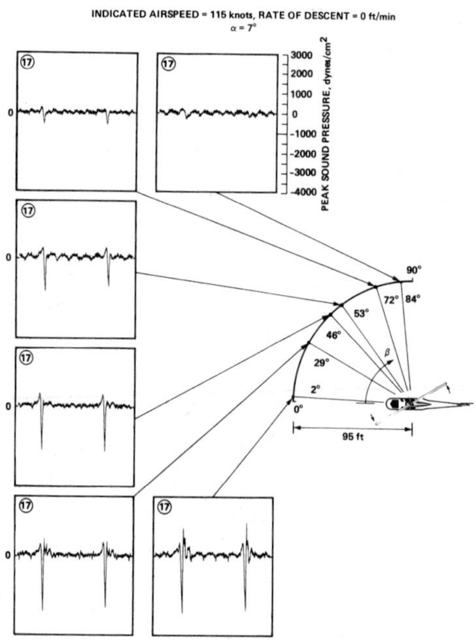

Figure 21. UH-1H lateral acoustic directivity, level flight, 115 knots IAS ($\alpha = 7°$).

We shall show later that this rapid increase in pressure is associated with a radiating shock wave.

The overwhelming success of the in-flight measurement technique encouraged its further refinement to more quantitatively examine blade-vortex interaction (BVI) noise. A specially built *quiet airplane* developed by the Army for Vietnam, called the YO-3A, was employed to gather additional acoustic data on a whole range of helicopters. Of course, one of the first tasks was an expansion of the original UH-1H matrix of flight conditions (Boxwell and Schmitz, 1982). The YO-3A could fly more slowly and thus make it possible to investigate fully the conditions of BVI during landing approaches. To emphasize further the signal-to-noise ratios during BVI, the microphone was positioned 30° below the rotor tip-path-plane, where BVI noise becomes more intense than high-speed impulsive noise (see Figure 22).

A sketch of typical BVI noise is shown in Figure 23 for this second test profile with the microphone located 30° below the tip-path-plane of the rotor. The wide negative-pressure pulse is indicative of high-speed impulsive noise, and the predominately positive-pressure pulses depict impulsive noise resulting from blade-tip vortex interactions. Tail-rotor compressibility noise is also depicted in the same figure by the higher-frequency, smaller-amplitude negative-going pulses.

A matrix of test conditions for the 30°-down microphone position for the UH-1H is shown in Figure 24. It is quite apparent that BVI noise plays a larger role in the noise radiation of the UH-1H at this 30°-down position.

3.1 Some Impulsive-Noise Power Spectra

Although not the only, or for that matter the most important, method of analyzing impulsive noise sources, a power spectral density representation of the characteristic waveforms does indicate in what band of energy the noise is predominant. For this reason, the energy content in each of the three basic waveforms is presented (Schmitz and Boxwell, 1976).

Figuer 25 illustrates the sound-pressure level of the acoustic signature measured in the far-field near the tip-path-plane ($\alpha = 3°$) for the 80 knot level-flight condition. Frequencies up to 800 Hz are

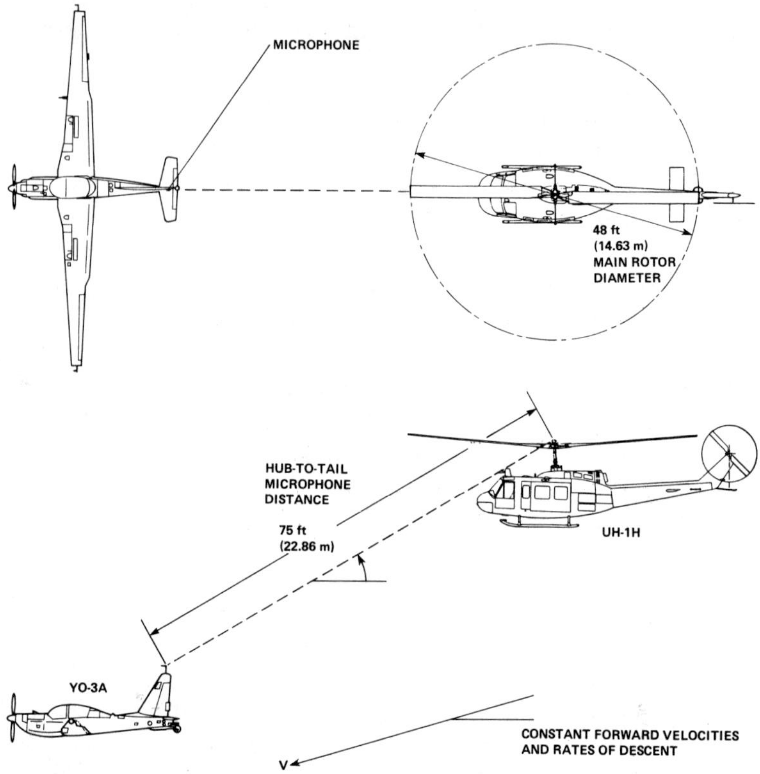

Figure 22. Relative orientation of aircraft for acoustic measurements of UH-1H blade-vortex interaction noise.

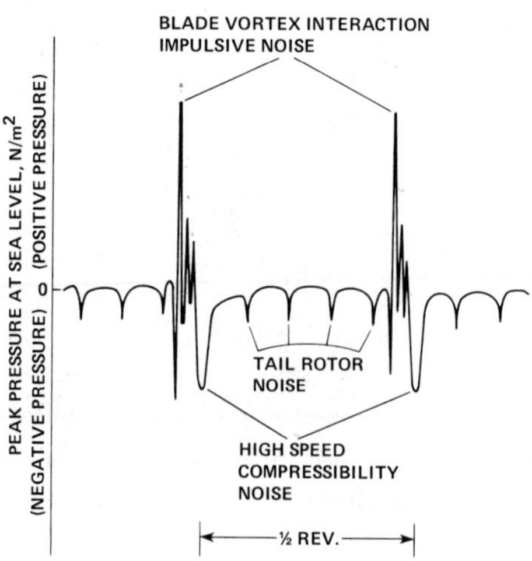

Figure 23. Composite drawing showing dominant acoustic waveform features.

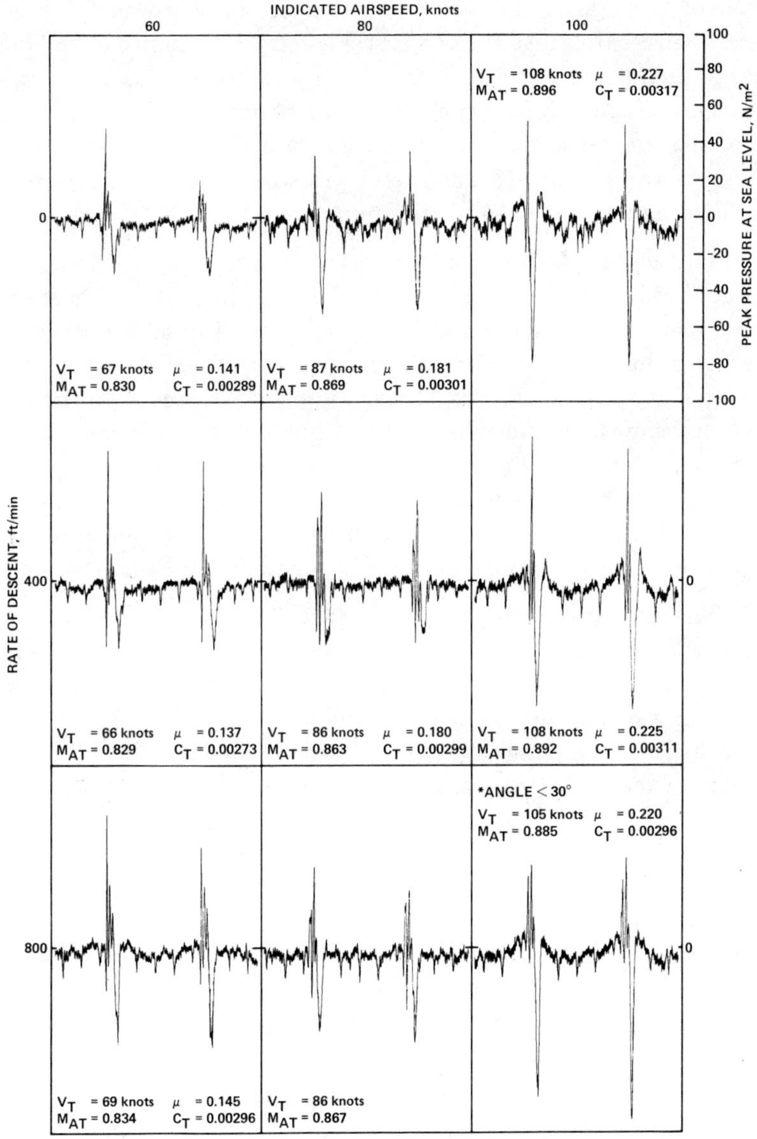

Figure 24. Unaveraged acoustic signature of UH-1H vs forward airspeed and rate of descent.

shown. At higher frequencies, the UH-1H acoustic signature drops into the background noise levels. For reference, the corresponding pressure time-history is also reproduced in the upper right-hand corner of the figure. The dominant feature of this waveform is the negative-pressure pulse that causes the many harmonics of the blade-passage frequency. Most of the energy in this pulse is contained in the low-frequency harmonics, up to about 150 Hz. The higher harmonics decrease at a more rapid rate, being 20 dB down at 300 Hz. Subjectively, this negative-pressure pulse signature can be described as a smooth, loud thumping sound.

The additional harmonics generated as a result of blade-vortex interaction are shown in Figure 26. The waveform, which was measured during descending flight (400 ft/min) at the 80 knot airspeed, illustrates the occurrence of positive-pressure pulses that precede the larger negative-pressure rise. Their major contribution to the power spectrum is in the higher harmonics of blade-passage frequency (300–600 Hz). Although the energy content of these pulses is considerably less than that of the negative-pressure disturbance, their dominant presence at the higher frequencies is believed to be responsible for the crisp popping or slapping character in the audible acoustic signature.

The last power spectrum, Figure 27, illustrates the sound-pressure level of the in-plane far-field acoustic signature for the high-speed, 115 knot level-flight condition together with the measured waveform. A striking increase in acoustic levels at these forward airspeeds is shown out to 1600 Hz. Even though the UH-1H acoustic signature exceeded background noise levels by at least 10 dB beyond these frequencies, the data began to exceed the dynamic range of the tape-recording equipment and became lost in electronic noise. Although it is difficult to separate rigorously the energy contribution of the large negative-pressure pulse from the following sharp increase in positive pressure, some general observations can be made. The energy at the low harmonics, as in the low-speed cases, is believed to result primarily from the presence of the large negative-pressure pulse. The sharpness and magnitude of the following increase in pressure causes most of the higher frequency noise shown. As noted previously, the resulting acoustic signal is very intense and definitely dominates all other noises generated by the helicopter.

All of the data shown so far were taken on the UH-1H helicopter. The key operational and design characteristics are shown in Table 1. The main point to note is the high main-rotor tip speed.

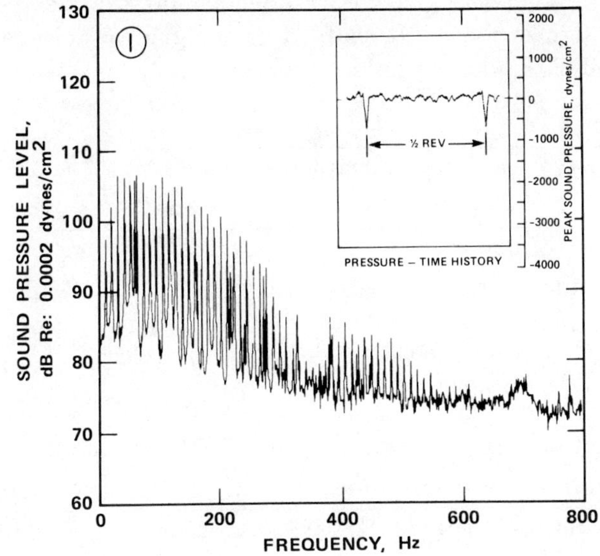

Figure 25. In-plane UH-1H acoustic power spectrum for level flight, 80 knots IAS: $\alpha = 3°$, $\beta = 0°$.

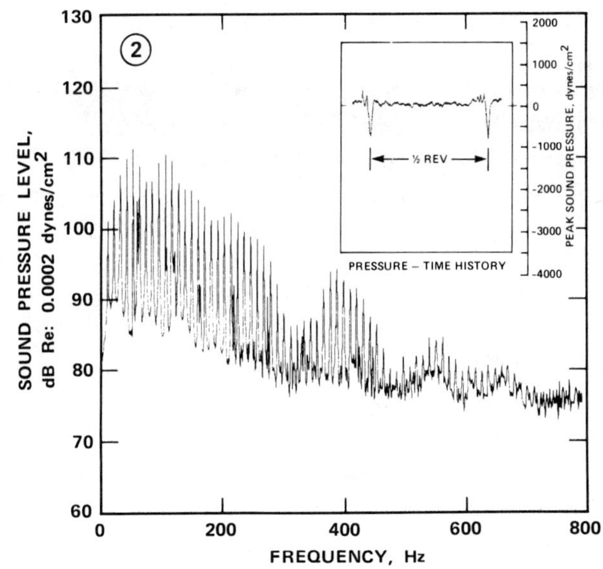

Figure 26. In-plane UH-1H acoustic power spectrum, descending flight, 80 knots IAS, and 400 ft/min R/D: $\alpha = 3°$, $\beta = 0°$.

In fact, the high tip speed is responsible for the clean high-level impulsive signals shown. It enabled good signal-to-noise levels to be maintained throughout the test program.

Rotor system design variables	Main rotor	Tail rotor
number of blades	2	2
rotor diameter, ft	48	8.5
rotor solidity	0.0464	0.105
blade chord, in.	21	8.41
blade airfoil	NACA 0012	NACA 0015
blade twist (root to tip), deg	−10.9	0
Aircraft operational limits	Maximum	Minimum
main-rotor tip speeds, ft/sec	813.8	740.0
tail-rotor tip speeds, ft/sec	736.1	669.0
forward flight airspeeds, knots	115	0
gross weight, lb	9500	6600

Table 1. UH-1H operational characteristics.

Over the years, the technique of using a *flying microphone* has been used by the Aeromechanics Laboratory to measure the noise of many helicopters. More recently, Ames Research Center acquired a YO-3A and has dedicated it to the helicopter-noise measuring role. The various acoustic signatures of all kinds of single-rotor helicopters have more similarities than differences. For example, acoustic data from a modern four-bladed helicopter in the in-plane in-trail position is shown in Figure 28. The helicopter is in level flight at 130 knots IAS. Notice that the mechanisms are quite similar to those of two-bladed rotors. A sequence of positive BVI impulses precedes the high-speed thickness-noise expansion. However, instead of two characteristic waveforms per rotor revolution, we now see four, corresponding to the four-bladed rotor system.

This particular helicopter has a much lower main-rotor hover tip speed than the UH-1H ($\Omega R \simeq 700$ft/sec). However, as in most modern helicopters, it also cruises at much higher velocities. The net result is a quite comparable advancing-tip Mach number on both

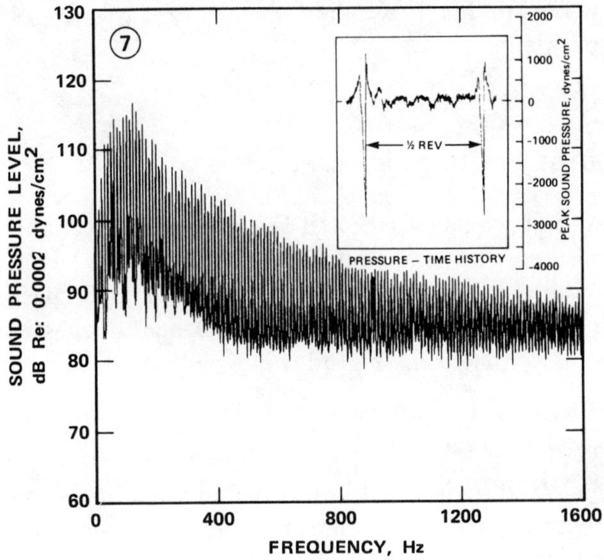

Figure 27. UH-1H acoustic power spectrum, level flight, 115 knots IAS: $\alpha = 6.6°$, $\beta = 0°$.

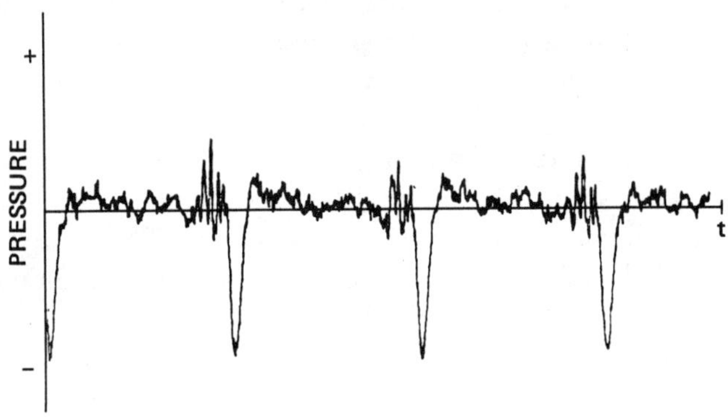

Figure 28. Acoustic signature of a modern four-bladed helicopter.

helicopters in cruise. Therefore, it is not too surprising to find that the acoustic radiation is quite similar in character. As we have seen, the advancing-tip Mach number plays a key role in all helicopter acoustic problems.

4. High-Speed Impulsive Noise

Now that we have reviewed the necessary mechanisms for impulsive noise on two- and four-bladed helicopters, let us look more closely at the detailed aerodynamic origins of high-speed impulsive noise. In particular, we pose four fundamental questions and attempt to answer them in the following sections:

1. Can high-speed impulsive noise be scaled, and if so, what are the scaling parameters?
2. Is the rapid high-speed impulsive noise compression which follows the expansion region at high advancing-tip Mach numbers a radiating (delocalized) shock wave, and can it be measured?
3. Can high-speed impulsive noise be duplicated in a steady environment (i.e., how important are the unsteady effects)?
4. How accurately can high-speed impulsive noise be predicted? What degree of modeling is required?

Our theoretical results so far have shown that a very few key nondimensional variables control much of the acoustic radiation. The fact that high-speed impulsive noise is predominately a noncompact (source and sinks do not completely cancel for an in-plane observer) high-Mach number (compressible) event would suggest that small models could be made to duplicate the full-scale acoustic phenomena. This fact was demonstrated in two separate wind-tunnel experiments (Schmitz et al., 1977; Splettstoesser et al., 1983). In the first, a 1/7 scaled UH-1H model rotor was fabricated and run in an acoustically treated wind tunnel. The data were compared with full-scale data taken at similar nondimensional conditions. As illustrated in Figure 29, model-to-full-scale comparisons are quite straightforward. There are no Doppler corrections and data records up to 1 min in duration are possible at steady-state flight conditions.

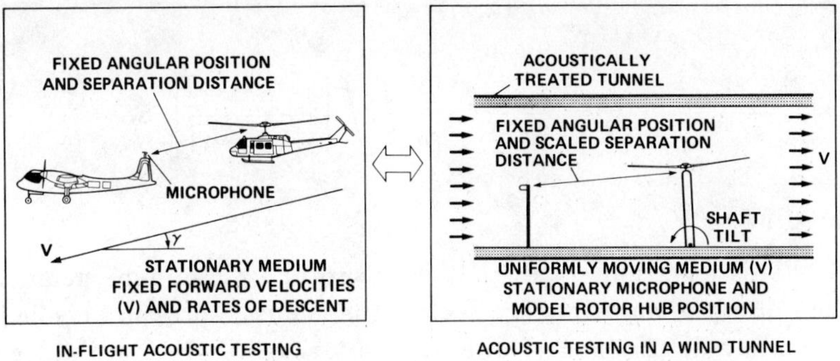

Figure 29. Model and full-scale acoustic testing.

From dimensional analysis considerations, we know that there exist nondimensional variables which govern acoustic radiation. If we consider linear thickness noise,* we have (equation 2.1)

$$P'(\vec{x},t) = \frac{1}{4\pi}\frac{\partial}{\partial t}\int\int\left[\frac{\rho_o v_n}{r|1-M_r|}\right]_\tau dS(\vec{\eta}) \tag{4.1}$$

where $t = \tau + r/a_o$. But $v_n = V \cdot \lambda$, where λ = local blade surface slope, and V = velocity of the undisturbed flow at each blade station. If we assume geometric scaling and introduce the nondimensional unit of time $\bar{t} \equiv t/(2\pi/\Omega)$, then

$$\begin{aligned} \overline{dS} &= \frac{dS}{R^2}, \\ \bar{r} &= \frac{r}{R}, \\ \bar{t} &= \frac{t}{2\pi/\Omega}, \\ \bar{t} &= \bar{\tau} + \frac{\bar{r}M_t}{2\pi} \end{aligned} \tag{4.2}$$

when $M_t = \Omega R/a_o$. Defining $C_p(\bar{x},\bar{t}) \equiv P'(x,t)/(\rho_o a_o^2)$ (i.e., a nondimensional acoustic pressure), we have in nondimensional notation,

*A nondimensionalization of the complete integral equation is given in Splettstoesser et al. (1983).

$$C'_p(\bar{x},\bar{t}) = \frac{P'(\bar{x},\bar{t})}{\rho_o a_o^2} = \frac{1}{8\pi^2} M_T \frac{\partial}{\partial \bar{t}} \int\int \left[\frac{M\lambda}{\bar{r}|1 - M_r|}\right]_{\bar{\tau}} d\bar{S} \tag{4.3}$$

where $M = V/a_o$.

This equation states that if the rotor, microphone geometry, Mach number, and time are scaled, the acoustic-pressure coefficient is uniquely determined. However, most acoustic data are not compared on a nondimensional basis. Instead, all pressures are normally referred to sea-level standard conditions where comparisons of pressure time-histories are made. For full-scale data taken at altitude, the referred pressure becomes

$$\frac{P'_{SL}(\bar{x},\bar{t})}{\rho_{o_{SL}} a^2_{o_{SL}}} = \frac{P'(\bar{x},\bar{t})}{\rho_o a_o^2} \tag{4.4}$$

$$P'_{SL}(\bar{x},\bar{t}) = \frac{\rho_{o_{SL}} a^2_{o_{SL}}}{\rho_o a_o^2} P'(\bar{x},\bar{t}) = \frac{P_{o_{SL}}}{P_o} P'(\bar{x},\bar{t}) \tag{4.5}$$

For model-scale data, the referred pressure becomes

$$P'_{m_{SL}}(\bar{x},\bar{t}) = \frac{P_{o_{mSL}}}{P_{o_m}} P'_m(\bar{x},\bar{t}) \tag{4.6}$$

Therefore, by matching the above nondimensional parameters, the referred model-scale acoustic pressures should duplicate the referred full-scale acoustic measurements. This is confirmed in Figure 30 for the UH-1H helicopter in high-scale flight. Notice that the time for one rotor revolution of model-scale data is Ω/Ω_m times the time for one revolution of full-scale data. (It should be noted that in this early research, changes in ρ_o were accounted for when comparing full-scale pressures to wind-tunnel model acoustic pressures; however, variations in a_o were not. It is estimated that corrections for a_o could increase the full-scale levels shown by 6%—thus making the comparison less favorable. However, more recent experimental research with very high quality model and full-scale data shows improved agreement (see Splettstoesser et al., 1983).

It should be noted that the full-scale data were acquired at larger scale nondimensional distances than the model-scale results. If this is accounted for, the peak negative-pressure comparison between model- and full-scale data is quite good, as shown in Figure 31. However, the model-data peak amplitudes are in generally slightly smaller than the full-scale data. The nonanechoic properties of the treated wind tunnel are believed to be partially responsible for this discrepancy.

More recent scaling tests on the AH-1 series helicopter (Splettstoesser et al., 1983) have resolved most of the model-scale/full-scale comparison differences. A nearly perfect match of peak amplitudes and pulse shapes is shown in Figure 32. The model-scale acoustic data were taken in the world's largest anechoic wind tunnel (DNW) and are of very high quality. The in-flight acoustic data which are shown were taken with the YO-3A aircarft. The pulse shapes of both the model and full scale have been averaged for comparison purposes. (For more details, see Splettstoesser et al. (1983).)

The model-scale/full-scale comparison is made even more convincing by considering other than the in-plane microphone position. Longitudinal and lateral directivity comparisons are explored in Figure 33 for one advancing-tip Mach number (Splettstoesser et al., 1983). As shown, agreement is the best at the in-plane microphone positions but is still very good at positions under the horizontal plane of the helicopter. Data from these UH-1 and AH-1 model-scale/full-scale tests demonstrate that carefully designed and nondimensionally tested small-scale models can duplicate the high-speed impulsive noise generated by full-scale rotors.

It is also interesting to note at this juncture that thrust plays only a secondary role in the in-plane high-speed impulsive noise radiation process. Figure 34 presents the variation of peak pressure levels with changes in thrust coefficient for a microphone positioned nearly in-plane with the plane of the rotor for the UH-1 model rotor. The data indicate a remarkable insensitivity to changes in thrust. Notice that the small variations about the mean line of zero slope that do exist are caused by variations in tip Mach number about $M_{AT} = 0.9$.

The rapid increase in the magnitude of the negative-pressure pulse with increasing advancing-tip Mach number is reemphasized in Figure 35. The solid curve was obtained at varying advance ratios (0.091 to 0.264), and the dashed curve was obtained at a constant advance ratio (0.229). The similarity of the two curves over a range of advancing-tip Mach numbers (0.87 to 0.93) shows

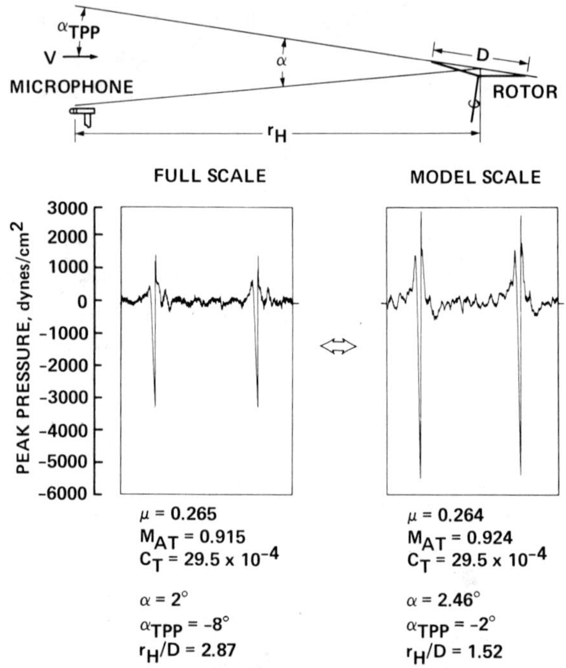

Figure 30. Waveform comparison: full-scale and model-scale high-speed data.

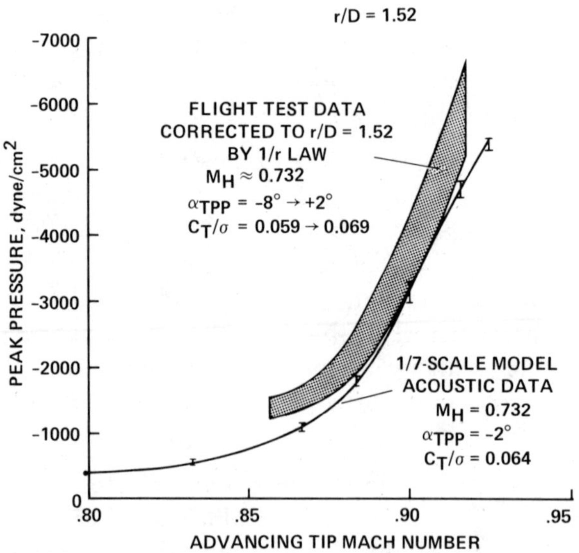

Figure 31. Peak negative-pressure amplitude comparison: full-scale and model-scale high speed data.

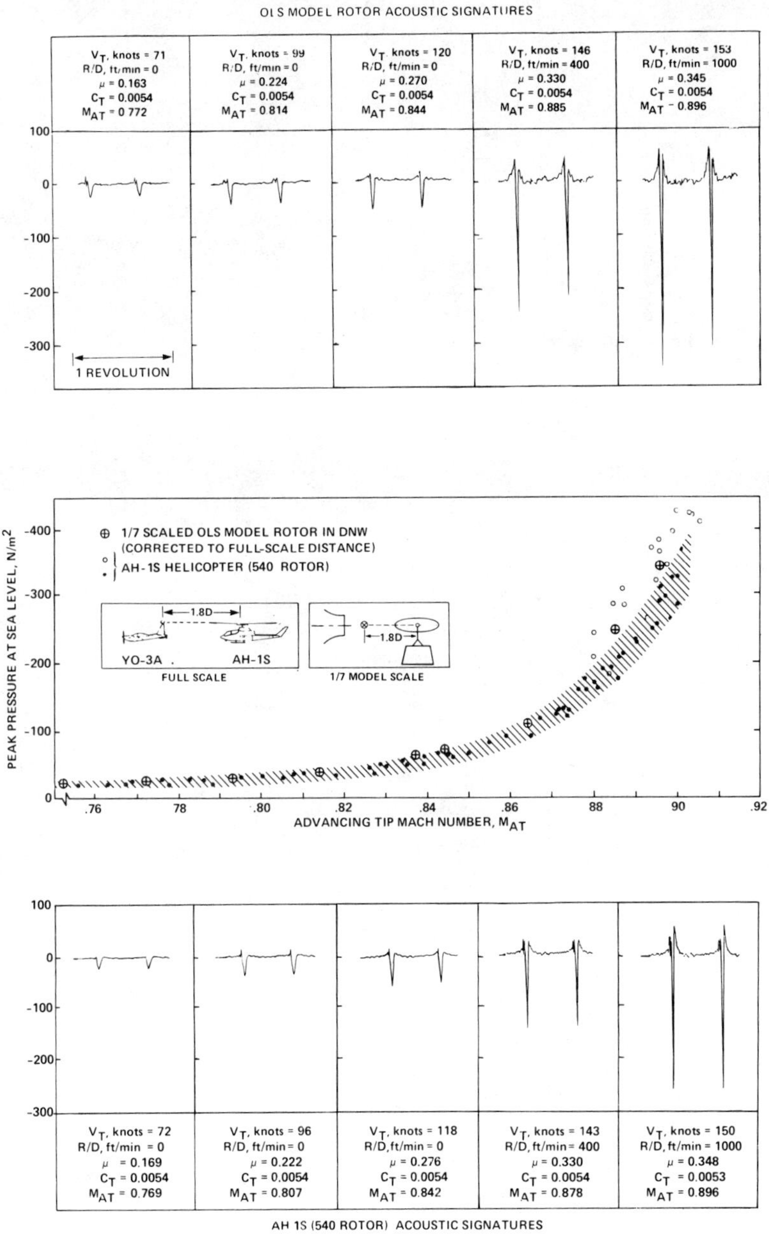

Figure 32. Comparison of model and full-scale acoustic pressure for an in-plane microphone 1.8 rotor diameters ahead.

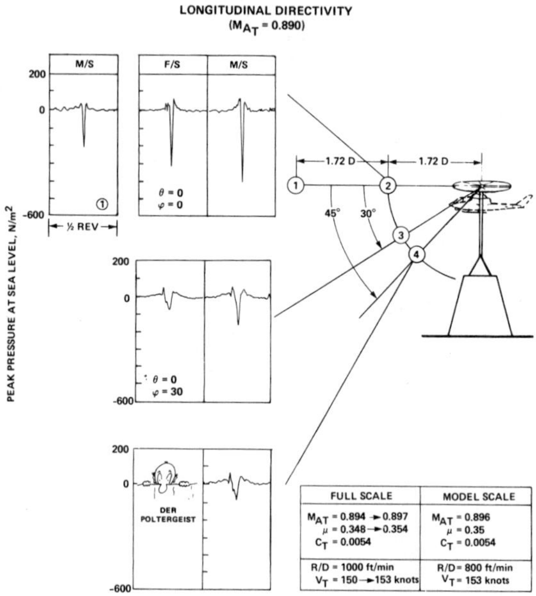

Figure 33a. Comparison of model and full-scale impulsive noise directivity. Longitudinal directivity ($M_{AT} = 0.896$).

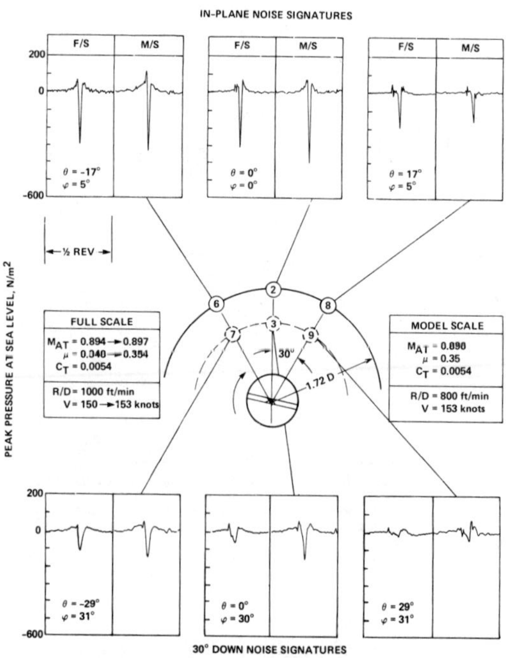

Figure 33b. Lateral directivity.

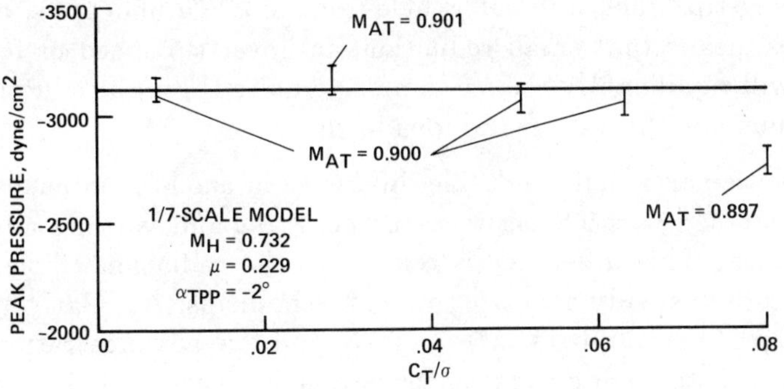

Figure 34. Peak pressure versus changes in thrust coefficient (UH-1H model rotor).

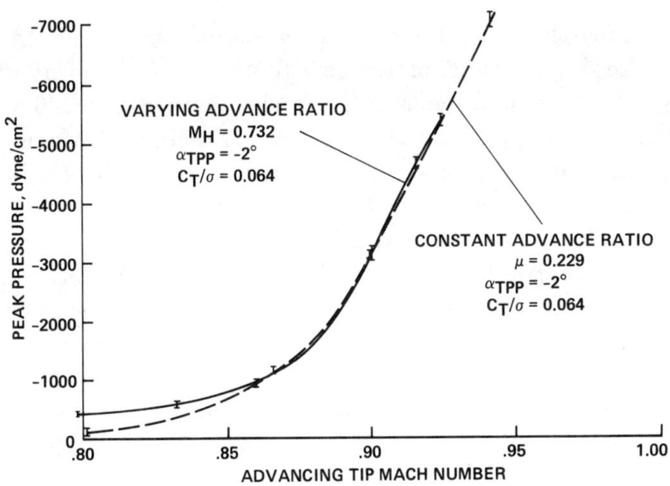

Figure 35. Peak pressure versus advancing-tip Mach number: wind-tunnel data (UH-1H model rotor).

that for high-speed impulsive noise, advancing-tip Mach number is the dominant parameter. Small changes can occur in advance ratio without significantly altering in-plane acoustic radiation, if advancing-tip Mach number is held constant. To helicopter designers, this means that small reductions in hover-tip speed or forward speed will significantly reduce impulsive noise through reductions in M_{AT}, but not through reductions in μ.

Notice also that a notable difference in acoustic intensity does exist near $M_{AT} = 0.8$ between the constant and varying advance-ratio cases. This difference is related to the radiation efficiency of steady and unsteady acoustic source mechanisms. At relatively high hover tip Mach numbers, $M_H = 0.73$, and low advance ratios, $\mu = 0.09$, each rotor blade experiences only small variations in the local Mach number around the azimuth, implying that the steady compressible source mechanisms are the most likely radiators of far-field acoustic energy. At the same M_{AT}, but at the lower hover-tip Mach number ($M_H \approx 0.65$) and the higher advance ratios ($\mu = 0.229$), significant variations in the rotor-blade local Mach number occur and can be expected to exert more influence on the radiated acoustic field. In this case, unsteady acoustic-source mechanisms become more pronounced. The data shown indicate that the steady radiators of acoustic energy are more efficient generators of impulsive noise than the unsteady ones for similar advancing-tip Mach numbers.

Perhaps the most interesting aspect of high-speed helicopter noise is the development of the saw-toothed waveform at high advancing-tip Mach numbers. This is shown in Figure 36 together with a plot of the peak pressure versus advancing-tip Mach number (M_H is constant and μ is varying.) (Schmitz et al., 1977). In case A ($M_{AT} = 0.867$), a near-symmetrical pulse is observed; the subjective qualities could be described as a loud thumping. As the advancing-tip Mach number is increased, the symmetrical pulse becomes sawtooth in character (case B, $M_{AT} = 0.90$); the waveform consists of a large decrease in pressure followed by an extremely sharp increase in pressure ($\Delta P/\Delta t \simeq 4 \times 10^6$ dynes/cm^2/sec). Crispness (many harmonics) and intensity of the acoustic signature are its dominant features. At still higher advancing-tip Mach numbers (case C, $M_{AT} = 0.925$), the peak negative pressure becomes very large, and the sudden rise in pressure becomes nearly instantaneous ($\Delta P/\Delta t \simeq 10^7$ dynes/cm^2/sec). Some overshoot can be seen, part of which is real and part of which is due to instrumentation bandwidth limitations. The noise generated by this latter waveform is rich in higher harmonics and can be subjectively classified as harsh and extremely intense.

It is known that this rapid increase in pressure (case C) is a radiating shock wave. Early indications of its formulation can be seen in case B at the lower advancing-tip Mach number. Of course we know that local shock waves do exist near the tip of the rotor blade throughout this Mach number range. However, this acoustic plot suggests that these local shock waves *delocalize* at a certain *delocalization Mach number* and propagate to the acoustic far-field. Below the delocalization Mach number (~0.9 for the NACA 0012 airfoil), all shock waves are confined to the blade. Above the delocalization Mach number, shock waves on the surface of the blade radiate as shock waves to the acoustic far-field (see Schmitz and Yu, 1981 and Yu et al., 1978).

Although it is obvious that such processes as transonic shock generation and propagation are nonlinear, much can be learned by looking at the geometry of the linear aerodynamic/acoustic process. Consider the space-fixed trajectory of a simple point source near the tip of a rotor blade. Its trajectory is the epicycloid pattern normally associated with the shed tip-vortex, as shown in Figure 37. If at regular azimuth angles (ψ) a pulse is emitted in space and allowed to propagate at the ambient speed of sound, depicted as a circle (a sphere in three dimensions), these pulses form the crescent-shaped wave shown. In effect, disturbances are propagating away from a source moving at almost the ambient speed of sound. As a result, disturbances accumulate and cause the *Doppler amplification*. As the advancing-tip Mach number increases (higher rpm or forward speed), the accumulation of disturbances becomes so great as to form local shocks and eventually the radiating shock wave. This accumulation process is represented as a singular integral, as we have seen in equations (1.1) and (2.1). The factor $1/(1 - M_r)$ is an integrable singularity for typical rotor geometry.

When these waves coalesce and a radiating shock wave develops, shadowgraph and Schlieren techniques can be used to visualize the radiation process. Using short-duration, high-intensity light pulses, the rotor's motion can be photographically frozen in space. A top view of the radiation process is shown in Figure 38 for a rotor with an advancing-tip Mach number of 0.92. These excellent photos were taken by J. Tangler of an approximately 2-ft diam model helicopter rotor (see Tangler, 1977).

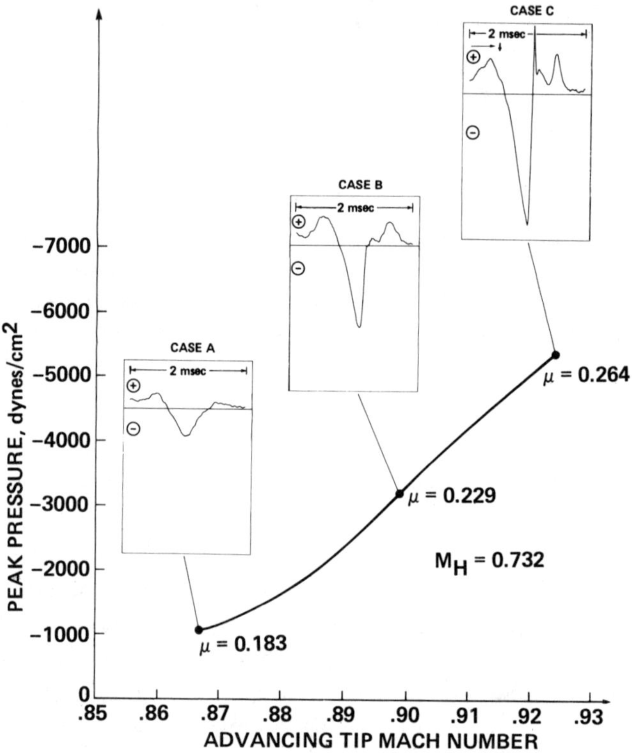

Figure 36. Waveform shape versus advancing-tip Mach number.

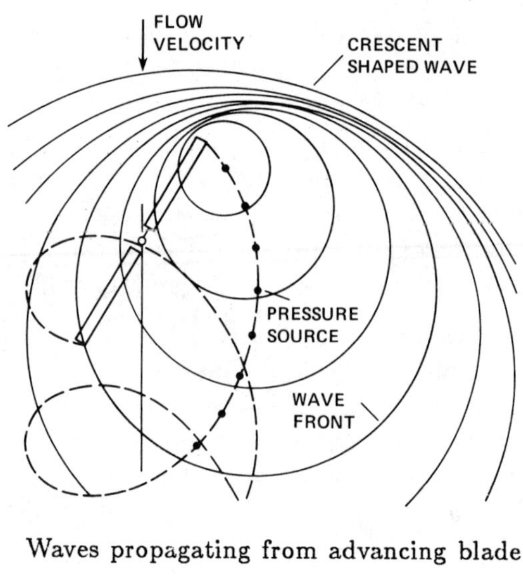

Figure 37. Waves propagating from advancing blade (from Tangler, 1977).

The top view of this small model-rotor experiment dramatically illustrates how local disturbances (in this case shock waves) are propagated to the observer in the far-field. At $\psi = 48°$ (Figure 38), a small disturbance field, which is quite diffuse, is seen off the tip of the rotor. When $\psi = 92°$, the field has increased and moved slightly aft along the chord line. At $\psi = 151°$, the waves are distinctly seen to move forward again and are traced into the region off the tip of the blade. Finally, at $\psi = 159°$, the wave move forward off the blade tip and escape to the acoustic far-field.

It should be noted that more than one wave appears to emanate from the tip of the rotor in this top view. It is believed that the two waves are really the extensions of the upper and lower surface shock systems on the blades. It is most likely that the stronger of the two radiating shocks is the second wave emanating from the tip of the rotor. Notice also, however, that both waves coalesce into one wave as the disturbance propagates into the far-field. The additive nature of these radiating waves suggests the following explanation for the weak role of thrust changes on the measured in-plane acoustic disturbances. As thrust increases, the upper surface shock strength increases while the lower surface shock strength decreases. The net result is a near-constant acoustic level as the radiating waves coalesce far from the rotor tip. Some of the aerodynamic details of the tip flow fields themselves can be seen more clearly in two-dimensional in-plane views of the same three-dimensional event (Schmitz et al., 1977).

Because most of the radiation process appears to be governed by Doppler amplification, it is logical to ask whether some of the key parameters of high-speed impulsive noise can be studied in hover. A *top view* of the linear acoustic wave behavior is sketched in Figure 39 for a simple hovering impulse with a rotational Mach number of 0.9.

As seen from this sketch, a crescent shaped wave is formed which is quite similar to the one formed in the forward-flight testing. In the hovering environment, the acoustic radiation processes normally associated with high-speed impulsive noise can only be attributed to steady (when viewed on a rotor) aerodynamic events. Therefore, the mechanism of linear amplification can be studied without regard to the effects of periodic force or thickness variations normally encountered in high-speed helicopter forward flight.

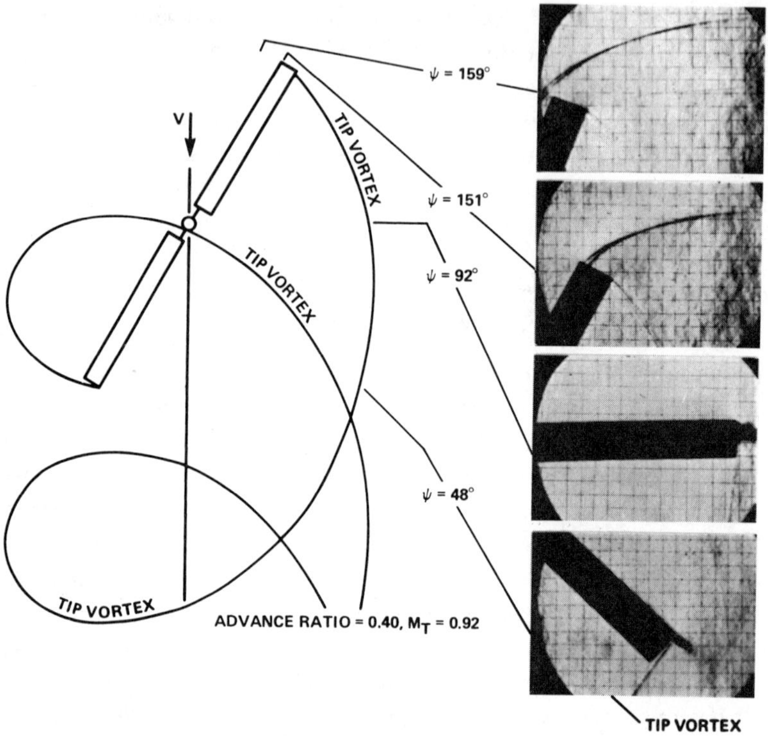

Figure 38. High-speed advancing-blade shock formation (from Tangler, 1977).

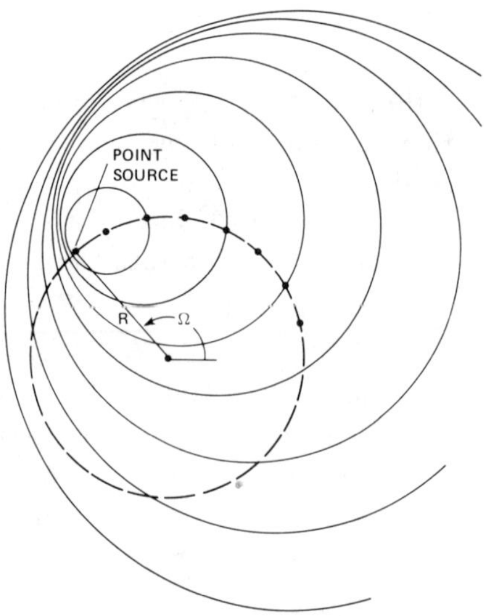

Figure 39. Linear wave amplification (Doppler effects) of a rotating point source: $M_T = 0.9$.

To obtain data of sufficiently high quality, the high-speed hover acoustic data were taken in the Aeromechanics Laboratories' Anechoic Hover Chamber. The chamber was lined with polyurethane foam and designed to be anechoic (without acoustic reflections) down to 110 Hz. As illustrated in Figure 40, flow recirculation is avoided by allowing quiescent air to be drawn into the room through acoustically lined ducts, collecting the wake of the hovering rotor through an annular diffuser, and exhausting the wake to the outside. In its current configuration, the test chamber can accommodate rotors from 1.5 to 2.4 m in diameter.

One of the same rotors used in the scaling tests (a 1/7 scale of a UH-1H main rotor) was run for these high-speed hover tests (Boxwell et al., 1978). The geometrically scaled rotor has a NACA 0012 airfoil section with a root-to-tip washout of 10.9°. Because thrust appeared to be unimportant in the high-speed noise generation process, a second set of untwisted but geometrically scaled rotor blades was run at near-zero net thrust. Some small positive net thrust was required to avoid shed-wake interference effects. The data were taken with

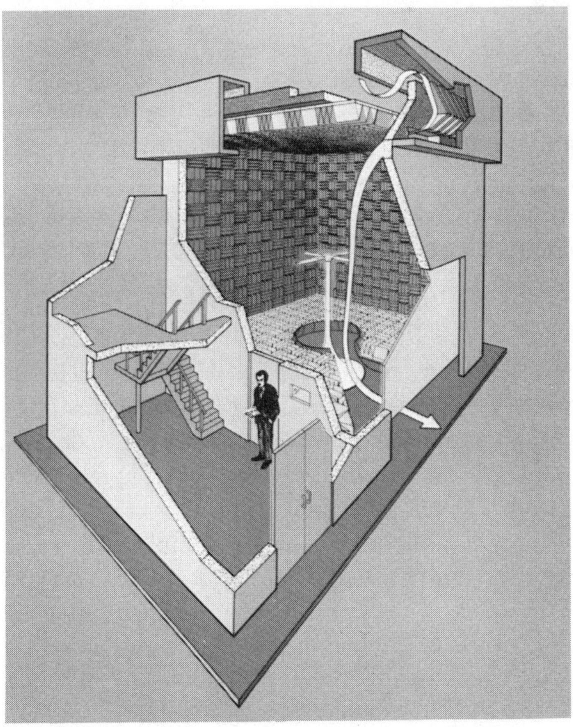

Figure 40. Anechoic hover chamber.

a microphone located within the tip-path-plane of the rotor at a distance of 1.5 rotor diameters ($r_H/D = 1.5$) from the hub. This in-plane microphone position is consistent with that used in previous in-flight and wind tunnel tests and is in a position to measure the most intense high-speed impulsive signature.

Figure 41 from Boxwell et al. (1978) presents the measured acoustic signature at a hover tip Mach number (M_T) of 0.8. Two time scales are presented. Figure 41a depicts two blade passages, approximately one-half a complete rotor revolution. Figure 41b is an expanded scale of the first acoustic pulse. The latter is used to emphasize the detailed waveform characteristics of the measured pulse. The most striking feature of the waveform at $M_T = 0.8$ is its almost symmetrical character. This same character has been observed in full-scale and model-scale forward-flight testing at tip Mach numbers below the delocalization Mach number.

Figure 42 illustrates the pressure time-history at a hovering tip Mach number of 0.9. The peak negative-pressure amplitude of the measured pulse has increased dramatically and the pulse shape has now lost its symmetry. The resulting saw-toothed waveform is known to generate large amounts of high-intensity, higher-frequency noise. Again, this same type of waveform was measured on the same rotor system operating in forward flight at an advancing-tip Mach number of 0.9. In this previous test, Schlieren photographs were used to correlate the discontinuous increase in pressure with a radiating shock wave. It is apparent that a similar phenomenon is occurring in this controlled hover test.

At a hover tip Mach number of 0.962 (Figure 43), the saw-toothed pulse shape is firmly established and the negative pressure peak level has doubled from the $M_T = 0.9$ condition. The large, discontinuous rise in pressure resulting from a radiating shock wave exhibits some variability from blade to blade. One particularly interesting aspect of the waveform shown in Figure 43b is the pulse width. At lower hover tip Mach numbers, the pulse width was observed to narrow with increasing rotor-tip speed up to the point of waveform transition from symmetrical to saw-tooth. Above this transition point, for example at $M_T = 0.962$, the pulse width has become larger. Figure 43 also shows that a positive-pressure wave (bow wave) begins to form. At still higher tip Mach numbers, the classical N-wave of sonic boom research is likely.

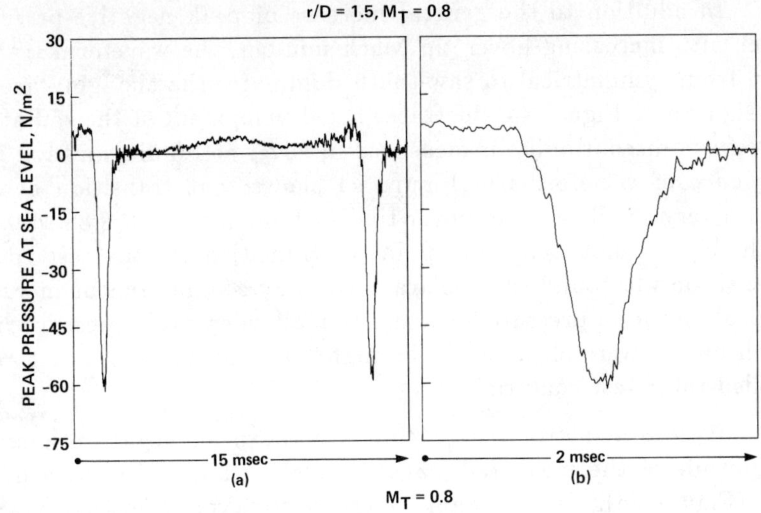

Figure 41. UH-1H model blade acoustic pressure time-history, in-plane, $r_H/D = 1.5$, $M_T = 0.8$.

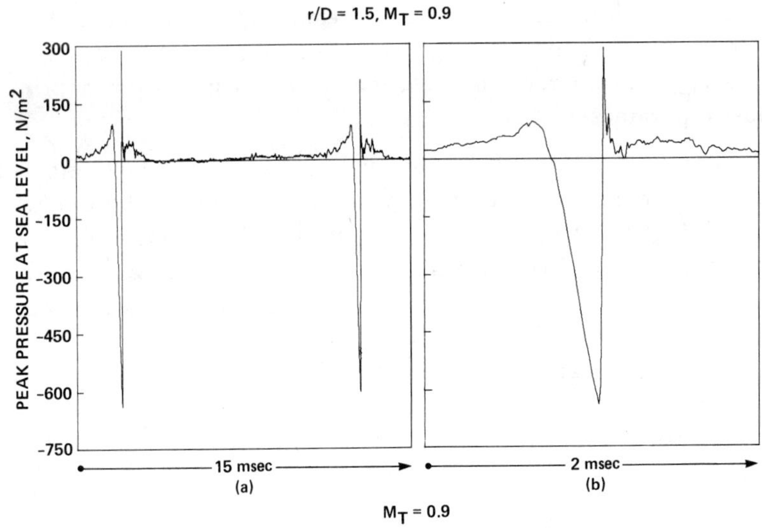

Figure 42. Acoustic pressure time-history, in plane, $r_H/D = 1.5$, $M_T = 0.9$.

In addition to the general increase of peak negative-pressure level with increasing hover tip Mach number, the waveform transition from symmetrical to saw-tooth dominates the changing acoustic signature. Figure 44 illustrates the development of the radiating waveform discontinuity as measured in hover at 1.5 diam away. The sequence of waveforms in Figure 44 shows that transition occurs over a very small range in hover tip Mach number from 0.88 to 0.90, with $M_T = 0.89$ being the point of transition for the test rotor. Transition was found to be characterized by a simultaneous increase in peak negative-pressure level and the following rapid pressure rise. Both events were observed to be highly unsteady, even under controlled rotor test conditions.

It is also instructive to compare the peak negative-pressure amplitude of the measured waveform versus hover tip Mach number (Figure 45). A very rapid increase in level is noticed as M_T approaches 0.9. However, as M_T increases beyond 0.9 to $M_T = 1.0$, the increase in peak level is not as rapid, or the rate of increase of this peak negative-pressure level with Mach number becomes smaller. As noted in Figure 45, the shaded area depicts the degree of unsteadiness in the measured data. The vertical solid bars reflect data taken with the UH-1H model twisted blades, and the dashed vertical bars are for the same dimension model rotor using untwisted blades. The correlation between twisted and untwisted results is good. This result gives further support to the idea that at the in-plane microphone position, rotor thrust is not a primary acoustic parameter.

The most remarkable observation about the hovering rotor data is their similarity to the model-scale acoustic data taken in forward flight. Although the amplitude of the hovering data is larger than that of the wind-tunnel data, its shape and character are remarkably similar. As in forward flight, the wave changes character at a tip Mach number of 0.9—radiating shock-like disturbances to the acoustic far-field. The major mechanisms of the high-speed impulsive-noise phenomenon are present in this simple hovering rotor experiment. This fact will become more and more evident as the discussion proceeds.

It is natural to question the accuracy of the model-rotor experiment. A major concern is the accurate measurement of the rotational tip Mach number. Fortunately, the experiment has been repeated many times in at least three different facilities throughout

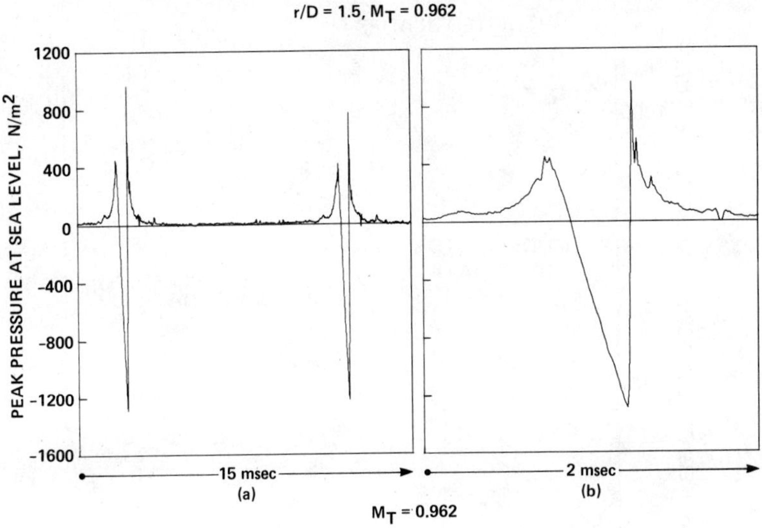

Figure 43. Acoustic pressure time-history, in-plane, $r_H/D = 1.5$, $M_T = 0.962$.

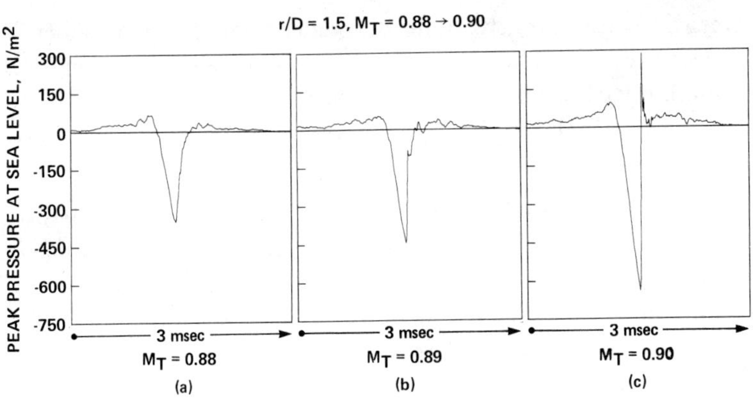

Figure 44. Waveform transition; the development of a radiating discontinuity, in-plane, $r_H/D = 1.5$, $M_T = 0.90$.

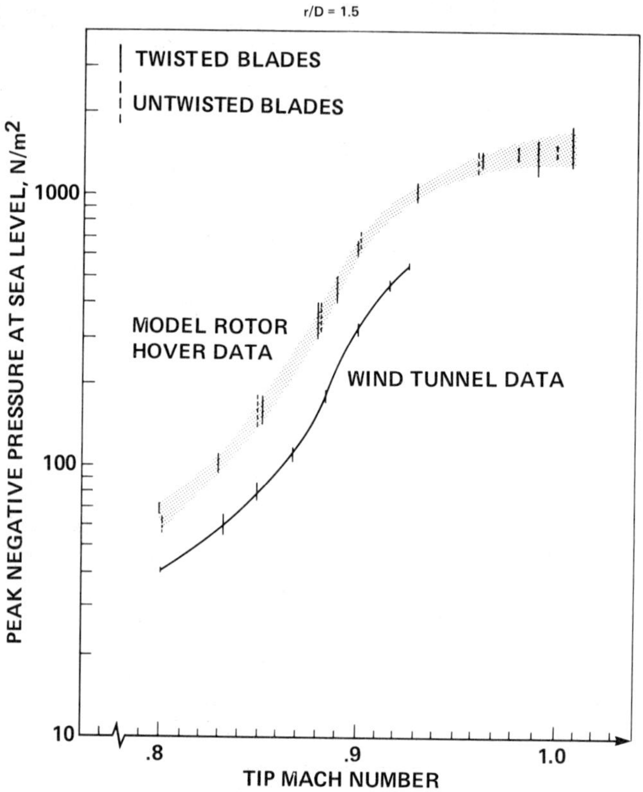

Figure 45. Peak negative-pressure level versus tip Mach number for rotor in hover and forward flight, in-plane, $r_H/D = 1.5$.

the world. More recently, humidity effects on Mach number have been accounted for. The results shown here have been consistent and reproducible within about a $\pm10\%$ error in absolute amplitude. All delocalization Mach-number trends are also reproducible to within 0.005 of a Mach number.

The simple hovering rotor is a natural place to begin to compare acoustic theory with experiment. Experimentally, it has been shown that thrust plays a small role (if any) in the in-plane noise radiation. If all nonlinear effects are also neglected ($T_{ij} = 0$), we can calculate the linear thickness noise equating the first two terms of equation (1.1) to zero and solving the resulting integral equation:

$$P'(\vec{x},t) = \frac{\partial}{\partial t}\int\int\left[\frac{\rho_o v_n}{r|1-M_r|}\right]_\tau dS(\vec{\eta}) \tag{4.7}$$

Fortunately, the entire right-hand side of this equation is known.

The major complications are the correct handling of the retarded time which is complicated by the geometry of the rotor problem (Schmitz and Yu, 1979). However, the tip Mach number of normal helicopter rotors is less than 1.0, facilitating a simple and straightforward calculation of equation (4.7).

The numerator of the right-hand side of equation (4.7) represents the equivalent acoustic sources (monopoles) that are created because each segment of the finite-thick rotor blade must displace mass as it moves through the medium. Thus, the portions of the blade that push fluid away from the blade are represented as acoustic monopole sources; those that flow over a two-dimensional airfoil section are given in Figure 46. From small perturbation theory for a slender two-dimensional body, the normal component of surface velocity is expressed as follows:

$$v_n \approx V \cdot \text{ (local surface slope)}$$

where V is the local free-stream velocity of the blade element (ΩR for the hovering rotor).

As shown in the simple geometrical patterns of a hovering rotor (Figure 39), exactly how the acoustic waves are collected in space is an important part of this linear acoustic problem. Correct treatment of the retarded time is essential, for it governs the amplification factor, $1/(1-M_r)$ of equation (4.7), as well as keeping track of when the wave propagation from the singularities in motion actually reach the observer.

Consider a simple hovering point singularity (top view) which emits a disturbance at position P that travels at the ambient speed of sound in the medium and arrives at the observer's position O at

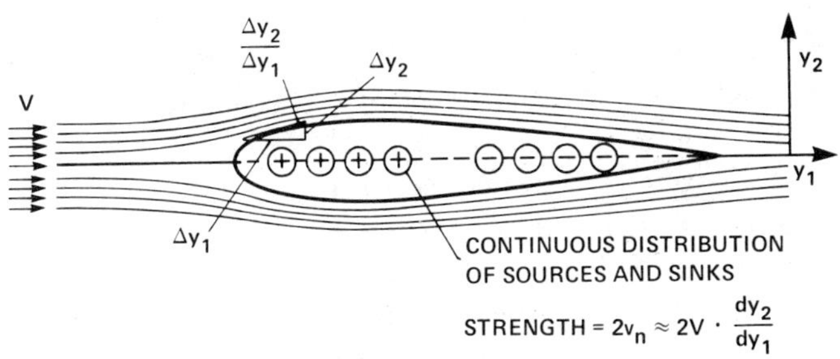

Figure 46. Monopole acoustic sources.

some time later, r/a_o (Figure 47). The velocity of the point P at any instant of time is ΩR, and M_r is the component of that velocity vector along the r direction divided by the speed of sound. From the given geometry, observer time is related to emission time τ by the implicit relationship

$$\tau = t - \frac{r(t,\tau)}{a_o} \tag{4.8}$$

where

$$r^2 = (r_H + R\cos\psi)^2 + (R\sin\psi)^2 \tag{4.9}$$

Therefore, a point source which was emitted at the retarded time τ takes time $r(t,\tau)/a_o$ to travel the distance r arriving at time t. If the rotor is rotating at angular velocity Ω, then pulses that are emitted at selected azimuthal positions, $\psi(\psi = \Omega\tau)$ obey the expression

$$\Omega\tau = \psi = \Omega t - \frac{\Omega r(t,\tau)}{a_o} \tag{4.10}$$

$$\frac{d\tau}{dt} = \frac{1}{1 - M_r} \tag{4.11}$$

which is plotted in Figure 47. Figure 47 shows that as the Mach number approaches 1, a large region of blade azimuth contributes to a narrow pulse width. The result is an inherent amplification of local source effects by the factor $1/(1 - M_r)$ which is illustrated in Figure 48.

Part of this amplification of acoustic energy is explicitly accounted for in equation (4.7). The Doppler factor, $1/(1 - M_r)$, which is the Jacobian of a coordinate transformation, already appears in the acoustic source term and represents the formation of a velocity potential wave with respect to the medium. A second Doppler factor is implicit in eqaution (4.7) and does not enter until the time-derivative of the potential function is taken. In each case, large-amplification effects are evident as advancing tip Mach numbers approach 1. These simple expressions show that the blade tip is acoustically the most efficient portion of the rotor, with acoustic efficiency decaying rapidly for inboard radial positions. It can also be concluded from Figure 48 that the dominant azimuthal source position is near the tangency point of a straight line drawn from

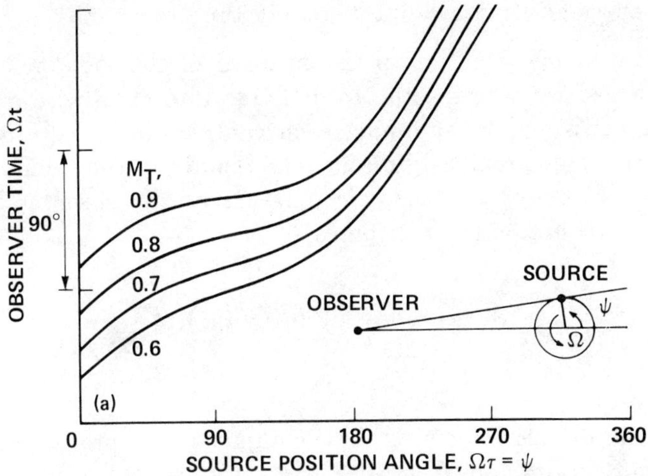

Figure 47. Retarded time factor.

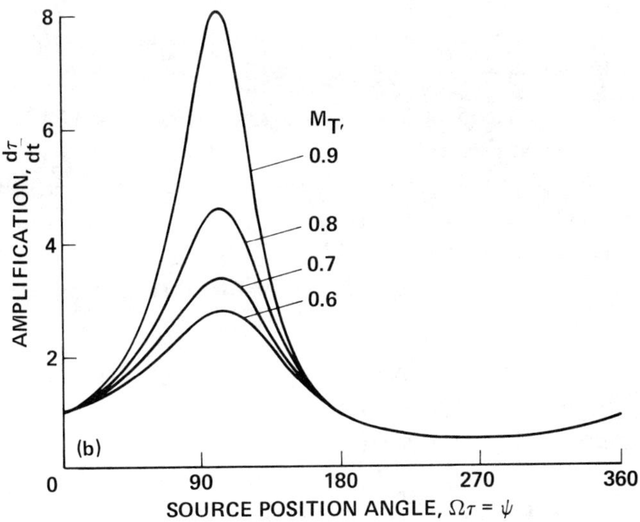

Figure 48. Retarded time factor.

the observer to the circle described by the tip of the rotor. At this azimuthal position, all line sources along the rotor blade arrive at the observer position at approximately the same time.

The final evaluation of the integral of the monopole sources is performed by dividing the rotor blade into chordwise and spanwise elements and by summing each contribution according to the geometric rules previously discussed. Finally, a time differential is performed yielding the acoustic pressure at the observer location. Therefore, equation (4.7) becomes

$$P'(x,t) \approx 2\frac{\partial}{\partial t}\sum_{k=1}^{\ell}\sum_{i=1}^{n}\rho_o\left[\frac{V\,(dy_2/dy_1)_{ki}}{r|1-M_r|}\right]dy_{1_i}dy_{3_k} \qquad (4.12)$$

Additional details describing these computational procedures can be found in Schmitz and Yu (1979).

A comparison between linear thickness calculations for hover versus the experimental data previously presented is shown in Figures 49–52 for several different hover tip Mach numbers (Boxwell et al., 1978). The striking features of the comparison between theory and experiment in hover at $M_T = 0.8$ (Figure 49) are the similarity in pulse shape and the discrepancy in peak pressure levels. As in forward flight at advancing-tip Mach numbers below 0.9, thickness-noise theory misses the measured negative-pressure peak levels by a factor of about 2.

The comparison of theory and experiment as M_T is increased to 0.88 (Figure 50) remains similar to that made at $M_T = 0.8$. The waveform shape is still generally symmetrical but the peak negative-pressure level is underpredicted by slightly more than a factor of 2. As was noted previously, $M_T = 0.88$ is slightly less than the delocalization Mach number for waveform transition, at least as measured at $r_H = 1.5$ with the test rotor.

At hover tip Mach number of 0.9, the situation becomes even worse (Figure 51). The amplitude of the peak negative-pressure pulse is again underpredicted, by a factor of about 2. However, as indicated previously, there is also a dramatic change in the waveform of the experimental data that is not predicted by the linear theory.

The comparison becomes even more intriguing at a hover tip Mach number of 0.962 (Figure 52). The theoretical waveform is still symmetrical and generally smooth in shape and thus does not

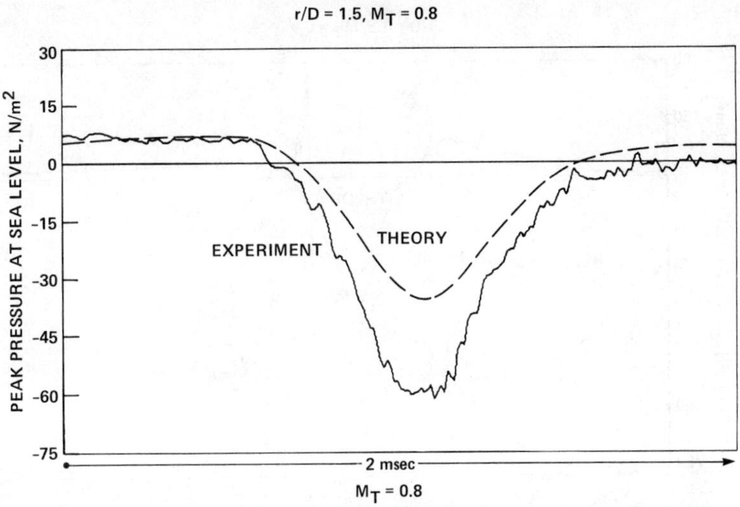

Figure 49. Comparison of theory with experimental pressure time-history, in-plane, $r_H/D = 1.5$, $M_T = 0.8$.

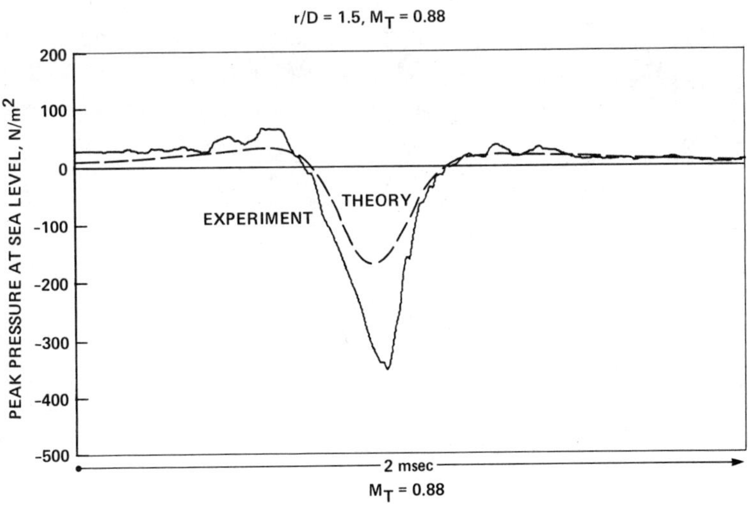

Figure 50. Comparison of theory with experimental pressure time-history, in-plane, $r_H/D = 1.5$, $M_T = 0.88$.

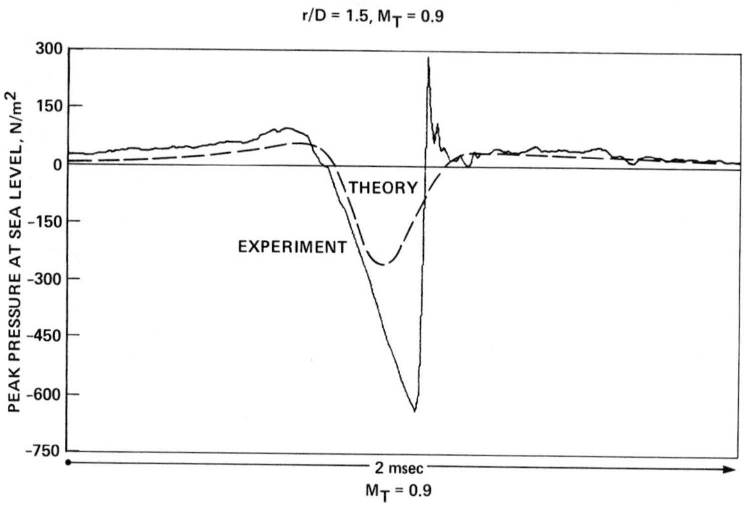

Figure 51. Comparison of theory with experimental pressure time-history, in-plane, $r_H/D = 1.5$, $M_T = 0.9$.

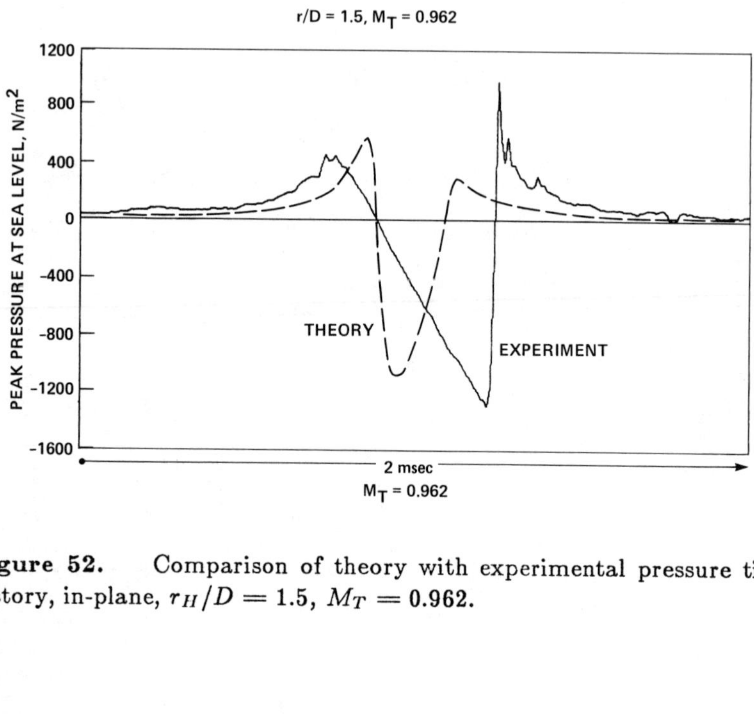

Figure 52. Comparison of theory with experimental pressure time-history, in-plane, $r_H/D = 1.5$, $M_T = 0.962$.

compare favorably with the measured data. In addition, theory now only slightly underpredicts the peak negative-pressure amplitude of the pulse. Also, as previously noted, the measured pulse width is becoming wider, whereas the linear theory predicts a more narrow pulse width with increasing hover tip Mach number. In fact, the experimental pulse width (measured at zero pressure) exceeds by at least 50% the width expected (by linear theory) from an airfoil of chord equal to that of the model rotor tested and traveling at sonic velocity. This pulse-widening effect suggests that aerodynamic events off the rotor blade trailing edge are contributing to the measured acoustic signature.

The difference in peak negative pressure levels between linear monopole theory and experiment can be seen more clearly in Figure 53. The theoretical model does not predict the rate of increase of the peak negative-pressure level.

It is tempting to attribute the lack of correlation with experiment to the simplicity of the linear theoretical model. Perhaps if the remaining linear *dipole* terms were included (the second term in equation (1.1)), the agreement with measured data might be better. This was tried in Farassat et al. (1979) for a rotor with predicted values of loading and skin friction suitably chosen to act as pressure dipoles. The resulting theoretical time-histories are almost indistinguishable from the simple linear monopole calculations previously presented. The comparison is shown in Figure 54 for hover tip Mach numbers of 0.8 and 0.9.

These results were not too surprising. Similar conclusions were drawn in Schmitz and Yu (1979) for a high-speed rotor in forward flight from order-of-magnitude and pulse-shape arguments. In addition, most of the experimental data had suggested that the high-speed noise phenomenon was not dependent on thrust or torque at constant Mach number.

The major conclusion from all of these linear acoustic analyses is that they do not adequately describe the in-plane noise radiation processes from a high-tip-speed hovering rotor. There have been some efforts to correct this situation by improving the modeling (improving the specification of the boundary conditions) in the linear formulation. It can be argued that rotor-tip end-plate (Farassat and Martin, 1983) and boundary-layer separation efforts increase the amplitude of the symmetrical pulse. The importance of the latter effect can be seen in Figure 55. This end view of a NACA 0012

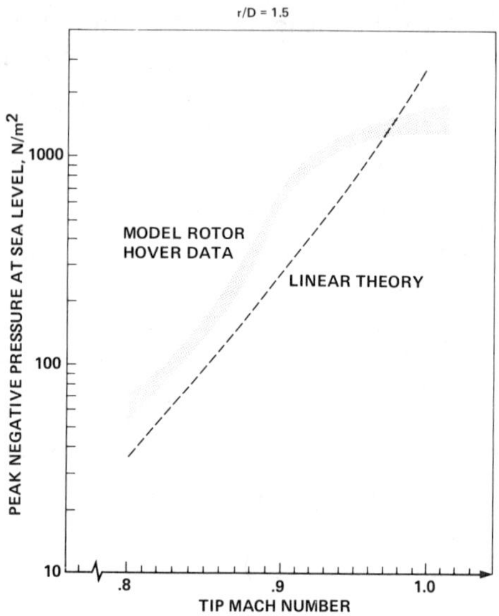

Figure 53. Comparison of theory and experiment in hover, in-plane, $r_H/D = 1.5$.

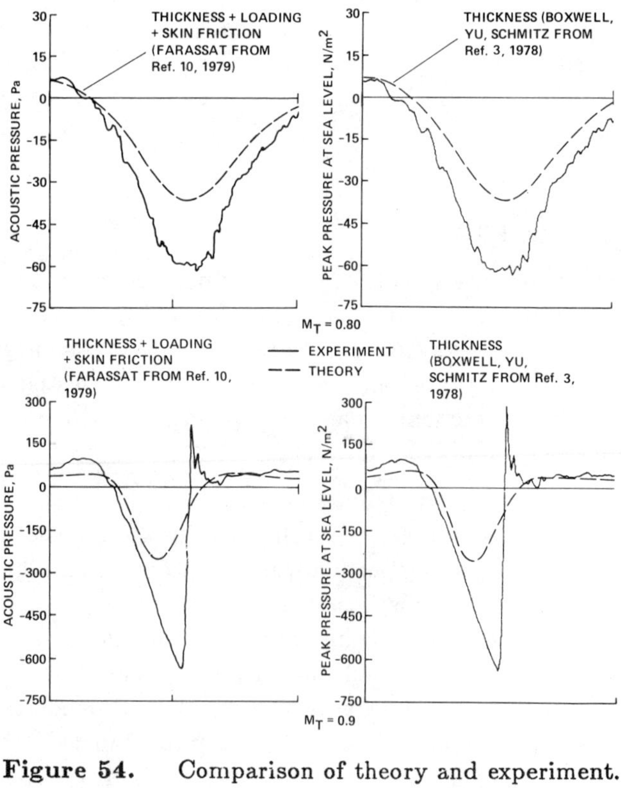

Figure 54. Comparison of theory and experiment.

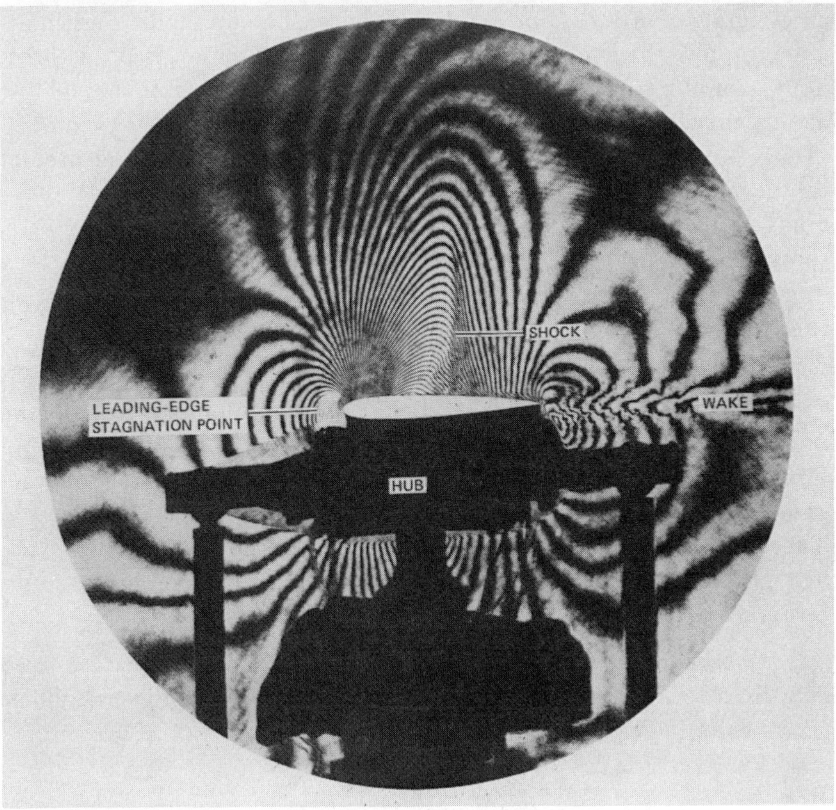

Figure 55. Interferogram recorded at 180° (chordwise view).

rotor at near zero lift (Kittleson and Yu, 1982) is a holographic interferogram of the integrated three-dimensional flow field surrounding the tip of a hovering UH-1H model rotor at the tip Mach number of 0.9. It is quite apparent that local shock waves on the surface of the blade interact with the boundary layer causing an enlarged separated flow region. A rigorous treatment of this problem is not usually attempted, for it would be necessary to model the boundary layer and separated-flow effects in equation (4.7). Instead, an *equivalent airfoil* comprising the original airfoil plus the outer edges of the separated-flow region is defined. This new equivalent airfoil is then used in equation (4.7) to define the strength of the distributed acoustic sources. If this is done, it is relatively easy to show that the peak negative-pressure calculations would increase substantially (they approximately double for each doubling of the effective airfoil chord at constant thickness). Although this effect has been known for many years, most researchers do not like to incorporate such an estimation in a *first-principles* analysis. The methods of estimating just how thick or extended the separation region is on a three-dimensional rotor in the transonic regions and how to model the equivalent airfoil for noise purposes are not well defined or even completely understood. In addition, none of these corrections predict the development and radiation of the delocalized shock wave above a hover tip Mach number of 0.9 for a scaled UH-1H rotor. Clearly, the radiation processes at these high tip speeds are governed to a large extent by transonic effects. These must be accounted for in the theoretical modeling.

The most straightforward approach to the nonlinear, high-speed acoustic problem might appear to be to simply include the *quadrupoles* in equation 1.1. However, equation 1.1 is in reality an integral equation which has no simple analytical solution. Some degree of approximation is necessary to proceed with this approach. These approximations will rely on what we know about the problem physically. Such insight can be gained by switching *hats* and formulating the problem as a transonic aerodynamist.

We begin with the classical potential equation in a space-fixed coordinate system. Assuming constant specific heats and weak shocks (i.e., negligible entropy increases), we have

$$\frac{\partial^2\phi}{\partial t^2} - a^2\nabla^2\phi + 2\text{grad }\phi\cdot\text{grad }\frac{\partial\phi}{\partial t} + \frac{1}{2}\text{grad }\phi\cdot\text{grad }[(\text{grad }\phi)^2] = 0 \tag{4.13}$$

where ϕ represents the velocity potential and a is the local speed of sound. Fortunately, the aerodynamics of a hovering rotor are basically steady when viewed from a blade-fixed frame. Therefore, following the work of Isom (1979), Schmitz and Yu (1981), Hawkings (1979), and Yu et al. (1978), the governing potential equation can be transformed to blade-fixed cylindrical coordinates and expanded to second-order yielding the following:

$$\begin{aligned}\left[\omega^2 - \frac{a_o^2}{r^2} - (\gamma+1)\frac{\omega}{r^2}\phi_\theta\right]\phi_{\theta\theta} - 2\omega\phi_r\phi_{r\theta} - 2\omega\phi_z\phi_{z\theta} \\ = [a_o^2 + (\gamma-1)\,\omega\phi_\theta]\left(\phi_{rr} + \frac{\phi_r}{r} + \phi_{zz}\right)\end{aligned} \tag{4.14}$$

where

ω angular rotation rate.

a_o undisturbed speed of sound.

r radial distance from the axis of the cylindrical coordinate system.

γ ratio of specific heats.

This nonlinear but steady second-order partial differential equation governs the transonic aerodynamics of the hovering rotor. In addition, it governs how disturbances propagate away from this rotating coordinate system (acoustic waves). At the present time, no closed-form solutions to this equation exist. A procedure adopted by some researchers is to numerically solve limited regions of the aerodynamic flow field (Caradonna and Isom, 1976; Caradonna, 1978). Others have chosen to solve the nonlinear acoustic far-field, using weak-shock theory (Isom, 1980). As we shall see, neither is a completely satisfactory solution, for the nonlinear aerodynamic and acoustic fields are interwoven.

The choice of a cylindrical coordinate system whose axis is aligned with the rotor is sketched in Figure 56. An observer riding in this coordinate system sees a free-stream velocity that increases linearly from zero at the region to ωr at r. As indicated, this increasing free-stream velocity continues out past the tip of the rotor; it will be shown to be important to many of the arguments to come.

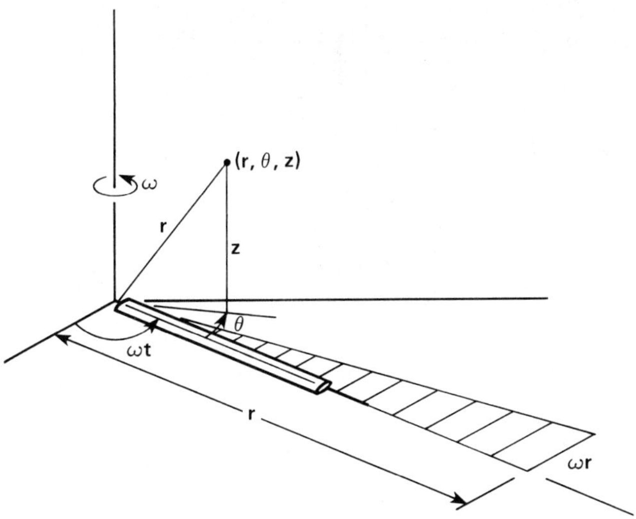

Figure 56. Cylindrical coordinate system.

Before attempting to solve equation (4.14), it is instructive to follow the approach of Isom (1979) and Hawkings (1979) and explore the behavior of the governing equation. It is known from the theory of partial differential equations that the coefficient of $\phi_{\theta\theta}$ governs the general character of the potential equation. For when

$$\begin{aligned} A &\equiv \omega^2 - \frac{a_o^2}{r^2} - (\gamma+1)\frac{\omega}{r^2}\phi_\theta \\ &< 0: \text{ elliptic behavior} \\ &> 0: \text{ hyperbolic behavior} \end{aligned}$$

However, A takes a more recognizable form after some further manipulation:

$$A = \omega^2 - \frac{a_o^2}{r^2} - \frac{(\gamma-1)\omega}{r^2}\phi_\theta - \frac{2\omega}{r^2}\phi_\theta$$

Using the energy equation, this becomes

$$A = \omega^2 - \frac{a^2}{r^2} - \frac{2\omega\phi_\theta}{r^2}$$

Defining $U_\infty = \omega r$, $\phi_\theta = -ur$, we obtain

$$A = -\frac{a^2}{r^2}\left[1 - \frac{(U_\infty + u)^2}{a^2} + \frac{u^2}{a^2}\right] \tag{4.15}$$

The last term in the brackets of equation (4.15) can be neglected. It is second order to A and a third-order correction to the second-order potential equation. If we now define

$$M_\ell \equiv \frac{U_\infty + u}{a} = \text{local Mach number} \tag{4.16}$$

the coefficient of $\phi_{\theta\theta}$ (equation (4.14)) becomes

$$A = -\frac{a^2}{r^2}(1 - M_\ell^2) \tag{4.17}$$

Therefore, the general behavior of the second-order transonic potential equation is governed by the local Mach number of the flow. If $M_\ell < 1.0$, then $A < 0$, and the governing equation is elliptic. In this case, no wave-like structure is possible. However, if $M_\ell > 1.0$, then $A > 0$, and the governing nonlinear partial differential equation is hyperbolic. Then, characteristics are formed along which disturbances can propagate in a wave-like manner. It is also important to realize that the local Mach number M_ℓ is dependent on the free-stream velocity ωr, the local speed of sound a, and the local perturbation velocity $u = -\phi_\theta/r$. Each contribution is separately identifiable in equation (4.16).

These ideas are quite useful when one is attempting to explain the phenomenon of delocalization for the hovering rotor experiment described previously. This connection was theoretically suggested in Yu et al. (1978) and Isom (1980), numerically calculated in Schmitz and Yu (1981) and Hawkings (1979), and experimentally confirmed in Yu et al. (1978). In the following paragraphs, the relationships are experimentally shown to depend on the local Mach number of flow. Three distinct cases are considered: $M_T = 0.85$, 0.88, and 0.90. Some freedom has been taken with the graphics in the interest of presenting a clear picture of the basic relationships involved. The data are the same as those reported in Schmitz and Yu (1981). In the figures that follow, the top views are sketches of events pieced together with limited experimental data, and the aft views are, for the most part, interpolations of experimental data.

Figure 57 depicts the top and aft views of a rotor operating at a free-stream tip Mach number (M_T) of 0.85. A locally supersonic region exists near the tip of the rotor. For this region $M_\ell = (\omega r + u)/a > 1.0$, even though $\omega r/a_o$ all along the blade span is less than 0.85. The hyperbolic nature of this pocket of supersonic flow is a result of local aerodynamic nonlinearities (i.e., changes in the local speed of sound, a, and local perturbation velocity, u). Surrounding this locally supersonic flow is a subsonic flow region, $M_\ell < 1.0$, in which the governing potential equation is elliptic. No wave-like behavior is possible through this compressible elliptic region. However, as r increases beyond the tip of the blade, M_ℓ again becomes greater than 1 because of the linearly increasing free-stream velocity field of the blade-fixed cylindrical coordinate system. For this region $u \approx 0$, $a \approx a_o$, so

$$M_\ell \approx \frac{\omega r}{a_o} > 1.0.$$

The surface where this first happens has been called the sonic cylinder (Yu et al., 1978; Isom, 1980). At radii larger than the sonic cylinder, the equation again becomes hyperbolic, and wave-like propagation is certain. The acoustic implications of this 0.85 case begin in the hyperbolic pocket of flow near the blade tip. Wave-like disturbances in this region terminate on the boundary of an elliptic region where they no longer propagate in characteristic directions. The wave-like character of the inner pocket is thus broadened as information passes through the elliptic region to the sonic cylinder. These broadened disturbances are then propagated in a wave-like manner throughout the outer hyperbolic region. The result is a smoothly varying, near-symmetrical acoustic signature in the far-field.

The competing phenomena become even more interesting when the undisturbed free-stream tip Mach number of the rotor (M_T) is increased to 0.88 (Figure 58). The inner supersonic (hyperbolic) region grows and extends off the tip of the rotor—again being driven by local aerodynamic nonlinearities. At the same time, the higher free-stream tip Mach number of the rotor decreases the radius of the sonic cylinder, thus moving the outer hyperbolic region toward the rotor tip. In addition, the proximity of the linear sonic cylinder to the blade tip introduces aerodynamic nonlinearities. These tend to warp the sonic cylinder inward, bringing the two hyperbolic regions even closer together. However, the hyperbolic regions do not overlap, thus ensuring that locally generated waves in the inner region do not propagate along characteristics to the hyperbolic far-field. Instead, the wave-like disturbances are forced to pass through a small elliptic region where they are broadened before entering the outer hyperbolic region for propagation to the far-field. The resulting acoustic signature becomes more saw-toothed in character, but does not contain radiating shocks.

The last and most interesting condition, in which the local free-stream tip Mach number is increased to 0.9, is sketched in Figure 59. The localized inner hyperbolic and outer hyperbolic regions connect off the blade tip, forming one continuous supersonic region ($M_\ell > 1.0$). In this case, shock waves that are generated on the surface of the rotor now propagate uninterruptedly in a radially outward characteristic direction to the acoustic far-field. The resulting delocalization phenomenon is quite striking, for the character and the intensity of the acoustic signature change dramatically. At all three of these conditions, measured values of local Mach number support and explain the phenomenon of *transonic delocalization*.

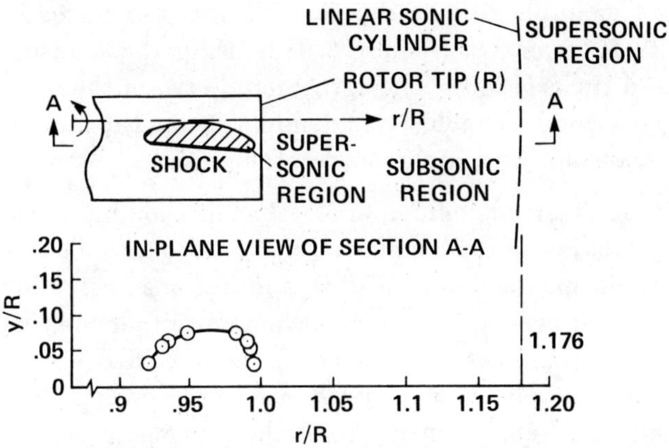

Figure 57. Top and aft views of shock boundaries of a rotor.

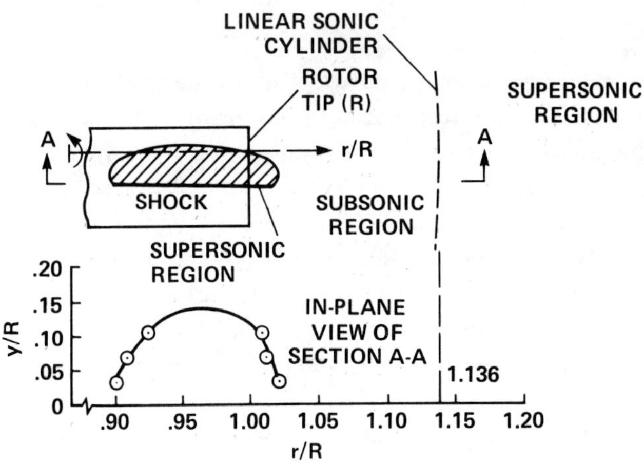

Figure 58. Top and aft views of shock boundaries of a rotor, $M_T =$ 0.88.

Further confirmations of the mechanism of delocalization can be made theoretically using existing transonic aerodynamic codes (see Caradonna and Steger, 1980; Tauber and Hicks, 1980; Chattot, 1980). An example calculation for this rotor in hover is shown in Figure 60, the work of Shenoy (1982) using the transonic code of Tauber and Hicks (1980). The agreement between theory and experiment is quite good—conclusively demonstrating the interrelationship between transonic and high-speed rotor noise.

Although the phenomenon of delocalization has been explained by simply looking at the coefficient of $\phi_{\theta\theta}$ in equation (4.14), predicting the radiating acoustic field is another matter. The explanations presented are themselves functions of either measured or calculated flow properties. In essence, a near-field description of the aerodynamic flow field is required before the events in the acoustic far-field can be explained. Even then, acoustic nonlinearities in equation (4.13) may alter the waveform of the propagating wave (Isom, 1980). Precise calculations of the radiating sound field are dependent on the full solution of the nonlinear potential equation (Isom, 1980).

On the other hand, the successful explanation of the delocalization phenomenon suggests that local aerodynamic nonlinearities strongly influence the acoustic radiation problem. Therefore, a logical step in the calculation of the acoustic field is the incorporation of the near-field aerodynamic nonlinearities in the acoustic radiation equation.

At least two alternative ways of implementing this idea have been presented in the literature. In one proposed method (Hawkings, 1979) the nonlinear near-field is mapped to a nonrotating control surface, where Kirchoff's theorem is applied. As reported, the control surface must be chosen to be large enough to capture the nonlinear aerodynamic behavior of the problem, but not to be so large as to make numerical computation impractical. Calculations using this procedure, coupled with an existing near-field numerical code, have shown improvement in peak amplitude levels, but not much improvement in waveform characteristics above the delocalization Mach number. As discussed in the paper, this is most likely a result of the numerical insensitivity of the transonic code at the boundary of the nonrotating control surface.

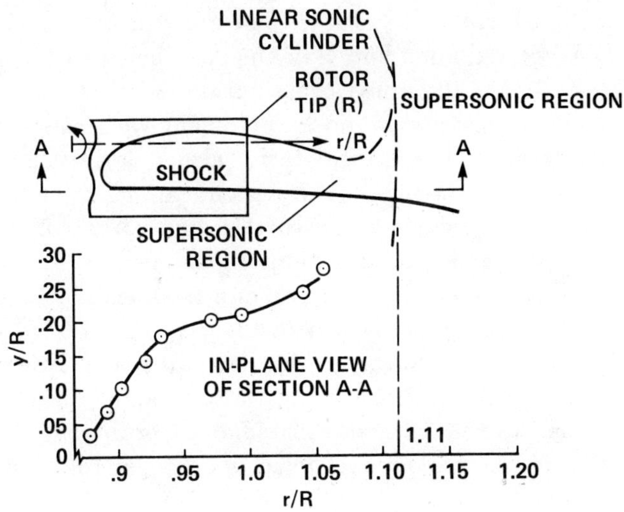

Figure 59. Top and aft views of shock boundaries of a rotor, the development of delocalization phenomenon, $M_T = 0.9$.

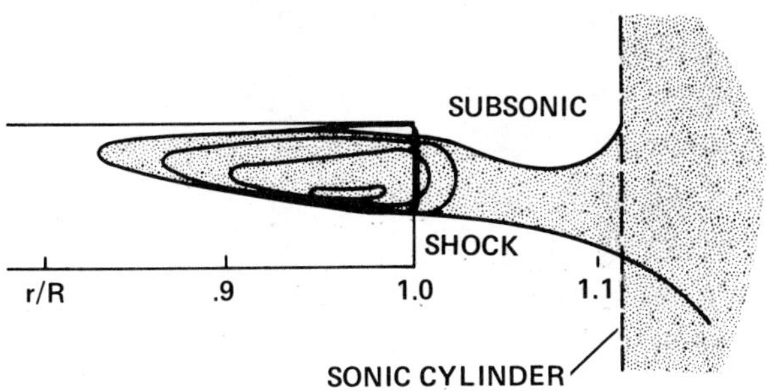

Figure 60. Tip flow field of a hovering rotor ($M_T = 0.9$).

In the second method, the well-known *acoustic analogy* procedures are used to evaluate the volume distributions of local aerodynamic nonlinearities or quadrupoles (Schmitz and Yu, 1979; 1981; Yu et al., 1978; Hanson and Fink, 1978). The analysis begins with equation (1.1) (Ffowcs Williams and Hawkings). The third term in this equation is the linear thickness contribution to the radiated noise, which by itself has shown the poor agreement with experiment, as previously discussed. The second term, also a surface integration over the blade, can be classified as either a linear or a nonlinear term, depending on how the surface pressure is approximated. However, for most of the in-plane computations reported here, its contribution to the radiated noise field is small and thus has been neglected. A strict evaluation of the nonlinear quadrupole (first) term in equation (1.1) requires a volume integration over all space. However, the interest in this paper is in capturing those nonlinear aerodynamic terms that may govern the acoustic radiation problem. Therefore, the evaluation of the quadrupole term is confined to a volume integration within the aerodynamic nonlinear flow field of the blade.

The quadrupole term assumes a simpler form if we restrict our attention to the acoustic far-field. Then the spatial differentiations can be easily converted to time differentiations. The first term of equation (1.1) becomes

$$\frac{\partial^2}{\partial x_i \partial x_j} \int \left[\frac{T_{ij}}{r|1-M_r|}\right] dV \simeq \frac{1}{a_o^2}\frac{\partial^2}{\partial t^2} \int \left[\frac{T_{rr}}{r|1-M_r|}\right] dV \tag{4.18}$$

where $T_{rr} = T_{ij}\hat{r}_i \cdot \hat{r}_j$ and $\hat{r}$ is the unit vector from the source at the retarded time to an observer in the acoustic far-field. It is known from transonic computations and experimentation that the primary quadrupole regions are confined within a few chord lengths normal to the rotor plane.

For in-plane far-field radiation, the unit vector $\hat{r}$ is nearly in the blade rotational plane and is nearly parallel to the blade chordwise direction when the acoustic pressure reaches its peak level. If isentropic flow is assumed and the perturbation velocities are measured in the coordinate system given in Figure 56, T_{rr} becomes

$$T_{rr} = \rho_o\left(v_\theta^2\cos^2\theta + 2v_r v_\theta \cos\theta\sin\theta + v_r^2\sin^2\theta\right) + \frac{\gamma-1}{2}\rho_o\left(\frac{\omega r}{a_o}\right)^2 v_\theta^2 \tag{4.19}$$

where the z-component of the perturbation velocity does not appear because of the choice of an in-plane far-field microphone position. For simplicity in the resulting calculations, it has been assumed that $\sin\theta \approx 0$ and that $u \approx v_\theta$ near the integration region of interest. This is true as long as the quadrupole field is in fact localized to a region near the rotor tip. Then,

$$T_{rr} \simeq \rho_o u^2 \cos^2\theta + \frac{\gamma - 1}{2}\rho_o\left(\frac{\omega r}{a_o}\right)^2 u^2 \tag{4.20}$$

where u represents the perturbation velocity along the blade chord and ωr is the free-stream velocity of the point in the flow field being evaluated. The two terms represented in equation (4.20) arise from similar properties of the flow already discussed in the potential formulation. Changes in the local speed of sound and local streamwise perturbation nonlinearities are included, although the equation forms do not permit a one-to-one correspondence of terms.

Equations (1.1), (4.18), and (4.20) described the nonlinear far-field acoustic radiation of the transonic hovering rotor. For subsonic tip Mach numbers, numerical evaluation of the surface integrals presents no real problems. However, the volume integration of quadrupoles is not as straightforward.

As discussed previously, the accuracy of this nonlinear formulation is totally dependent on the detailed knowledge of the flow field surrounding the blade. In this paper, the flow fields of interest were computed by the three-dimensional transonic numerical code described in Caradonna and Isom (1976) and Caradonna (1978). This code solves the near-field transonic small-disturbance potential equation in a blade-fixed reference frame. The solution uses a conservative, mixed-difference relaxation scheme. Because this code was developed to predict blade-surface aerodynamics, some simplifying assumptions were made to equation (4.14); namely, the nonlinearities known to be small near the blade surface were neglected. In particular, the terms $2\omega\phi_r\phi_{r\theta}$ and $2\omega\phi_z\phi_{z\theta}$ on the left-hand side and the factor $(\gamma - 1)\omega\phi_\theta$ on the right-hand side of equation (4.14) were not programmed.

The nature of the calculated chordwise pressure distributions as a function of blade radius is shown in Figure 61 for a tip Mach number of 0.9 at zero angle of attack. On the blade surface, a shock wave forms at about 90% of the span and persists over the outer portions of the blade and beyond. These calculated pressure

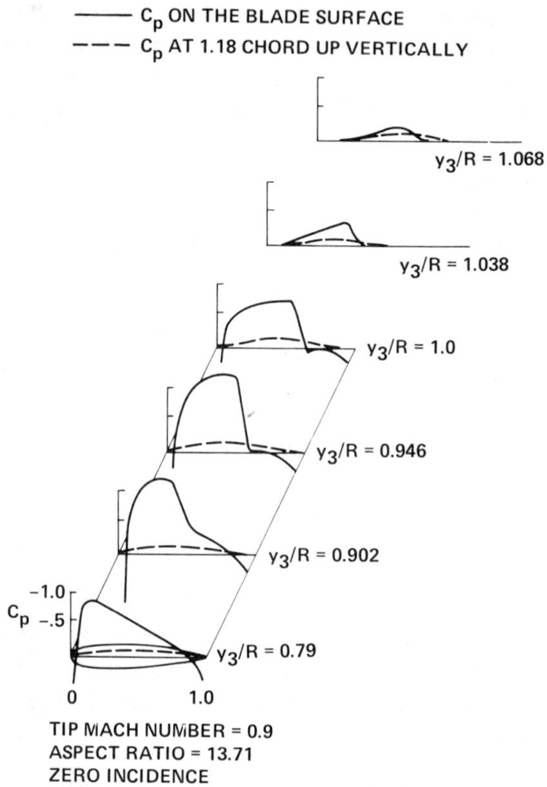

Figure 61. C_p distribution on the blade surface and at 1.18 chords up, aspect ratio = 13.7, NACA 0012, near zero angle of attack, M_T = **0.9.**

distributions are consistent with the experimentally sketched flow regions of Figure 59. The maximum value of shock strength is calculated to occur at 95% radius. It is also important to note that the shocks that are calculated with this code do not look like measured shock waves. The numerical solution has built into it a numerical viscosity which tends to smooth discontinuities over several mesh points. This affects the final acoustic waveforms to some degree and is discussed later.

The final problem in the evaluation of the quadrupole integral is the development of a calculation procedure for equations (1.1), (4.18), and (4.20), that is valid when $M_r = 1.0$. This problem occurs when the volume integration is extended up to and beyond the linearized sonic cylinder. The integrand in equation (4.18) then contains the product of two terms which compete to decide the eventual magnitude of the quadrupole radiation. The first is the decaying source field represented by the T_{rr} term in equation (4.18). This is multiplied by the $1/(1 - M_r)$ term which goes to ∞ as M_r approaches 1. Fortunately, the singularity is integrable, but it must be handled quite carefully. In the results summarized here, the acoustic planform technique was chosen to perform the numerical integration near $M_r = 1.0$. A complete discussion of the procedures and pertinent references is given in Schmitz and Yu (1981).

An evaluation of the prediction accuracy is presented below by comparing theory with the same UH-1H hover rotor data. Figure 62 presents the monopole and quadrupole contributions to the radiated noise at $M_{\text{tip}} = 0.88$ (slightly before delocalization). At this Mach number, the shape of the quadrupole term is basically still symmetrical, however, some asymmetrical character is present on the pressure recovery side of the quadrupole calculation. When the monopole and quadrupole contributions are added, good correlation in amplitude and pulse shape is observed (Figure 62). The overall shape of theory and experiment are still basically symmetrical in character, but the local shock structure of the transonic flow field is acting to destroy this symmetry.

At the slightly higher hover tip Mach number of 0.90, localized transonic effects cause large changes to the radiated noise field (Figure 63). Although the linear term (monopole, Figure 63a) remains quite symmetrical in shape and substantially underpredicts the measured data, the nonlinear (quadrupole, Figure 63b) term changes shape dramatically and increases in amplitude. This is a reflection of the fact that local shocks are propagating to the acoustic far-field (delocalization in Isom (1979) and Yu et al. (1978)). When the monopole and quadrupole terms are compared with experimental data (Figures 63c and 63d), good agreement in pulse shape is observed.

The relative accuracy with which the peak negative amplitude of the high-speed impulsive noise phenomenon can be predicted is illustrated in Figure 64. Much better agreement between theory and experiment is demonstrated for Mach numbers up to 0.9. At $M_T > 0.9$, amplitudes are overpredicted. Additional time-history comparisons can be found in Schmitz and Yu (1981).

The most exciting part of the high-speed impulsive noise problem is that we have come so far in understanding and predicting the basic physical phenomena. Nevertheless, the computation of the

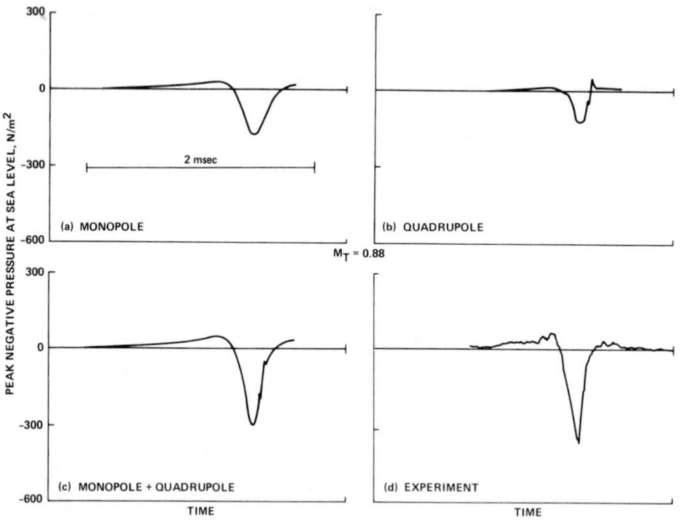

Figure 62. Theory-experiment comparison, $M_T = 0.88$.

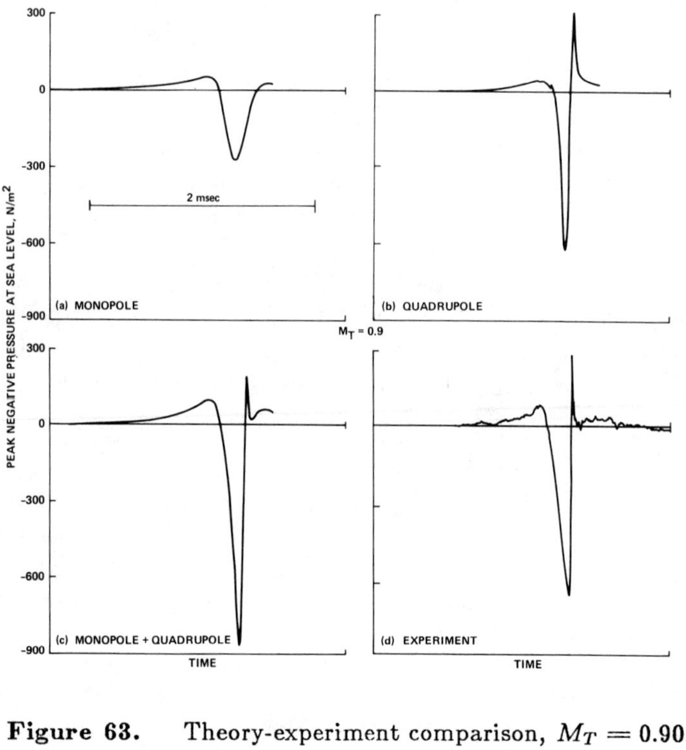

Figure 63. Theory-experiment comparison, $M_T = 0.90$.

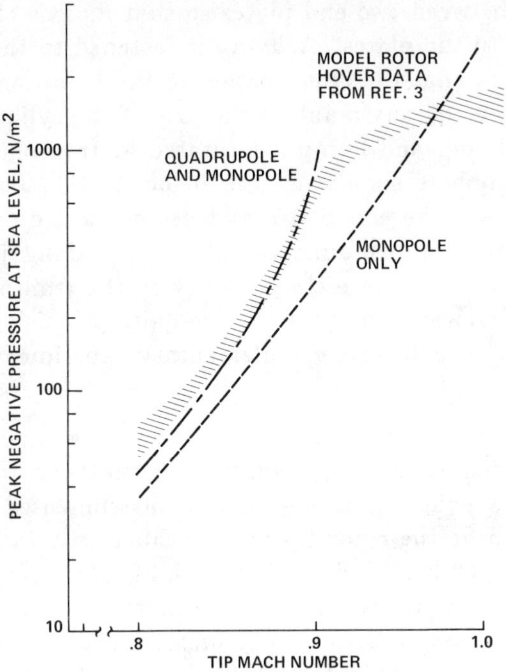

Figure 64. Comparison of theory and experiment in hover, in-plane, $r_H/D = 1.5$.

noise is not routine. It remains a difficult task involving a matching between a near-field transonic code and a suitable acoustic code. Several areas where improvements can and are being made (in some cases) are suggested in the following paragraphs.

The high-speed impulsive noise prediction process is dependent on the quality of the transonic near-field aerodynamic data. At the present time, existing numerical codes contain *numerical viscosity* which tends to smear the local shock structure. This smearing is most pronounced off the blade surface, past the blade tip, and can cause numerical errors in either the Kirchoff (Hawkings, 1979) or quadrupole acoustic formulations. Some alternative transonic formulations that use shock-capturing methods may help better define these important regions of the flow field.

It is also possible to study numerical and physical aspects of the delocalization process by considering a related two-dimensional problem—wave propagation from disturbances emanating from a physical rotating cylinder (Rutherford, 1983). In essence, a cylinder is mounted between two end plates so that its axis of rotation is at right angles to the plates. A *bump* is fastened to the cylinder surface aligned so that the cross section of the bump and the cylinder are the same along any point on the axis of the cylinder (see Figure 65). The cylinder and bump are rotated at transonic circumferential Mach numbers simulating the delocalization flow field of the hovering rotor. The end plates and geometrical designs make this simulated problem two-dimensional, thus making it much easier to investigate the numerical sensitivity of the transonic delocalization process. Initial computations are quite promising (Rutherford, 1983). In the near future, complementary experimental testing will commence.

In addition to numerical considerations, new experimental methods are needed to develop quantitative methods to measure the rotor tip flow field. Of course, blades instrumented with pressure transducers near the rotor tip yield invaluable data on the surface of the blade (Thibert and Philippe, 1982). However, the flow field off the rotor blade tip is perhaps even more important for transonic rotors. In the past, hot-wire anemometry has been successfully used to measure the transonic flow field (Yu et al., 1978). More recently, these measurements were charted using laser velocimetry. In each case the velocity field surrounding the rotor-blade tip extending past the blade's sonic cylinder has been extensively mapped and used as input to equation (1.1) to predict the acoustic far-field. The major problem with either the hot-wire or the laser velocimeter approaches is that they are time consuming.

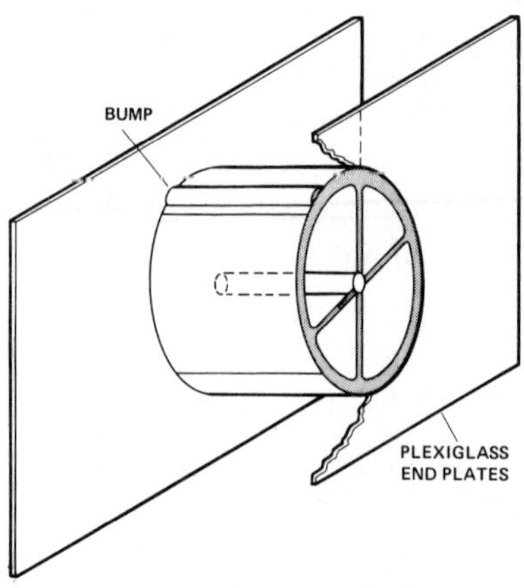

Figure 65. Schematic of a rotating cylinder.

A quite promising new method for obtaining the flow-field measurements employs holographic interferometry (Kittleson, 1983; Kittleson and Yu, 1982). A laser beam is expanded and passed through the rotor blade very quickly, mixed with a reference wave, and stored as a hologram of the three-dimensional flow field. A second beam is passed through the same path with the rotor at rest and a second hologram is made. These two holograms are then superimposed, creating an interferogram of the integrated three-dimensional flow field. Figures from Kittleson (1983) present a holographic interferogram of a hovering transonic UH-1H model rotor (see Figure 55). By simply counting fringes in the interferogram it is possible to quantitatively define the integrated density of the flow field. However, if many photographs similar to this are gathered at many different angles, it is possible to use the known techniques of tomography to define quantitatively the density of any point in the flow field. This truly exciting prospect has been simulated by successful research at our laboratory, at Stanford University, and at the University of Michigan. The next few years should see the development of holographic interferometric tomography for compressible rotor problems.

Finally, the acoustic calculation process, or coding, still leaves room for improvement. In quite a few cases of interest, including the evaluations of the quadrupole near the linear sonic cylinder region, it is necessary to evaluate an integral whose integrand is singular but integrable. This is normally done by using *acoustic planform methods* (Schmitz and Yu, 1981; Hanson, 1976; Nakamura and Azuma, 1978). However, it is the authors' opinion that the codes that calculate the far-field acoustics near singularities could be simplified and improved by using more of the physics of the transonic rotor. These improvements are being investigated by researchers at the Aeromechanics Laboratory.

5. Blade-Vortex Interaction Impulsive Noise

The blade-vortex interaction impulsive noise story is not as complete as that of high-speed impulsive noise, although neither is totally understood. As discussed at the beginning of this article, the problem is quite complex because it depends strongly on the local aerodynamic state of the helicopter. There have been many excellent experimental and theoretical approaches to this difficult problem over the past 15 years. Although each effort showed progress in understanding, only a few rather qualitative design changes have resulted.

It is the authors' opinion that more definite progress has not been made because we are not sure to what degree we are in fact modeling (theoretically or experimentally) the full-scale aerodynamic sources of interest. For example, it is becoming common practice to gather impulsive-noise acoustic data on a model rotor in an acoustically treated wind tunnel. However, it is often not known quantitatively how these data relate to measured noise on the full-scale vehicle. It is not too surprising then if the design changes based on these model results do not prove effective on the full-scale helicopter.

The review of blade-vortex interaction (BVI) impulsive noise presented below is based on the following three fundamental questions:

1. Can BVI impulsive noise be scaled, and what are the known scaling parameters?
2. What are the known physical origins of BVI impulsive noise?
3. How accurately can BVI impulsive noise be predicted?

All of the data and arguments to be presented are for the two-bladed, single-main-rotor helicopter class. Because it has been, historically, the most notable contributor to BVI impulsive noise, much of the BVI research has concentrated on this class of helicopters. However, it should not be concluded that three- and four-bladed helicopters do not generate impulsive noise. On the contrary, as shown in Figure 28, four-bladed helicopters can generate impulsive noise as well. In many instances, these noise sources are the dominant source of radiated acoustic energy. (It should be noted that tandem-rotor helicopters also generate BVI impulsive noise by some of the same source mechanisms as the single-rotor helicopter (Sternfield et al., 1972; Sternfeld and Schaeffer, 1982).

Another reason for focusing on the two-bladed BVI impulsive noise problem is the relative simplicity of the two-bladed teetering system. The rotors themselves are often large with minimal dynamic-aerodynamic coupling effects. The number of potential blade-vortex interactions for a given flight condition is directly proportional to the number of rotor blades. The choice of two rotors minimizes the number of encounters and facilitates the interpretation of the time-history of the acoustic pulse.

The scaling of balde-vortex interaction noise was recently addressed in Schmitz et al. (1982). Full-scale data taken by the in-flight method described previously were compared with model-rotor data gathered in an anechoic wind tunnel.

The full-scale acoustic data were taken on the AH-1S helicopter in 1978 and 1979 in two separate test programs and reported in Boxwell and Schmitz (1982; 1979). The data are repeatable and clearly define the blade-vortex interaction noise phenomena of interest. To optimize the signal-to-noise ratio of BVI impulsive noise, selected conditions of forward velocity and rate of descent were flown with the microphone positioned directly ahead of the helicopter but 30° below the rotor's tip-path-plane (Figure 66). As discussed, the major advantages of the in-flight technique are the long data records, the absence of ground reflections, and the ability to fly conditions normally associated with terminal-area operations.

Model-scale acoustic data were gathered in one of the largest anechoic wind tunnels in the world (CEPRA-19, located near Paris, France). The model rotor was 1/7 geometrically scaled (blade aspect ratio, twist rate, and sectional properties were duplicated) and mounted on a two-bladed teetering hub—as in the full-scale helicopter. A photograph of the rotor installed in CEPRA-19 is shown in Figure 67. Dynamic scaling was not attempted. Fortunately, the rigid, large-mass blades appeared to render the aeroelastic and dynamic effects to a secondary status. Details of the testing procedure are given in Schmitz et al. (1982). (There were some small-section geometry differences between the model and full-scale blades. Details can be found in Schmitz et al. (1982).)

Besides blade geometry and relative orientation between the rotor and the microphone, there are at least four nondimensional parameters which should be duplicated if model-scale acoustic data are to be expected to match full-scale data. As we have discussed, the in-plane geometry (see Figure 68) between the rotor blade and the tip-vortex structure is a dominant BVI interaction parameter. The first nondimensional parameter, **advance ratio** μ determines these large-scale geometrical interaction patterns. Advance ratio μ is defined to be the ratio of the helicopter's forward velocity of translation divided by the main rotor hover tip-speed. The in-plane projection of the spatial tip-vortex positions is governed by

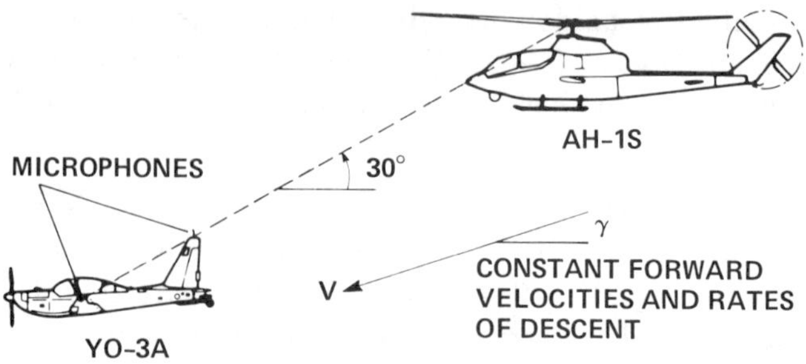

Figure 66. Full-scale blade-vortex interaction acoustic measurement technique.

Figure 67. 1/7 scale model rotor and test stand installed in French CEPRA-19 anechoic wind tunnel.

the free-wake geometry of the rotor-wake system. A very good approximation to these geometrical in-plane projections can be made if self-induced wake distortion is totally neglected (rigid wake). In this case, the in-plane projection simply becomes two epicycloid tip-vortex patterns shed from each rotor blade tip, as shown in Figure 68. The major differences between this calculation and the one given in Figure 68 are in the tip region after the individual vortices begin *rolling together* to form the classical fixed-wing-like vortex system. However, the BVI encounter often occurs before this distortion becomes effective. In these regions above $\mu = 0.1$, the rigid and free-wake in-plane projections are almost indistinguishable.

Because advance ratio governs the large-scale BVI geometry, it plays a key role in the acoustic radiation. When viewed from above, the rotor appears to slice through the epicycloid pattern of previously shed tip vortices. The resulting loci of interactions determines the number and strength of the blade-vortex interaction encounters and thus strongly influences the radiated noise.

For scaling purposes, let us assume that only changes in pressure on the surface of the blades causes BVI impulsive noise. As before, the second term of equation (1.1) becomes

$$P'(x,t) = -\frac{1}{4\pi}\frac{\partial}{\partial x_i}\int\int\left[\frac{P_{ij}n_j}{r|1-M_r|}\right]_\tau dS \qquad (5.1)$$

where P_{ij} is the local surface-pressure tensor on the surface of the rotor blade. Nondimensionalizing this pressure tensor, we obtain

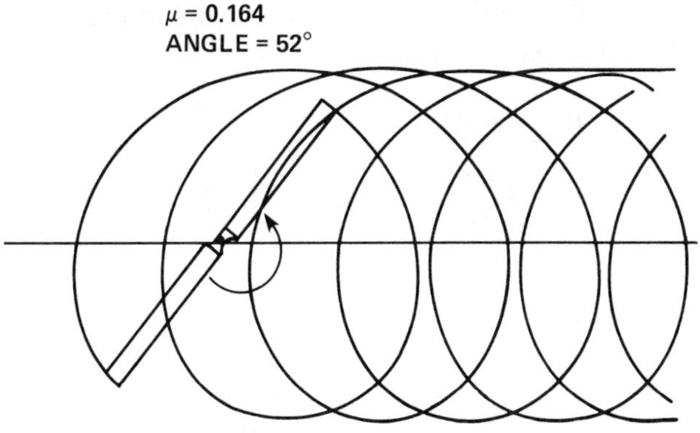

Figure 68. Epicycloid tip-vortex pattern for $\mu = 0.164$.

$$P'(x,t) = -\frac{1}{4\pi}\frac{\partial}{\partial x_i}\int\int\left[\frac{C_{P_{ij}} n_j \rho_o V^2}{r|1-M_r|}\right]_\tau dS \tag{5.2}$$

where

$$C_{P_{ij}} \equiv \frac{P_{ij}}{\rho_o V^2}$$

As in the high-speed scaling arguments, if we assume geometric scaling and introduce the nondimensional unit of time $\bar{t} \equiv t/(2\pi/\Omega)$, then

$$\begin{aligned} d\overline{S} &= \frac{dS}{R^2}, \\ \bar{r} &= \frac{r}{R}, \\ \bar{x} &= \frac{x}{R}, \\ \bar{t} &= \bar{\tau} + \frac{\bar{r}}{2\pi} M_T \end{aligned} \tag{5.3}$$

and

$$C'_P(\bar{x},\bar{t}) = \frac{P'(\bar{x},t)}{\rho_o a_o^2} = \frac{-1}{4\pi}\frac{\partial}{\partial \bar{x}_i}\int\int\left[\frac{C_{P_{ij}} n_j M^2}{\bar{r}|1-M_r|}\right]_{\bar{\tau}} d\overline{S} \tag{5.4}$$

where $M = V/a_o$. So if rotor, microphone geometry, Mach number, and time are scaled, the acoustic pressure coefficient is uniquely determined. Notice that this result is similar to the high-speed scaling arguments presented earlier. All comparisons between model and full-scale results are done in referred acoustic pressure—as was done for the high-speed scaling.* Therefore,

$$P'_{SL}(\bar{x},\bar{t}) = \frac{P_{o_{SL}}}{P_o} P'(\bar{x},\bar{t}) \tag{5.5}$$

$$P'_{m_{SL}}(\bar{x},\bar{t}) = \frac{P_{o_{m_{SL}}}}{P_{o_m}} P'_m(\bar{x},\bar{t}) \tag{5.6}$$

*The full-scale pressure time-history data shown here are from Schmitz and Boxwell (1976). The data were corrected for density but not for speed-of-sound effects to an effective pressure at sea level. As mentioned previously, direct comparisons should account for a_o changes in both the model and full-scale results. The effect can increase the amplitude disparity between model and full-scale by as much as 6%.

These scaling arguments highlight the importance of the second key nondimensional parameter, **rotor Mach number.** Because the geometry of the problem is scaled and μ (advance ratio) is also duplicated, choosing the hover tip Mach number (M_{HT}) completely specifies the advancing-tip Mach number [$M_{AT} = (1 + \mu)M_{HT}$] and the Mach number in the radiation direction, M_r. In effect, all Mach numbers associated with the large-scale geometry of BVI are governed by two nondimensional parameters, μ and M_{HT}.

Judicious matching of the third and fourth nondimensional parameters–**thrust coefficient,** C_T, and **nondimensional inflow**– is necessary to duplicate the pressure coefficients ($C_{p_{ij}}$) of the model and full-scale experiments. For a geometrically scaled rotor, the thrust coefficient governs the local angle of attack of the rotor blade and thus the steady-pressure field. In addition, it affects the average strength of the shed tip-vortex, thus directly influencing the unsteady pressure field as well. The nondimensional inflow $\mu(-\alpha_i + \alpha_{TPP})$ also affects the magnitude of the unsteady pressures by governing the vertical separation between the vortex and the rotor blade at the time of an encounter. In a rigorous sense, this parameter should scale over the portion of the rotor disk where blade-vortex interactions occur. However, it is often assumed that by scaling geometric properties and C_T, an average value in space and time of the induced angle (α_i) at the rotor disk governs the interaction problem ($\alpha_i \sim C_T/\mu$). Therefore, if C_T and μ are duplicated in a model-to full-scale test, the nondimensional inflow or **tip-path-plane angle** (α_{TPP}) becomes the fourth nondimensional test variable.

In normal, unaccelerated level flight, the helicopter pilot must tilt the rotor tip-path-plane ($\alpha_{TPP} \equiv$ the angle between the plane of the rotor tips and the incoming velocity vector; positive for rearward tilt) to balance the drag of the vehicle at each velocity. The result is an increasingly negative tip-path-plane angle with increasing forward velocity. In a climb, the rotor must be tilted farther forward ($-\alpha_{TPP}$) both to balance drag and to oppose gravity, whereas in a descent the rotor must be tilted rearward ($+\alpha_{TPP}$). The strongest blade-vortex interactions are known to occur in the descent condition, when the tip-vortices close to or into the rotor's tip-path-plane (Figure 68).

The most rigorous test of the scalability of impulsive noise is the most direct: that of simply comparing the character of the model- and full-scale acoustic time-histories on a one-to-one basis.

In addition to being a straightforward comparison, it is also helpful in identifying the occurrences of blade-vortex interactions in the acoustic signatures. This phenomenological approach is illustrated in Figure 69 for the AH-1G helicopter for a microphone located approximately 30° beneath the plane of the rotor tips. This relative orientation of the microphone and the rotor is known to maximize the blade-vortex interaction noise and to reduce the intensity of high-speed impulsive noise. In Figure 69a a measured acoustic time-history is shown for one rotor revolution as measured using the full-scale, in-flight technique. The helicopter and measurement aircraft were flown in formation at a 60 knot (IAS) partial-power descent (400 ft/min rate of descent), a condition known to produce strong blade-vortex interaction noise. The four important nondimensional scaling parameters are listed in Figure 69a. At this 30° microphone position, both the blade-vortex interaction noise and high-speed impulsive noise are discernible. During advancing blade-vortex interaction, a sequence of narrow, small-negative and large-positive spikes occurs in the waveform just before the broader negative-pressure pulse. As discussed previously, the latter is the high-speed thickness noise and is characteristic of acoustic radiation below the delocalization Mach number at a microphone position 30° under the rotor plane. It should be noted that the full-scale data in Figure 69a show a *snapshot* of a long data record. Time-averaging of the in-flight signals is prohibited by small but discernible changes in the relative position between the microphone and the helicopter.

The model-scale data, at approximately the same scaled geometric distance, are shown in Figures 69b and 69c. Figure 69b presents one instantaneous snapshot of one rotor revolution and Figure 69c presents an average of 100 revolutions. The similarity of the model- and full-scale pulse shapes is evident.

A more complete picture of the comparison of model- and full-scale blade-vortex interaction noise is shown in Figure 70 for similar flight conditions. Good general agreement between model-scale and full-scale is apparent at all descent conditions. The averaged model-scaled blade-vortex interaction amplitudes, as well as pulse shapes, match those of the full-scale data. The figure also shows that the acoustic pulses resulting from a BVI with an older vortex tend to decay, and that those pulses resulting from a BVI with an older vortex tend to grow with increasing descent rate (more positive tip-path-plane tilt). As explained before, as the tip-path-plane of the rotor is tilted rearward, simulating descending flight, a vortex-interaction

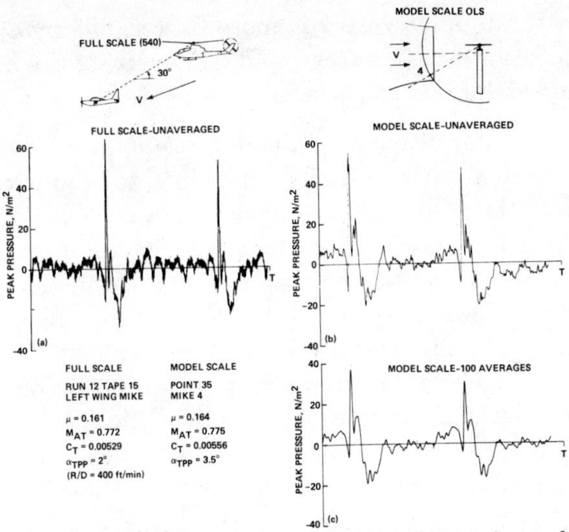

Figure 69. Time-history comparisons of full-scale and model-scale helicopter blade-vortex interaction noise.

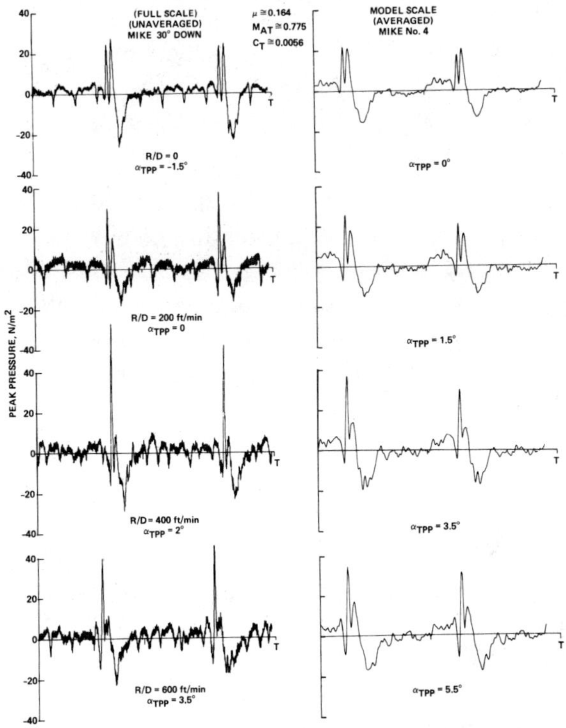

Figure 70. Time-history comparisons of full-scale and model-scale helicopter blade-vortex interaction noise for several descent conditions.

encounter with an older vortex is more likely. The model- and full-scale data substantiate this trend and demonstrate the continuity of the blade-vortex interaction process.

These BVI impulsive-noise scaling results are quite encouraging. They demonstrate that it is possible to duplicate many of the features of the full-scale BVI phenomena on model-scale rotors. However, there remain some discrepancies in this first BVI scaling comparison. Most notably, the amplitude of the model-scale BVI is consistently less than that of the full-scale data. In addition, the pulse widths of the model scale seem wider than those of the full-scale data. Both discrepancies suggest that the tip-vortex strength of the model may be less intense than the full-scale data. Further analysis of these tests and follow-on testing in the DNW wind tunnel (Splettstoesser et al., 1983) should help clarify these concerns.

One of the most interesting aspects of the blade-vortex interaction phenomenon is the directivity of the resulting noise. In general, we have shown (Figure 71) that the BVI noise is radiated forward and down approximately 30° beneath the rotor plan (Boxwell and Schmitz, 1982). Figure 71 confirms this fact for the $\mu = 0.164$ and $\alpha_{TPP} = 3.5°$ case but also indicates that the rotor noise decay is less gradual from the 30° position toward the rotor's tip-path-plane than farther below this position.

The lateral-directivity sweep shown in Figure 72 at the $\psi = 30°$ elevation angle reveals many interesting phenomena as well. The impulsive noise spike is quite weak at the $\theta = 90°$ position of the microphone and appears to consist of one event. It grows as the microphone is moved to the 45° position and grows even larger at the $\theta = 30°$ position. Here it is apparently joined by another pulse. The amplitude of the first pulse increases further at the straight-ahead microphone position and remains high at the microphone retreating side positions. At the $-45°$ position, a decrease in amplitude occurs, some of which may be attributable to blockage by the rotor stand.

Some caution should be exercised when interpreting the data recorded on the microphone located near the 3-m nozzle. The reverberant/reflected field could have distorted the results somewhat. Shear-layer effects can also be important at these microphone locations because the sound path enters the shear layer at shallow angles. Nevertheless, the general trends illustrated in Figures 71 and 72 are believed to be correct.

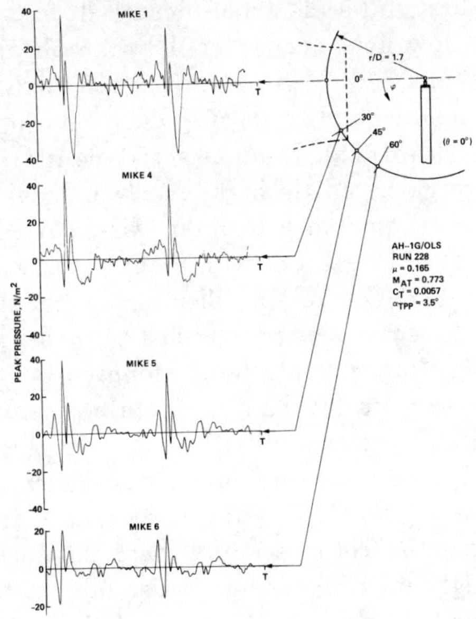

Figure 71. Longitudinal directivity of blade-vortex interaction noise (averaged data).

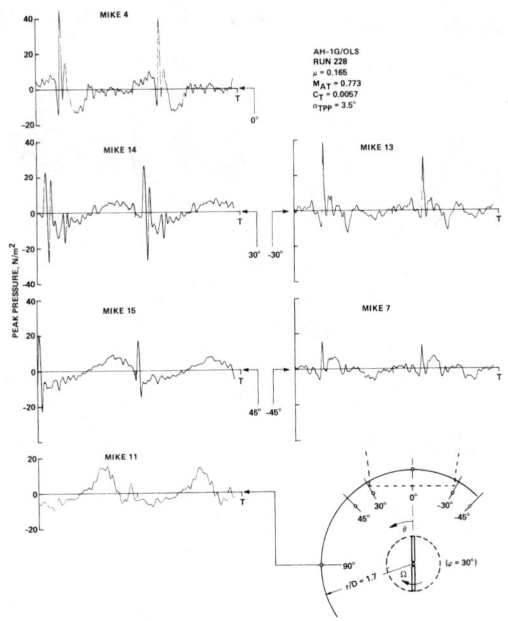

Figure 72. Lateral directivity of blade-vortex interaction noise (averaged data, $\phi = 30°$).

One of the best full-scale experimental attempts to document the aerodynamic flow field encountered by a helicopter rotor was the joint U. S. Army/Bell Helicopter Textron flight-test program called the *Operational Loads Survey (OSL)* (Shockey et al., 1976; Sakowski and Charles, 1976). In-flight aerodynamic, structural, and acoustic data were gathered throughout a wide range of operating conditions. The instrumented helicopter was an AH-1 Cobra, quite similar to the one used to gather the in-flight acoustic data discussed in the previous section. One minor difference is that the Operational Loads Survey blade sections were modified to incorporate extensive rotor-blade pressure instrumentation. Mounted on the upper and lower airfoil surfaces were 110 Kulite absolute-pressure transducers to measure the static-pressure distribution over the airfoil at five spanwise stations. The blade-pressure measurement system had a frequency response of 400 Hz. Thus, events occurring within approximately 5° steps in rotor azimuth were measured. As shown in Figure 73, in-flight microphones were also used to measure simultaneously the near-field noise at three fixed positions. Additional experimental details can be found in Sakowski and Charles (1976) and Shockey et al. (1976).

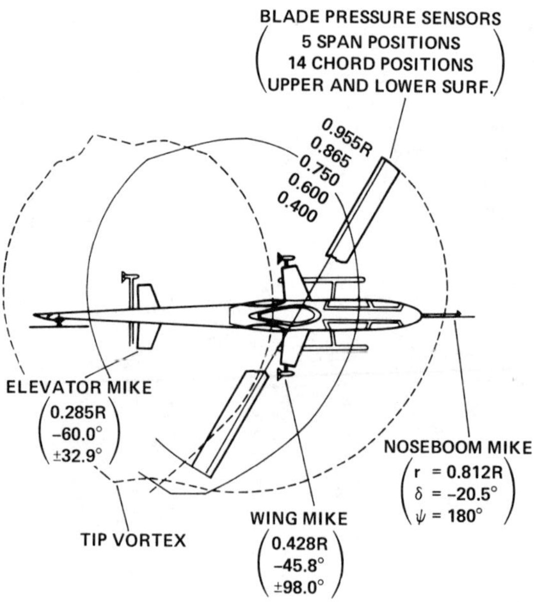

Figure 73. Simultaneous measurement of blade surface-pressure and noise.

Figures 74 and 75 present an example of the measured pressure fluctuations for a typical BVI descent case ($\mu = 0.147$, rate of descent = 200 ft/min) at several spanwise locations near the blade tip identify the aerodynamic BVI phenomena qute nicely. Rapid pressure fluctuations from $\psi \approx 45°$ to $\psi \approx 90°$ are seen corresponding to the BVI locations previously noted on the advancing side of the rotor disk. Relatively strong pressure fluctuations are also seen on the retreating side ($\psi \approx 270°$). These also generally correlate with the BVI locations indicated in the top view of the free-wake sketched in Figure 68. Unfortunately, the 400-Hz upper frequency limit on the blade pressure transducers apparently tends to smooth the pressure variations that are shown. More recently (Schmitz et al., 1982) similar model-scale tests of a scaled OLS rotor have confirmed these general findings and have indicated that the acutal BVI pressure transducer response does contain higher frequency data.

One of the most important findings of this experiment is shown in Figure 75 where the upper-surface pressure transducer responses are plotted versus azimuth angle for several chordwise stations. The BVI phenomenon is seen to be concentrated near the very leading edge of the blade chord. In fact, the dominant BVI pressure changes are confined to the first 10% of the blade chord. At 20% of the chord, very few BVI pressure changes are noted. The same pattern (see Sakowski and Charles, 1976) also exists for the lower-surface pressure transducers. In the more recent model-scale tests (Splettstoesser et al., 1983; Schmitz et al., 1982), these findings were substantiated, indicating that the aerodynamic source location for BVI noise is the very leading edge of the rotor blade. It was also noticed that the pressure disturbances on these model-rotor tests were *shock-like*, again indicating that the aerodynamic phenomena near the rotor's leading edge were quite sharp (nearly discontinuous). These experiments clearly show that the BVI is an impulsive-like phenomenon located near the leading edge of the rotor blade.

Simultaneous in-flight near-field acoustic measurements were also recorded during these experiments. As expected, the advancing-blade/vortex interactions radiated most of their energy forward and were measured on the right wing-tip and nose-boom microphones. The retreating-blade/vortex interactions radiated to aft directions and were measured on the left wing tip and elevator microphones. The largest acoustic pulses were recorded on the right-wing-tip and nose-boom microphones, even though Figure 74 shows that the

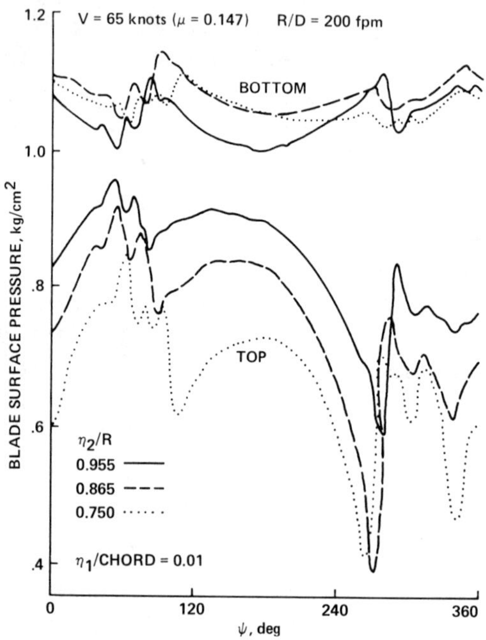

Figure 74. Example of blade surface-pressure history at different span positions (Shockey et al., 1976).

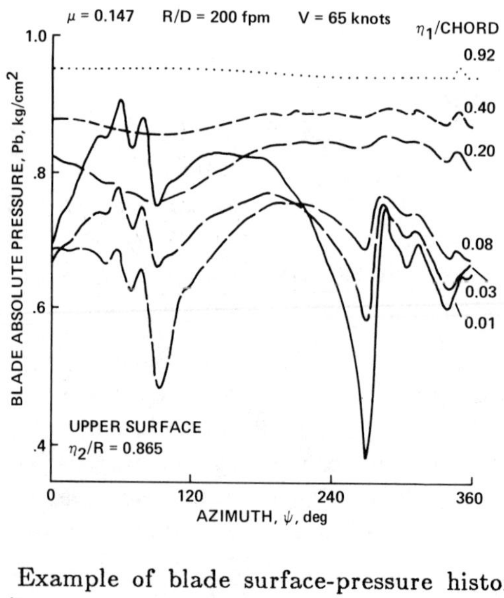

Figure 75. Example of blade surface-pressure history at different chord positions (Shockey et al., 1976).

advancing-blade interaction pressure variations are less than those measured on the retreating side. The conclusion to be drawn is that advancing-side BVIs are apparently *acoustically* more efficient than the retreating side BVIs.

The near-discontinuous nature of these measured pressure variations is supported by some earlier Schlierin photographs of advancing-side BVI taken by Tangler (1977) on a small model rotor in a wind tunnel. The photographs shown in Figure 76 are admittedly qualitative in many testing respects, because parameters such as thrust, blade design, and tip-path-plane tilt were adjusted to obtain clear photographs. In addition, the Schlierin system presents only two-dimensional representations of a three-dimensional phenomenon. Nevertheless, the BVI interaction shown is believed to be the $1\frac{1}{2}$ old (No. 3) *near broadside* encounter illustrated in Figure 76 which in a local sense is nearly two-dimensional. A lower-surface shock wave is seen in these photographs which begins near the blade leading edge and propagates forward and downward into the direction of the oncoming flow. Judging from these remarkable photographs, the compressible nature of the BVI phenomena at these typical full-scale tip Mach numbers is clear. The photographs also support the full-scale and model-scale experimental findings that the aerodynamic source location is the leading edge of the rotor blade.

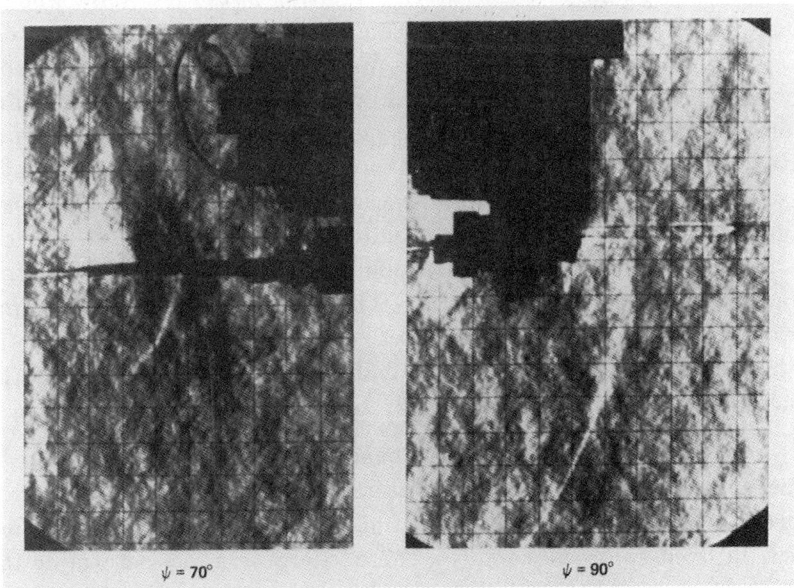

Figure 76. Shock propagation from blade-vortex interaction.

It is tempting to conclude that shock waves play a role in many of the blade-vortex interaction encounters on full-scale helicopters. However, the full-scale experimental evidence to support such a conclusion is sketchy and inconclusive at best. The measured full-scale in-flight acoustic data (Schmitz and Boxwell, 1976) have shown that the waveform exhibits a *shock-like* character. However, a discontinuous radiating shock wave similar to high-speed delocalization has never been quantified either in ground-based or in-flight acoustic measurements. What seems certain is that compressibility plays a significant role in the formation of the sharp pressure disturbance. On full-scale helicopter measurements to date, these disturbances radiate to the acoustic far-field as saw-tooth waves.

It has always been the acoustician's dream to be given accurate pressure measurements on the surface of the rotor blade and then to be asked to compute the acoustic pressure at some remote microphone location. Theoretically, equation (2.2) can be used to sum pressure disturbances to calculate the acoustic far-field. Such calculations were performed by Nakamura (1981) utilizing the full-scale OLS data with remarkably good results. However, these results were achieved after many of the experimentally observed phenomena previously described were carefully accounted for in the analysis.

The first assumption in this analysis was that only known surface-pressure variations contributed to the radiated noise (eq. 2.2). Of course, the noncompact nature of the aerodynamic and acoustic problems was realized and included in the acoustic formulation. However, the first attempt at just *plugging in* all of the interpolated blade-vortex interaction pressures totally failed. The acoustic prediction only accounted for the very low frequency loading noise. All higher frequency data were inadvertently filtered by the necessary interpolation between data points (Nakamura, 1981). The important lesson to be learned from this exercise is that just *plugging in* measured or calculated pressures to an existing computer code to calculate far-field acoustics is probably a waste of time unless the phenomenon of interest is carefully represented in the input data.

Representation of the BVI phenomena from the measured data was not a trivial task. A computer model was constructed which carefully fit the measured data and kept track of the interaction loci. In fixed-space coordinates, these interaction loci determine the *trace Mach number* and thus govern the relative efficiency of the BVI. Figure 77 from Nakamura (1981) shows some of the difficulties of constructing continuous BVI interaction loci from the measured data. All of the expected BVI pressure pulses were not measured at each radial station. Instead, it appeard as if the blade-tip region encountered fewer interactions on the advancing side than did the more inboard regions. These findings are probably a result of the nonplanar character of the BVI and are aggravated by the 400 Hz upper frequency limit of the pressure instrumentation. The intelligent estimates of the interaction structure (shown in Figure 77) were necessary to arrive at a successful BVI model.

Figure 78 shows the comparison of the measured OLS and computed acoustic waveforms for the right-wing microphone and nose-boom microphone. The general waveform, including the impulsive shape, is well predicted in both observer positions; however, the peak amplitudes are underestimated and the pulse widths overestimated. At the present time it is not known how much of this discrepancy is due to experimental inaccuracies, to numerical interpolation of the measured data, or to the omission of nonlinear terms in the governing equation (1.1). The wider pulse shapes of the theoretical predictions seem to suggest that the measured pressures did not have a high enough frequency response. New data recently taken on model-scale OLS blades are currently being analyzed to address these concerns.

Based on the model- and full-scale experiments discussed so far, it is strongly suspected that compressibility plays a role in the aerodynamics of BVI. For the near *broadside* BVI encounter depicted in Figure 8, the mean-flow velocity $(\omega R + V \sin \alpha)$ plus the local velocity induced by the vortex (v_i) can exceed sonic velocities over small regions near the leading edge of the rotor blade. Unfortunately, this local aerodynamic flow field is transonic and unsteady and thus governed by nonlinear equations. As in the high-speed problem, numerical solution of the governing equations is the only known method of capturing these effects.

At the present time, the local transonic flow field which exists during BVI is being addressed by several researchers. The problem has initially been simplified to a two-dimensional encounter with the vortex passing by a stationary airfoil. The governing equations are then numerically solved to yield the aerodynamic, as well as the radiated noise. A first attempt at such a simulation of BVI was recently performed in George and Chang (1983). Even though only the vertical upwash disturbance of the BVI encounter was considered, these preliminary results are encouraging. They reaffirm the

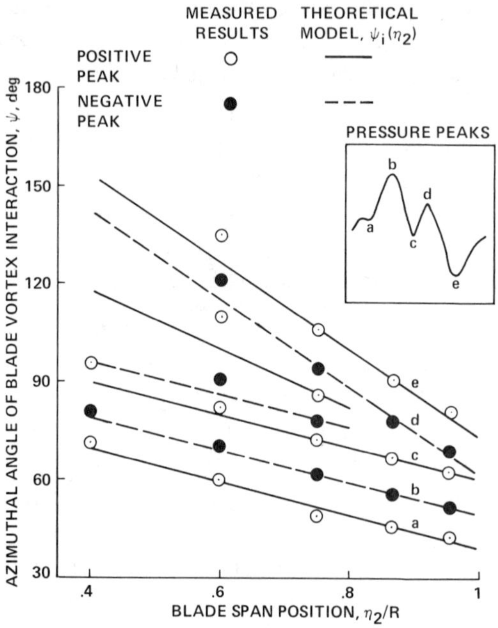

Figure 77. Modeling of blade-vortex interaction lines from measured results.

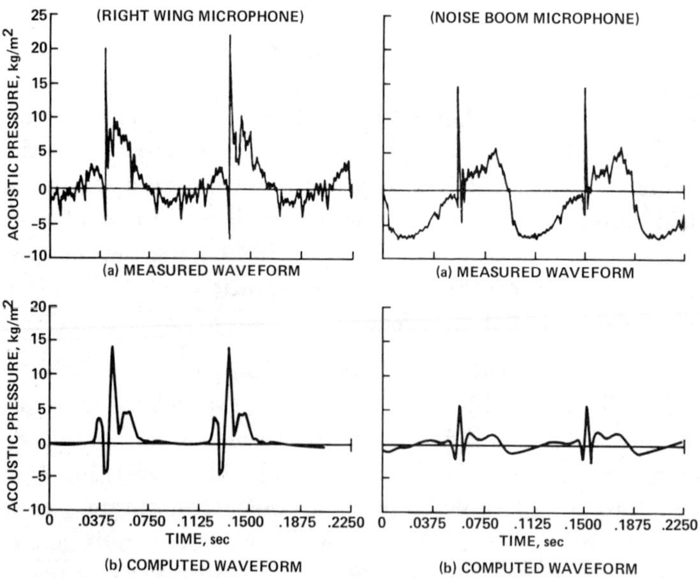

Figure 78. Comparison of acoustic waveform between measurement and computation.

importance of leading-edge blade geometry(Tangler,1977; Nakamura, 1981). When developed, they should be able to suggest leading-edge geometries that tend to minimize the airfoil pressure disturbances that can radiate to the acoustic far-field. Experimental work to support these efforts is also under way.

As mentioned previously, the BVI impulsive-noise phenomena are not as well understood as high-speed impulsive noise. Nevertheless, substantial progress is being made. Several areas in which improvements can and are being made (in some cases) are suggested in the following paragraphs.

The first attempt at scaling BVI impulsive noise (Schmitz et al., 1982), although generally good, is not completely satisfactory. There remain the problems of reduced amplitude and wider pulse shapes of the model scale data. Fortunately, more recent tests (Splettstoesser et al., 1983) have documented BVI under more controlled aerodynamic and acoustic conditions. The detailed analysis of these data (both blade pressure and radiated noise) should more fully answer the scaling question.

It would also be helpful to conduct some new experiments. In particular, laser velocity measurements on model- and full-scale rotors to define the vortex strength and its geometry during a BVI encounter at full-scale tip speeds are suggested.

Another very important question that needs to be addressed is: How applicable are these normal operating tip-speed results to lower tip speeds? From our experience, it appears that the mechanisms of the problems that we have discussed exist on full-scale helicopters with hover tip Mach numbers of 0.6 to 0.73. However, much of the model-scale BVI experiments that have been run are at lower Mach numbers (Hubbard and Harris, 1979; Leighton and Harris, 1983; Leverton and Taylor, 1966). How do these lower-Mach-number results relate to the phenomena which we have been discussing? New testing over a wide range of Mach number is needed.

The BVI research to date has focused on a very limited amount of measured acoustic data. In this summary, it has been shown that there are directions in which BVI impulsive noise is efficiently radiated. It remains to calculate these radiation directions and to correlate these findings with high-quality pressure-instrumented rotor blades run in an anechoic wind tunnel. The same high-quality pressure data could be used in equation (2.2) (similar to that used in Nakamura (1981)) to predict the radiated noise.

Acknowledgements

The support and contributions of our many friends and colleagues at the Aeromechanics Laboratory are gratefully acknowledged. The inspiration and support of Mr. H. Andrew Morse and Dr. Irving C. Statler, the technical contributions of Mr. Donald Boxwell and Mr. C. Rande Vause, and the experimental support of Mrs. Georgene Laub and Mr. Robert George made the entire effort over the past ten years possible. In addition, colleagues at the Research and Technology Laboratories, various universities and colleges, the helicopter manufacturers, the European and United States Army Research Offices, ONERA, DNW, DFVLR, NLR, and NASA also helped with direct as well as indirect support—bring the talents of many researchers all over the world to bear on the problem of helicopter impulsive noise.

References

[1] Boxwell, D. A. and Schmitz, F. H. "In-Flight Acoustic Comparison of the 540 and K747 Main Rotors for the AH-1S Helicopter," *Appendix, U. S. Army Aviation Engineering Flight Activity Report 77-38,* Edwards AFB, CA, 1979.

[2] ————————————. "Full-Scale Measurements of Blade-Vortex Interaction Noise," *J. Am. Helicopter Soc., 27, No. 4,* Oct. 1982.

[3] Boxwell, D. A., Yu, Y. H., and Schmitz, F. H. "Hovering Impulsive Noise: Some Measured and Calculated Results," *NASA CP-2052,* 1978; also *Vertica, 3, No. 1,* 1979.

[4] Caradonna, F. X. "The Transonic Flow on a Helicopter Rotor," *Ph.D. dissertation,* Stanford Univ., Stanford, CA, 1978.

[5] Caradonna, F. X. and Isom, M. P. "Numerical Calculations of Unsteady Transonic Potential Flow Over Helicopter Rotor Blades," *AIAA J., 14, No. 4* (1976), 482–88.

[6] Caradonna, F. X. and Steger, J. L. "Implicit Potential Methods for the Solution of Transonic Rotor Flows," presented at the 1980 Army Numerical Analysis and Computers Conference, Moffett Field, CA, 1980.

[7] Chattot, J. J. "Calculation of Three-Dimensional Unsteady Transonic Flows Past Helicopter Blades," *NASA TP-1721,* 1980.

[8] Farassat, F. "The Acoustic Far-Field of Rigid Bodies in Arbitrary Motion," *J. Sound vibration, 32* (1974), 387–405.

[9] Farassat, F. and Martin, R. M. "A Note on the Tip Noise of Rotating Blades," *J. Sound Vibration, 86, No. 3* (1983), 449–53.

[10] Farassat, F., Nystrom, P. A., and Morris, C.E.K., Jr. "A Comparison of Linear Acoustic Theory With Experimental Noise Data for a Small Scale Hovering Rotor," *AIAA Paper 79-0608,* Seattle, WA, 1979.

[11] Ffowcs Williams, J. E. and Hawkings, D. L. "Philos. Trans. R. Roc. London, Ser. A," *264,* May 1969, 321–42.

[12] George, A. R. "Helicopter Noise—State of the Art," *AIAA Paper 77-1337,* Atlanta, GA, 1977.

[13] George, A. R. and Chang, S. B. "Noise Due to Transonic Blade-Vortex Interactions," *Paper A-83-39-50-D000,* 39th Annual National Forum of the American Helicopter Society, 1983.

[14] Hanson, D. B. "Near-Field Noise of High Tip Speed Propellors in Forward Flight," *Paper 76-565,* Third Aero-Acoustics Conference, July 1976.

[15] Hanson, D. B. and Fink, M. R. "The Importance of Quadrupole Sources in Predictions of Transonic Tip Speed Propeller Noise," Spring Meeting of the Institute of Acoustics, Cambridge Univ., England, 1978.

[16] Hawkings, D. L. "Noise Generation by Transonic Open Rotors," *Research Report,* Westland Helicopter Limited, June 22, 1979.

[17] Hawkings, D. L. and Lowson, M. V. "Theory of Open Supersonic Rotor Noise," *J. Sound Vibration, 36, No. 1,* 1974.

[18] Hubbard, J. E. and Harris, W. L. "An Investigation of Model Helicopter Rotor Blade Slap at Low Tip Speeds," *AIAA Paper 79-0613,* Seattle, WA, 1979.

[19] Isom, M. P. "The Theory of Sound Radiated by a Hovering Transonic Helicopter Blade," *Poly-AE/AM Report No. 75-4,* Department of Aerospace Engineering and Applied Mechanics, Polytechnic Inst. of New York, 1975.

[20] ————. "Some Nonlinear Problems in Transonic Helicopter Acoustics," *Report No. 79-19,* Dept. of Mechanical and Aerospace Engineering, Polytechnic Inst. of New York, 1979.

[21] ————. "Acoustic Shock Waves Generated by a Transonic Helicopter Blade," *Paper 63,* 36th Annual National Forum of the

American Helicopter Society, 1980.

[22] Kittleson, J. K. "A Holographic Interferometry Techniques for Measuring Transonic Flow Near a Rotor Blade," *Paper B,* Ninth European Rotorcraft Forum, Sept. 13–15, 1983.

[23] Kittleson, J. K. and Yu, Y. "Holographic Interferometry Technique for Rotary Wing Aerodynamics and Noise," *Proc. of 1982 Army Science Conference, Vol. II,* 1982, 209–22.

[24] Leighton, K. and Harris, W. L. "A Parametric Study of Blade Vortex Interaction Noise for Two-, Tree-, and Four-Bladed Model Rotors at Moderate Tip Speeds: Theory and Experiment," *Paper A-83-39-50-D000,* 39th Annual National Forum of the American Helicopter Society, 1983.

[25] Leverton, J. W. and Taylor, F. W. "Helicopter Blade Slap," *J. Sound Vibration, 4, No. 3* (1966), 345–57.

[26] Lowson, M. V. "Helicopter Noise: Analysis-Prediction and Methods of Reduction," *AGARD Report LS-63,* 1973.

[27] Nakamura, Y. "Prediction of Blade-Vortex Interaction Noise From Measured Blade Pressure," *Paper 32,* Seventh European Rotorcraft and Powered Lift Aircraft Forum, Garmisch-Partenkirchen, Federal Republic of Germany, 1981.

[28] Nakamura, Y. and Azuma, A. "Improved Methods for Calculating the Thickness Noise," *NASA CP-2052,* 1978.

[29] Rutherford, J. "The Aerodynamics and Aeroacoustics of Rotating Transonic Flow Fields," *Paper A-83-39-50-D000,* 39th Annual National Forum of the American Helicopter Society, 1983.

[30] Sakowski, P. and Charles, B. "Noise Measurement Test Results for AH-1G Operational Loads Survey," *Vols. I and II, Report 299-099-831,* Bell Helicopter Co., 1976.

[31] Schmitz, F. H. and Boxwell, D. A. "In-Flight Far-Field Measurement of Helicopter Impulsive Noise," *J. Am. Helicopter Soc., 21, No. 4,* Oct. 1976.

[32] Schmitz, F. H., Boxwell, D. A., and Vause, C. R. "High-Speed Helicopter Impulsive Noise," *J. Am. Helicopter Soc., 22, No. 4,* Oct. 1977.

[33] Schmitz, F. H. and Yu, Y. H. "Theoretical Modeling of High-Speed Helicopter Impulsive Noise," *J. Am. Helicopter Soc., 24, No. 1,* 1979.

[34] ——————————. "Transonic Rotor Noise—Theoretical and Experimental Comparisons," *Vertica, 5* (1981), 55–74.

[35] Schmitz, F. H., et al. "A Note on the General Scaling of Helicopter Blade-Vortex Interaction Noise," presented at the 38th Annual Na-

tional Forum of the American Helicopter Society, Anaheim, CA, 1982.

[36] Shenoy, K. R. "Semiempirical High-Speed Rotor Noise Prediction Technique," presented at the 38th Annual National Forum of the American Helicopter Society, Anaheim, CA, 1982.

[37] Shockey, G. A., Williamson, J. W., and Cox, C. R. "Helicopter Aerodynamics and Structural Loads Survey," presented at the 32nd Annual National V/STOL Forum of the American Helicopter Society, Washington, DC, 1976.

[38] Splettstoesser, W. R., et al. "Model Rotor High-Speed Impulsive Noise—Parametric Variations and Full-Scale Comparisons," *Paper 83-39-53,* 39th Annual Forum of the American Helicopter Society, 1983.

[39] Sternfield, H. and Schaeffer, E. "An Investigation of Rotor Harmonic Noise by the Use of Small Scale Wind Tunnel Models," *NASA CR-166-338,* 1982.

[40] Sternfield, H., et al. "An Investigation of Noise Generation on a Hovering Rotor, Part II." *Report D210-10550-1,* the Boeing Co., Vertol Div., Philadelphia, PA, 1972.

[41] Tangler, J. L. "Schlieren and Noise Studies of Rotors in Forward Flight," *Paper 77, 33-05,* 33rd Annual National Forum of the American Helicopter Society, Washington, DC, 1977.

[42] Tauber, M. E. and Hicks, R. M. "Computerized Three-Dimensional Aerodynamic Design of a Lifting Rotor Blade," *Proceedings of the 36th Annual National Forum of the American Helicopter Society,* Washington, DC, 1980.

[43] Thibert, J. J. and Philippe, J. J. "Studies of Aerofoils and Blade Tips for Helicopters," presented at AGARD/FDP Meeting on Prediction of Aerodynamic Loads on Rotorcraft, London, England, 1982.

[44] White, R. P. "The Status of Rotor Noise Technology," *J. Am. Helicopter Soc., 25, No. 1,* Jan. 1980.

[45] Yu, Y. H., Caradonna, F. X., and Schmitz, F. H. "The Influence of the Transonic Flow Field on High-Speed Helicopter Impulsive Noise," *Paper 58,* 4th European Rotorcraft and Powered Lift Aircraft Forum, Italy, 1978.

Part III
Generation and Propagation

The Development of Contemporary Views of Flow-Tone Generation

William K. Blake and Alan Powell
David Taylor Naval Ship R&D Center
Bethesda, Maryland 20084

Abstract

It is now commonly recognized that there are basic elements required for the occurrence of any self-sustained flow tone. The flow must be intrinsically capable of generating orderly disturbances in response to some applied (small) stimulus. Such stimuli may be in the form of either pressure or velocity fluctuations to which the flow may respond. The primary ordered motions must thus generate secondary pressure or velocity disturbances. This generation also accounts for the mechanism of sound radiation. The pressure or velocity fluctuation must be transmitted back to the upstream boundary that initiates the primary flow disturbance and it must have an adequate magnitude and phase to provide a new initial condition that reinforces the primary motion. The development of all these notions will be traced from their beginning in the nineteenth century to the present for flow tones in general, and jet tones in particular.

It will be shown how these views have matured and that now there is a common physical basis for the generation of all jet and cavity tones which involve impinging shear layers.

1. Introduction

1.1 Brief Overview

Our views of the physical mechanisms of the generation of flow tones have evolved for well over 100 years and the evolution is really a mirror of the evolution of our understanding of the role of unsteady fluid mechanics in aerodynamic sound generation. This development progressed from a gradual recognition that the sound sensitivity of jets and the creation of various types of jet whistles were somehow related and that both phenomena involved the stability of parallel-flow fluid layers. This intimate connection of sound radiation, fundamental fluid mechanics, and creation of orderly flow structure by the appropriate juxtaposition of fluid boundaries were all recognized before 1900 through the efforts of Tyndall, Le Conte, and Rayliegh. Thus, many notions and beliefs that are fundamental to our physical understanding of flow tones have really been around for quite a while. The passage of time and the results of volumes of reported research have served to formalize the early views which were at the same time somewhat elementary and sophisticated. Elementary because many of the mathematical and experimental tools of the time were often too crude to make measurements over broad enough parameter ranges, and because the notions of these physical interconnections perhaps lacked sufficient formalism. Sophisticated because fluid mechanics and acoustics were considered not as separate fields of endeavor, but as a single subject area which had to be handled separately only to suit the analytical and experimental tools at hand. Also, sophisticated because the sensitivity of the existence of certain tones to details in the geometry of the flow and impinging surfaces was recognized for many types of flow tones.

Since 1900, our physical perception of flow tones have developed in an evolutionary manner. The first half century was a period of spasmodic observation of vortex structures and the conditions (i.e., ranges of velocity and geometry) for the existence of tones. It would appear that the prevailing views had lost some of the focus which could be found in the earlier work, that is, the intimacy of fluid mechanics and acoustics seems to have been less appreciated that it might have been. The role of hydrodynamic stability in the creation of embryonic disturbances from which would emerge the eddy structures that could be so readily observed was apparently not recognized until after 1950.

During the last 30 years, the contemporary view of flow tone generation emerged and now its connection with the early work is more clear. A recognition that eddy structures grow through an evolutionary process that begins with natural flow instability is now rather common. Furthermore, it is now taken for granted that flow tones are generated whenever these eddy structures are regulated through the appropriate selection of a reinforcing geometry, that the creation of these vortices is primarily responsible for sound generation, and that these phenomena are not restricted to flows at low Reynolds number. The interconnection of these features is still somewhat vague, and many of the limitations of analytical and observational tools which were apparent (and recognized) in the early work are still apparent, although much advancement is obvious. A few of these persisting problem areas are worth noting. Parallel flow stability models are inadequate to properly describe growing disturbances in diverging viscous flows, yet only the simplest techniques are available to account for divergence. Most analyses are linear, yet the creation of the characteristic vorticity formations is fundamentally nonlinear phenomenon. The validity of boundedness, or *Kutta* conditions, at edges typically used in analyses have a doubtful application at high frequency for finite-thickness shear flows. Shear layer dynamics in slit-edge gaps, and orifices, although coupled with and modified by the sound field produced have not been really comprehensively dealt with.

1.2 Organization of the Paper

The various types of flow tones which have received extensive investigation are illustrated in Figure 1. They are displayed in two groups: those requiring a shear-flow surface interaction with a simple fluid path feedback and those for which the motion is reinforced by an acoustic resonance or a flow entrainment involving solid surfaces. The various flow oscillators cover a broad range of geometries, but fall into these two easily identifiable classifications according to the fundamental flow involved. Most of the geometries shown involve the impingement of a shear flow, a jet, or a single shear layer on a surface which plays a key role in stablizing the oscillatory flow. Vortex shedding from a bluff body (not illustrated), either in free space or in a resonator, involves flow separation and this generally occurs without the stabilizing assistance of another body. Although certain axisymmetric jet modes may be excited by nozzle separation, the tones generated are not nearly so intense as in the case of bluff body vortex shedding. Thus, jet and cavity tones form one roughly-defined classification, vortex shedding from bluff bodies, the other. In this paper, we shall follow the development of work on subsonic and choked jet tones and, to a much lesser extent, cavity tones, as these phenomena now are regarded from a common physical basis that may be traced to the same shear layer dynamics. Vortex shedding will not be discussed.

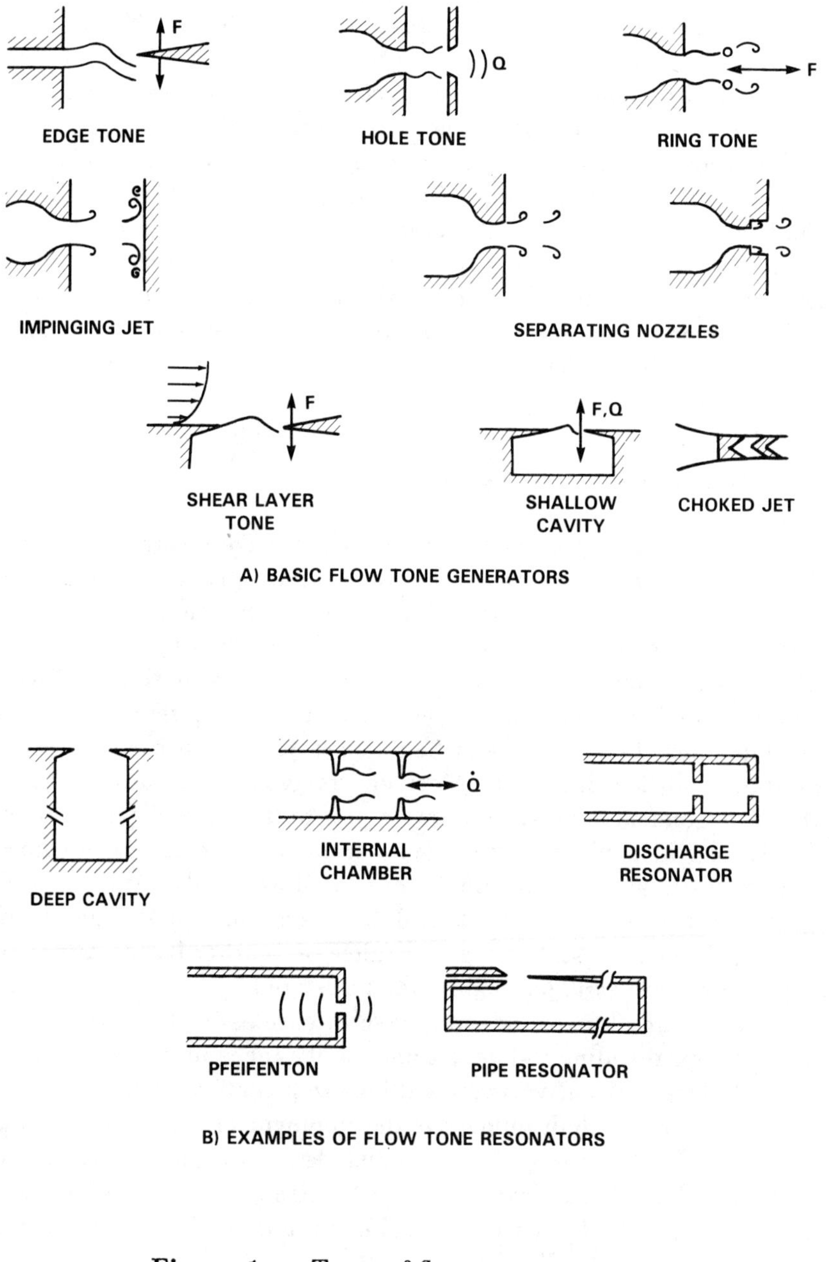

Figure 1. Types of flow tone generators

The paper will be separated into three eras. First the pre-1961 era will cover the evolution of alternative views of jet tones. Research in this era was marked by extensive flow visualizations and observations of tone frequencies. In this era, the importance of hydrodynamic stability as an initiating phenomenon was first emphasized. In the second era, 1961–1970, the foundations of contemporary views of flow tones are formalized. The importance of hydrodynamic stability, the development of a secondary disturbance at the impinging body,* the fact that this secondary disturbance determines both the sound field and self-excitation of the jet, all emerged in this era. Also, the development of these notions paralleled the advent of the *vortex sound* theory. Thus, the notion of flow-body impingement, vorticity creation, and sound radiation were all fundamentally connected. The present era since 1970 is marked by more sophisticated experimental and analytical tools, extension of flow visualization and measurements to higher Reynolds number, as well as much greater attention to the fluid mechanics of cavity tones. Also, there is a wider recognition of the importance of high Reynolds number jet tones, largely as a result of the emergent knowledge that ordered large scale eddy structures occur naturally in developing jet flows.

2. Early Developments in Flow Tones and Sound Sensitivity of Jets (Pre-1960)

2.1 Sound Sensitivity and Development of Large Scale Structures

The existence of jets sensitive to sound have been known for over a hundred years, Le Conte's investigation having been stimulated by observation of a gaslight flickering in response to music. Tyndall (1861) found that the flame was unnecessary, the fluid flow of the jet itself being sensitive to sound. Closer observation showed that, if a small steady jet of smoke-laden air issuing into the air is disturbed in some way by normally audible sound, then the jet

*In this context, the terms *primary* and *secondary* are meant to denote the notion of disturbances which are intrinsic to fluid dynamic stability (primary) and those resulting from the sources (secondary). In practice it is not possible to make such a differentiation.

becomes sinuous. This sinuosity grows along the jet when it is *sensitive* until filaments of smoke are carried from the wave peaks to form discrete vortices arranged in an alternate manner. Even when not deliberately disturbed, smoke jets generally show some development, attributable to random disturbances and sooner or later there is always transition to a random, turbulent motion downstream.

Lord Rayleigh (1896) became interested in both the theoretical and the experimental aspects of the subject. He found that an inviscid vortex sheet is unstable for all wavelengths of a sinusoidal disturbance, and the shorter the wavelength, the greater the growth rate. This was not in accord with the sensitive jet and so he proceeded to investigate the stability of inviscid parallel jet flows, of simple velocity profile. For a sinuous bending of the jet, relevant to sensitive jets, he found a region of marked instability bounded by the definite stability of short wavelengths (which correspond to high frequencies in the sensitive jet), and a fall to neutral stability at infinite wavelength.

In his theories Rayleigh made use of a parameter bf/V, where b is the streamwidth, f the frequency, and V the jet velocity. This parameter has come to be known as the Strouhal number but its importance in sensitive jets, as distinct from an instability theory, appeared to be lost until it was reemphasized by Powell (1953).

In continuation of Rayleigh's investigation, Rosenhead (1931) retained second-order terms in the inviscid vortex sheet equation to show the increasing importance of higher-order terms with the passage of time. Perturbation of a row of point vortices resulted in a waveshape sequence terminating in the rolling up of the pattern into concentrated vortices. Andrade (1941) used the electromagnetic analogy to inviscid vortex motion to discover the results of applying an initially sinusodial disturbance to two rows of point vortices, representing a jet. The patterns which developed have many similarities to those of real jets.

Until very recently, all theoretical investigations pertaining to the instability of jet flows had been based on Rayleigh's (1896) linearized equation, valid for inviscid flows, or its development including viscous effects, i.e., the Orr-Sommerfeld equation. As it stands, the latter equation is one presenting considerable difficulty, even though in its derivation the postulate is made that the basic flow is parallel. This simplification appears to be nearly unavoidable so one has to be content to study an artificial parallel jet flow, a *pseudolaminar jet*. In recent years, however, methods have been developed to account for the gradual spectral growth, e.g., Crighton and Gaster (1976), and numerical methods offer great promise (see Ohring, 1983).

The steady laminar-jet flow issuing from a linear slit in an infinite wall was discussed by Schlichting (1933) and later by Bickley (1937). The latter found that the viscous jet spreads like x, the velocity falling off like $x^{-1/2}$, where x is measured from the line source. The instability theory based on parallel flow clearly cannot be directly applied to such a flow; but the velocity profile can be used in the corresponding pseudolaminar jet.

Savic (1941) used Rayleigh's equation for a pseudolaminar jet with this velocity profile. His solution for neutrally stable sinuous and initially small disturbances yielded streamlines looking rather like those of stable vortex rows and like the smoke patterns photographed by Brown (1935), although these smoke patterns are not of streamlines but of streak lines of fully developed highly nonlinear vortex formations.

Although one might have expected Bickley's velocity profile to apply best to jets at low Reynolds numbers, it is unfortunate that this is just the region where the spread of the jet is most pronounced. Consequently, even prior to 1960, the parallel-flow assumption remained a somewhat doubtful proposition. On the other hand, at higher Reynolds numbers, the jet spreading is reduced but the velocity profiles differ significantly from those found by Bickley because of the real jet issuing from a slot of finite width (see Sato, 1960). Thus, too close a quantitative agreement between theory and experiment was not to be expected.

Another factor which may have caused and may yet cause some disparity is the manner in which the jet is disturbed. In theory, the stream is assumed perturbed along its infinite length at a specified instant with an infinitesimal sinusoidal velocity or displacement and its subsequent behavior as time passes is estimated. In practice, the jet was disturbed temporarily near the orifice, by a not necessarily small disturbance in which velocity and displacement are inseparable; the subsequent motion then develops in space. Although it had been fairly well demonstrated that sound waves do not normally disturb the two-dimensional jet along its length, e.g., Brown (1935), the possibility exists that the vigorous vortex development itself (i.e., spatial amplification) can materially effect the stream at the orifice. The existence of certain jet tones (Powell, 1951) lends support to the idea that such feedback may sometimes occur naturally, in which case it is possible a crucial element in the mechanism of the transition from laminar to turbulent flow.

Brown (1935) performed some classical experiments on the sensitive jet. As Rayleigh's work would lead one to expect, he found that the velocity of efflux, the frequency, and the slit width are all important, and that there was a frequency above which a given jet would not respond. He also noted that the amplitude of the disturbing sound was not immaterial. Actually, his results were found to display dynamical similarity provided the comparison is made on the basis of both Reynolds and Strouhal numbers, and they were clearly subject to influences from resonances in his apparatus.

In this and in an earlier work, Brown (1932) found well-defined and fixed frequencies where the jet response was maximum or minimum quite independently of the other variables.

2.2 Early Work on the Two-Dimensional Edge Tone

The edge tone is the sound resulting from the action of a jet emerging from a slit orifice and impinging upon a fixed cylinder (traditionally a wedge but called an edge), symmetrically placed and parallel to the length of the slit. The mechanism by which this tone is generated has been the subject of much controversy over the years since its discovery over a century ago.

The action was supposed to take place near the frequency for which the jet is most unstable with respect to sinuous disturbances. The plausibility of this had the fundamental backing of Lord Rayleigh's classical studies (1896) of jet instability and what was probably the most important of Brown's (1937, 1935) observations, that the edge tone occurs at frequencies for which the edgeless jet is most sensitive to acoustic disturbances (1937) (the latter aspect having not been free of controversy (see Brown, 1935 and Andrade, 1941). The amplified asymmetric disturbances interact with the edge, resulting in oppositely phased sources of sound on each side of it.

This was shown quite clearly in the schlieren visualization of the pressure field of the edge tone published by Powell (1953) and reproduced here as Figure 2. This illustration shows that pressure gradients exist across the jet, distorting it as it leaves the orifice with the distortions being amplified as they are convected to the edge. A feedback loop was developed to account for these interactions. Figure 3 shows the various elements of the feedback as it applied to the edge tone. A major feature of the loop was the inclusion of an acoustic path of communication between the slit and the edge. In that paper, it was suggested that for jets of high speeds the sources at the edge would be of the simple type—as indeed schlieren photographs indicated—but that at low speeds the stream would offer little or no impedance to them with a consequent degeneration of the pressure fields across it, resulting in a tendency for an overall dipole effect. Since for the traditional very low speed edge tone, the whole system is in the very near field where the speed of sound is inconsequential, the description *hydrodynamic* is as fully applicable as is *acoustic*. In other words, at these low speeds the actual sound radiation plays no essential role in the feedback mechanism (and so the validity of this theory does not hinge on such things as the audibility of the tone). This is certainly not the case at high speeds; so the description *acoustic* has the advantage of direct generality, though there had been a tendency to place it in opposition to what has been called by Richardson (1955) the hydrodynamic outlook.

Working simultaneously and quite independently of Powell, this hydrodynamic or vortex view was furthered considerably by Curle (1953). He followed F. Kruger's (1920) suggestion that the edge tone operated in the same frequency range as does the *jet-tone*, the weak sound of the edgeless jet, combined this with an idea hinging on Richardson's (1931) observation that another series of vortices spring from the edge of the wedge in sympathy with those already developed in the stream (these vortices having also been studied by Mokhtar and Youssel (1952)). Because of the variable crossflow at the edge, flow separation must occur there, this resultant free vortex in turn causing a crossflow at the orifice where a small new vortex is created. Then owing to the inherent instability of the jet flow, this new vortex gathers strength as it carried towards the edge, and so on. Thus Curle also emphasized the role of jet stability, and produced a frequency formula closely analogous to Powell's (1953); the stage having to be chosen by reference to experimental data. The nature of the sound source was not dealt with; the view that it is a purely incidental side effect to the mechanism of self-excitation was also held by Mokhtar and Youssef (1954) and Lenihan and Richardson (1940).

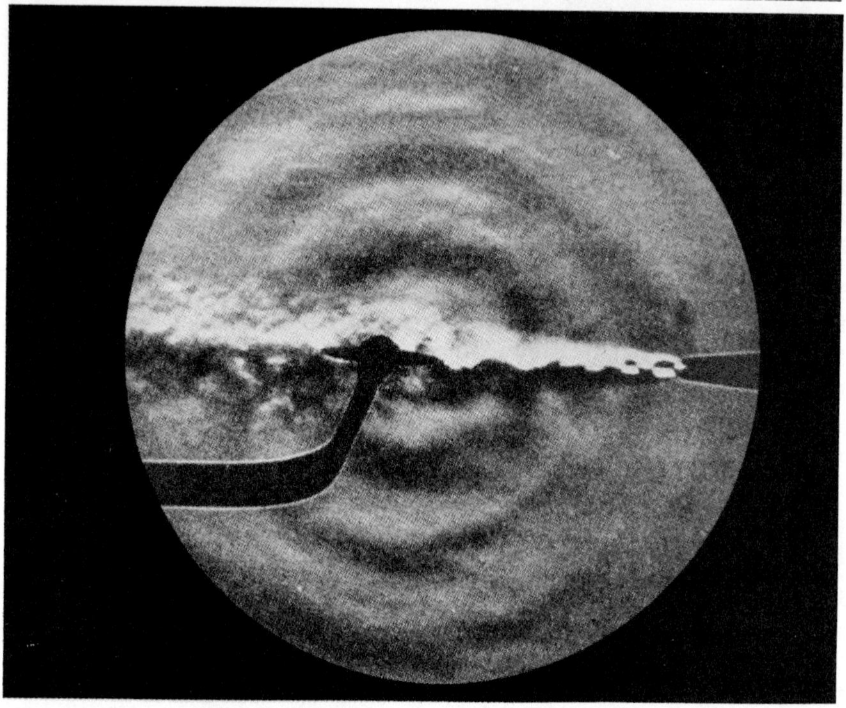

Figure 2. Schlieren photograph of flow structure and sound wave fronts of a high speed edge tone. Knife edge perpendicular to jet.

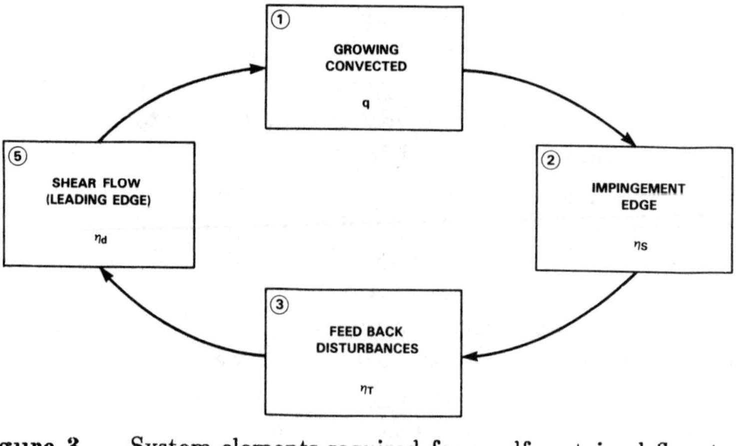

Figure 3. System elements required for a self-sustained flow tone.

After 1953, controversy about the nature of the edge tone persisted. Nyborg (1954) gave his dynamical theory following a series of observations on edge tones with and without resonators (Nyborg et al., 1952, 1953). This is also a feedback mechanism, but the jet is supposed to act as if it were composed of a line of unconnected discrete mass particles acted on by an uniformly distributed lateral force field, this being associated with the lateral displacement of the particle trajectory at the edge. With a very elementary form for this force, the resultant integral equation yielded a set of eigenfrequencies bearing ratios to each other very much like Brown's empirical values, and particle trajectories not unlike the stream deformation for certain edge tone motions occuring in water (Buoyoucos and Nyborg, 1954). Though the viscosity has been deliberately varied (Brackenridge and Nyborg, 1961), the basic nature of this model-remarkable for its total omission of all hydrodynamical considerations had been retained for still future embellishments (Brackenridge, 1960). Finally Gross (1959) argued that a feedback of the pressure (held attributable to a volume distribution of quadrupoles) is much effective than a vortex field and that Nyborg's theory might be a valid first approximation except that at certain speed the fluid dynamics must be taken into account.

In a review paper, Powell (1958) made the following suggestions relevant to edge tone theory. First, at low speed the sound field should be of dipole nature directly attributable to the fluctuating force exerted by the stream upon the edge, according to Lighthill's (1952) theory of aerodynamic sound production when expressed in the form by Curle (1953). Second, stress was laid again upon essential role nonlinearity in connection with the amplitude of the steady edge tone. Third, a new parameter connected with the necessary amplification of the stream disturbances was introduced providing a *gain criterion*. It was also suggested that for the low-speed case there really exists no essential conflict between the so called acoustic and *hydrodynamic* theories, but that Nyborg's is best applied when the jet stream instability is very weak (although even there the disturbance convection speed—and consequently the frequency—is clearly incorrect). More generally, it was questioned whether or not any connection existed between the dominant eddy formations in turbulent jets and the creation of vortex structures form instabilities in laminar flow jets.

2.3 Other Jet Tone Phenomena

A. Hole Tone

Sondhauss (1854) first reported on the hole tone and found that the tone frequency f_s increased with increasing jet velocity V_j, and the decreased with increasing distance L between orifices. Rayleigh later discussed this phenomenon in its relation to sound sensitive jets. The sound was though to be due to periodic jet flow through the second orifice. The internal edge of the downstream plate must be sharp, and the hole must not be too small. From the observed uniformity of the sound field, he (correctly) deduced that the disturbances at the second plate were "$\cdots$ *thrown back, probably with the speed of sound, to the first plate* $\cdots$," where they gave rise to further jet disturbances, which then were amplified by the unstable jet as they progressed downstream. This apparently was the first hint that jet-edge systems were hydrodynamic oscillators controlled by *acoustic* feedback. From dynamical considerations, he concluded that according to Sondhauss' law, the number $S_D = fD/U_J$ (Strouhal's!), was a function only of the Reynolds number $R_D = DU_s/\nu$ and the dimensionless-plate distance (ν is the kinematic viscosity and D is the jet-orifice diameter). He noted the sensitivity of the system to reflections from downstream of the second plate and commented that the tone might be augmented by a reflector when the "*distance is an (even or odd) multiple of the half wavelength.*" It is not clear from this statement whether both even and odd were meant, or whether the issue was unresolved.

B. Pipe Tone, Pfeifenton

When an orifice plate with a sharp edge is used to terminate a pipe a jet tone is emitted due to the periodic flow separation at the lip which induces an oscillatory flow from the pipe. An extensive review of the 19th century work done by Mason, Strohual, Kruger, and Sondhauss has been given by Anderson (1952). The tone emitted back into the pipe has been given the special name *Pfeifenton* by Strouhal. The sound generated to the outside field has been called *jet tone* by Richardson (1947). In a series of researches between 1952 and 1955 Anderson evaluated the pipe tones generated over a wide range of Reynolds numbers and orifice geometries. These will be discussed more fully in Section 3.3.

von Gierke (1950) studied various jet-edge geometries and shows that there was a relationship between jet, hole, and slot tones; the slot-tone system is the planar version of the hole-tone system. For the hole tone, he confirmed Rayleigh's observation that the sound was uniform over a wide angle. The Strouhal number was not used by him explicitly in presenting his data, although he mentions it in the form used by Rayleigh. His hole-tone apparatus had a sharp-edged orifice, so separated flow may have occurred within the nozzle. In two sketches of his equipment, he shows the hole in the second plate to have a level that faces the jet in one case and faces away in another. A cursory examination of plate reversal showed that there was a measurable change in frequency and sound pressure; this was probably due to the change in the edge distance from the orifice.

C. Choked Jets and Screech

As the velocity through a slit nozzle is increased the propensity for edge tone generation becomes less apparent so that tones may be generated only under relatively controlled situations. Accordingly, the acoustic emissions are less intense and broadband in nature. When the velocity in the nozzle becomes sonic, i.e., at the above the critical pressure, the characteristics of both noise and the jet structure change markedly. Superimposed on the broadband sound is an acoustic tone which is associated with the existence of a stable cellular pattern of expansion and compression zones along the jet axis. A schlieren photograph of such a pattern induced in a two-dimensional jet is shown in Figure 4. These cells are formed from periodic compression and expansion zones in the fluid as it passes down the jet axis; the downstream extent of this pattern is idealized in the upper part of Figure 4 as a series of conical wedges which evolve into the cells. The process is initiated at the nozzle exit as jet disturbances initiate an oscillation of the Mach lines originating there. Although generally studied in two-dimensional jets, the phenomenon occurs in high speed circular jets. Generally three to five cells occur until they are dispersed by downstream turbulent mixing. In the case of circular jets as many as twelve cells have been observed. The number of cells formed, and their axial dimension, is dependent on both pressure ratio and diameter.

The conditions required for the existence of these screech tones and, as well, an explanation of the directivity of the sound emitted was given by Powell (1951, 1953b, c, d). These conditions will be quantified in Section 3.4 so that only the essential qualities will be described here. The sound sources are stationary with respect to the jet exit and the source of oscillation arises from the passage of orderly vortex structures through the cells. Accordingly, the strongest tones are generated when the phases of the individual pressure radiated form each cell is such that the contributions best reinforce each other at the nozzle exit. These feedback pressures drive the jet shear layer producing the aerodynamic disturbances which *pulse* the shock cells. These sound waves are clearly visible in the schlieren photograph of the slit jet in Figure 4. The sound on opposite sides of the jet is 180° out of phase. The effective source center depends on conditions and may be situated three to four cells downstream. This location will be governed not only by geometrical aspects of acoustic reinforcement but also by the weighting of source strengths of adjacent cells down the jet axis. Near the nozzle the source strength will be relatively small, as the jet instabilities are still growing there, and a few cells downstream the strength will diminish because of the generation of disorder and decorrelation induced by turbulence. The most important acoustic sources are included between these extremes.

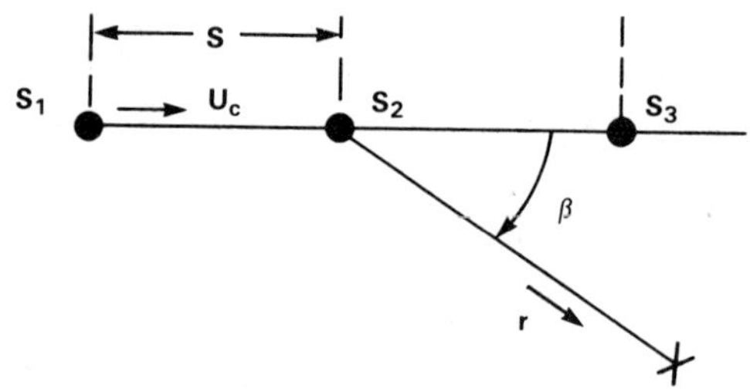

(A) IDEALIZATION OF SOURCE GEOMETRY

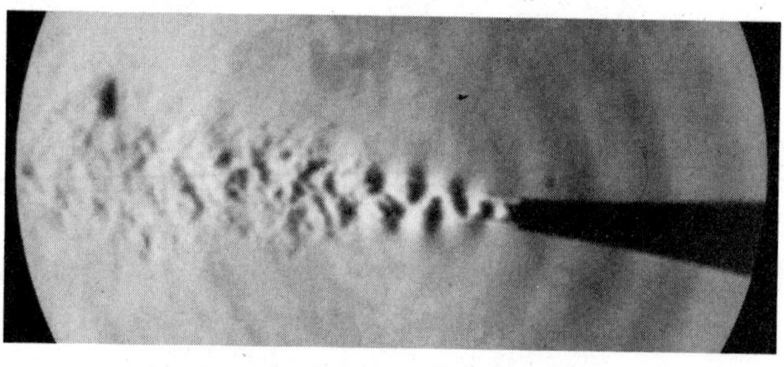

(B) SCHLIEREN PHOTOGRAPH OF CHOKED JET AND UPSTREAM-PROPAGATING SOUND WAVES, R = 3.05

Figure 4. Shock cell patterns and source geometry for the sound field of a choked rectangular jet with breadth normal to page.

The directivity of the sound is determined by the acoustic wavelength, the cell dimension along the axis of the jet, and the convection speed of the jet vortices through the shock cells since this determines the relative phases of cell undulations. The primary directivity of sound, i.e., at the fundamental frequency, is upstream back to the jet exit. This directivity is required to determine the coupling of the sound to aerodynamic disturbances in the exit plane of the jet.

2.4 Summary

The early era work on jet tones disclosed a number of features which relate closely to the contemporary state of understanding of flow tones:

- The sound sensitivity of jets was observed by Tyndall and Le Conte to affect not only the vortical (*fringed by luminous tongues*) structure, but also the jet spreading (*shrinkage*) and this behavior depended on the frequency of the sound.
- The stage character of hole, edge, and pipe tones was well known, but not altogether precisely quantified.
- The Rayleigh analysis of the jet showed the wave number range for unstable operation to be between $0 < k_c D < 4$, and for shear layers between $0 < 2k_c\delta < 2$, where 2δ is the shear layer thickness, $k_c = \omega/U_c$, $\omega = 2\pi f$, and U_c is the phase velocity of disturbances.
- The importance of jet spreading in controlling the behavior of instabilities was well regarded.
- The dipole nature of the edge tone had been observed, but it was not widely recognized as such.
- A sort of vortex pairing was noted by von Gierke (1950) in referring to the rotational symmetrical rolling up on the shear layer into structures which approach circular vortices (which occurred at $R_D \simeq 2600$) and to the unification of small vortices to form larger eddies.
- A general physical similarity exists between all forms of jet tones which involve the interconnection of both sound-sensitive instability modes and the location and directivity of sound sources. Moreover, strong tones may be generated

when the sound source location (effective or physical) is fixed relative to the position of the exit plane, as a fixed geometry is essential to the establishment of necessary feedback phase.

3. Foundations of Contemporary Views—Jet Tones (1960–1970)

3.1 Requirements for Self-Sustained Oscillations—the Edge Tone

Although the first elements of the requirements for two-dimensional self-sustained edge tones were set down by Powell (1953), it was not until some time later Powell (1961) that each element of the feedback circuit was clearly quantified. The 1953 paper set a format for determining, either by theory or measurement, each stage of the feedback circuit. Thus, building on the first paper, Powell (1961) determined the dipole strength at the edge and the loop gain for entire ranges of jet-edge separations and jet speed of any particular laminar flow jet-edge tone.

Three overall requirements must be met in any self-sustained oscillation. First, the primary shear flow must be fundamentally capable of generating orderly unsteady motion, i.e., it must be unstable to small environmental disturbances. Second, the unsteady motion created by the primary flow must be capable of generating a second disturbance which can propagate upstream back to the location at which the primary disturbance was generated. This means that secondary disturbance must have sufficient intensity. Third, the phase between the primary and secondary (feedback) disturbances, at the point of initiation of the primary disturbance, must be such that reinforcement (rather than cancellation) occurs. The total phase around the loop shown in Figure 3 must include all the propagation times and any phase shifts which are appropriate to the physical mechanisms involved. To emphasize the importance of these elements in general, we shall review the various elements of the two-dimensional jet-edge tone process as illustrated in Figure 5. The elements are numbered consistently in Figures 3 and 5 and they shall be discussed in order based on these early papers. In the following, we shall be examining the mechanics of the velocity fluctuation

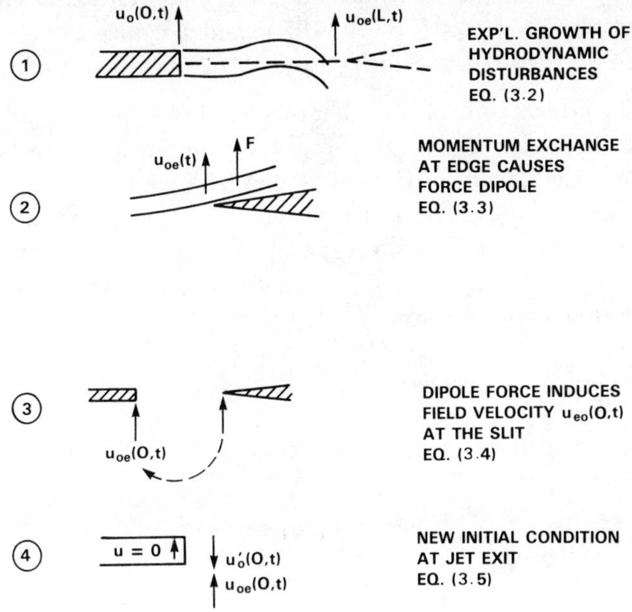

Figure 5. Illustrations of events which control the operations of edge tones.

directed perpendicular to the jet axis and all disturbances are pure-tone of frequency $f = \omega/2\pi$ and so the $e^{-i\omega t}$ time dependence is suppressed.

In the growth process from 1 to 2, the hydrodynamic disturbances $u_{0_e}(x,t)$ may be considered to a first approximation to grow according to

$$u_{0_e}(L,t) = u_0 \exp\left(\int_0^L \alpha_i \, dx\right) \exp\left[i \int_0^L (\alpha_r \, dx) - \omega t + \phi\right] \tag{3.1}$$

where $k_c = \alpha_r - i\alpha_i$ is the complex propagation wave number of the disturbance in the jet. It changes with x in accordance with the spreading of the jet. Although equation (3.1) represents the result of a linear stability analysis of the shear flow, it is not necessary to be this specific. All that is necessary to say at this point is that the disturbance $u_{0_e}(L,t)$ at the edge and the initial disturbance $u_0(0,t)$ at the jet opening must be related as

$$\frac{u_{0_e}(L,t)}{u_0(0,t)} = q e^{i\phi_N} \tag{3.2}$$

where $q \geq 1$ and ϕ_N is a phase angle. This generalization was in fact used at the outset (Powell, 1953) and permits a view of either wave like or vortical motions (Powell, 1962) in the slit-edge gap.

The interaction of the shear layer flow with the edge, interaction 2, depends on the geometries of the shear layer (since this determines the details of the entire disturbance field in the presence of the edge, say $u_{0e}(L,t)$) and the edge. The geometry thus controls both the magnitude and the type of the flow-edge interaction. In the edge tone, this interaction is manifested as a force, F exerted on the fluid that depends on $u_{oe}(L,t)$ as

$$F(t) = a_1 \rho_o U_J W \delta_o u_{0e}(L,t) \tag{3.3}$$

where W is the width of the jet, δ_o its height, and U_J its exit velocity, and

$$a_1 \simeq 3,$$

or

$$F(t) = \eta_s u_{0e}(L,t)$$

This force can be considered to generate the secondary field, which again depends on the geometry of the fluid region. Since the fluid region is constrained by no other surfaces but the free edge, Powell (1961) and Chanaud and Powell (1962) show that in event 3 of Figure 5 this force generates a velocity of the fluid across the jet u_{e0} which drives the jet. The new field velocity (which is directed parallel to the force on the fluid and which is really the direct result of the induced pressure gradient across the jet) is related to F by the well-known relationship for the *near-field* of a dipole, i.e., when $L \ll \lambda_o$, yet still far enough away to be outside the influence of local aerodynamic disturbances,

$$u_{e0}(0,t) = a_2 \frac{F}{4\pi} (\omega \rho_o L^3)^{-1} e^{+i\pi/2}$$

or

$$u_{e0}(0,t) = \eta_t F(t) \tag{3.4}$$

where a_2 generally lies somewhere between unity and 2 depending on the relationship among the jet-edge distance L, the jet width h, and the acoustic wavelength λ_o. The most important feature of equations (3.3) and (3.4) is the resultant $\pi/2$ phase shift between jet disturbance, $u_{oe}(L,t)$ incident on the edge and the induced disturbance $u_{e0}(0,t)$ on the leading edge of the jet. The sound pressure generated by the edge tone is wholly controlled by this force dipole and it accounts for the $\cos\theta$ directivity of sound about the force axis $\theta = 0, \pi$ as is apparent in Figure 2. The independent measurement of sound pressure magnitude, and directivity and of the edge force presented by Powell (1961) represent an early unambiguous confirmation of the Curle (1955) result for the force dipole on a rigid surface.

The initial condition on disturbances at the jet orifice is that $u = 0$ there. Thus the combination of the incident secondary disturbance $u_{e0}(0,t)$ and the hydrodynamic disturbance in the jet that it induces $u_0'(t)$ must be such that $u = 0$ just at the jet exit, thus

$$u_0'(t) + u_{e0}(0,t) = 0$$

so that the final condition in the loop at event 4 is

$$\begin{aligned} u_0'(t) &= -u_{e0}(0,t) \\ &= |u_{e0}(0,t)|\, e^{+i\pi} \end{aligned} \tag{3.5}$$

or

$$u_0'(t) = \eta_d u_{e0}$$

Combining equations (3.2) through (3.5) we can find the overall loop gain as

$$\frac{u_0'}{u_0} = \frac{u_{0_e}(L,t)}{u_0(0,t)} \cdot \frac{F(t)}{u_{0_e}(L,t)} \cdot \frac{u_{e_o}(0,t)}{F(t)} \cdot \frac{u_0'}{u_{e_o}(0,t)}. \tag{3.6a}$$

or

$$\frac{u_0'}{u_0} = q \cdot \eta_s \cdot \eta_t \cdot \eta_d \tag{3.6b}$$

In order for the feedback and initial disturbance to reinforce they must be in phase, i.e.,

$$u_0', u_0 = 2\pi n$$

where n is an integer. Summing the phases around the loop, we obtain the condition on ϕ_N of equation (3.2) that

$$\phi_N = 2\pi\,(N + \frac{1}{4}) \qquad \text{where } N = 1, 2, 3, \ldots \tag{3.7}$$

The physical interpretation on ϕ_N has been that it represents the number of complete hydrodynamic waves which are present between the slit and the edge. In the context of equation (3.1) ϕ_N is just the L-dependent phase, or

$$\left(\frac{L}{\overline{\lambda}}\right)_N = N + \frac{1}{4} \tag{3.8}$$

where $\overline{\lambda}_N$ is average over L of the hydrodynamic wavelength which may be defined as

$$2\pi\left(\frac{L}{\overline{\lambda}}\right)_N = \int_0^L \alpha_r(x)\,dx \tag{3.9}$$

for the Nth mode of hydrodynamic oscillation.

All of the above notions are, by this time, fairly well known. Reviewing them here for this very special case of the edge tone emphasizes many issues which are still of considerable importance in generally determining the existence and stages of flow-tones. The feedback condition is determined by phase changes not exclusively caused by propagation times but also by delays associated with the operation of disturbances at the impinging surface and at the originating edge.

An important generalization which may be considered and which alters the form of ϕ_N regards the relationship between $u_{e0}(0,t)$ and F. Suppose that L/λ_o is no longer much less than unity. Then since the jet orifice will no longer be in the hydrodynamic near field of the edge dipole force, we must account for the acoustic propagation of the disturbance. Accordingly, instead of equation (3.4), the propagating component of $u_{e0}(0,t)$ evaluated at the slit outlet is

$$u_{e_o}(L,t) = \frac{a_3}{4\pi}\frac{F}{\rho_o c_o L^2}\cdot e^{i2\pi L/\lambda_o} \tag{3.4a}$$

so that replacing equation (3.8) we have

$$\frac{L}{\lambda_o} + \left(\frac{L}{\overline{\lambda}}\right)_N = N + \frac{1}{2}, \qquad N = 1, 2, \ldots \tag{3.8a}$$

The difference between equations (3.8) and (3.8a) lies in the appearance of the acoustic propagation time delay, given here by the term L/λ_o, and the fact that the far-field velocity is an acoustic particle velocity which is in phase with the force on the $y = o$ plane and far enough from the edge such that $L/\lambda_o > 1$.

The Strouhal number of the tone, fh/U, is given in the first case by

$$S_h = \frac{fh}{U_\infty} = \left(\frac{\overline{U}_c}{U_\infty}\right)\frac{h}{L}\left(N + \frac{1}{4}\right) \tag{3.10}$$

As pointed out by Powell (1961) S_h should not be regarded as necessarily proportional to $N + \frac{1}{4}$ or to L/h because one should expect the convection velocity $\overline{U}$ to be dependent on both frequency and the slit-wedge distance. This caveat applies to *all* types of shear-flow induced tones.

The general form of equation (3.10) may be deduced from a comparison of equations (3.8) and (3.8a) to be

$$S_h = \frac{fh}{U_\infty} = \left(\frac{U_c}{U_\infty}\right)\cdot\left(\frac{h}{L}\right)\cdot\left(\frac{N+\gamma}{1+\overline{M}_c}\right) \tag{3.11a}$$

or

$$\frac{fL}{U_\infty} = \frac{U_c}{U_\infty}\cdot\left(\frac{N+\gamma}{1+\overline{M}_c}\right) \tag{3.11b}$$

where the average convection Mach number is

$$\overline{M}_c = \overline{U}_c/C_o \tag{3.12}$$

and the factor γ is a fraction which may be added (as above) or subtracted from the stage integer N. Equation (3.11) is of a form which quite generally applies to all flow-tones that do not involve as surface vibration.

With respect to edge tones, the factor γ has received a great deal of attention and in the past was the subject of some controversy.

As we shall see in Section 4.3 the question of measurement of γ is now being pursued for the case of cavity tones, with much the same sort of discussion that occurred previously. Attempts by Karamcheti and Bauer (1963) and later Karamcheti, Bauer, Shields, Stegan, and Wooley (1969) to measure the phase ϕ_N in equation (3.2) to give stage orders that differ by exactly $N + \frac{1}{4}$ were largely unsuccessful. They correctly noted that Strouhal numbers for a given stage could be only roughly predicted using equation (3.10). This caveat had been given originally by Powell (1961) who later (1965) confirmed that empirical values of $\phi_N/2\pi$ could take on values such that γ in equation (3.10) ranged roughly from 0 to $\frac{1}{2}$. Both Powell (1965) and Karamcheti et al. (1969) found that phases deduced from hot wire traverses along the slit-edge gap would give different values depending on the mean position of the shear layer relative to the traverse locus.

Finally, in the work of Powell and Unfried (1964) the rectangular jet edge tones were examined as a function of wedge angle with from 15° to 180° included angle. The 180° angle is just the rectangular jet impinging on a flat surface. For a given velocity, it was found that for angles of 15° through 90° the dependence on slit-edge distance was very similar. For the flat surface, however, somewhat greater separation was required for the onset of a stage 1 tone.

3.2 The Role of Hydrodynamic Instabilities

Perhaps the major distinction of the second edge tone paper (Powell, 1961) is the identification of the importance of shear layer dynamics in generating the edge tone and in controlling the persistence of stages. At about the same time Sato (1960) and Sato and Sakao (1964) experimentally and analytically examined the modes of a two-dimensional jet at low Reynolds number. The antisymmetrical mode of such jets was found to be less stable and more sensitive to acoustic excitation from the side than the symmetric mode. Such a sensitivity implied that the jet responded well to transverse velocities (or pressure gradients) thus providing a link between the pressure differential across the jet disclosed in the Schlieren photographs such as shown in Figure 2.

This point was further elaborated on by Chanaud and Powell (1962) who determined the phase and amplitude relationship between an imposed transverse velocity and the appearance of vorticity at the slit of a two-dimensional jet. They further found that instability could be observed down to very low excitation Strouhal numbers as well as being bounded by high values as expected from stability theory and the characteristics of neutral stability were dependent on the amplitude of excitation.

When compared to other types of shear layers, it can be seen that the antisymmetric mode of a two dimensional jet is theoretically more unstable than any other. Figures 6 and 7 show a range of shear layer profiles together with their linear stability characteristics. In all cases the Reynolds number may be taken as virtually infinite. The diagram in Figure 7 gives the relationship between the α_r and α_i that appear in equation (3.1). As shown by Sato (1960), as the gradient of mean velocity across the jet, $d\overline{U}/dy$, increases at the edges, the jet becomes less stable. In any case the range of $\alpha_r\delta$ for which the α_i is positive (denoting exponential growth) is limited to $0 < 2\alpha_r\delta < 4$ (where 2δ is a measure of jet width). Such pass bands of unstable wave numbers are typical of all the shear layers with finite shear layer thickness 2δ. Sato (1956) and Michalke (1965) found that the parallel-flow shear layer (C) in Figure 6 had an instability range between $0 < 2\alpha_r\delta < 1$ to 2. It was noted in Section 2.4 that this qualitiative behavior had been calculated for both jets and shear layers by Rayleigh.

In postulating the onset of stages of the edge tone, Powell (1961) reasoned that onset and cut off of a given stage for which $(L/h)_N$ is a constant was determined by a range of wave numbers $(2\alpha_r\delta)_\ell$ to $(2\alpha_r\delta)_u$, say, for which the jet was particularly responsive to excitation. This sensitivity would be achieved by large enough values of $\alpha_i L$. It is to be noted that 2δ is not the slit height h, rather it is a measure of the thickness of the laminar jet. This thickness, as noted previously, has long been recognized as dependent on distance from the jet slit and on the mean velocity. The Strouhal number S_h and $(2\alpha_r\delta)$ may be related by

$$S_h = \frac{1}{2\pi}\left(\frac{h}{2\delta}\right)\frac{c_r}{U_J}\left[2\alpha_r\delta\right] \tag{3.13}$$

According to equation (3.10), for a given stage N as L/h increases from 0 the required upper limit value consistent with $(2\alpha_r\delta)_u$ for the onset of the tone establishes a lower limit on L, i.e., when

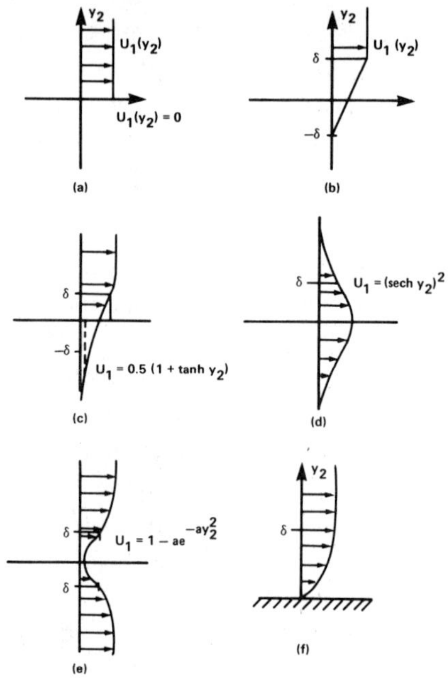

Figure 6. Classical types of two-dimensional shear flows, (a) discontinuous, (b) linear, (c) hyperbolic tangent, (d) jet, (e) wake, and (f) laminar boundary layer.

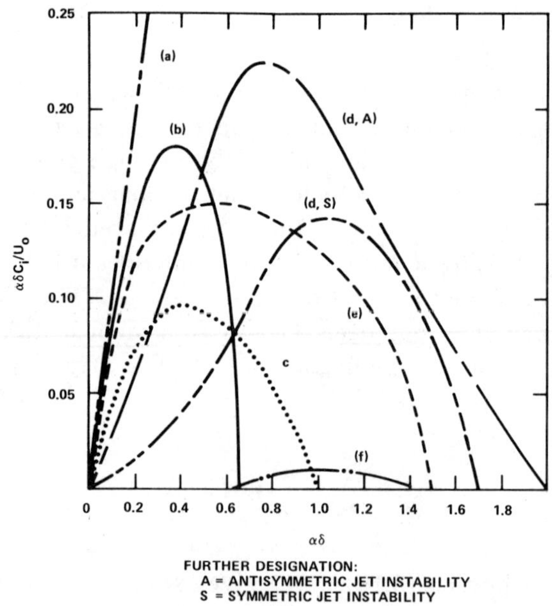

Figure 7. Theoretical growth rates based on temporal instabilities in the two-dimensional shear layers of Figure 6.

$$\left(\frac{L}{h}\right)_{\min} \simeq \frac{2\pi(2\delta/h)(\overline{U}_c/c_r)(N+1/4)}{[2\alpha_r\delta]_u}$$

the growth constant $\alpha_i L$ is large enough for the necessary system gain. At this value of L/h the onset of the tone develops and continues as L/h increases until the limit corresponding to $(2\alpha_r\delta)$, i.e., until

$$\left(\frac{L}{h}\right)_{\max} \simeq \frac{2\pi(2\delta/h)(\overline{U}_c/c_r)(N+1/4)}{[2\alpha_r\delta]_\ell}$$

Meanwhile, the sequence is repeated for each value of N until a value of L/h is reached for which coherent shear layer motion breaks down before it reaches the wedge. This range of allowable L/h for stage N may be expressed as

$$[2\alpha_r\delta]_u^{-1} < b\frac{L/h}{(N+1/4)} < [2\alpha_r\delta]_\ell^{-1} \tag{3.14}$$

where b is a constant. Up to four stages generally can occur unless the jet drives a resonator. Figure 8 utilizes some of Brown's (1937) measurements for three stages to demonstrate the ranges of L/h. In the above, jet spreading across the slit-edge gap has been ignored; this effect could alter the ranges slightly.

Similar stages occur as speed is increased for a fixed value of L/h. Typical behavior consists of the cut-on of a low order stage (not always stage 1) at some threshold velocity followed by a linear increase of frequency with velocity up to some point at which the frequency jumps to the next higher stage. Thus dependence of frequency on speed then assumes another linear relationship with an increased slope. The jump velocities may be seen to be connected with the threshold wave numbers $(2\alpha_r\delta)_\ell$ and $(2\alpha_r\delta)_u$. Since the laminar jet thickness decreases with increasing Reynolds number in the fashion of

$$\frac{2\delta}{h} \propto \frac{1}{\sqrt{\frac{U_J h}{\nu}}} = \frac{1}{\sqrt{R_h}} \tag{3.15}$$

The range of R_h for which each stage is self-excited is limited according to equation (3.13) by

$$[2\alpha_r\delta]_\ell < C\frac{S_h}{\sqrt{R_h}} < [2\alpha_r\delta]_u$$

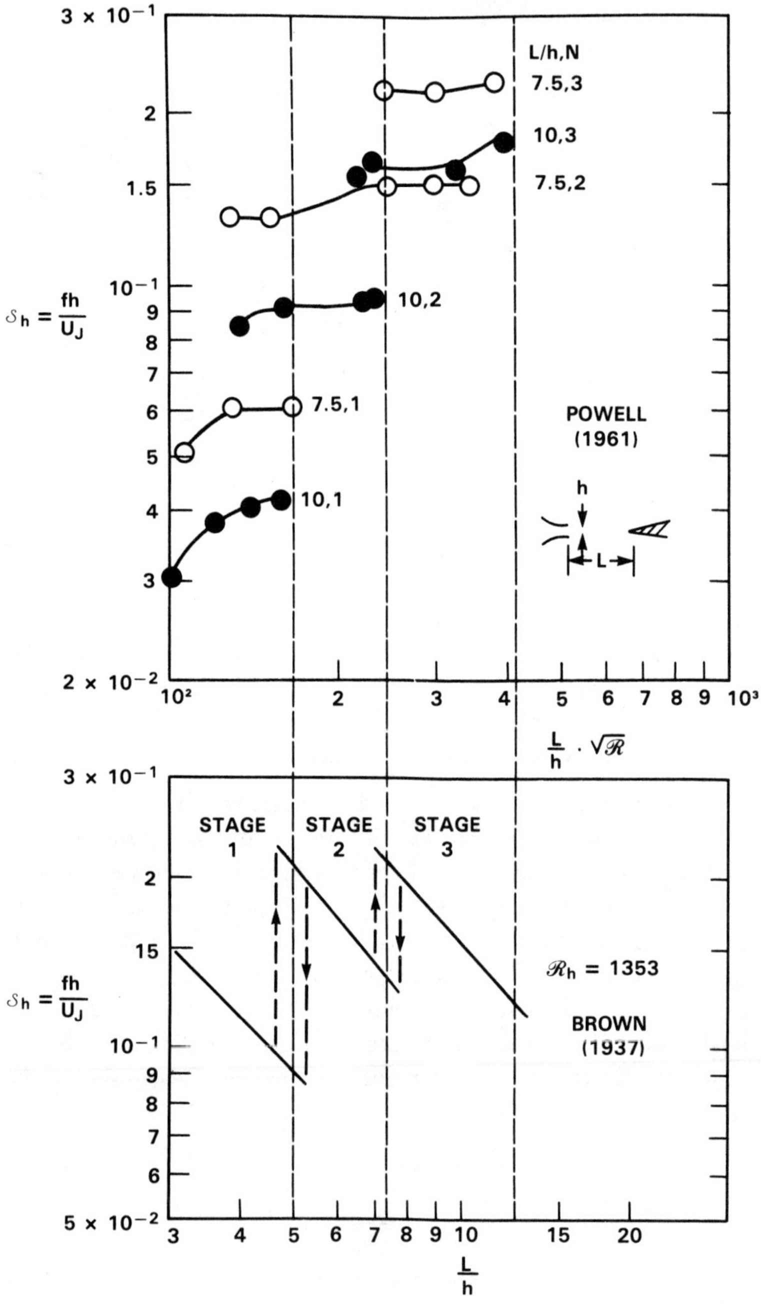

Figure 8. Stages of jet tones correspondence between stages observed with changes of velocity and wedge distance.

where C is a constant. Accordingly, equation (3.10) gives each stage N to be self-excited over a range

$$\frac{1}{[2\alpha_r\delta]_u} < \frac{d(\sqrt{R_h})\,L/h}{(N+1/4)} < \frac{1}{[2\alpha_r\delta]_\ell} \tag{3.16}$$

where d is a constant. Thus, the tone at a given stage will persist for a range of (velocities)$^{1/2}$, or $\sqrt{R_h}\cdot\frac{L}{h}$, which will cover the same proportional interval as the region of L/h given by equation (3.14). Figure 8 uses some data from Powell (1961) to illustrate this correspondence for three stages and for values of $L/h = 7.5$ and 10.

There is a general increase of Strouhal number with Reynolds number that follows the pattern of maximum sound sensitivity of the two-dimensional jet. Furthermore, Chanaud and Powell (1962) found experimentally that for neutrally stable oscillations the Strouhal number/(Reynolds number)$^{1/2}$ was significantly dependent on the level of oscillation.

3.3 Circular Jets—the Hole, Ring, and Pipe Tones

From the point of view of flow tone generation, the acoustic activity of circular jets received somewhat less attention in this decade than the two-dimensional edge tone. These tones all involve axisymmetric modes of the jet and the arrangements are illustrated in Figure 9.

It is interesting to note the variety of commonplace observations of these tones and related jet instabilities afforded by acoustically excited flames (Tyndall, 1875 and Rayleigh, 1896) bird calls and whistling tea kettles (Rayleigh, 1896 and Chanaud and Powell, 1965), human whistles (Wilson, Beavers, DeCosler, Holger, and Regenfuss, 1971).

The hole tone, is generated when a plate with a sharp-edged orifice is placed coaxially with the jet. Rayleigh (1896) discusses in great detail the aspects of construction of this oscillator which is the source of bird calls, tea kettle whistles, and human whistling. Axisymmetric disturbances are caused at the hole which reinforce the initiating disturbances at the efflux. Sound radiation results from the pulsating efflux at the second plate. In the experiments of Chanaud and Powell (1965), the breadth of the second plate

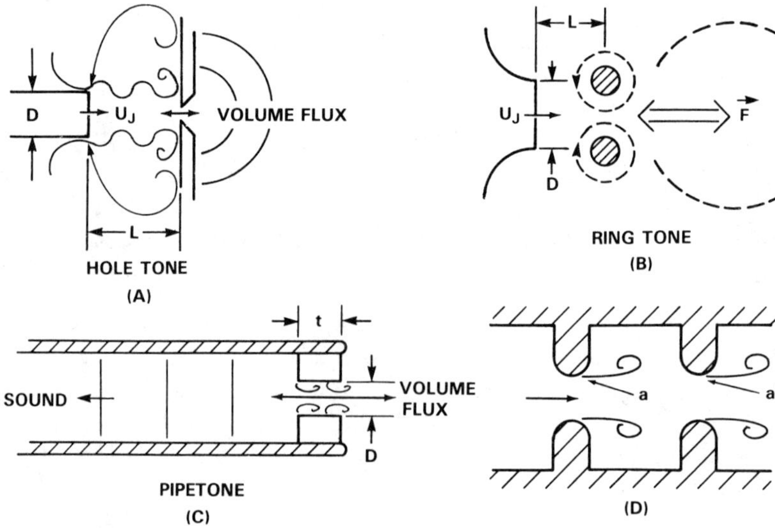

Figure 9. Hole, rig, pipe tones, and cavity tone.

was larger than the wavelength of the sound emanated so that the radiation was omnidirectional. The ring tone (1965) works on the same principle, however, the hole is replaced by a coaxial ring. Toroidal vorticies shed from the ring cause alternating forces on the ring. These dipoles radiate in the direction of the jet axis.

The pipe tones are generated in a pipe when a square or sharp-edged orifice is used as a termination. In this case the jet formed at the pipe terminus is backed by the resonant column of air in the pipe. Such tones do not require the backing of a pipe, although the resonance permits the generation of tones of greater intensity.

The stages of hole and ring tones have been shown to be governed by Chanaud and Powell (1965)

$$\left(\frac{L}{\lambda}\right)_N = N - \frac{1}{4} \tag{3.17}$$

where the average convection velocity across L varies from $0.55\,U_J$ to $0.9\,U_J$ with a typical value of $0.7\,U_J$. These stages are determined by the integer number of complete disturbance wave lengths between the orifice and the hole or ring. Figures 10 and 11 summarize the operating ranges of the hole and ring tones and put these ranges in perspective of the two-dimensional edge tone. It is to be noted that the predominant Strouhal numbers of antisymmetric (edge) tones and symmetric (hole) tones differ in the same direction that is suggested by the stability diagrams in Figure 7.

The pipe tones measured by Anderson (1953, 1954, and 1955), as well as the axisymmetric disturbances in a jet emerging from a knife edge orifice plate occur at about the same value of Strouhal number. Strouhal numbers based on the orifice diameter and efflux velocity are shown in Figure 12 together with a schematic of the experimental arrangement used. The sloped lines show that the numbers are functions of the ratio of thickness to diameter. For each parameter

$$S_t = \frac{ft}{U_J}$$

the corresponding values of S_D generally range from 0.4 to 1.0. This range corresponods roughly to the observed range of numbers observed for hole and ring tones in Figure 11. The limit of $t/D = 0$ corresponds to the knife edge orifices used by Beavers and Wilson (1970) and Johansen (1929). Parametric dependence of S_D on t/D shown in Figure 12 may be explained by noting the observatioins of Kurzweg (1933) or of Beavers and Wilson (1970) which disclosed a train of axisymmetric ring vorticies downstream of a sharp-edged orifice plate. For thicker orifices a *standing-wave* train of n such vorticies generated along the orifice thickness would satisfy the condition

$$t = n\Lambda$$

where

$\Lambda =$ vortex spacing

$t =$ thickness of the orifice plate

$n =$ integral number of vorticies along the orifice axis

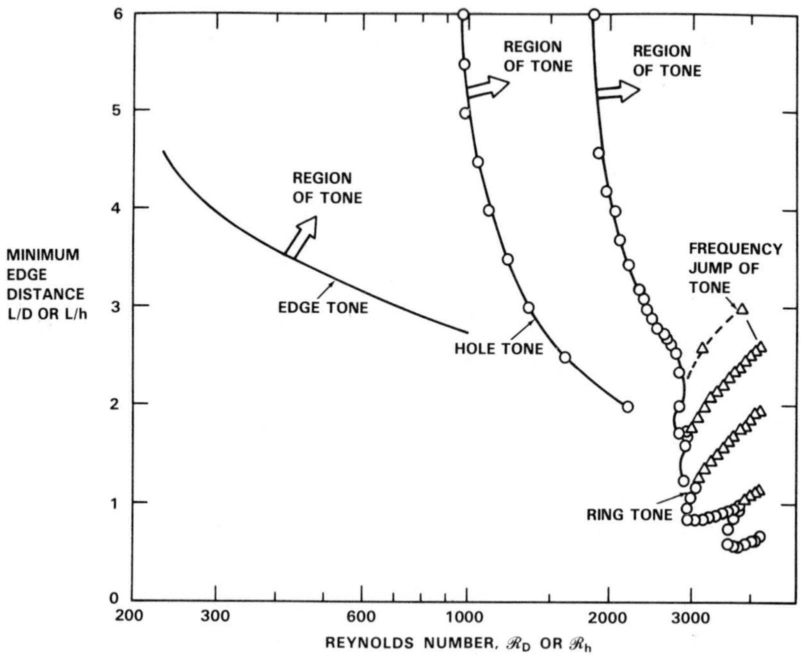

Figure 10. Minimum edge distance for an audible tone was different for the various geometries. The edge tone had asymmetric disturbances to a planar jet; the hole and ring tone had symmetric disturbances to a circular jet. Indirect feedback caused the ring tone contour to be irregular.

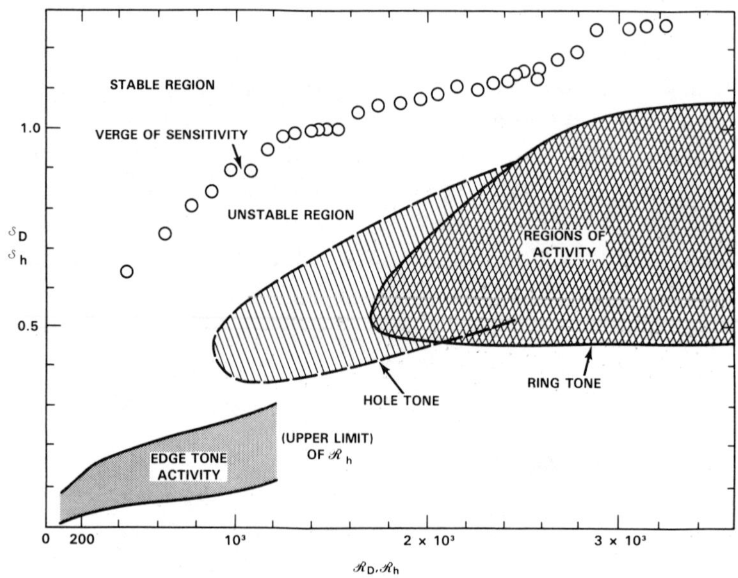

Figure 11. Comparison of regions of tone activity of hole and edge tones as a function of Reynolds number.

Letting

$$\frac{f\Lambda}{U_J} = \Omega = \text{constant}$$

where U_J is the convection velocity of the orificial eddies, so that

$$S_t = \frac{(ft)}{U_j} = n\Omega \qquad \text{for } n \geq 1$$

describes the Strouhal numbers for the various stages of tones in analogy to equation (3.10). If we let the allowable range of S_D for the jet to be between $0.4 < S_D < 1$ (as suggested by Figure 11), at the appropriate values of S_D, then the dependence of stages on D/t is given by

$$0.4 < S_D = \left(\frac{D}{t}\right)(S_t) < 1 \tag{3.18}$$

which corresponds to the diagonal lines shown in Figure 12 with the indicated values of S_t.

The observations of Becker and Massaro (1968) and of Beavers and Wilson (1970) began to extend the techniques of flow visualizations of shear layer dynamics up to fairly large Reynolds numbers. The observations of Becker and Massaro (using a contoured short

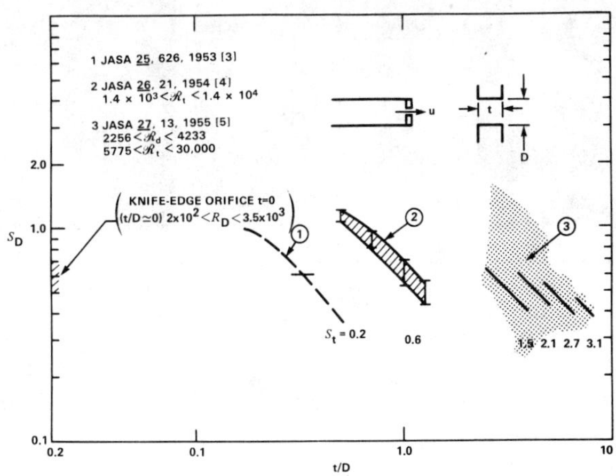

Figure 12. Tone frequencies for square—edge orifice plates terminating a pipe, from Anderson, compared with δ_D for knife–orifices from Johansen, Beavers, and Wilson. Reprinted with permission of Cambridge University Press.

nozzle) extended throughout the entire zone of hole and ring tone activity. A sample of their photographs, which appears in Figure 13, shows a number of features which would be examined more fully in later years. These are: initial shear layer motions at the nozzle exit, a developing axisymmetric vortex structure which results from these initial disturbances, and vortex-pairing of downstream disturbances. It can be seen that regular vortex structure may be observed up to R_D of order 8000. Since a similar structure is first apparent at $R_D \simeq 1000$ (which is in the hole tone range) a potential for extension of the range of R_D for hole tone activity up to 8000 is suggested. It is to be noted that many of the qualities of jet vortex structure for $R_D < 3000$ had been observed earlier by von Gierke (1950).

Becker and Massaro excited their jets acoustically from 45° above the axis and facing the nozzle. Chanaud and Powell (1965) ensonified the jets from various orientations around the jet and found that axisymmetric disturbances were generated even with the sound source positioned at 90° to the axis. This was in contrast to the behavior of the two-dimensional jet as noted earlier, and indicated an apparently greater pressure sensitivity of the circular jet than the slit jet.*

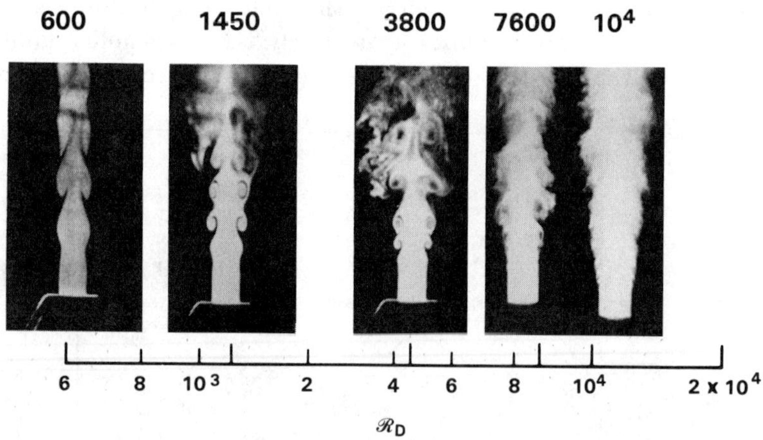

Figure 13. Photographs of smoke jets from laminar flow exiting an ASME short-radius circular nozzle; (from Becker and Massaro, 1968). Reprinted with permission of Cambridge University Press.

*In this context the terms *pressure* and *velocity* sensitivity are invoked to distinguish between primarily scalar (directionless) and vector (directional) forms of excitation. In principle it is not possible to impose experimentally purely pressure or velocity disturbances because each will induce the other upon interaction with the nozzle exit.

3.4 The Screech Tones of Choked Jets

As mentioned previously, the generation of tones by choked jets is facilitated by an acoustic path feedback from the undulating shock cells to the jet exit where the driving disturbances are initiated. A fixed phase relationship between the emitted sound and the aerodynamic disturbances in the exit plane is required just as in the case of the subsonic edge tone. As shown in Figure 4, the source distribution is stationary with respect to the nozzle exit and this essentially provides a fixed *acoustic center* at a distance x_s from the jet which is analogous to the slit-edge distance of the edge tone. In contrast to the edge tone however, the effective source location will be a variable that depends on pressure ratio and diameter. Dealing with the case of two-dimensional jets, the condition analogous to equation (3.11b) which is required for self-excitation of a tone of frequency f_s is Powell (1953d, 1962).

$$\frac{f_s x_s}{U_J} = \frac{U_c}{U_J}\left(\frac{N+k}{1+M_c}\right)$$

The effective source location x_s is not a fixed quantity; it depends on the particular jet and the pressure ratio and it must be deduced experimentally from schlieren photographs as the acoustic center of cylindrically spreading wave fronts. In cases reported by Powell (1953d) it appears to range typically from 1.7 to 2.3 λ_c, where λ_c is the effective eddy disturbance wavelength, but the generality off this relationship has not been established. Similarly, the convection velocity and the parameter k depend on the specific properties of the jet. Typically $U_c \sim 0.6\,U_J$. The parameter k is a nonintegral measure of the number of eddy disturbance wavelengths along the axis which participate in the tone and the index N is number of complete acoustic wavelengths between the source and the jet exit. Thus the sum $N + k$ represents the total number of acoustic and aerodynamic cycles around the feedback loop. There are as yet no general rules for precisely determining x_s, k, or U_c/U_J. However for the jets examined by Powell (1953d), the parameter $N + k$ was a constant over a broad range of pressure ratios for a given jet and the frequency followed the rule

$$\frac{f_s h}{c_0} = \frac{0.2}{\sqrt{R - R_c}} \tag{3.19}$$

This rule was subsequently also observed by Krothapalli et al. (1983a) and it thus appears to be reasonably general. R is the pressure ratio, R_c is its critical value, and h is the height of the jet.

The directivity of the sound is determined by the geometry of the cells in relation to the wave length of sound. The dominant parameters are given in the top of Figure 4. Each cell represents a source and the number of sources depends on Reynolds number and turbulence level in the jet since turbulence will eventually determine the desintegration of the cells downstream. The cells are pulsed by eddy disturbances which convect at speed U_c; as with subsonic edge tones U_c is an effective value for a given jet exit velocity. The phase between adjacent sources is, thus, $2f_s s/U_c$. Although the number of participating cells and their relative strengths indefinite, the directivity pattern of the main lobes are given in a representative fashion by only three sources of equal strength by which

$$D_f = \frac{1}{3} + \frac{2}{3}\cos\left[2\pi\frac{s}{\lambda_c}(1 - M_c\cos\beta)\right]$$

for the fundamental frequency, and

$$D_h = \frac{1}{3} + \frac{2}{3}\cos\left[4\pi\frac{s}{\lambda_c}(1 - M_c\cos\beta)\right]$$

for the first harmonic of $f = 2f_s$. Higher numbers of sources add additional factors of $1 - M_c\cos\beta$ that effect the side lobes, but note the main lobe directivity. Even though M_c may be a variable of the jet, the cell width s and λ_c are given by the empirical relationship

$$\frac{s}{h} \simeq 1.9\sqrt{R - R_c}$$

and

$$\frac{\lambda_c}{h} = \frac{U_c}{f_s d} \approx 2.9\sqrt{R - R_c}$$

so that λ_c/s is a fixed number about equal to 0.66. For a range of typical values of M_c, then, the main lobe (greatest intensity of sound) will be nearly upstream (toward the jet) for the fundamental. The directivity of the first harmonic will be nearly broadside with much reduced intensity upstream. Accordingly this harmonic is incidental to the tone generation and its magnitude in relation to the fundamental is dependent on the details of the pulse shape of the eddy-shock interaction and the time stationarity of the flow. Thus, no further definition of characteristics of the fundamental and its harmonic is possible at this time.

Whether or not a tone of a given stage N is generated depends on principally the same factors which control the edge tone. The factors η_d and η_t, in equation (3.6b), being pertinent to both the generation of the feedback disturbances and the interaction at the jet exit are continuously variable functions of frequency, but the frequency η_s pertains to the net disturbance at the jet exit from the transmission of pressures from the individual sources to the jet exit. Complete reinforcement, and therefore a maximum value of η_s, is determined by the condition (Powell, 1962b)

$$\frac{s}{\lambda_c}(1+M_c)=1$$

Thus, reinforcement of the aerodynamic and acoustic disturbances will be facilitated within a band

$$\Delta\left[\frac{s}{\lambda_c}(1+M_c)\right]$$

centered on $s(1+M_c)/\lambda_c = 1$.

In the case of the two-dimensional jet the range of allowable tone frequencies is relatively large, with the frequency varying continuously with pressure ratio. In the case of the circular jet the observed dependence is discontinuous and hysteric effects occur much as they do in the edge tone. Thus, the band $\Delta[s(1+M_c)/\lambda_c]$ for circular jets appears to be much narrower than for two-dimensional jets. A prior prediction of the stages for the tones of circular jets is not yet possible, however, the frequencies observed generally discontinuously follow the rule

$$\frac{f_s D}{c_0}=\frac{0.33}{\sqrt{R-R_c}}$$

and involve a spiral mode of the jet disturbances. Follow-on work through 1970 on sounds from choked jets have further quantified some of these relationships and demonstrated that the feedback may be broken by the insertion of a shield forward of the exit plane, e.g., Hammitt (1961), two-dimensional jets, and Davies and Oldlfield (1962), circular jets.

3.5 Dynamics of the Free Shear Layer and Cavity Tones

Progress in the mechanics of free shear layers and in the generation of cavity tones continued with an apparent independence of respective researches that persisted into the 1970's.

The most extensive theoretical analysis of free shear layers per se, which focused on the linear instability and the exponential growth region, was that of Michalke (1964) using the hyperbolic tangent profile shown in Figure 6c. Experimental confirmation in the linear region was made by Sato (1958) and Freymuth (1966). Freymuth examined the shear layer near the lip of a laminar flow nozzle; the layer was acoustically excited by a loud speaker situated inside the nozzle. The Strouhal number based on free stream velocity and a value of the momentum thickness θ_m measured just downstream of the edge, i.e., $S_{\theta m}$, was used to characterize the measured shear layer dynamics for comparison with theory. Fluctuating velocities, both transverse to and aligned with the flow direction, were determined at downstream distances over a broad range of Strouhal numbers. For $R_{\theta m} < 330$ the range of exponential growth extended to 60–110 θ_m downstream of the lip depending on the amplitude of sound. Higher levels of excitation promoted a longer exponential growth range. Figure 14 shows a distribution of transverse velocity amplitude u_2/U_0 somewhat downstream of the lip, but still in the exponential growth range. Superimposed on this is a plot of $e\alpha_i x$ adjusted to arbitrary vertical scale and this may be seen to correspond well with the measured values of the disturbances. The length of the exponential growth range and the value of S_θ for maximum growth are all significant parameters in subsequent evaluation of shear layer tones. Generally speaking, the shear layer with a hyperbolic targent profile becomes unresponsive when excited at Strouhal numbers, much exceeding $S_{\theta m} = 0.04$.

The characteristics of the developed shear layer downstream of the zone of linear fluctuations were observed nearly simultaneously by both Freymuth (1966) and Browand (1966) to be characterized by the generation not only of disturbances of the fundamental wave but also of subharmonics of that wave. Browand (1966) observed also that disturbances at combinations both of multiplies of the fundamental and fractions of the fundamental could be generated when the shear layer was driven strongly enough. Kelly (1967) analytically examined large-amplitude nonlinear fluid motions finding natural instabilities to exist in the shear layer at combinations of frequencies,

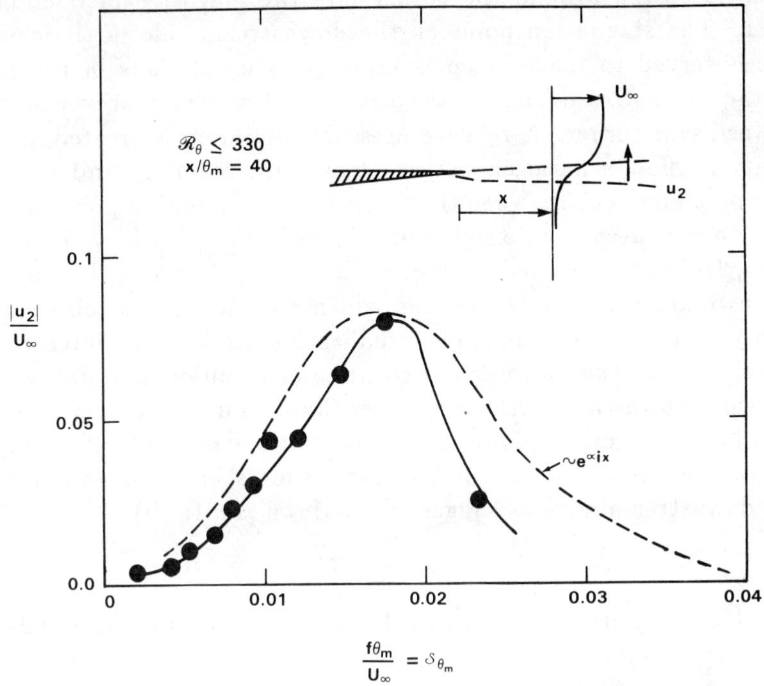

Figure 14. Fluctuating transverse velocity in developing parallel-flow shear layer (from Freumuth, 1966). Reprinted with permission of Cambridge University Press.

such as those observed experimentally by Browand.

The analysis of cavity tones during this period was generally based on fairly elementary views of shear layer motions in the cavity opening. Rossiter (1964) and East (1966), for example, working with shallow rectangular air cavities, found that the stages of the tones may be predicted to reasonable accuracy with a formular such as equation (3.11) for which $\gamma = -1/4$ and $\overline{M}_c \simeq 0.5 M_\infty$.

Dunham (1962), working at higher Reynolds numbers in water found the formula

$$fL/U_\infty \propto N$$

where $N = 1$ or 2 with a proportionality constant which appeared to be around 0.4. Although his paper has gone largely unnoticed, flow visualizations by Dunham disclosed a convected vortex across the mouth of the cavity which impacted on the downstream corner. The vortex, as it encountered the corner, entrained fluid from the cavity when it passed around the corner into the downstream boundary layer. The stagnation point on the downstream side of the cavity was observed to move in consonance with oscillations in pressure at the downstream face of the cavity and vortex motions at the downstream corner. A relative pressure minimum is created in the cavity as flow separation occurs at the upstream lip and a pressure maximum occurs when the downstream stagnation point has its maximum deflection into the cavity. Thus in the early flow visualizations, Dunham observed a unique relationship between the up and downstream shear layer motions which contributed to self-excited cavity tones. In particular, his visualization disclosed an interaction of vorticity at the downstream corner and resultant fluctuations in the close downstream boundary layer that could be regarded as support for more contemporary analytical notions of cavity shear layer interactions that provide for high amplitude shear layer motions at the downstream edge as suggested by Howe (1981a, b).

4. Contemporary Work on Flow Tone Generation (1970–present)

Most experimental work since 1970 has dealt with flows at higher Reynolds numbers. There is also a noticable shift of emphasis away from jet tone generation to cavity tones. Work on jets was devoted to the nature of large scale structures and the relationship that these structures have to the generation of high-Reynolds number flow tones. The views of cavity tone generation began to fall in step with those of jet tones. The integration of the fluid dynamics of unsteady shear flows in the cavity opening with the kinematic requirements of cavity oscillations followed very much the pattern of edge tones. Also, interest in acoustic resonators excited by unsteady shear flows was reemphasized. Thus, the contemporary views of jet flows (at high Reynolds numbers particularly), cavity tones, and shear-flow tones in general began to develop from a common fluid dynamic basis. All these types of flow which originate with an upstream laminar boundary layer develop as a result of linear instabilities in the shear layer at the upstream edge or nozzle lip which progressively develop a dominant vortex structure. The dynamics of these vortices are controlled by the specifics of the flow geometry (i.e., free or impinging jet, cavity, etc.). In comparison to this work on shear layer mechanics, the interest in the primary acoustic aspects of flow tone generation has been relatively lacking. Of particular note in this regard is the limited work on the acoustic intensities emitted, the nature or control of acoustic excitation of the shear layer, or the dominant mechanisms of sound production.

4.1 Development of Disturbance Modes in Shear Layers

The observations of fundamental instabilities of two-dimensional shear layers by Freymuch (1966) and of separated flow behind a step by Brown and Roshko (1971) at high Reynolds number disclosed more clearly an identifiable ordered vortex structure which coexists with the general turbulence level. In this context, the Reynolds number is based on the greater mean velocity outside the shear layer (of two co-moving fluids at velocities U_1 and U_2 with $U_1 > U_2$) and distance from the generating edge. It was noted in Section 3.4 that the exponential growth of initial instabilities extended to roughly $X/\theta \sim 100$ for $R_\theta \sim 330$. This corresponds to $R_x \sim 3.3 \times 10^4$. Ordered vortex structures were apparent in the free shear flows up to $R_x \sim 3 \times 10^6$ (see Roshko, 1976). Figure 15 (taken from Brown and Roshko, 1974) illustrates the type of developing layer and its vortex structure under consideration. The Reynolds number indicated is that for the value of x, corresponding to the right hand extremity of the picture. The existence of large eddy structures in turbulent flows had, of course, been implied by the well known characteristics of space-time correlations of velocity and pressure fields in boundary layers, wakes, and jets. They were also obvious in various schlieran photographs of jets at moderate to high Mach numbers (see e.g., von Gierke (1950), Darlington (1949), Powell (1951, 1977a, b)). A form of pairing had been previously noted in the cited works of von Gierke and Powell. Three examples are shown in Figure 16 of the large-scale orderlines superimposed on a fine grain turbulent structure and existing 1–2 diameters downstream of the jet exit. It is apparent from this and other supporting evidence that there is a close connection between the space-time convection of fluctuating variables in turbulent shear flows and these physical vortex structures.

In subsequent work, Winant and Browand (1974) observed that the developed vortices appear to interact with each other in pairs, with the vortices of each pair rotating about one another in relative motion. These pairs merge to one larger vortex. Subsequently, pairs of larger vortices again interact forming still larger vortices.

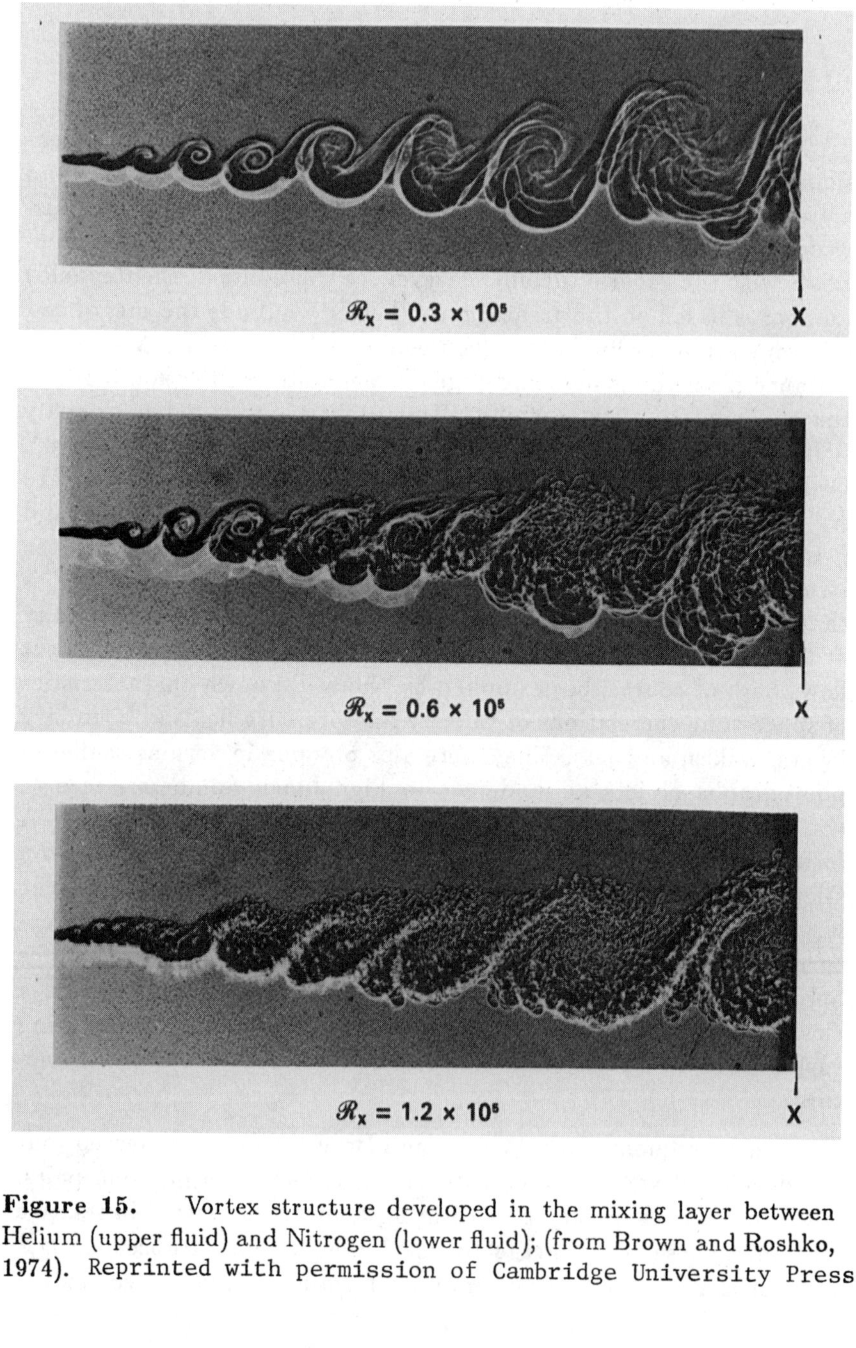

Figure 15. Vortex structure developed in the mixing layer between Helium (upper fluid) and Nitrogen (lower fluid); (from Brown and Roshko, 1974). Reprinted with permission of Cambridge University Press.

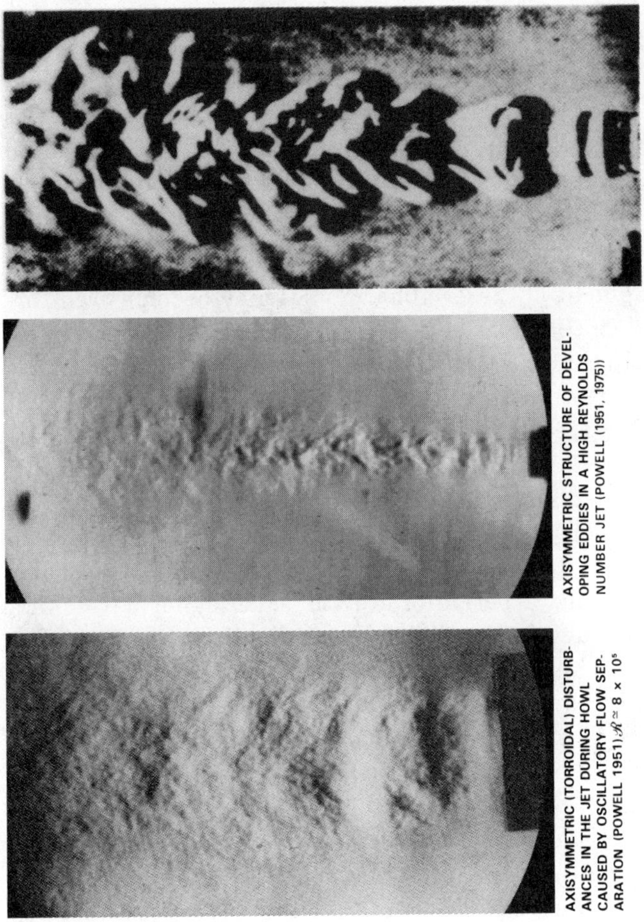

AXISYMMETRIC (TORROIDAL) DISTURBANCES IN THE JET DURING HOWL CAUSED BY OSCILLATORY FLOW SEPARATION (POWELL 1951) $\mathscr{R} \simeq 8 \times 10^5$

AXISYMMETRIC STRUCTURE OF DEVELOPING EDDIES IN A HIGH REYNOLDS NUMBER JET (POWELL (1951, 1975))

THE FLOW OF JET-TONES. FLOW SEPARATION OCCURS WITHIN THE ORIFICE AND THE PERIODIC SHEDDING OF THE VORTICES WITH THE ASSOCIATED FLUCTUATION IN FLOW RESULTS IN THE CHARACTERISTIC TOROIDAL VORTEX FORMATIONS. (FROM A PHOTOGRAPH BY W. H. DARLINGTON (1949))

Figure 16. Early visualizations of large EDDY structures in circular jets at large Reynolds number.

As each pairing occurs, up to three to four times, the streamwise spacing of paired vortices doubles and the width of the shear layer gradually increases. To velocity probes situated at varying distances downstream of the generating edge of the shear layer, the frequency of passage of the vortices is halved for each pairing. Thus, letting S_1 be the fundamental Strouhal number that is characteristic of initially formed vortices, pairing results in a succession of Strouhal numbers $S_1/2$, $S_1/4$, etc. at increasing distances from the generating edge. Browand and Weidman (1976) then found that net production of Reynolds stresses was associated with each vortex pairing event. The production is gradual, occuring continuously over time scales that are equivalent to the duration of a pairing event. This gradual production of Reynolds stresses is in contrast to the more impulsive *bursting* production that has been so well characterized in turbulent wall layers.

The development of these subharmonics may become even further organized when the shear layer is either acoustically or mechanically driven. In Figure 17 is shown the growth of the various harmonics of disturbances in the shear layer shown at the upper right of the figure. The measurements are Miksad's (1972). Each harmonic or subharmonic was separately driven. The fundamental, β is 2π times the Strouhal number of least stable naturally-occurring disturbance just downstream of the edge. The excited fundamental saturates at $x/\theta_m \simeq 27$ while the first subharmonic $\beta/2$ saturates at $x/\theta_m \simeq 85$ following a period of growth. The higher harmonics, even driven, do not develop as markedly as the disturbances at frequencies β or $\beta/2$. Thus the fundamental and its subharmonics have clearly identifiable growth regions, each may be separately excited, and a unique relationship between frequency and the downstream growth zone for each mode may be identified. The same or similar behavior has been observed recently by Ho and Huang (1982). In their experiment, which also involves the development of parallel-flow shear layer, the fundamental and the successive subharmonics were all observed to have a much more clear exponential growth range than was shown in Miksad's experiments. The location of saturation or of limitation to exponential growth moves to greater distances from the generating edge as the subharmonic frequency decreases. Flow visualization by Ho and Huang disclosed that *vortex pairing* occurred at these points of saturation and that the momentum thickness of the shear layer increased in stages which roughly corresponded to these pairings. In the same experiments, reported

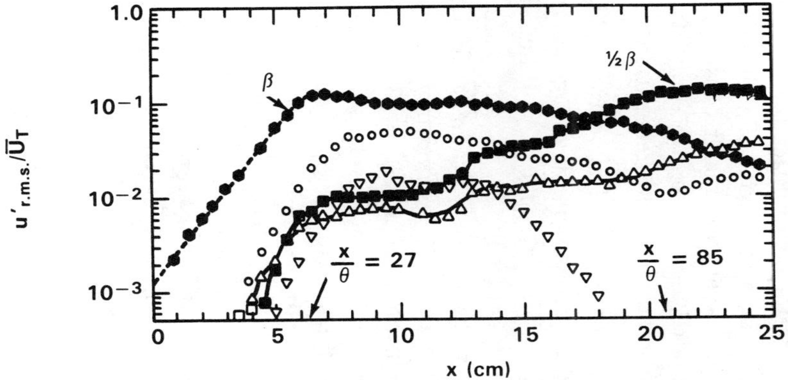

DOWNSTREAM GROWTH OF $U'_{r.m.s.}$ MAXIMA. EXCITATION AT $\beta = 0.222$; $\mathscr{R}(x_0) = 145$. ● β; • , 2β; ■, $\frac{1}{2}\beta$; △, $\frac{3}{2}\beta$. — — —, THEORETICAL LANDAU EQUILIBRIUM CURVE.

$\beta = 2\pi\, \mathcal{S}_\theta$ $\qquad$ $\mathscr{R}_\theta = 374$

$\mathcal{S}_\theta = 0.035$

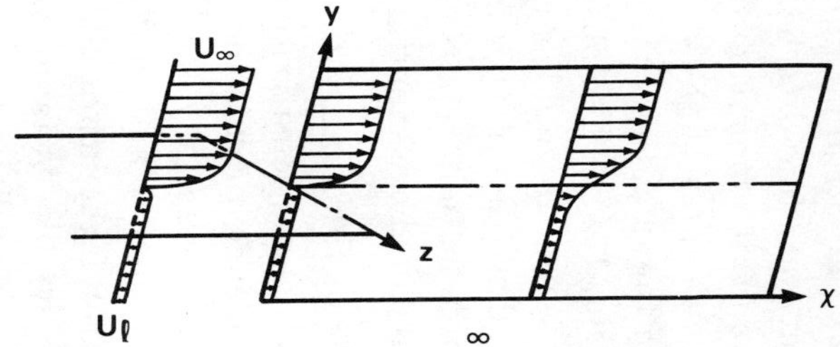

$$\theta_m = \int_{-\infty}^{\infty} \frac{U}{U_\infty}\left(1 - \frac{U}{U_\infty}\right) dy$$

$$\frac{U_\ell}{U_\infty} = 0.16$$

Figure 17. Development of subharmonics in a free shear layer (Miksad, 1972). Reprinted with permission of Cambridge University Press.

by Ho (1981) and Ho and Huang (1982), the Strouhal numbers for the various stages of vortex development were found to be bounded between narrow limits when based on the *local* momentum thickness. These characteristics as well as the growth rates of the driven subharmonic disturbances, closely resemble the qualities that would be predicted with linear stability theory (Ho, 1981). The observations of acoustically forced motions of jets at high Reynolds number by Crow and Champagne (1971) (also to be discussed further in Section 4.2) disclosed similar behavior analogous to that reported for the parallel-flow shear layer and predated all the above-mentioned work.

Collectively, these results thus began to quantify the integration of shear layer instability, vortex production, sensitivity to acoustic excitation, and the spreading turbulence production in shear layers. It is also evident that free shear layers may be more sensitive to excitation at the fundamental frequency and subharmonics of the fundamental frequency of the shear layer instability than at multiples of it. Furthermore, the instability-like behavior of the ordered subharmonics also suggested the possibilities for other flow-excited forces. It is apparent that the appearance of pairing or agglomeration of vortices in the shear layers is a result of the gradual diminishment of higher order modes and the maturation of lower order modes with distance downstream from the generation point.

As pointed out in Section 3, the modes of acoustic sensitivity of jets were found to be somewhat dependent on the geometry of the jet. At low Reynolds numbers ($R_h < 4000$) two-dimensional jets respond best (although not exclusively) in an antisymmetric mode to an imposed cross velocity, or pressure gradient. Morkovin and Paranjape (1971) attempted a controlled experiment in which they intended to drive two-dimensional jet at $R_h = 3900$* either symmetrically or unsymmetrically with two sound sources located close to the jet and on opposite sides of it. They could drive the sources either in phase or antiphase. It was found that the unsymmetrical modes nearly always dominate the response whether or not the applied excitation was symmetric or antisymmetric. Clear symmetric response of two-dimensional jets is only rarely observed, but such results are in some contrast to that observed by Sato (1960) who reported a relative insensitivity to the orientation of the acoustic source placed remotely to the jet for both types of modes. One

*This Reynolds number is comparable to that used by Sato (1960).

interpretation of this discrepancy is that, owing to the somewhat greater instability of the antisymmetric mode (see Figure 6) it is difficult to experimentally create sufficiently symmetric excitation levels without contamination from nonsymmetries. It is also probable that the near field disturbances in Morkovin and Paranjape's induced nonsymmetric three-dimensional disturbances. Conversely, circular jets at low Reynolds number have been found to be *pressure* sensitive and to respond more frequently in axisymmetric modes than in nonaxisymmetric modes. Theoretically, the symmetric mode is more sensitive to the details of the velocity profile than the antisymmetric mode (Sato, 1960). Furthermore Chanaud and Powell (1965) observed that the existence or nonexistence of hole tones depended on the velocity profile of the jet and also they reported on the greater *pressure* sensitivity compared with *velocity* sensitivity of circular jets. However, in spite of this, high speed choked circular jets undergo primarily a spiral mode. Thus it is apparent that the acoustic responsiveness and tone propensity of circular jets will be very much dependent on the nozzle details and, perhaps, as well as on the geometry of acoustic driving.

The acoustic sensitivity of planar shear layers was traced by Miksad (1972) in his experiment to a transverse oscillation of the stagnation point at the trailing edge of the upstream splitter plate between the two shear layers. This oscillation caused the generation of vorticity the near wake. That the shear layer was responsive to a transverse pressure gradient was further suggested by its relative unresponsiveness to sound propagation in the plane of the layer and its responsiveness to sound incident in the transverse direction. This result suggests that the drive mechanism is somehow related to the establishment of a Kutta condition at the edge. The spreading of sound waves emerging around the edge of a nozzle, or diffracting around the edge of a flat plate, causes significant gradients in velocity and pressure particularly in directions normal to the edge regardless of the direction of incidence of the sound field. Upon interaction of these gradients with the unstable mean flow, it is not difficult to comprehend that a wide variety of driving parameters might always result in the generation of the same or similar modes and that from time to time axisymmetric and nonaxisiymmetric responses may be more or less easily excited. In work by Bechert and Pfizenmaier (1975) this connection is refined through measurements of the displacement amplitude of the free shear in near wake of a nozzle lip under the influence of acoustic excitation from an upstream point in the nozzle at various Strouhal numbers. An example is shown in Figure 18 for the shear layer displacement normalized on the amplitude factor $p_a/\rho_0\omega^2 D$ where p_a is the amplitude of the fluctuating drive pressure. At low frequencies it appeared that a full Kutta condition (i.e., $\eta = 0$ and $\delta\eta/\delta x = 0$) was satisfied while at high Strouhal numbers a rectified condition ($\eta = 0$, $\delta\eta/\delta X \neq 0$) was apparently met. Since the amplitude of the displacement increased parabolically with distance from the trailing edge in the near wake. The bound between the high and low frequency behavior is dependent on both Reynolds number and the shear layer thickness, but for their experiments ($R_\theta \sim$ 100–200) a useful upper limit of S_θ is 10^{-3}. There is an implication in this result (Bechert and Michel, 1975) that analytical modeling of shear layer mechanics which do not account for the finite thickness of the shear layer, may only be valid at suitably low frequencies.

4.2 Ordered Structures and Jet Tones

4.2.1 Elucidation of Large Eddy Structures and Turbulence Mechanics

Whether or not disturbances of a given qualitative form occur in developing jets is dependent on the character of the mean velocity profile of the efflux, and therefore somewhat on the type of nozzle used. We shall now summarize the response characteristics of various nozzles over a range of Reynolds numbers. For short, potential-flow nozzles (a) in Figure 19, the efflux contains a modestly-sized potential core with an annular shear layer so that $2\delta/D \ll 1$, where δ is the shear-layer thickness. When the nozzle, or slit in the case of a $2-D$ nozzle, is made many diameters longer, the flow in the tube is fully sheared so that a velocity profile with a parabolic shape occurs (in these profiles the shear layer incorporates a half-diameter). The developing flow outside the nozzle then resembles the profile in Figure 6d. The efflux of these jets will be laminar for Reynolds numbers (U_0D/ν) less than 1400 when the inlet to the nozzle is also well formed to avoid separation of flow inside the nozzle. Such circular jets formed from long enough nozzles that turbulent flow develops at the exit do not generate well-developed hole tones (Chanaud and Powell, 1965). For jets ensuing from long square-edge orifices such as (c), the efflux is laminar only for $R_D <$ 600 because vortices caused by separation of flow are generated

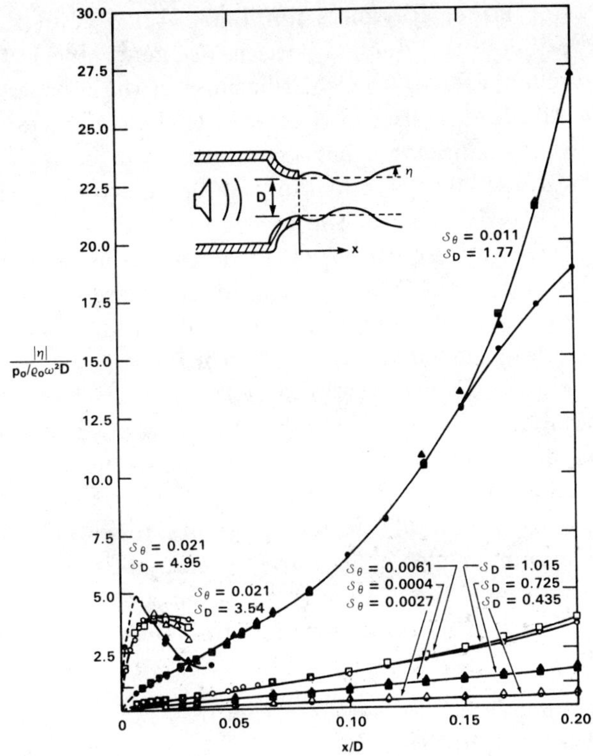

Figure 18. Optical compensation measurements: envelope of the jet border deflexion downstream. $Rc = 10^4$, S_D varied. ○, ●, 104db; ■, □, 94db; ▲, △, 84db. $\theta/D = 0.0061$, $\mathcal{R}_0 = 10^4$. (Bechert and Pfizenmaier, 1975). Reprinted with permission of Cambridge University Press.

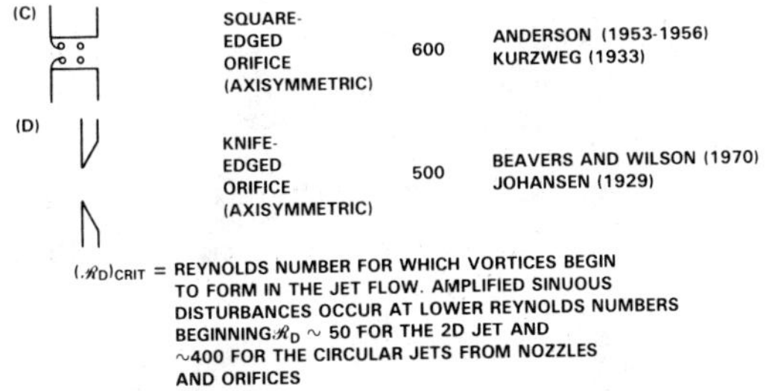

Figure 19. Illustrations of nozzles and orifices that produce jet tones.

at the inlet for larger Reynolds numbers. Finally, in the case of knife-edged orifices, well-defined vortices are generated in the jets at Reynolds numbers greater than 500 because of the rather thin shear layer in the efflux. As noted in Section 4.1, the saturation of initial instability waves in planar shear layers occurs in only 20–40 shear layer momentum thickness, and thus the linear stability region will breakdown relatively close to the nozzle exit. The references cited in Figure 19 are representative of those for which extensive flow visualizations and quantitative measurements were obtained over a wide range of Reynolds number. The critical Reynolds numbers cited are the minimum values for which growing sinuous disturbances give way to clearly-defined vortex structures. This critical value of Reynolds number is not well defined since it is often influenced by the presence of extraneous disturbances and its identification also depends on experimental detail and on the manner of observation. This extreme sensitivity of the jet instability to its velocity profile and Reynolds number also contributes to many of the discrepancies in early observations of jet tones. As noted at the bottom of the figure, jets are disturbance sensitive at Reynolds numbers that are as low as 50.

The connection that vortex structures such as visualized in Figure 13 have to the behavior of the parallel-flow shear layer is now reasonably obvious. However, the measurements of Crow and Champagne (1971) predate the bulk of work which has been done on describing this behavior. They obtained measurements of the shear layer instabilities in the lip of a short nozzle at Reynolds numbers between $R_D = 10^4$ and $R_D = 5 \times 10^4$, and photographed the formation of toroidal vortices in the downstream jet flow. They made the connection between the disturbance evolution that they observed to the prooduction of subharmonics which had been previously reported by Browand (1960) and Freymuth (1966). The vortex structures had the visual appearance of *puffs* which formed from an apparent pairing process identical to that observed earlier by von Gierkie (1950). The final vortex train had a Strouhal number of about 0.3 and it developed within about eight diameters downstream of the jet. Acoustical excitation of the jet from inside the nozzle at $S_D = 0.3$, and with the lip flow tripped to turbulence to remove any Reynolds number dependence, resulted in an ordering of the vortex formmation and a contraction of the formation zone of large scale eddies to four diameters. Growth rates of the driven disturbances were found to resemble those of linear spatial hydrodynamic instabilities just as found later by Browand and coworkers for the planar shear layer. A minimum amplitude of acoustical excitation was required to noticeably excite the vortex structure and greatest sensitivity of the jet to excitation in the range of $0.15 < S_D < 0.50$ was $S_D = 0.3$.

An examination of the space-time statistics of near-field pressure and velocity fluctuations in the mixing layer of a turbulent jet by Lau, Fisher, and Fuchs (1972) further disclosed the existence of ordered vortex structures at Strouhal numbers near $S_D \simeq 0.6$. The subsequent expeprimental programs of Chan (1974) and Peterson, Kaplan, and Laufer (1974) generally confirmed the emerging picture of the cascade of shear layer instability to vortex structure events with the finally emerging vortex structure controlling the general space-time statistical behavior of turbulent variables in the jet. The development generally proceeded in stages over the first three to four orifice diameters and culminated in Strouhal numbers ranging from roughly 0.3 to 0.6. The process is illustrated in Figure 20 taken from Browand and Laufer (1975). In all the methods of measurement or observation, the resulting phase or convection velocities of turbulence properties of visualized vortex structures ranged from 0.6 to 0.7 of the jet efflux velocity when x/D exceeded roughly 2. Finally, a connection between this subharmonic vortex development and the production of Reynolds stress was implied by measurements of Hussain and Zaman (1977). The variation in Strouhal number S_D was from 0.3 to 0.7 generally. These variations in S_D or U_c/U_j are probably due mostly to differences in the methods of measurement (or of the variable measured) and the scatter of measurements, but it is clear that these numbers are well bounded.

These ranges of disturbance frequencies which may be observed in circular jets at all Reynolds numbers are summarized in Figure 21. All the jets featured are issued from either a short nozzle or from a knife edge. Accordingly, the mean velocity profile will contain a cylindrical flow region surrounded with a shear layer of momentum thickness θ. For R_D less than 2000 to 3000, the wavelength of oscillation is larger than oor comparable to the diameter of the jet. At higher Reynolds numbers, the oscillations which account for the disturbances at higher Strouhal numbers and which increase as

$$S_D \sim (R_D)^{1/2}$$

in accordance with the decrease in shear layer thickness

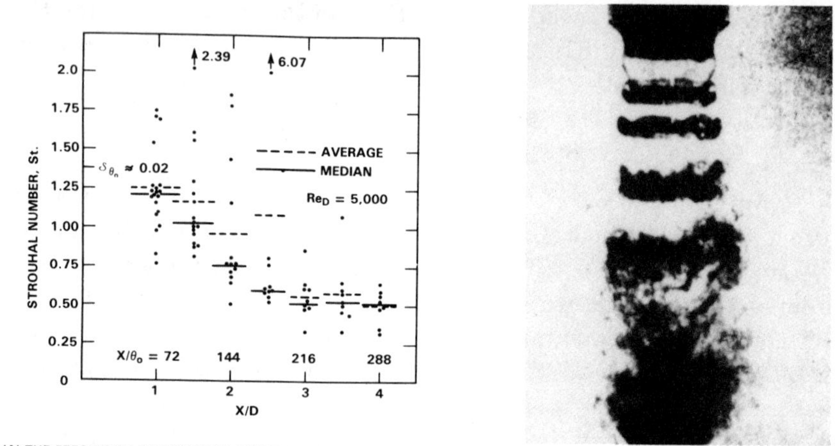

(A) THE FREQUENCY OF INDIVIDUAL VORTEX PASSAGES AS A FUNCTION OF DOWNSTREAM POSITION. $St = D/TU_{jet}$, WHERE T IS THE TIME PERIOD BETWEEN ARRIVAL OF ADJACENT VORTICES.

$\theta_o/D = 0.98/\sqrt{\mathscr{R}_D} = 0.014$

(B) TRANSITION OF RINGS TO PUFFS AT TERMINATION OF POTENTIAL CORE. JET IS FORCED SLIGHTLY AT St ≃ 1.0.

Figure 20. The development of vortex structure in a jet at high Reynolds number (Browand and Laufer, 1975).

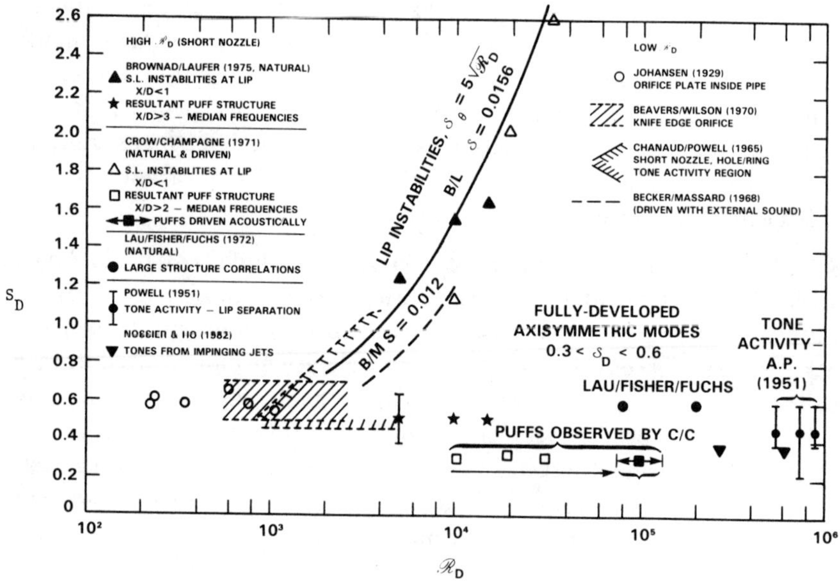

Figure 21. Strouhal numbers for disturbance-sensitive circular jets from orifice plates and short nozzles.

$$\frac{\theta}{D} \sim \frac{1}{(R_D)^{1/2}} \tag{4.1}$$

This behavior along this leg should persist until the shear layer becomes turbulent thus preventing the natural development of instability waves, i.e., until R_θ increases to roughly 300–400. For most jets studies, θ/D lies from 0.003 to 0.01, which places this limit in the range of R_D from 3×10^4 to 10^5 roughly. For higher values of R_D, a train of small axisymmetric vortices is probably generated right at the lip without the generation of sinuous instability waves although visualizations have not been carried out in this region. The lower leg of S_D in the range of 0.3 to 0.6 is determined by the large eddy structure which is developed within a distance $X/D \simeq 3$ to 4 from the jet exit. For R_D in excess of 2×10^5 the reported frequencies are those of sounds emitted by jets. These shall be discussed below.

4.2.2 Tones Generated by Jets at High Reynolds Number

A major portion of work since 1970 has involved the tones of jets at high Reynolds number. As Reynolds number increases, the tones from both two-dimensional and circular jets (e.g., edge and hole tones) generally become feeble and disappear. Such tones may reappear, as conditions permit, at higher Reynolds number. Then, when the jet becomes choked the screech tone may be generated. Following the illustration in Figure 1, tones by jets at high Reynolds number may be generated with the separating nozzle jets, by choked jets and by circular and two-dimensional jets impinging on a flat surface.

A notable example of sound radiation from a separated flow nozzle is the study of Powell (1951) in which sound from choked and subsonic circular jets was measured for a variety of nozzle geometries. Figure 22 shows the sound radiated in a direction 90° to the jet axis for a nozzle with a lip fashioned to induce separation. The lip expanded the exit radius from 1 inch to $1\frac{1}{8}$ inch in 1/2 inch as shown in the inset. The acoustic measurements were made in octave frequency bands so that the tonal quality of the sound is not apparent in the measured spectra. The levels shown in Figure 22 are equivalent to 1/3 octave band levels and they have a pronounced peak occurring at $S_D \sim 0.4$ to 0.6, although the analysis bandwidth was actually too large (one octave) to precisely identify it. The frequencies of these peaks are shown in Figure 21 in the range of 5×10^5 to 9×10^5 Reynolds number. The sound radiated by the jet with lip separation substantially exceeded that of conventional short nozzles. Figure 23 compares the sound radiated from both types of nozzles in a dimensionless form that follows directly from the similitude of subsonic and quadrupole radiation. The sound levels from the short nozzle without separation compare favorably with more contemporary levels obtained by Karchmer et al. (1974). In more recent work, Hasan and Hussian (1979, 1982) examined the tones emitted and the modes of unsteady jet flow over a range of R_D extending from 6×10^4 to 3×10^5 which may be induced by separation at the nozzle lip. The separation in this case is generated by a concentric step formed by a collar placed at the nozzle.

A supersonic version of the separating flow into a collar is that of a rapidly expanding supercritical jet into a duct. As discussed by Meir et al. (1980) and Muller (1980), the periodic interaction of the shock cell formed at the jet efflux with the bound annular stagnant fluid surrounding it causes the pulsations. The shock cell oscillates such that a fluctuating counter flow is generated in the contiguous *dead-air* region.

Little work on the tones from two-dimensional (or high-aspect ratio) impinging jets has followed the work of Powell and Unfried (1964) in which they observed relatively erratic behavior of edge tones, as part of their study of edge tones at lower Reynolds numbers. Although difficult to detect when the jet becomes fully turbulent, the tones reappear at higher speeds and may occur for a wider range of slit-edge spacing than in the laminar case and with reduced hysterises effects. With a still greater increase of exit velocity beyond critical, the screech tone appears. Though not reported, they speculated an interaction between shock cells and the edge which could give rise to another class of tones. Recently, Krothapalli (1983) observed dual tones emitted by a choked jet impinging on a flat plate. In addition to the high frequency screech tone, a lower frequency tone was emitted as a consequence of interaction of the jet with the plate which produced a lateral (edge tone-like) oscillation of the jet. Thus, the two-dimensional impinging choked jet may generate both the classical screech tones and the *180° wedge* edge tone simultaneously.

A few further investigations into the classical laminar two-dimensional edge tone offered additional analysis and measurements to look at details of the frequencies and hysteresis conditions near stage jumps (Stegen and Karamcheti, 1970); the quantification of jet

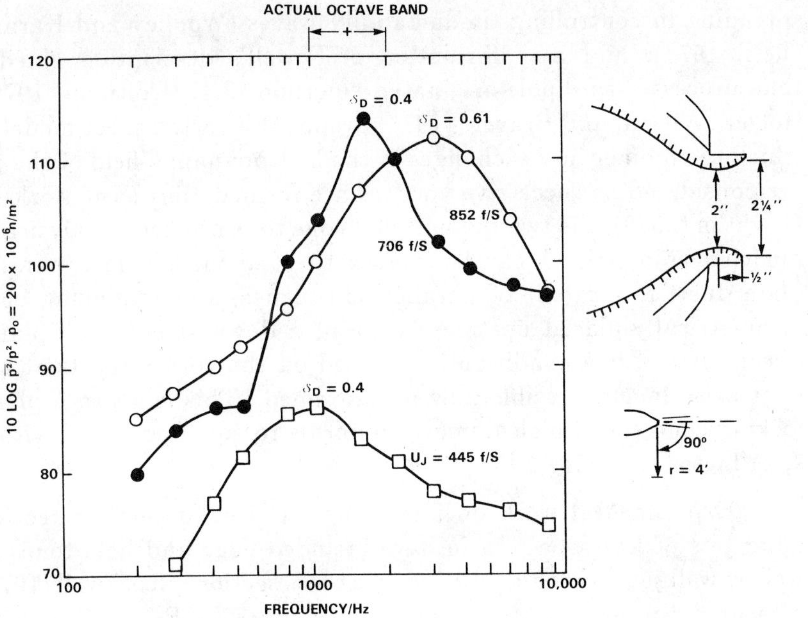

Figure 22. Sound pressure levels from *break away* nozzle. One-third octave band, adjusted ($-5db$) from octave band levels (Powell, 1951).

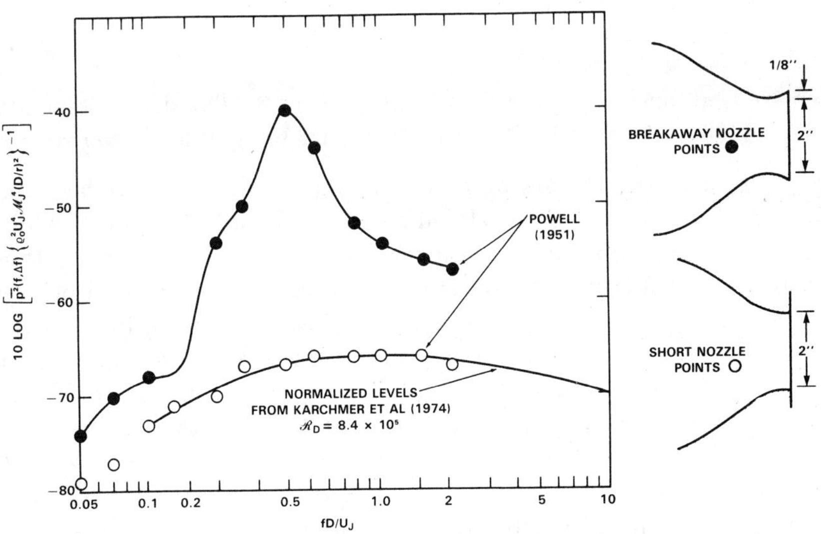

Figure 23. Jet noise in one-third octave bands at 90° nondimensionalized on quadrupole sound parameters.

spreading in controlling the instability waves (Woolley and Karamcheti, 1974); and a reaffirmation of Powell's description of edge tone processes as dipole-originated equation (3.4) (Coltman, 1976). Holger, Wilson, and Beavers (1977) proposed a vortex sheet model in which the momentum exchanges in the hydrodynamic field of the jet are considered as successive vortices are formed. In recent work by Crighton (1983), the two-dimensional edge tone has been analytically modeled as a pair of parallel vortex sheets issued from a parallel plate duct, i.e., a pair of parallel half-planes, and impinging on a symmetrically-placed upstream edge of a down stream half-plane. A complete Kutta condition is imposed on the vortex sheets at the duct exit. Incompressible flow is assummed. This leads to a phase locking condition which roughly conforms to equation (3.10) with a 7/24 factor replacing 7/28.

Experimental work on impinging on circular jets, or rectangular jets of low aspect ratio have included edge and hole tones as well as wall jets. Using a jet of square cross sections, Rockwell (1972) measured the first stage ($N = 1$) edge tones to $R_h = 7500$ with $\frac{h}{L}$ up to 10.8. The lowest Reynolds number for tone generation at any jet-edge separation was found to be 600 with Reynolds number independence for $R_h > 3000$. The Strouhal number was found to depend on L/h as

$$S = \alpha \frac{L}{h}$$

where $\alpha = 0.51$ at $\frac{L}{h} = 2$ to 0.31 at $\frac{L}{h} = 10.8$. The decrease in α is due to the reduction in $\overline{U}_c$ across the jet-edge gap as $\frac{L}{h}$ increases.

Experimental work on hole-tone like oscillations have been undertaken in a variety of unrelated and sporadic instances all of which involved the use of an enclosed resonant cavity of the general type illustrated in Figure 9d. Wilson et al. (1971), using curved orifices at a and a' (see Figure 9d) to promote flow separation, found tones to be generated in the range of S_D from 0.29 to 0.55 in the range of R_D from 5600 to 15,000. Flow visualization disclosed the formation of ring vortices at the upstream orifice which lost their identity before impinging the downstream orifice separated by fairly large L/D which ranged from 5.4 to 18. The mechanism of sound production was thus not considered to involve the impingement of vortex structures on the downstream orifice plate. Insertion of a porous baffle close to the downstream orifice to inhibit direct impingement of the jet did not remove the tone. Accordingly the primary acoustic source was viewed as generated by the upstream orifice alone, but enhanced by the Helmhooltz resonance of the chamber. Thus, it would appear that the Wilson's work would have some relationship to the sound generated by the separated flow nozzle as shown in Figures 22 and 23. Furthermore, Wilson et al. did not observe the generation of staged tones. Morel (1978) investigated a similar arrangement but with an upstream nozzle and a downstream orifice plate for a range of smaller $\frac{L}{D}$ from 0–8 and over a Reynolds number range from 2×10^4 to 10^5. In this case, staged tones were generated and the most intense tones occurred at the Helmholtz resonance frequencies of the chamber. Depending upon the jet-orifice separation, Strouhal numbers of the tones ranged from $S_D \simeq 0.2$ to 1.0. For a given stage, $S_D \simeq 1.0$ would occur at small separations, and $S_D \simeq 0.2$ at larger separations. Weak tones would occur whenever $\frac{L}{D}$ exceeded 4.5. The stages were ordered much as those of hole tones, equation (3.17). Isaacson and Marshall (1982), generated tones with two equal knife-edge rectangular slits of height h across a wind tunnel at $R_h =$ from 4.5×10^4 to 9×10^4 and separated $L/h = 2$. They observed fundamental tones at S_D between 0.5 and 1.0 which coupled with acoustic resonance of the cavity. In the same vein, Stull, Curron, and Velkoff (1983) generated tones with a resonating chamber slit jet and found that the primary jet mode was of the transverse type rather than the axisymmetric type. These two studies of *slot tones* are probably the first since von Gierke's (1950). Similar problems of tone generation in a pipe terminated by an axisymmetric cavity was examined by Shacknmann and Rockwell (1980) and of a jet crossing a cavity by Rebuffet and Guedel (1982). DeMetz et al. (1979) examined tones emitted from multi-holed orifice plates.

Sound radiation from high Reynolds number subsonic circular jets impinging on flat surfaces show similar features to edge tones. Although the first observations of such sounds was made by Marsh (1961) the similarities to the laminar jet tones could not be made until relatively recently. Marsh's measurements disclosed a tone generated at large enough exit velocities when the nozzle to plate distance was two diameters. For a fixed exit velocity, as the nozzle to plate distance was increased from two to five diameters, a pronounced peak in the 1/2-octave sound spectrum decreased in frequency and in intensity. The sound was more broadband for distances greater than five diameters. No visualizations of the flow were reported. The next reported measurements on the impinging jet were those of Powell and Unifred (1964), but as discussed above, they were on a rectangular jet. Our present knowledge of the stage behavior of the jet, and the occurrence of feedback is due to the work of Quick (1971) and Ho and Nosseir (1980).

Quick (1971) published a series of sound measurements and schlieren visualizations for a jet impinging on a rigid wall set 90° to the jet axis. Figure 24a shows that an axisymmetric vortex structure exists between the nozzle and the plate in qualitative similarity to jet modes active in hole tones.

The sound observed was intense and tonal; it was most pronounced for Mach numbers in excess of 0.6 and for L/D between 1 and 5 where L is the distance from the nozzle to the plate. Furthermore, the tone frequency coincided with the vortex impingement frequency on the plate, and as many as two tone stages were observed. The tone was also generated when the impinging plate was inclined to 60° from the jet axis. The creation of the tone and its dependence on Mach number and separation (L/D) were explained as due to an acoustic feedback from the pressure field on the plate and the exit conditions at the nozzle.

Recently, the experiment was repeated by Nosseir and Ho (1980, 1982) and Ho and Nosseir (1981) who examined more fully the vortex dynamics which controlled the pressure field on the impinging plate as well as the various stages. Figure 25 shows the succession of stages which follows an ordering of the type given by equation (3.8), i.e.,

$$\phi_N = 2\pi\left[\left(\frac{L}{\lambda_c}\right)_N + \left(\frac{L}{\lambda_o}\right)\right] \tag{4.2}$$

for a range of $2 < \frac{L}{D} < 8$ and a band of Strouhal numbers S_D

$$0.3 < S_D < 0.6$$

which has been shown to be controlled by the large eddy structure (see Figure 20). Visualization disclosed the same type of transition from shear wave instability through stages of vortex pairing. Nosseir and Ho deduced that acoustic radiation emanated from a zone near the impingement location. Subsequently, Ho (11983) has determined that following impingement, the vortices spread radially inducing a circular separation region on the plate as shown by the dye streak visualization in Figure 24b. Thus it appears that this phenomenon holds a very close analogy to the laminar flow hole tone with similar jet modes and feedback involved.

(a)

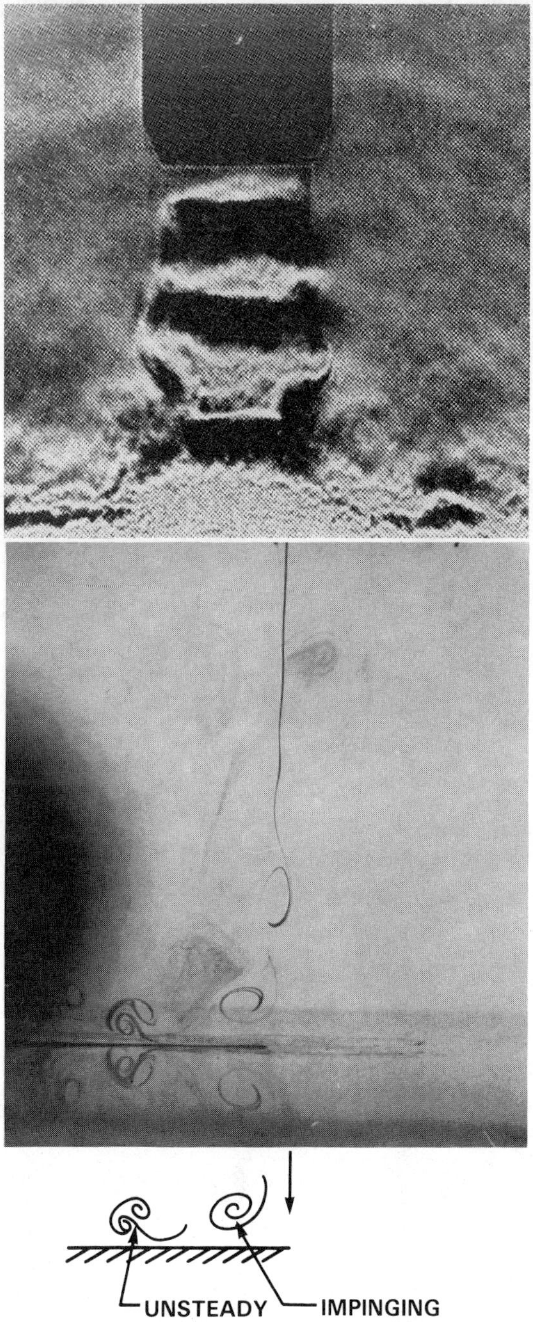

(b)

Figure 24. Flow visualization of vortex structures generated by impinging jets.

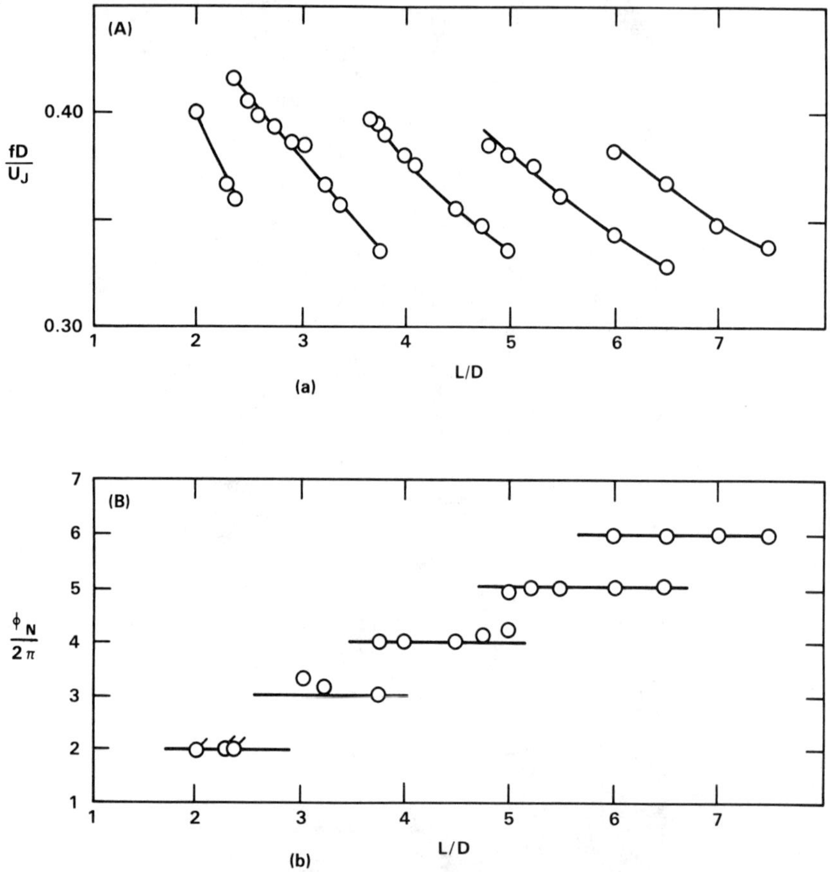

Figure 25. (a) Variations of the nondimensional resonance frequency parameter $f_r D/U$ with plate locations, $M = 0.9$. (b) Resonance frequency stages, $N = L/\lambda_C + L/\lambda_O$, o, measured; σ, calculated from equation (21a) with f_r measured from a surface pressure transducer and $C_1/U = 0.62$. (Ho and Nossier, 1981). Reprinted with permission of Cambridge University Press.

All these researches deal with an important class of jet tone problems at high Reynolds numbers. While tones are not generally excited by free turbulent jets with conventional nozzles, it is clear that the tendency of jets even at high Reynolds number, $R_D > 3000$, say, to form ordered vortex structures makes them candidates for creating forces under favoroable conditions. These conditions are:

flow separation at the nozzle lip, impingement of the jet on a plate (or very likely an orifice) for $1 < \frac{L}{h} < 7$, or the passage of two- or three-dimensional flow from a separated-flow nozzle through a chamber that will sustain a Helmholtz resonance with no impingement necessary.

4.3 Shear Layer/Edge Interaction Tones

4.3.1 Cavity Tone Stages

As discussed in Section 3, prior to 1970 work on cavity tones was much more fragmented with regard to the coupling of fluid mechanics and acoustics than was work on jet tones. Work on cavity tones per se continued with Heller, Holmes, and Covert (1971), Bilanin and Covert (1973), Martin et al. (1975), DeMetz and Farabee (1977), Ethembabaoglu (1978), and Sarhoia (1977) providing basically empirical evaluations of flow induced cavity oscillations. Of these, only that of DeMetz and Farabee dealt with the acoustically resonant cavity; all others dealt with the so-called *shallow cavity*. As time progressed, the interpretation of results would reflect the developing notions of shear layer instabilities and large scale vortices (Section 4.1). Heller et al. found Rossiter's (1964) formula to apply to the stages. In the case of small shallow cavities the stages do not depend much on the volume of the cavity, only on the breadth of its opening. With the work of Bilanin and Covert (1973) the coupling of the leading and trailing edges of the opening was considered in developing a formula for successive stages of tones. These involved the envisioning of a monopole line source at the downstream edge to induce an unsteady flux inside the cavity which would propagate upstream to deflect the shear layer. A formula for the stages like Rossiter's (1964), as equation (3.11) was accordingly rederived, but with an additional phase shift proposed to account for delays in the response of the shear layer to excitation. More recently Tam and Block (1978) used a similar model to describe volume-source-like trailing edge boundary condition.

The development of the shear layer across the cavity opening must be considered in any scheme to either predict or interpret the tone frequencies. DeMetz and Farabee (1977) and Sarhoia (1977) found that laminar shear layers over either resonant (deep) or non-resonant (shallow) cavities all generated tones at $S_\theta \simeq 0.017$ to 0.025

for L/θ between 7 and 50, in the case of DeMetz et al. and between 35 and 70 in the case of Sarhoia (1977). For turbulent flow over the openings of resonant cavities, the stages of DeMetz and Farabee (1977) as well as of other investigators followed the general rule

$$\frac{fL}{U_\infty} = \frac{\overline{U}_c}{U_\infty}\gamma_N \tag{4.3}$$

where γ_N is really $(L/\lambda)_N$, as shown in Figure 26. As the ratio δ/L increases, where δ is the thickness of the upstream boundary layer, the mean convection velocity and γ_N are expected to change in the direction of decreasing $\overline{U}_c/U_\infty$ with decreasing L/δ because a less of the faster-moving outer flow may be entrained into relatively small cavities. Depending on the selection of convection velocity, $\gamma_N = N \pm 1/4$ are possible alternates. In the case of laminar flow, Sarhoia's interpretation gave $\gamma_N = N + 1/2$, while a resonance condition of DeMetz gave $\gamma_N = N - 1/2$ with $\overline{U}_c/U_\infty \simeq 0.56$.

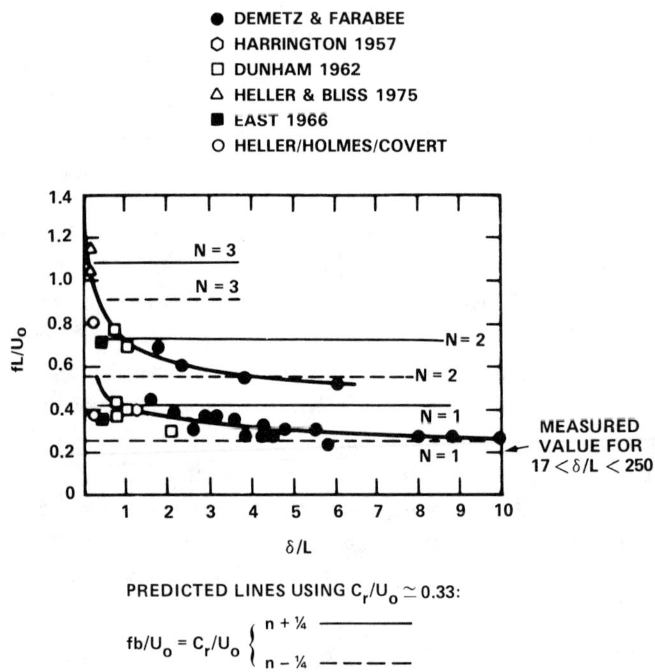

Figure 26. Cavity resonance frequencies for turbulance flow past the opening. (from DeMetz and Farabee, 1977). Copyright American Institute of Aeronautics and Astronautics, and reprinted with permission of the AIAA.

The notion of the average convection speed and wavelength has been pursued in the formulations proposed by Martin et al. (1975), Woolley and Karamcheti (1976), and Rockwell (1977), who suggested that the feedback was determined by a phase condition between the pressure oscillations in the cavity and the deflection of the shear layer streamlines into and out of the opening. Specifically, an average internal deflection caused a relative pressure maximum which induced an outward transverse deflection of the streamlines at the upstream lip.

All of these treatments of the cavity oscillations gloss over much of the fluid mechanics and acoustics of the coupling between cavity dynamics and shear layer mechanics. The presumed existence of a volume source at the downstream edge of the cavity opening as the exclusive originator of the feedback disturbance has not really been verified until recently in spite of its apparent usefulness in ad hoc accounting for the observed stages of tones.

Recent studies by Rockwell and Knisely (1979a, b) and Knisely and Rockwell (1982) begin to satisfy this need for physical understanding. In this work the development of the unsteady flow in a cavity opening has been studied in detail. The geometry involved a slot with 90° corners at the up and downstream edges and observations result in the following picture. The laminar shear layer (for $R_\theta < 330$) issued from the upstream edge, the layer undergoes a period of oscillation which culminates in the formation of a vortex, the vortex convects across the opening and impinges on the downstream corner. Upon impingement, the vortex is either directed into the cavity or swept downstream in the boundary layer. It appears that the impingement event couples with the fluctuation of the streamwise velocity at the leading edge. Specifically, the pressure on the downstream surface in the cavity is a minimum just as the vortex rounds the corner and is a maximum just before impingement when the streamline moves just inside the cavity. The upstream velocity is maximum between the minimum and maximum pressure as the ejected vortex is swept away by the downstream boundary layer. This is also the instant of maximum average deflection of the shear layer into the cavity. Thus the premise that feedback in small shallow cavities is controlled by a coupling of a volumetric influx and an outward velocity perturbation at the leading edge has some fluid dynamical support.

4.3.2 Free Mixing Layers

In this class of tones a shear layer impinged on a sharp edge, but it is not reinforced by a cavity. Figure 27 shows the arrangement used by Hussain and Zaman (1978). The fluctuating velocity near the apex of the wedge is the measured variable, radiated sound has not yet been reported for this tone. The frequency of the tone generated is normalized on the momentum thickness of the shear layer at the wedge, θ. The momentum thickness at the upstream edge is denoted by θ_0. As the distance between the generation point of the shear layer and the downstream edge is increased the shear layer oscillation frequency goes through stages according to the approximate rule.

$$\left(\frac{L}{\lambda_c}\right)_N \simeq N + \frac{1}{2}$$

The fifth stage actually has a *reduction* in frequency to 1/2 of the fundamental. This is attributed to vortex pairing in the shear which, as noted previously, may occur at distances downstream of L/θ_0 from 80 to 100. For smaller values of L/θ_0, the increase in S_θ with L/θ_0 matches the increase of θ with the spreading of the shear layer. Strouhal numbers S_{θ_0} would be much more constant for small values

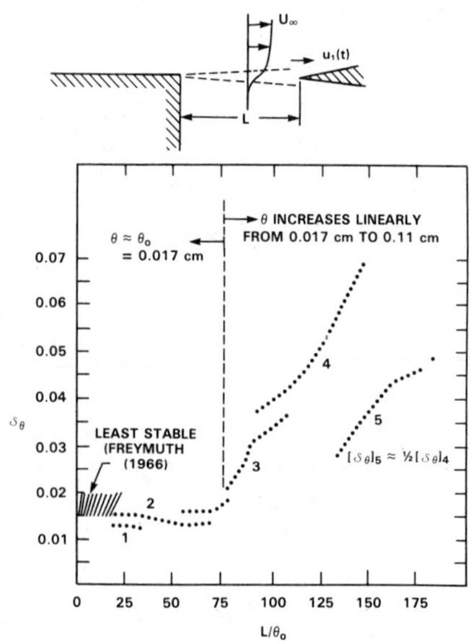

Figure 27. Shear layer tones, numbers denote stages. Strouhal nos. based on local θ, $\mathcal{R}_{\theta_0} = 404$ (Hussain and Zaman, 1978). Reprinted with permission of Cambridge University Press.

of L/θ_0, the S_θ thus appear to be determined by the least stable wave numbers in the shear layer as would be predicted by the theory (see also Figure 14).

The problem of determining the phase of the shear layer disturbance as it is convected across the opening is as complex as it has been in the case of edge tones. Attempts have been made at measuring the convection velocities across the gap by Zaida and Rockwell (1982a, b), Hussain and Zaman (1977), and across the cavity opening by DeMetz and Farabee (1977). These measurements are prone to inaccuracy because of phase variations of disturbances across the shear layer and because the shear layer undulates. Thus as velocity probes traverse the shear layer the measured phase shift is a mixture of contributing causes, only one of which is the actual convection of the disturbances.

4.4 Flow-Excited Resonators

Interaction of a shear flow with an acoustic cavity resonance or of a jet flow with a Helmholtz resonance was already touched on in Sections 4.2.2 and 4.3.1 when the results of Morel (1978) and DeMetz and Farabee (1977) were discussed. Flow excited resonances are possible whenever the resonance frequency of a mode of the resonator coincides with one of the possible instability modes of the flow. The acoustic sensitivity of shear flows makes it possible for the resonator to cause an ordering or regulation of the disturbance so that the ordered modes of the flow may actually extend beyond the parametric ranges for the flow without a resonator. Examples are shown in Figure 28 for the edge tone (Powell, 1963) excitation of a double ended organ pipe and for the axisymmetric jet oscillations in a resonator. In both cases six to nine stages are observed and these are bounded by the ranges of Strouhal number over which natural oscillations are possible. Note that the tones coincident with the first and second pipe resonances shown in Figure 28a are discontinuous.

When a resonator is present the character of the loop again described in Section 3.1 is changed because the physical constraint of the resonator may alter both the nature of the disturbance which may be generated at the downstream edge (or η_s and η_t in equations (3.3) and (3.4)) and the effectiveness with which the feedback disturbance drives the initial conditions on the flow, (or η_d in equation (3.5)). In the case of cavity tones, for example, the acoustic source is a volumetric pulsation of the cavity which radiates sound as an acoustic volume source.

An example of the alternative manifestations of speed dependence in different source mechanisms is shown in Figure 29. Dependence on speed of the sound intensity radiated by the double organ pipe shown in Figure 28 differs significantly from the dependence shown by the free-field edge tone. The intensity from the organ pipe follows a U^4 dependence which suggests monopole-like source mechanism. This contrasts with the velocity dependence of the free-field edge tone which generates a U^6 behavior with the dipole ($\cos^2 \phi$) directively measured by Powell (1961). Measurements of DeMetz and Farabee (1977) also demonstrated the volumetric source mechanism of a cavity tone emitted by a orifice in a wall by obtaining an omindirectional directivity pattern.

Most analytical work which has been developed deals with organ pipe resonances generated by the jet edge interaction. Typically the compressible fluid in the resonator and the shear layer mechanics are modeled as lumped parameter systems. Positive feedback to the shear layer is then generated because of the acoustical resonance in the pipe. The foundations of this vein of current analysis originated with Cremer and Ising (1967, 1968) who viewed the jet to generate an unsteady volumetric flow in the slit-edge opening. The volumetric pulsation, then, coupled to the acoustic field in the duct. An alternative view of the jet motion as generating an unsteady pressure oscillation due to momentum transfer in the opening was taken by Coltman (1968, 1969) shortly thereafter. It has been revised recently by Coltman (1976).

A recent series of papers by Elder (1973, 1978a, b, 1980) and Elder, Farabee, and DeMetz (1982) extend these notions to cavity resonators such as those studied by DeMetz and Farabee (1977). The drive effect of the shear layer oscillations is modeled as a simple lumped parameter with general behavior characteristics determined by elementary stability theory supported by measurements of shear layer disturbances in the openings of resonanting cavities. The method gives resonably good agreement with observed frequencies of flow excited resonance and the thesholds of onset and cutoff of various modes. Fletcher's (1974, 1976a, b, c) extension to the approach of Cremer and Ising has been devoted to the jet-edge excitation of the flute and the organ pipe. Various refinements are made to the modeling of the jet-edge dynamics and the generation

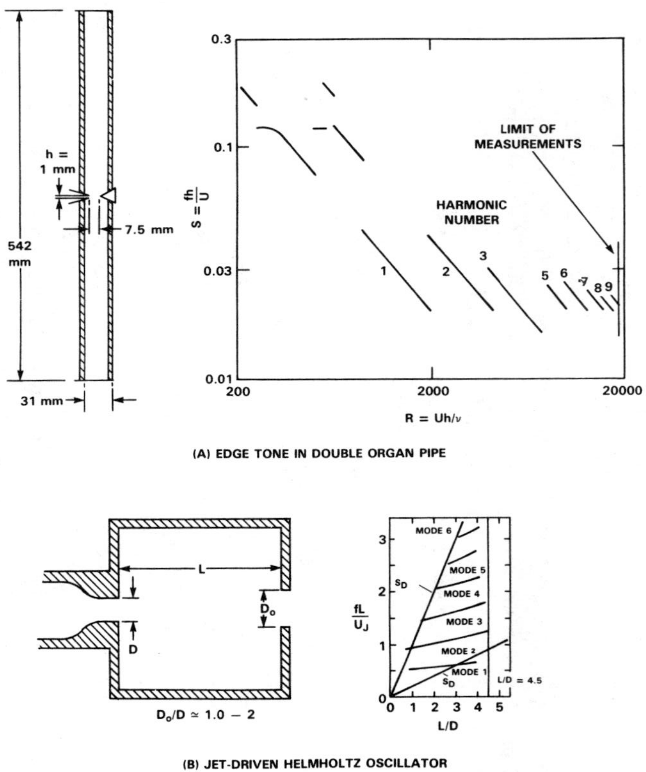

Figure 28. Examples of flow—excited resonators and the stages of flow tone.

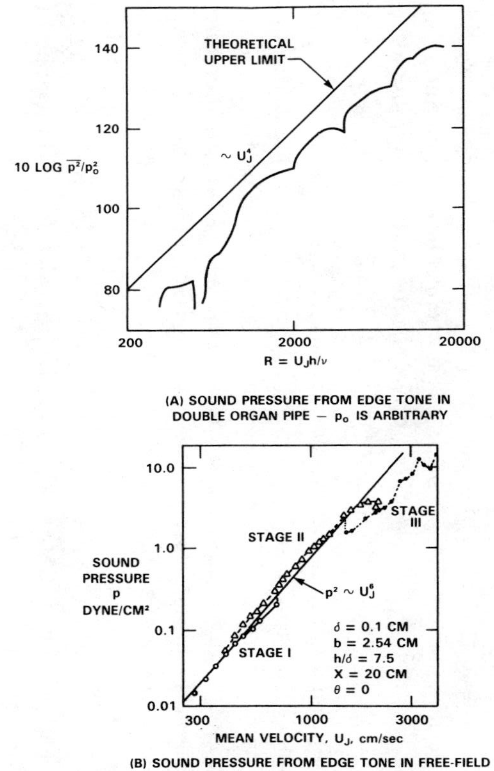

Figure 29. Variation of sound pressure with velocity for jet-edge in resonator and jet-edge in free space.

of transients as well. In both Elder's and Fletcher's approaches the shear layer mechanisms are not regarded as modified by the coupling to the resonator other than a simple linear relationship between acoustic motion and shear layer amplitude. Furthermore the acoustic impedance of the opening is not considered to be modified by the shear flow. These interactive effects may be significant as shown in the work of Ronneberger (1972, 1980).

The modeling of alternative approaches to shear flow excited resonances has been recently adopted by Howe (1981a,b) and Mironov (1982) who regard the shear layer motion as modified by the imposed oscillations of the acoustic resonator. This necessitates a modeling of the shear layer as a vortex sheet. The driving of the shear layers is coupled to the acoustical motions in the opening by applying a complete Kutta condition, i.e., $\eta(0) = 0$ and $\delta\eta(0)/\delta x = 0$ at the leading edge of the opening. This specification differs from the modeling of Elder and Fletcher in that it not only provides for interactions of the fluid motions, but also applies a condition of tangential flow at the opening. They also ignore the effect of flow on the acoustic impedance of the openings. The shear layer is interpreted to control the stages of oscillation by acting as a transductor of energy from the mean flow into the acoustic field. Questions remain, however, which concern the influence that the finite shear layer thickness and its spatial growth across the opening might have on the results and on the removal of certain inconsistencies between measurement and theory which are discussed in the papers. The limit thickness of the shear layer may also influence the validity of the Kutta condition at all frequencies of excitation, as demonstrated by Bechert and Pfizenmier (1975) (see Figure 18 and Section 4.1).

5. Concluding Remarks

Occurrence of jet tones may thus be expected at all Reynolds numbers from 100 to 10^6 depending on the geometry of the jet and the impingement surface. The flow structures which are responsible for these tones share in common the appearance of prominant eddies. Throughout a significant range of Reynolds numbers (1000 to 30,000), the axisymmetric eddies of circular jets (with nearly uniform exit velocity profiles) are now widely regarded as having emerged via a growth of well ordered vortex structures, the smallest of which is controlled by the naturally-occurring instability waves at the lip shear layer and the largest of which by the growth of subharmonics which mature at greater distances downstream as the order of the subharmonic cases decreases. The development has the visual appearance of a pairing in some cases. In nondriven jets this pairing per se is not as clear. Rather, as the jet spreads, the vortex diameter and spacing grows (in roughly similar proportion) in a manner analogous to the wake development behind a cylinder, but with a notable difference: that the Strouhal number decreases downstream of the jet exit. The view shown in Figure 15 of the free shear layer is qualitatively similar to this behavior.

The linear stability range of shear layer motions generally does not exist for R_θ greater than 300 to 400 (roughly corresponding to $R_D \simeq 3 \times 10^4$), yet regular eddy structures (and jet tones) do exist at such high Reynolds numbers. Clearly the fluid mechanics responsible for this behavior is fully nonlinear, and so the question of how there can be any connection to linear stability models remains as is occasionally suggested.

The observed jet tones do, however, fall into neat categories since behavior at all Reynolds numbers shows qualitative similarities for both circular and rectangular jets. The major distinction between these two geometries seems to be that circular jets undergo primarily (though not exclusively) axisymmetric mode oscillations while the rectangular jets undergo primarily (though not exclusively) antisymmetric mode oscillations. In both cases the laminar flow tone involves an oscillation of the jet column which disappears when turbulence sets in as $R_h \sim 1000$ or $R_D \sim 3000$. Tones are largely absent, or at least feeble over a succeeding range of Reynolds numbers until the turbulent jet tones are reestablished. These tones appear to involve the large eddy structures which we have been referring to as subharmonics of the principle shear layer modes near the jet exit and which seem to originate through a growth mechanism analogous to the linear instability of laminar jets. A clear connection between this stability analogy and the notion of vortex pairing has yet to be made particularly with regard to setting requirements for feedback loop gain.

When the nozzle flow becomes choked, the shock cell undulations generate screech tones apparently quite independently of whether or not the jet is impinging on a surface. It appears that jets impinging on flat surfaces are essentially special cases of the edge tone for rectangular jets or the hole (or ring) tone for circular jets.

The generation of cavity tones and requirements for shear layer mechanics in the orifice and the physics of feedback appear to be qualitatively analogous to the jet phenomena.

Questions still remain that concern the method of driving shear layers, principally jets, with regard to both the intensity of sound and the orientation of the source with respect to the nozzle that is required to create a given mode of disturbances. This is largely neglected topic, yet it relates to the response sensitivity of jets in various geometries. Very few controlled measurements have been conducted, and some discrepancies still persist particularly with regard to controlled acoustic excitation of axisymmetric and symmetric modes. Such experiments are very difficult, being hampered by undesired and uncontrolled reflections and by acoustically-induced and flow-induced vibration of nozzles and impinging surfaces. Similar questions remain in the case of cavity oscillations.

Numerical methods are now being developed which may be used to examine mathematically some of these questions. It may be anticipated that future work will include *canonical* flow tone problems which are addressed both experimentally and mathematically as parts of a single program of evaluation. In the program reported by Ohring (1983) the configuration under study is the jet issuing from a channel impinging on a wedge shaped body. The numerical modeling of the flow is being done at DTNSRDC on the Cray-1S computer. The complete Navier-Stokes equations are used in a finite difference scheme to compute vorticity and stream lines in a two-dimensional fluid space as a function of time following a perturbation in flow symmetry introduced at the exit plane. A parallel measurement program, using the same geometry, is underway at Lehigh University by Rockwell and Lucas. Figure 30 shows computed equivorticity lines for $R_h = 250$ which compares well to the experimentally observed streaklines. The basic features that are numerically simulated are the generation of vorticity in the shear layer of the nozzle, the distortion of the vortex by the edge, the production of additional vorticity by the edge-vortex interaction, the convection of vorticity downstream, and the phase between the force at the wedge and the incident vortex-leading edge interaction. Thus, all of the factors in equation (3.6) may be deduced. The development of such mathematical tools will permit parallel mathematical and physical experiments in which initial conditions, forms of flow excitation, and the effects of impingement surfaces may be varies systematically.

Another new trend in the analytical treatment of acoustically reinforced shear flows is a modeling of both the shear layer dynamics and the acoustic field as a coupled problem. An essential feature of this analysis is the provision of an upstream Kutta condition on the unsteady shear layer motion. However, some additional experimental investigations are required to determine what motions of the shear layer occur at the separating edges in order to establish any limitations that might exist in the analysis.

Investigations of jet and cavity tones which have been reported to date deal with aerodynamic sound radiation and fluid-born feedback paths. Another class of problems involves feedback through flow-induced vibration. Little work has been done to determine the types of vibration which may influence different flow geometries outside of that done on vortex shedding from vibrating cylinders and hydrofoils. Such properties as surface geometry, material properties, and structural dynamic fluid loading should all be examined. As noted above, such influences may also have caused confusion between measurements made by investigators over the years.

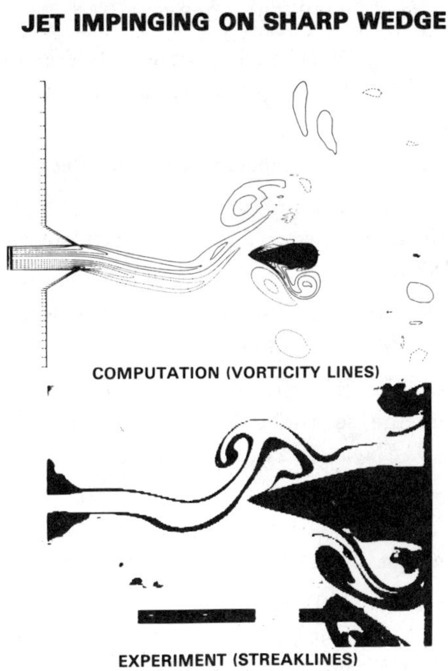

Figure 30. Numerically computed and experimentally visualized of the motion of a jet impinging on a sharp wedge (provided courtesy of S. Ohring).

References

[1] Anderson, A.B.C. "Dependence of Pfeifenton (Pipe Tone) Frequency on Pipe Length, Orifice Diameter, and Gas Discharge Pressure," *J. Acoust. Soc. Am. 24* (1952), 675–681.

[2] ——————. "Dependence of the Primary Pfeifenton (Pipe Tone) Frequency on Pipe-Orifice Geometry," *J. Acoust. Soc. Am. 25* (1953), 541–545.

[3] ——————. "A Circular-Orifice Number Describing Dependency of Primary Pfeifenton Frequency on Differential Pressure, Gas Density, and Orifice Geometry," *J. Acoust. Soc. Am. 25* (1953), 626–631.

[4] ——————. "A Jet-Tone Orifice Number for Orifices of Small Thickness-Diameter Ratio," *J. Acoust. Soc. Am. 26* (1954), 21–25.

[5] ——————. "Mestable Jett-Tone States of Jets From Sharp-Edged, Circular, Pipe-Like Orifices," *J. Acoust. Soc. Am. 27* (1955a), 13–21.

[6] ——————. "Structure and Velocity of the Periodic Vortex-Ring Flow Pattern of a Primary Pfeifenton (Pipe Tone) Jet," *J. Acoust. Soc. Am. 27* (1955b), 1048–53.

[7] ——————. "Vortex-Ring Structure-Transition in a Jet Emitting Discrete Acoustic Frequencies," *J. Acoust. Soc. Am. 28* (1956), 914–921.

[8] ——————. "The Sensitive Flame," *Proc. Phys. Soc., (London) 53* (1941), 329–355.

[9] Beavers, G. S. and Wilson, T. A. "Vortex Growth in Jets," *J. Fluid Mech. 44* (1970), 97–112.

[10] Becker, H. A. and Massaro, T. A. "Vortex Evolution in a Round Jet," *J. Fluid Mech. 31* (1968), 435–448.

[11] Bechert, Dietrich. "Die Steuerung eines ebenen Turbulenten Freistrahls durch eine seittliche Wechselstroomung, erzeugt in einem Schallfeld," *Z. Flugwiss, 24* (1976), 25–33.

[12] Bechert, Dietrich and Michel, V. "The Control of a Thin Free Shear Layer With and Without a Semi-Infinite Plate by a Pulsating Flow Field," *Acoustica, 33* (1975), 287–307.

[13] Bechert, Dietrich and Pfizenmaier, E. "Optical Compensation Measurements on the Unsteady Exit Condition at a Nozzle Discharge Edge," *J. Fluid Mech. 71* (1975), 123–144.

[14] Bickley, W. "The Plane Jet," *Phil. Mag. 23* (1937), 727–731.

[15] Bilani, A. J. and Convert, E. E. "Estimation of Possible Excita-

tion Frequencies for Shallow Rectangular Cavities," *AIAA J. 11 (3)* (1973), 347–351.

[16] Bouyoucos, J. V. and Nyborg, W. L. "Oscillations of the Jet-Edge System II," *J. Acoust. Soc. Am. 26* (1954), 511–514.

[17] Brackenridge, J. B. "Transverse Oscillations of a Liquid Jet," *J. Acoust. Soc. Am. 32* (1960), 1237–42.

[18] Brackenridge, J. B. and Nyborg, E. L. "Transverse Oscillations of a Liquid Jet II," *J. Acoust. Soc. Am. 33 (1961), 1078–84.*

[19] Browand, F. K. "An Experimental Investigation of the Instability of an Incompressible, Separated Shear Layer," *J. Fluid Mech. 26* (1966), 281–307.

[20] Browand, F. K. and Laufer, J. "The Role of Large Scale Structures in the Initial Development of Circular Jets," Dept. Aerospace Engrg. (1975), 333–344.

[21] Browand, F. K. and Weidman, P. D. "Large Scales in the Developing Mixing Layer," *J. Fluid Mech. 76,* 127–144.

[22] Brown, G. B. "On Sensitive Flames," *Phil. Mag. 13* (1932), 161.

[23] ——————. "On Vortex Motion in Gaseous Jets and the Origin of Their Sensitivity to Sound," *Proc. Phys. Soc., (London) 47* (1935), 703–732.

[24] ——————. "The Vortex Motion Causing Edge Tones," *Proc. Phys. Soc., (London) 49* (1937), 493–520.

[25] Brown, G. L. and Roshko, A. "The Effect of Density Difference on the Turbulent Mixing Layer Turbulent Shear Flows," *AGARD-CP-93* (1971a), 1–12.

[26] ——————————————. "On Density Effects and Large Structures in Turbulent Mixing Layers," *J. Fluid Mech. 64* (1974), 775–816.

[27] Chan, Y. Y. "Spatial Waves in Turbulent Jets," *Phys. Fluids, 17* (1974), 46–53.

[28] Chanaud, R. C. and Powell, A. "Experiments Concerning the Sound Sensitive Jet," *J. Acoust. Soc. Am. 34* (1962), 907–911.

[29] ——————————————. "Some Experiments Concerning the Hole and Ring Tone," *J. Acoust. Soc. Am. 37* (1965), 902–911.

[30] Clenshaw, C. W. and Elliot, D. "A Numerical Treatment of the Orr-Summerfeld Equation in the Case of a Laminar Jet," *J. Mech. Appli. Math. 13* (1960), 300.

[31] Coltman, J. W. "Sounding Mechanism of the Flute and Organ Pipe," *J. Acoust. Soc. Am.* (1968), 983–992.

[32] ——————. "Sound Radiation From the Mouth of an Organ Pipe," *J. Acoust. Soc. Am. 44* (1969), 477.

[33] ——————. "Jet Drive Mechanisms in Edge Tone and Organ Pipes," *J. Acoust. Soc. Am. 60* (1976), 725–733.

[34] Cremer, L. and Ising, H. "Die Selbsterregte Schwingungen yon Orgelpfeifen," *Acoustica, 19* (1967/1968), 143–153.

[35] Crighton, D. G. and Gaster, M. "Stability of Slowly Diverging Jet Flow," *J. Fluid Mech. 77* (1976), 397–413.

[36] Crighton, D. G. To be published, 1983.

[37] Crow, S. C. and Champagne, F. H. "Orderly Structure in Jet Turbulence," *J. Fluid Mech. 48* (1971), 547–591.

[38] Curle, N. "The Mechanics of Edge Tones," *Proc. Roy. Soc., (London), A216* (1953), 412–424.

[39] ————. *Proc. Roy. Soc., (London), A231* (1955), 505.

[40] Darlington, W. H. "Air Jets for Use in Gas Turbulent Combustion Chambers," Univ. of London, *Ph.D. thesis,* 1949.

[41] Davies, M. G. and Oldfield, D.E.S. "Tones From a Choked Axisymmetric Jet, Part I: Cell Structure, Eddy Velocity and Source Locations; Part II: the Self-Excited Loop and Mode of Oscillation," *Acoustica, 12* (1962), 257–277.

[42] DeMetz, F. C. and Farabee, T. M. "Laminar and Turbulent Shear Flow Induced Cavity Resonances," *AIAA 77-1293,* 4th Aerospace Conference, Atlanta, GA, 1977.

[43] DeMetz, F. C., et al. "Noise Radiation From Subsonic Airflow Through Simple and Multiholed Orifice Plates," *DTNSRDC Report SAD 237E-1942,* 1979.

[44] Dunham, W. H. "Flow-Induced Cavity Resonance in Viscous Compressible and Incompressible Fluids," 4th Symposium on Naval Hydrodynamics, Washington, DC, 1962.

[45] East, L. F. "Aerodynamically Induced Resonance in Rectangular Cavities," *J. Sound Vib. 3* (1966), 277–287.

[46] Elder, S. A. "On the Mechanism of Sound Production in Organ Pipes," *J. Acoust. Soc. Am. 54* (1973), 1554–64.

[47] —————. "Self-Excited Depth-Mode Resonance for a Wall-Mounted Cavity in Turbulent Flow," *J. Acoust. Soc. Am. 64* (1978a), 877–890.

[48] —————. "Edge Tones Versus Pipe Tones," *J. Acoust. Soc. Am. 64* (1978b), 1721–23.

[49] —————. "Forced Oscillations of a Separated Shear Layer With

Application to Cavity Flow-Tones Effects," *J. Acoust. Soc. Am. 67* (1980), 774–781.

[50] Elder, S. A., Farabee, T. M., and DeMetz, F. C. "Mechanisms of Flow-Excited Cavity Tones at Low Mach Nummber," *J. Acoust. Soc. Am. 72* (1982), 532–549.

[51] Ethembabaoglu, S. "Some Characteristic of Unstable Flow Past Slots," *J. Hydraulic Div. 104* (1978), 649–666.

[52] Fletcher, N. H. "Nonlinear Interaction in Organ Flue Pipes," *J. Acoust. Soc. Am. 56* (1974), 645–652.

[53] ———————. "Transients in the Speech of Organ Flue Pipes—A Theoretical Study," *Acoustica, 34* (1976a), 224–233.

[54] ———————. "Jet-Drive Mechanism in Organ Pipes," *J. Acoust. Soc. Am. 60* (1976b), 481–483.

[55] ———————. "Sound Production by Organ Flue Pipes," *J. Acoust. Soc. Am. 60* (1976c), 926–936.

[56] Freymuth, P. "On Transition in a Separated Laminar Boundary Layer," *J. Fluid Mech. 25* (1966), 683–704.

[57] Fuchs, H. V., Fisher, M. J., and Lau, J. C. "The Intrinsic Structure of Turbulent Jets," *J. Sound Vib., 22* (1972), 379–406.

[58] Gierke, Henning von. "Uber Schneidentone an Kreisrunden Gasstrahlen und ebenen Lamellen," *Zeitschrift fur angewandte Phys. 3* (1950), 97–106.

[59] Gross, M. J. "Underwater Edge Tones," *Acoustica, 9* (1959), 164–172.

[60] Hasan, M.A.T. and Hussain, A.K.M.F. "A Formula for Resonance Frequencies of a Whistler Nozzle," *J. Acoust. Soc. Am. 65* (1979), 1140–42.

[61] ————————————————. "The Self-Excited Axisymmetric Jet," *J. Fluid Mech. 115* (1982), 59–89.

[62] Heller, H. H., Holmes, D. G., and Covert, E. E. "Flow-Induced Pressure Oscillations in Shallow Cavities," *J. Sound Vib. 18* (1971), 545–553.

[63] Hammitt, A. G. "The Oscillation and Noise of an Overpressure Sonic Jet," *J. Aero. Sci. 28* (1961), 673–680.

[64] Ho, C. M. "Free Shear Layers and Stability Analyses," *Lecture Notes of Phys. Springer Verlag Co. 1-6,* 1981.

[65] Ho, C. M. Private communication, 1983.

[66] Ho, C. M. and Huang, L. "Subharmonics and Vortex Merging in Mixing Layers," *J. Fluid Mech. 119* (1982), 443–473.

[67] Ho, C. M. and Nosseir, N. S. "Dynamics of an Impinging Jet, Part 1: The Feedback Phenomenon," *J. Fluid Mech. 105* (1981), 119–142.

[68] Holger, P. K., Wilson, J. A., and Beavers, G. S. "Fluid Mechanics of the Edge Tone," *J. Acoust. Soc. Am. 62* (1977), 1116–28.

[69] Howard, L. N. "Hydrodynamic Stability of A. Jet," *J. Math. Phys. 37* (1959), 283–298.

[70] Howe, M. S. "The Role of Displacement Thickness Fluctuations in Hydroacoustics, and the Jet-Drive Mechanism of the Flue Organ Pipe," *Proc. R. Soc. Am. 374* (1981a), 543–568.

[71] ————. "The Influence of Mean Shear on Steady Aperture Flow, With Application tto Acoustical Diffraction and Self-Sustained Cavity Oscillations," *J. Fluid Mech. 109* (1981b), 125–146.

[72] Hussain, A.K.M.F. and Zaman, K.B.M.Q. "Controlled Perturbation of Circular Jets," Symp. Turb., Berlin: Springer Verlag, 1977, 31–42.

[73] ————————————. "The Free Shear Layer Tone Phenomenon and Probe Interference," *J. Fluid Mech. 87* (1978), 349–383.

[74] Isaacson, L. K. and Marshall, A. G. "Acoustic Oscillations in Internal Cavity Flows: Nonlinear Resonant Interactions," *AIAA Journal 20* (1982), 152–154.

[75] Johansen, F. C. "Flow Through Pipe Orifices at Low Reynolds Numbers," *Proc. Roy. Soc. Am. 216* (1929), 231–245.

[76] Karamcheti, K. and Bauer, A. B. "Edge Tone Generation," *Dept. Aeronautics and Astro., Stanford Univ. Rep. 162* (1963), 1–76.

[77] Karamcheti, K. et al. "Some Features of an Edge Tone Flow Field," NASA Basic Aerodynamic Noise Research Conference, *NASA SP-207 (1969), 275–304.*

[78] Karchmer, A. M. et al. "Acoustic Tests of a 15.2-Centimeter-Diameter Potential Flow Convergent Nozzle," *NASA TM X-2980,* 1974.

[79] Kelly, R. E. "On the Stability of an Inviscid Shear Layer Which is Periodic in Space and Time," *J. Fluid Mech. 27* (1967), 657–689.

[80] King, J. L., Boyle, P., and Ogle, J. B. "Instability in Slotted Wall Tunnels," *J. Fluid Mech. 4* (1958), 283–305.

[81] Knisely, C. and Rockwell, D. "Self-Sustained Low-Frequency Components in an Impinging Shear Layer," *J. Fluid Mech. 116* (1982), 157–186.

[82] Krothapalli, A. "Discrete Tones Generated by an Impinging Underexpanded Rectangular Jett," *AIAA Paper 83-0729,* 1983.

[83] Krothapalli, A., Baganoff, D., and Hsia, D. Y. "On the Mechanism of Screech Tone Generation in Underexpanded Rectangular Jets," *AIAA Paper 83-0727*, 1983.

[84] Kruger, F. "Theorie der Schneidentone," *Ann. Physik. 62* (1920), 673.

[85] Kurzweg, H. "Neue Unterschaungen uber die Entsehung der Turbulenten Rohrtromung," *Ann. der Physik, 18* (1933), 193–216.

[86] LeConte, J. "On the Influence of Musical Sounds on the Flame of a Jet in a Cool Gas," *Phil. Mag. 15* (1858), 235.

[87] Lenihan, J.M.A. and Richardson, E. G. "Observations on Edge Tones," *Phil. Mag. 29* (1940), 400–406.

[88] Lighthill, J. "On Sound Generated Aerodynamically, One General Theory," *Proc. Roy. Soc., (London), A211* (1952), 564–587.

[89] Lucas, M. and Rockwell, D. "Self-Excited Jet: Upstream Modulation and Multiple Frequencies," in preparation.

[90] Marsh, A. H. "Noise Measurements Around a Subsonic Air Jet Impinging on a Plane Rigid Surface," *J. Acoust. Soc. Am. 33* (1961), 1065–66.

[91] Martin, W. W., Naudascher, E., and Padmanabhan, M. "Fluid-Dynamic Excitation Involving Flow Instability," *Proc. ASCE J. Hydraulics Div.* (1975), 681–698.

[92] Meier, G.E.A. et al. "Oscillations of the Supersonic Flow Downstream of an Abrupt Increase in Duct Cross Section," *AIAA Journal, 18* (1980), 394–395.

[93] Michalke, A. "On Spatially Growing Disturbances in an Inviscid Shear Layer," *J. Fluid Mech. 23* (1965), 521–544.

[94] Miksad, R. W. "Experiments on the Nonlinear Stage of Free-Shear-Layer Transition," *J. Fluid Mech. 56*, 695–719.

[95] Mironov, M. A. "Impedance of a Hole in a Wall Separating a Moving Medium From a Nonmoving Medium," *Sov. Phys. Acoust. 28* (1982), 310–313.

[96] Mokhtar, M. and Youssef, M. "Observations on Edge Tones," *Acoustica, 2* (1952), 135.

[97] Morel, T. "Experimental Study of a Jet-Driven Helmholtz Oscillator," *Am. Soc. Mech. Engrg. Paper 78-WA/FE-16*, 1978.

[98] Morkovin, M. V. and Parangape, S. V. "On Acoustic Excitation of Shear Layers," *Zietschrift for Flugwiss enschaftern 19 (1971), 328–337.*

[99] Muller, E. A. "Recent Progress in Understanding Basic Progress in Flow-Acoustics," invited lecture 10th International Congress on

Acoustics, Sidney, Australia, July 9–16, 1980.

[100] Nosseir, N. and Ho, C. M. "Pressure Fields Generated by Instability Waves and Coherent Sturctures in an Improving Jet," AIAA 6th Aero. Conf. 1-5, 1980.

[101] ———————. "Dynamics of an Impinging Jet, Part 2: The Noise Generation," *J. Fluid Mech. 116* (1982), 379–391.

[102] Nyborg, W. L. "Self-Maintained Oscillation of the Jet in a Jet-Edge System," *J. Acoust. Soc. Am., 26* (1954), 174–182.

[103] Nyborg, W. L., Burkhard, M. D., and Schilling, H. K. "Acoustical Characteristics of Jet-Edge Resonator System," *J. Acoust. Soc. Am., 24* (1952), 293–304.

[104] Nyborg, W. L., Woodbridge, C. L., and Schilling, H. K. "Characteristics of Jet-Edge-Resonator Whistles," *J. Acoust. Soc. Am., 25* (1953), 138–146.

[105] Ohring, S. "Numerical Solution of an Impinging Jet Flow Problem," to be presented at the GAMM Conference on Numerical Methods in Fluid Mechanics, Univ. of Rome, Italy, Oct. 5–7, 1983.

[106] Pai, S. I. "On the Stability of Two Dimensional Laminar Jet Flow of Gas," *J. Aeronaut. Sci., 18* (1951), 731–743.

[107] Peterson, K. A., Kaplan, R. E., and Laufer, J. "Ordered Structures and Jet Noise," *NASA CR-139733* (1974), 1–23.

[108] Powell, A. "A Schlieren Study of Small Scale Air Jets and Some Noise Measurements In Two-Inch Diameter Air Jets," *ARC 14726 FM 1694*, 1951.

[109] ————. "On Edge Tones and Associated Phenomena," *Acoustica, 3* (1953a), 233-243.

[110] ————. "The Noise of Choked Jets," *J. Acoust. Soc. Am. 25* (1953b), 385–388.

[111] ————. "On the Mechanism of Choked Jet Noise," *Phys. Soc. London, B66* (1953c), 1039–57.

[112] ————. "On the Noise Emanating From a Two-Dimensional Jet Above the Critical Pressure," *Aero. Quart., 4* (1953d), 103–122.

[113] ————. "Theory and Experiment in Aerodynamic Noise With a Critique of Research on Jet Flows and Their Relationship to Sound," Second Symposium on Naval Hydrodynamics, Washington, D.C., 1958.

[114] ————. "On the Edge Tone," *J. Acoust. Soc. Am. 33* (1961), 395–409.

[115] ————. "Vortex Action in Edge Tones," *J. Acoust. Soc. Am., 34* (1962a), 163–166.

[116] ————. "Nature of the Feedback Mechanism on Some Fluid Flows Producing Sound," 4th International Cong. on Acoustics, Copenhagen, 1962b.

[117] ————. "Mechanisms of Aerodynamic Sound Production," *AGARD Report 466,* 1963.

[118] ————. "Aspects of Edge Tone Experiment and Theory," *J. Acoust. Soc. 37 (1965), 535–536.*

[119] ————. "Flow Noise; A perspective on Some Aspects of Flow Noise and of Jet Noise in Particular, Part 1," *Noise Control Engineering, 8* (1977a), 69–80.

[120] ————. "Flow Noise; A Perspective on Some Aspects of Flow Noise and of Jet Noise in Particular, Part 2," *Noise Control Engineering, 8* (1977b), 108–120.

[121] Powell, A. and Unfried, H. H. "An Experimental Study of Low Speed Edge Tones," Dept. of Engineering, Univ. of Calif. Reg. 64-49, 1964.

[122] Quick, A. W. "Zum Schall-und Stromungsfeld eines axialsymmetrischen Freistrahs veim Auftreffen auf leine Wand," *Zietschrift fur Flugwissenschaftern 19* (1971), 30–44.

[123] Rayleigh, Lord. "Pure Tones From Sounding Observations," *Theory of Sound,* New York: Dover, 1945.

[124] Rebuffet, P. and Guedel, A. "Model Based Study of Various Configuration of Jet Crossing a Cavity," *Recherche Aerospatiale, N.1* (1982), 11–22.

[125] Richardson, E. G. "Edge Tones," *Proc. Phys. Soc., (London), 43* (1931), 394–404.

[126] ————. *Sound,* New York: Langman's Green, 1947.

[127] ————. "Acoustic in Relation to Aerodynamics," *J. Aeronaut. Sci., 22* (1955), 775.

[128] Rockwell, D. O. "Transverse Oscillations of a Jet in a Jet-Splitter System," *ASME J. Basic Engrg. 94* (1972), 675–681.

[129] ————. "Prediction of Oscillation Frequencies for Unstable Flow Past Cavities," *J. Fluids Engrg., 99* (1977), 294–300.

[130] Rockwell, D. and Knisely, C. "The Organized Nature of Flow Impingement Upon a Corner," *J. Fluid Mech., 93* (1979a), 413–432.

[131] ————. "Unsteady Features of Flow Past a Cavity," *ASCE J. Hydraulic Div. 105* (1979b), 969–979.

[132] Ronneberger, D. "The Acoustical Impendance of Holes in the Wall of Flow Ducts," *J. Sound Vib., 24* (1972), 133–150.

[133] ————. "The Dynamics of Shearing Flow Over a Cavity—

a Visual Study Related to the Acoustic Impendance of Small Orifices," *J. Sound Vib., 71* (1980), 565–581.

[134] Rosenhead, L. "The Formation of Vortices From a Surface of Discontinuity," *Proc. Roy. Soc., (London), 134* (1931), 170–193.

[135] Roshko, A. "Structure of Turbulent Shear Flows: a New Look," *AIAA J., 14* (1976), 1349.

[136] Rossiter, J. E. "Wind Tunnel Experiments on the Flow-Over Rectangular Cavities at Subsonic and Transonic Speeds," *Royal Aircraft Establishment Tech. Rep. 64037,* 1964.

[137] Sarhoia, V. "Experimental Investigation of Oscillations in Flows Over Shallow Cavities," *AIAA, 15* (1977), 984–991.

[138] Sato, H. "Experimental Investigation on the Transition of Laminar Separated Layer," *J. Phys. Soc. Japan, 2* (1958), 702–709.

[139] ———. "The Stability and Transition of a Two-Dimensional Jet," *J. Fluid Mech., 7* (1960), 53–80.

[140] Sato, H. and Sakao, F. "An Experimental Investigation of the Instability of a Two-Dimensional Jet at Low Reynolds Numbers," *J. Fluid Mech., 20* (1964), 337–352.

[141] Savic, P. "On Acoustically Effective Vortex Motion in Gaseous Jets," *Phil. Mag., 32* (1941), 245–252.

[142] Schlicting, H. "Z. Angew," *Math. U. Mech., 13* (1933), 260.

[143] Shachenmann, A. and Rockwell, D. "Self-Sustained Oscillations of Turbulent Pipe Flow Terminated by an Axisymmetric Cavity," *J. Sound Vib., 73* (1980), 61–72.

[144] Sondhauss, C. "Uber die beim Ausstromen der Luft ensthenden Tone," *Ann. Physik 91* (1854), 126–147, 214–240.

[145] Stegun, G. R. and Karamcheti, K. "Multiple Tone Operation of Edge Tones," *J. Sound Vib. 12 (3)* (1970), 281–284.

[146] Stull, F. D., Curran, E. T., and Velkoff, H. R. "Investigation of Two-Dimensional Cavity Diffusers," *AIAA Paper No. 73* (1973), 685.

[147] Tam, C.K.W. and Block, J. W. "On the Tones and Pressure Oscillations Induced by Flow Over Rectangular Cavities," *J. Fluid Mech., 89* (1978), 373–399.

[148] Tatsumi, B. T. and Kakutani, T. "The Stability of a Two Dimensional Laminar Jet," *J. Fluid Mech., 4* (1958), 261–275.

[149] Tyndall, J. "On the Action of Sonorous Vibrations on Gaseous and Liquid Jets," *Phil. Mag., 33* (1867), 375–391.

[150] ———. "The Science of Sound," reprinted by the Citadel Press, 1964.

[151] Wilson, T. A. et al. "Experiments on the Fluid Mechanics of Whistling," *J. Acoust. Soc. Am. 50,* (1971), 366–371.

[152] Winant, C. D. and Browand, F. K. "Vortex Pairing: The Mechanism of Turbulent Mixing Layer Growth at Moderate Reynolds Number," *J. Fluid Mech., 63* (1974), 237–255.

[153] Woolley, J. P. and Karamcheti, K. "Role of Jet Stability in Edge Tone Generation," *AIAA J.* (1974), 1457–58.

[154] ————————————. "The Two-Dimensional Development of Flow in the Presence of Periodic of Random Fluctuations With Application to the Calculation of Cavity Tones," Nielsen Engrg. Research Inc., 1–92, 1976.

[155] Zaida, S. and Rockwell, D. "Oscillations of an Unstable Mixing Layer Impinging Upon an Edge," *J. Fluid Mech., 124* (1982a), 307–334.

[156] ————————————. "Generation of Higher Harmonics in a Self-Oscillating Mixing Layer-Wedge System," *AIAA, 20* (1982b), 196–202.

A Simple Theory of the Sondhauss Tube

Nikolaus Rott
Federal Institute of Technology
CH-8092 Zurich

"Complex problems have simple, easy-to-understand wrong answers." Grossman's misquote of H. L. Mencken, from *Murphy's Law, Book II,* by Arthur Bloch, 1980.

Of the many fascinating thermoacoustic devices, in which heat is the direct cause of gas oscillations, probably the oldest known is the so-called Sondhauss-tube. The name (which is not universally used) does not refer to the discoverer (who is unknown), but to the first systematic experimenter, whose work (1850) is quoted in Rayleigh's famous "Theory of Sound". The earliest reference to the effect is probably a *Letter to the Editor* in Gilbert's *Annalen der Physik* of the year 1804. Since then, several descriptive reports were published *before* Sondhauss; there was an inventor who wanted to use the effect for a musical instrument and had a correspondence on this idea with Goethe in 1827.

The device itself is extremely simple. A hot bulb is connected to a cold tube; the temperature change occurs in a short section of the cold tube nearest to the bulb, which is to be called the *neck* (Figure 1). Such configurations were occasionally produced by glass-blowers, who sometimes noticed that when a hot bulb was blown, the air in the device came to spontaneous oscillations. Sound of a certain frequency was produced, which faded away while the bulb cooled. Sondhauss observed correctly that the shape of the bulb is of negligible importance. He also observed that the frequency is independent of the bulb temperature, and he gave a formula for the frequency as a function of bulb volume, tube length, and tube cross section. Rayleigh immediately recognized this frequency as the natural frequency of an oscillator with the bulb as spring and the cold gas in the tube as mass. Damping is known to influence this result only by small corrections (except for extreme cases); for the Sondhauss-tube, the *neck* is obviously an element of negative damp-

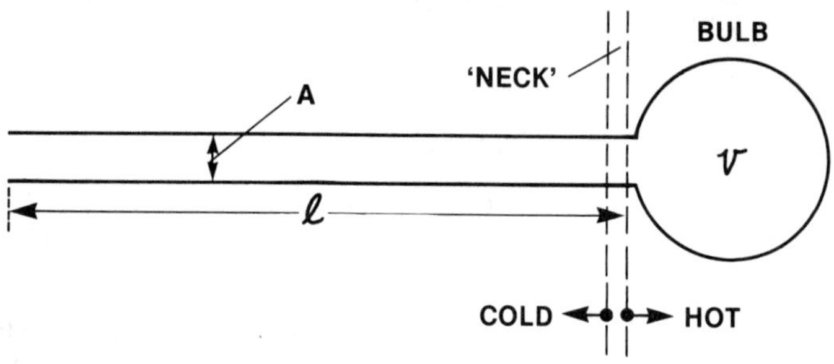

Figure 1. Sondhauss-tube.

ing, and the real part of the frequency is known to a high degree of accuracy. To ask for the imaginary part, i.e., to ask for the negative damping, that is to ask for the mechanism of thermoacoustic excitation.

With the problem cleared thus far, a strong temptation has to be fought, namely to think of a *model* which would produce the negative damping and thus provide a simple explanation. A model can be defined as a mechanical system in which at least one parameter is free to adjust for unknown mechanical features. It follows that any problem in which there is only one significant unknown is not a suitable object for modeling.

Rayleigh was cautious in his comment on the phenomenon and gave only a necessary condition for the maintenance of the oscillations. He pointed out that in the high pressure phase of the oscillation, gas is being displaced into the hot bulb and heated; conversely, cooling takes place in the low pressure phase. Thus, the maintenance of the oscillations is energetically possible, and a hot closed end is a necessary conditions for the excitation.

Attempts to get a sufficient condition for the stability limit from energy considerations alone were not successful. A new era started when the same kind of oscillation was observed in connection with quite different devices. Indeed, the new configurations were so different that the relations to the Sondhauss-tube were not recognized in the beginning. It started with the liquification of helium by Kammerlingh Onnes (1908). Thermoacoustic oscillations were observed from the beginning in configurations as shown in Figure 2. A tube leads into a dewar with liquid helium, and has an open end in the helium vapor region, at 4.2 K (for atmospherically vented containers). Outside the dewar, the tube takes on room temperature (300°K, say) and is typically closed by a valve. In this configuration the tube has (in contrast to the Sondhauss tube) a constant cross section; also, the temperature change occurs over a longer fraction of the tube. The temperature ratio between hot and cold in this case is naturally immensely high compared to the one for the tubes investigated by Sondhauss.

It will be shown that all these differences do not affect the basic elements of the forthcoming explanation. Nevertheless, the straight tube and the not-too-steep temperature gradient is more suggestive of a key element in the explanation of the instability, namely, that it is caused by the periodic expansion and contraction of the wall boundary layer in the gas.

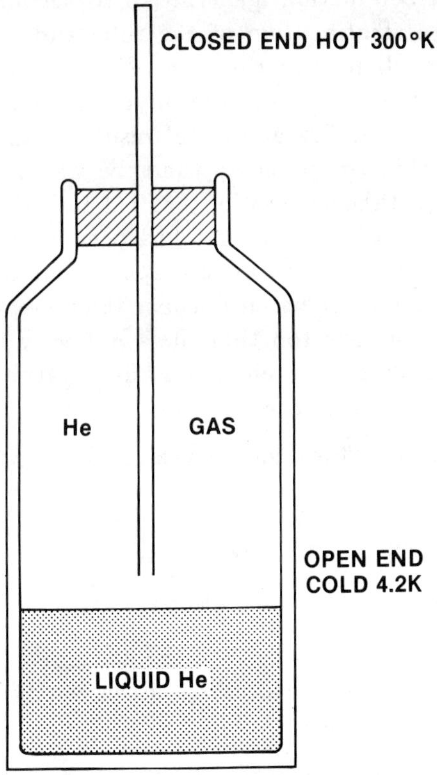

Figure 2. Taconis oscillator.

The first to propose such a mechanism was Taconis (1949). For the phase when the gas is displaced furthest towards the hot closed end, the following two facts have to be noted: first, the pressure is highest; second, the boundary layer is dragged furthest towards hotter wall regions and therefore expands. The expanding boundary layer *squeezes* (by a transverse motion) the gas core and thus increases the pressure, in the phase when it is already highest. With the converse happening at the phase of the lowest pressure, which is further diminished by a *yielding* boundary layer condensed on colder wall parts, the driving mechanism is clear. Actually, one has to realize that there is a phase lag between displacement and pressure; thus, velocity and pressure have in-phase components and excitation is possible.

Inspired by Taconis' idea, a calculation was made by Kramers (1949) based on the Kirchhoff theory of oscillating gas columns with friction and heat conduction, generalized to account for (common) axial temperature stratification of the tube and the gas, with the radial pressure gradient and the axial effects of friction and heat conduction neglected. One more simplifying assumption was important for Kramers to obtain analytical results, namely, to have the boundary layers thin compared to the tube wall radius. As can be seen in retrospect, this assumption has led to results not in agreement with experience, a story that has been told often before by the author. One can see, also in retrospect, that for the excitation of oscillations, there must be an *optimal thickness* of the boundary layer, because when it is too thin, it is not very effective; on the other hand, when it is thick enough to fill the tube, the driving is *choked* by the dissipative effects.

The thickness of the boundary layer δ is given, according to Stokes, by the formula

$$\delta \sim \sqrt{\frac{\nu}{\omega}}$$

where ν is the kinematic viscosity of the gas and ω is the frequency. The ratio of δ to the tube radius r_w is the essential fluid dynamic parameter of the problem.

There was one more simplification that Kramers made for his calculations in the limit $\delta/r_w \to 0$, namely, he assumed that the axial temperature distribution has two constant levels, and that there is a jump between the hot and the cold part. The two isothermal sections can be described by analytic solutions, and the problem is reduced to join the two solutions properly at the position of the temperature jump. The joining condition at the discontinuity has to express, as is well known in acoustics, that pressure and displacement must be continuous. The pressure is constant across the tube and poses no problems, but with the velocity distribution varying radially, the continuity of the displacement needs a careful investigation, which becomes difficult in case that the mean gas temperature has a jump. It has to be noted that Kramers did not face this difficulty, because with very small boundary layers the constant velocity part over the cross-section dominates and the classical joining conditions for displacements holds.

Here, a second temptation has to be fought, namely, to try a simple model for the temperature discontinuity. However, one has to realize that the problem of the gas moving over a wall with a *sharp* temperature jump leads to a nonlinear problem. Before investing effort into its solution, one also has to realize that an actual jump in practice will be spread, when it is real sharp, over 5 to 10 tube diameters (say), or maybe 20 to 50 boundary layer thicknesses. The reasonable approach is, therefore, to assume a *small* rate of change of the wall temperature per diameter and/or per boundary layer thickness, and to solve this *linear* problem. Then, in case it is convenient, the results of this theory can be considered for a sharp jump, knowing that a mathematical approximation is obtained for an actual jump spread over a length long enough to make the physical assumptions of the analysis valid.

Actually, problems with discontinuities represent a very important mathematical limit. First, any continuous solution can be considered (for these linear problems) as a superposition of small jumps. Second, the discontinuous solutions give, in many cases, all the essential results without insignificant and confusing details.

This is already well known for isothermal tubes with varying cross-sections. The linear theory is only valid when the cross-sectional variation is slow When solutions with a discontinuity are sought for problems with friction, the proper definition of the displacement or flux which has to be continuous is considered to be obvious. Actually, it should be derived with some care. It is discussed, by way of introduction, in the isothermal case first. The continuity equation suffices for this consideration. Let $ue^{i\omega t}$, $ve^{i\omega t}$ be the velocity components and $\rho e^{i\omega t}$ the acoustic density fluctuation;

ρ_m is the mean density. Then the continuity equation is

$$i\omega\rho + \rho_m \frac{\partial u}{\partial x} + \rho_m \frac{\partial(r^j v)}{r^j \partial r} = 0 \tag{1}$$

($j = 1$ for tubes, $j = 0$ for channels). Let the volume displacement ς be defined as

$$\varsigma = \frac{1+j}{i\omega} \int_0^{r_w} u r^j \, dr \tag{2}$$

It follows from (1) that

$$\frac{d\varsigma}{dx} + (1+j) \int_0^{r_w} \frac{\rho}{\rho_m} r^j \, dr = 0 \tag{3}$$

where use has been made of $u = v = 0$ for $r = r_w$. It is clear that ς has to be differentiable, i.e., continuous wherever the cross section has a jump, in order to avoid an infinity of $d\varsigma/dx$, which must have finite (but different) values at the two sides of the jump.

In case there is a temperature (and density) stratification, the continuity equation has an additional term and becomes

$$i\omega\rho + \rho_m \frac{\partial u}{\partial x} + u \frac{d\rho_m}{dx} + \rho_m \frac{\partial(r^j v)}{r^j \partial r} = 0 \tag{4}$$

and ς now fulfills the relation

$$\frac{d\varsigma}{dx} - \theta\varsigma + (1+j) \int_0^{r_w} \frac{\rho}{\rho_m} r^j \, dv = 0 \tag{5}$$

where

$$\theta = \frac{d \log T_m}{dx} = -\frac{1}{\rho_m} \frac{d\rho_m}{dx} \tag{6}$$

(as $p_m =$ const.). Now it is obvious that if θ has a jump, then ς also must have one to make (5) a valid equation. But what, then, has to be continuous?

To answer this question, the integral in (5) has to be investigated. The density fluctuations can be replaced by the acoustic pressure and temperature

$$\frac{\rho}{\rho_m} = \frac{p}{p_m} - \frac{T}{T_m} \tag{7}$$

Now it can be argued, without going into detailed calculations, that the radial distribution of T/T_m can be represented as a sum of two functions:

$$\frac{T}{T_m} = \frac{p}{p_m}\mathcal{F}_1(r) + \theta\varsigma\mathcal{F}_2(r) \tag{8}$$

The first (classical) term is the only one present for isothermal tubes, and is caused by the adiabatic acoustic fluctuations. The second term represents the essential thermoacoustic effect due to the axial thermal stratification. In the absence of oscillations, a common temperature distribution $T_m(x)$ for the gas and the wall is assumed, maintained by heat sources and sinks not further specified. By the displacement of the gas against the wall through oscillations, a transverse (radial) temperature difference Δ is caused which is equal to the product of the displacement ς and the temperature gradient dT_m/dx , and Δ related to T_m is $\Delta/T_m = \varsigma\theta$.

In order to evaluate the integral in (5), the radial distributions in (8) have to be determined. This requires lengthly calculations which have been carried out by the author (Rott, 1969); by inserting those results in (5), the present analysis can be kept simple and concise. The following relation between displacement and pressure gradient is needed to make use of these earlier results, which were expressed in terms of dp/dx:

$$\varsigma = r_w^{i+j}\frac{1-f_j}{\omega^2\rho_m}\frac{dp}{dx} \tag{9}$$

Here, f_j is defined as a function of the parameter

$$\eta_w = r_w\sqrt{\frac{i\omega}{\nu}} \tag{10}$$

which is recognized as a (complex) ratio between r_w and the Stokes layer thickness, and f_o (for channels) is

$$f_o(\eta_w) = \frac{1}{\eta_w}tanh\eta_w \tag{11}$$

while for tubes, f_1 is given by

$$f_1(\eta_w) = \frac{2J_1(i\eta_w)}{i\eta_w J_o(i\eta_w)} \tag{12}$$

Expressed by these functions ($f_j \equiv f$), the following result is obtained:

$$(1+j)\int_0^{r_w} \frac{\rho}{\rho_m} r^j \, dv = r_w^{1+j} \frac{p}{\gamma p_m}[1+(\gamma-1)f^*] + \theta\varsigma \frac{(1-f^*)-\sigma(1-f)}{(1-\sigma)(1-f)} \tag{13}$$

where σ is the Prandtl number, and

$$f^*(\eta_w) = f(\eta_w\sqrt{\sigma}) \tag{14}$$

Inserting (13) in (5) gives, after some elementary algebra,

$$\frac{d\varsigma}{dx} - \theta\varsigma\frac{f^*-f}{(1-\sigma)(1-f)} + r_w^{1+j}\frac{p}{\rho_m a^2}[1+(\gamma-1)f^*] = 0 \tag{15}$$

and now it becomes evident, how ς has to behave to compensate for possible jumps in T_m and make (15) a valid equation:

$$\frac{d\varsigma}{\varsigma} - \frac{dT_m}{T_m}\frac{f^*-f}{(1-\sigma)(1-f)} = 0 \tag{16}$$

which integrates to the condition

$$E\varsigma = \text{const.} \tag{17}$$

where

$$E = \exp\left[-\int \frac{f^*-f}{(1-\sigma)(1-f)}\frac{dT_m}{T_m}\right] \tag{18}$$

(The relation between ς and the quantity ψ used by the author previously is $\psi = \rho_m a^2 E\varsigma$.) The main result is (Rott, 1969) that a function $E(T_m)$ has been determined, given by (18), such that the product $E\varsigma$ is invariant across a temperature jump,

$$(E\varsigma)_{\text{hot}} = (E\varsigma)_{\text{cold}} \tag{19}$$

and (15), with E inserted, becomes

$$\varsigma \frac{d \log(E\varsigma)}{dx} + r_w^{1+j} \frac{p}{\rho_m a^2} [1 + (\gamma - 1) f^*] = 0 \tag{20}$$

Before discussing the E-function any further, it is shown how its use modifies the calculation of the Sondhauss frequency. The displacement ς_h at the hot side of the neck into the bulb produces a pressure rise (see Figure 1)

$$\frac{\Delta p}{p_m} = \gamma \frac{A \varsigma_h}{V} \tag{21}$$

which balances the inertia of the gas on the cold side, calculated with ς_c:

$$\Delta p = \rho_{m_c} \ell \omega^2 \varsigma_c \tag{22}$$

It follows from (21), (22) and (19) that

$$\omega^2 = a_c^2 \frac{A}{\ell V} \frac{E_c}{E_h} \tag{23}$$

It is seen that the imaginary part of E answers the question of the excitation in the Sondhauss-tube.

The calculation of E as a function of η_w is done either by series expansions or numerically. Here only the first significant term of a series is given which has been calculated by the author (Rott 1969) for thin boundary layers, i.e., $|\eta_w| \gg 1$:

$$E = 1 - \frac{(1+j)k}{\eta_w} = 1 - \frac{(1+j)k}{r_w} \sqrt{\frac{\nu}{i\omega}} \tag{24}$$

$E = 1$ is the value for $\nu = 0$, a normalization causing no loss of generality. The constant k in the second term was found to be equal to

$$k = \frac{2}{1+\beta} \frac{1}{\sigma + \sqrt{\sigma}} \tag{25}$$

where β is an exponent which gives the variation of the gas viscosity μ with the absolute temperature, $\mu \sim T_m^\beta$.

The magnitude of the complex term in (24) is small compared to 1, so that the change in the real part of E (representing the boundary layer influence on the spring constant) can be safely neglected; then (24) becomes

$$E = 1 + i\frac{(1+j)k}{r_w}\sqrt{\frac{\nu}{2\omega}} \tag{26}$$

The imaginary part of E determines the excitation in (23). It depends on the temperature through $\nu = \mu/\rho_m \sim T_m^{1+\beta}$. Inserting (26) in (23) gives, properly expanded

$$\omega^2 = a_c^2 \frac{A}{\ell V}\left\{1 - \frac{(1+j)k}{r_w}\sqrt{\frac{\nu_c}{2\omega}}\left(\sqrt{\frac{\nu_h}{\nu_c}} - 1\right)\right\} \tag{27}$$

Herein, the term

$$\sqrt{\frac{\nu_h}{\nu_c}} = \left(\frac{T_{m_h}}{T_{m_c}}\right)^{\frac{1+\beta}{2}} \equiv \alpha^{\frac{1+\beta}{2}} \tag{28}$$

contains the ratio α of the absolute temperatures on the two sides of the neck. Finally, with $\omega = \omega_r + i\omega_i$ and $\omega_i \ll \omega_r$, it follows from (27) that

$$\omega_r^2 = a_c^2 \frac{A}{\ell V} \tag{29}$$

(this is the Sondhauss frequency) and

$$\frac{\omega_i}{\omega_r} = -\frac{(1+j)k}{2r_w}\sqrt{\frac{\nu_c}{2\omega_r}}(\alpha^{\frac{1+\beta}{2}} - 1) \tag{30}$$

The negative sign of ω_i indicates excitations; its value (per cycle) is proportional to the ratio of boundary layer thickness to wall radius at the neck. The last factor shows that excitation is found whenever the temperature ratio is bigger than one. This result is obviously wrong; it is the consequence of the fact that no losses were considered. For the Sondhauss tube, two main sources of losses exist which together with the driving effect lead to a critical value of α for the beginning of the excitation (stability limit).

The first cause of losses is that the pressure fluctuations in the bulb are adiabatically connected with temperature fluctuations; the associated pulsating thermal wall boundary layer causes damping. The second source of losses is the friction due to the oscillating motion in the tube. It is interesting to note that both types of losses can be reduced by simple means: the first by making V big, and the second by replacing the gas mass in the cold part by a fluid with a low value of ν ; most liquids will have favorably low values of the kinematic viscosity. The theory of such gas-liquid oscillations was given by Zouzoulas and Rott (1976) and by Müller and Rott (1983). With these design principles, devices with a very low critical temperature ratio can be built; they were actually tested at the Institut für Aerodynamik of the ETH Zürich. It has to be noted, however, that with losses taken into account, the thin boundary layer approximation for E is no longer adequate for the determination of optimal conditions.

Much higher temperature ratios are needed for excitation in a straight tube, with one end closed and one end open, and a temperature jump. Actually, the theory of excitation gives the right order of magnitude with the gas in the hot part near the closed end taken as spring and the cold gas considered as mass. (The most effective driving is found near to a fifty-fifty partition of the tube by the temperature jump.) The losses, however, are particularly high in the hot part, which should only serve as a spring, and should have a big cross section to minimize losses. Thus, temperature ratios of over 5 are typically needed to produce oscillations in a straight tube under optimal conditions. Moreover, when the temperature ratio in a straight tube gets very high, then under certain conditions the oscillations can be *choked* by the very viscous gas in the hot part (Rott, 1973). This effect was verified by von Hoffmann, Lienert, and Quack in helium (1973).

It is seen that the Sondhauss tube configuration is the one in which it is easiest to keep the losses low; this remark is meant as an excuse for the simplified analysis presented here, and explains why this effect was discovered first and is still interesting today, for the possibility of thermoacoustic power generation.

References

[1] von Hoffmann, T., Lienert, U., and Quack, H. *Cryogenics 13* (1973), 490.

[2] Kramers, H. A. *Physica 15* (1949), 971.

[3] Müller, U. and Rott, N. *Z. Angew. Math. Phys. 34* (1983), 609.

[4] Rott, N. *Z. angew. Math. Phys. 20* (1969), 230.

[5] ———. *Z. angew. Math. Phys. 24* (1973), 54.

[6] Sondhauss, C. *Ann. Phys. Leipzig (2) 29* (1850), 1.

[7] Taconis, K. W., et al. *Physica, 15* (1949), 738.

[8] Zouzoulas, G. and Rott, N. *Z. angew. Math. Phys. 27,* (1976), 325.

A Review of Numerical Methods in Acoustic Wave Propagation

*Sébastien M. Candel**

Office National d'Études et de Recherches Aérospatiales
BP72, 92322 Châtillon, France

Abstract

In the last ten years, considerable progress has been accomplished in the numerical analysis of acoustic wave problems. Recent advances in this field are surveyed with emphasis on the more specific techniques, such as the parabolic and geometrical approximations, spectral, space-time finite difference and boundary integral methods. Finite element methods are not considered and the review is restricted to linear acoustics.

Nomenclature

A	wave action.
$c(\mathbf{x})$	local sound speed.
c_o	reference sound speed.
d	layer thickness.
F	Fourier transform operator.
$G(\mathbf{r}/\mathbf{r}_0)$	free space Green's function for the Helmholtz equation.
H_0^1	zero order Hankel function of the first kind.
k	local wave number $k = \omega/c(\mathbf{x})$.
k_0	reference wave number $k_0 = \omega/c_0$.
$\mathbf{k}$	wave-vector.
L	differential operator, spatial extension of the calculation domain.

*Research scientist at ONERA, 92320 Châtillon and professor of engineering at Ecole Centrale des Arts et Manufactures, 92290 Chatenay Malabry, France.

L_x, L_y, L_z	finite difference one-dimensional operators.
M	Mach number.
$\mathbf{M}$	Mach-vector.
N	refraction index, $N = c_0/c$.
p	pressure field.
$\tilde{p}$	Fourier transform of the pressure field.
(r, θ, z)	polar coordinates.
(r, θ, ϕ)	spherical coordinates.
(x, y, z)	cartesian coordinates.
α, β	spatial wave numbers, arguments of spatial Fourier transform.
γ	phase or damping factor.
$\delta(\ )$	delta function.
δa	elementary ray-tube cross section.
$\delta\Sigma$	elementary wave front cross section.
Δx	spatial sampling period.
$\Delta\alpha$	spatial wave number sampling period.
ς	argument of Fourier-Bessel transform.
η	interfacial displacement.
λ	wavelength.
$\underset{\sim}{\nu}$	unit vector in the wave vector direction $\underset{\sim}{\nu} = \mathbf{k}/k$.
ρ	density.
ϕ	velocity potential.
σ, μ	source and dipole singularity distributions.
ψ	complex field amplitude (parabolic approximation), phase function (geometrical approximation).
ω	angular frequency.
∇^2	Laplace operator.
$\nabla^2_\perp$	transverse Laplace operator.

Subscripts and Superscripts

$*$	reduced (dimensionless) quantity.
j	layer index.
j	spatial node number.
n	time-step index.

1. Introduction

Considerable progress has been accomplished in the numerical analysis of acoustic wave problems. In the last ten years a large number of methods have appeared, and a wide range of topics have been covered. Numerical wave propagation has found many practical applications in aeroacoustics, hydroacoustics, atmospheric and underwater sound, geophysical exploration and data processing, seismology, ultrasonic imaging, nondestructive testing and signal processing. Fundamental studies in physical acoustics also have benefited from these advances.

The literature dealing with numerical acoustic wave propagation is already extensive, but a review paper seems to be missing. To keep the present paper at a normal length, it is necessary to restrict the survey to the field of linear acoustics and only review some of the more specific numerical methods. As a consequence, certain general techniques like the finite element methods have not been covered.

The survey concentrates on five major classes of methods:

(1) the parabolic equation method,
(2) the geometrical approximation,
(3) spectral methods,
(4) finite difference time-dependent methods,
(5) boundary integral methods.

This selection of topics is somewhat arbitrary and reflects the author's own involvement in this area of research.

No attempt has been made to cite all the relevant literature, and the connections with other fields of physics, like fluid mechanics, hydrodynamics, optics, and electromagnetics are only briefly mentioned. Progress in these areas has paralleled, and sometimes preceded, that of acoustic wave propagation. Indeed, methods like the geometrical and parabolic approximations find their origin in these other domains of physics, while spectral, finite difference, and boundary integral methods have been strongly developed in fluid dynamics. Thus, an in-depth account of progress in the various numerical wave propagation disciplines would have been of interest, but remains outside the scope of the present paper.

We start by surveying the classical parabolic and geometrical approximations which form the basis of a large number of modern numerical methods (Sections 2 and 3). Spectral techniques are then considered with emphasis on wave extrapolation and migration, wave synthesis in layered media, and the numerical treatment of pole contributions (Section 4). Some time-dependent finite difference explicit and implicit schemes are then reviewed (Section 5). The last section gives a short account of the current boundary integral methods (Section 6).

2. The Parabolic Equation Method (PEM)

2.1 General Considerations

The parabolic approximation allows the numerical solution of a large class of wave problems in nonhomogeneous media. This approximation is an extension of the classical paraxial approximation of physical optics, and it is adapted to situations where the sound field propagates at a small angle with respect to a preferred direction. This is the case, for example, in underwater long range propagation, where the main part of the acoustic energy is transmitted by rays travelling at small grazing angles while energy radiated at higher angles is lost at each bottom reflection due to absorption and ground-leakage.

Physically, the parabolic approximation consists of neglecting the backscattered waves, and only considers the waves propagating forward about a preferred axial direction. It is a useful approximation if the inhomogeneity of the medium is considered to be weak and slowly varying in space. This is the case in underwater acoustic applications, where the sound-speed-relative variations do not exceed 5.5% (Urick, 1967, p. 114). Similarly in the lower atmosphere, temperature variations are of the order of 25 K and produce relative sound speed variations of about 5% (Gossard and Hooke, 1975, p. 20). Additional changes in phase speed may also result from wind and turbulence, but they do not generally exceed 10%.

Many aeroacoustic problems also belong to the weakly inhomogeneous class. This is clearly so in anechoic open wind tunnels developed in recent years to simulate forward flight effects on aircraft noise radiation. In these facilities the flow speed may reach 100 m/s, but the turbulent fluctuations encountered by the sound waves in their propagation through the jet shear layer are of the order of 12% of the main stream velocity, and they induce about 4% variations in the acoustic phase-speed (see Candel et al., 1976).

The parabolic approximation is also extensively used in geophysical seismic applications, where it is often referred to as the 15° approximation (Claerbout, 1976). In this case the speed of the longitudinal waves may exhibit notable variations associated with changes of rock composition and porosity. However, the medium may be modelled as a collection of slices, in which the wave speed changes weakly, and the parabolic approximation becomes applicable.

When compared to the classical geometrical approximation, the parabolic equation method has the important advantage of taking into account wave diffraction effects in the transverse direction. This is why it forms the basis of the majority of theoretical studies of scattering by turbulent media. (See Tatarski (1961) for a discussion of conditions of application of the geometrical and paraxial approximations to wave scattering problems.)

From the computational viewpoint, the main interest of the paraxial approximation is that it reduces the propagation problem of elliptic type to a Cauchy problem (initial value problem) of parabolic type.

Now, the solution of initial value problems is easily obtained by marching in the direction of propagation, and this is why the PEM has been so successful.

The method was introduced in electromagnetic wave studies by Leontovich and Fock (1946) and Fock (1965), and it has been extensively used in this field, and in optics and laser beam propagation. One of the first numerical studies of the last topic is due to Kelley (1965), and concerns nonlinear self-focusing of optical beams. Other examples are provided by Wallace and Lilly (1974) and by Lilly and Miller (1977).

In acoustics, the parabolic equation method was introduced by Hardin and Tappert (1973), and it forms the basis of a growing number of studies mainly dealing with underwater sound. McDaniel (1975a, b) considers normal mode propagation in the ocean, provides estimates of phase and group velocity errors associated with the parabolic approximation, and gives some interesting formal derivations of the parabolic equation based on sound field splittings. Fitzgerald (1975) compares the PEM and normal mode representations and presents estimates for the validity of the paraxial approximation. Various theoretical aspects are discussed by Palmer (1976) and De Santo (1977). De Santo et al. (1977) and Brock et al. (1977) describe modifications which improve the accuracy of the PEM. Tappert (1977) gives a broad review of the method together with some remarkable examples of calculations. Candel (1978, 1979) treats propagation in ducts, free space radiation, and wave scattering problems. Flatté and Tappert (1975) and Baer (1980) analyze sound interaction with internal waves and eddies in the ocean, while Myers and McAninch (1978) consider wave propagation in the atmosphere.

Numerical solution of the parabolic equation was initially obtained by Hardin and Tappert (1973) using an elegant and original split-step Fourier method (see Tappert, 1977). More recently, implicit finite difference methods or the method of lines were used by McDaniel (1975), Candel (1978, 1979), Lee and Papadakis (1980), Lee et al. (1981), and McDaniel and Lee (1982). Three-dimensional calculations, based on the PEM, are reported by Perkins and Baer (1982). Applications of the paraxial approximation to elastic waves are described in the early work of Landers and Claerbout (1972) (see also Claerbout (1976) and the more recent studies of McCoy (1977) and Hudson (1980)). Lee (1983) provides a recent tutorial review of the PEM.

2.2 Derivation of the Parabolic Equation

Most derivations of the parabolic equation assume a cylindrically symmetric geometry. This condition is well suited to underwater sound problems, but it is not essential and we shall first study other geometrical configurations.

Consider as a starting point the Helmholtz equation describing monochromatic wave motion in a medium characterized by a variable sound speed $c(\mathbf{x})$:

$$\nabla^2 p + k^2 p = 0 \tag{2.1}$$

with $k(\mathbf{x}) = \omega / c(\mathbf{x})$.

Now suppose that the wave motion takes place around the z-axis, i.e., the z-axis is the preferred direction of motion. Then replace p by the following expression:

$$p = \psi(x, y, z)\, e^{ik_o z} \tag{2.2}$$

where k_o designates an *average* wave number $k_o = \omega/c_o$, and ψ is the complex amplitude of the pressure field. The reference sound speed c_o is usually defined as some spatial average of $c(\mathbf{x})$ over the domain of propagation.

Making the substitution of (2.2) in equation (2.1) yields:

$$\frac{\partial^2\psi}{\partial x^2} + \frac{\partial^2\psi}{\partial y^2} + \frac{\partial^2\psi}{\partial z^2} + 2ik_o\frac{\partial\psi}{\partial z} + (k^2 - k_o^2)\,\psi = 0 \tag{2.3}$$

It is convenient to introduce the transverse Laplace operator $\nabla_\perp^2 = \frac{\partial^2}{\partial x^2} + \frac{\partial^2}{\partial y^2}$, and the local index $N(\mathbf{x}) = c_o/c(\mathbf{x})$. With these notations equation (2.3) becomes:

$$\nabla_\perp^2\psi + \frac{\partial^2\psi}{\partial z^2} + 2ik_o\frac{\partial\psi}{\partial z} + k_o^2\,(N^2 - 1)\,\psi = 0 \tag{2.4}$$

Now if the wave moves along the z-axis, the factor $e^{ik_o z}$ in (2.2) essentially describes the z-dependence of the wave field, and the second derivative of ψ with respect to z may be neglected when compared to the other terms of equation (2.4):

$$\left|\frac{\partial^2\psi}{\partial z^2}\right| \ll \left|2ik_o\frac{\partial\psi}{\partial z}\right|,\ |\nabla_\perp^2\psi| \tag{2.5}$$

Equation (2.4) then becomes parabolic with respect to the z coordinate:

$$\nabla_\perp^2\psi + 2ik_o\frac{\partial\psi}{\partial z} + k_o^2\,(N^2 - 1)\,\psi = 0 \tag{2.6}$$

Similar derivations may be performed in cylindrical and spherical geometries and the results are summarized in Table 1.

Most underwater sound studies assume cylindrical symmetry and the following field representation:

$$p(r,z) = \frac{e^{ik_o r}}{(k_o r)^{1/2}}\,\psi\,(r,z) \tag{2.7}$$

The corresponding parabolic equation takes the form:

$$\frac{\partial^2\psi}{\partial z^2} + 2ik_o\frac{\partial\psi}{\partial r} + k_o^2\left[N^2 - 1 + \frac{1}{4\,(k_o r)^2}\right]\psi = 0 \tag{2.8}$$

In many circumstances the range r is quite large and $1/4(k_o r)^2$ may be neglected from the last term of equation (2.8).

	Cartesian	Polar	Spherical
Field description	$p(x,y,z) = \psi(x,y,z)e^{ik_o z}$	$p(r,\theta,z) = \psi(r,\theta,z)\dfrac{e^{ik_o r}}{(k_o r)^{1/2}}$	$p(r,\theta,\phi) = \psi(r,\theta,\phi)\dfrac{e^{ik_o r}}{4\pi r}$
Parabolic equation	$\dfrac{\partial^2\psi}{\partial x^2} + \dfrac{\partial^2\psi}{\partial y^2} + 2ik_o\dfrac{\partial\psi}{\partial z} + k_o^2(N^2-1)\psi = 0$	$\dfrac{\partial^2\psi}{\partial z^2} + \dfrac{1}{r^2}\dfrac{\partial^2\psi}{\partial\theta^2} + 2ik_o\dfrac{\partial\psi}{\partial r} + k_o^2\left[N^2 - 1 + \dfrac{1}{4(k_o r)^2}\right]\psi = 0$	$\dfrac{1}{r^2}\dfrac{\partial^2\psi}{\partial\theta^2} + \dfrac{\cos\theta}{r^2\sin\theta}\dfrac{\partial\psi}{\partial\theta} + \dfrac{1}{r^2\sin\theta}\dfrac{\partial^2\psi}{\partial\phi^2} + 2ik_o\dfrac{\partial\psi}{\partial r} + k_o^2(N^2-1)\psi = 0$
Field description	propagation in two dimensions $p(x,z) = \psi(x,z)e^{ik_o z}$	rotational symetry and $k_o r \gg 1$ $p(r,z) = \psi(r,z)\dfrac{e^{ik_o r}}{(k_o r)^{1/2}}$	rotational symetry around the z-axis $\theta \simeq \pi/2$ $p(r,\theta) = \psi(r,\theta)\dfrac{e^{ik_o r}}{4\pi r}$
Parabolic equation	$\dfrac{\partial^2\psi}{\partial x^2} + 2ik_o\dfrac{\partial\psi}{\partial z} + k_o^2(N^2-1)\psi = 0$	$\dfrac{\partial^2\psi}{\partial z^2} + 2ik_o\dfrac{\partial\psi}{\partial r} + k_o^2(N^2-1)\psi = 0$	$\dfrac{1}{r^2}\dfrac{\partial^2\psi}{\partial\theta^2} + 2ik_o\dfrac{\partial\psi}{\partial r} + k_o^2(N^2-1)\psi = 0$
Geometry	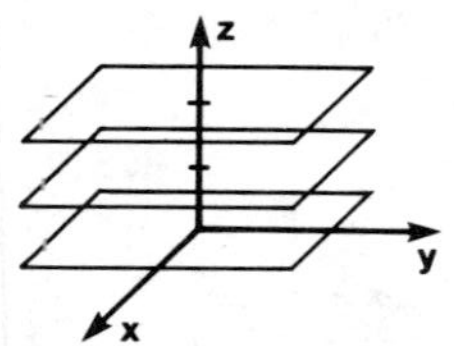	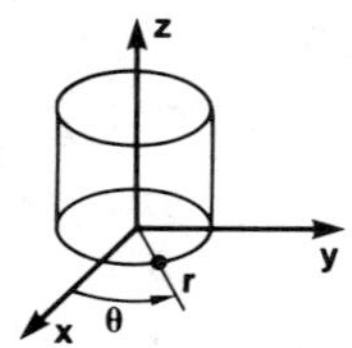	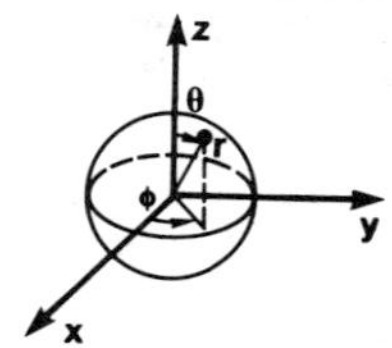

Table 1.

A somewhat more rigorous derivation of the parabolic equation may be based on classical techniques of singular asymptotic expansions. Consider again the Helmholtz equation (2.1) and define the following stretched coordinates:

$$\overline{x} = k_o x\epsilon, \quad \overline{y} = k_o y\epsilon, \quad \overline{z} = k_o z\epsilon^2 \tag{2.9}$$

Here ϵ designates a small parameter characterizing the weak inhomogeneity of the medium:

$$N^2 - 1 = \epsilon^2 \mu(\mathbf{x}). \tag{2.10}$$

Substituting the new coordinates (2.9) and definition (2.10) in equation (2.1) yields:

$$\epsilon^2 \frac{\partial \psi}{\partial \overline{z}^2} + \left(\frac{\partial^2 \psi}{\partial \overline{x}^2} + \frac{\partial^2 \psi}{\partial \overline{y}^2} \right) + 2i \frac{\partial \psi}{\partial \overline{z}} + \mu\psi = 0 \tag{2.11}$$

The outer expansion of ψ in the stretched coordinates (2.9) may be defined by the following series:

$$\psi = \psi^{(0)}(\overline{x}, \overline{y}, \overline{z}) + \epsilon^2 \psi^{(1)}(\overline{x}, \overline{y}, \overline{z}) + \epsilon^4 \psi^{(2)}(\overline{x}, \overline{y}, \overline{z}) + \cdots \tag{2.12}$$

Substituting this expansion in (2.11) yields:

$$\frac{\partial^2 \psi^{(0)}}{\partial \overline{x}^2} + \frac{\partial^2 \psi^{(0)}}{\partial \overline{y}^2} + 2i \frac{\partial \psi^{(0)}}{\partial \overline{z}} + \mu \psi^{(0)} = 0 \tag{2.13}$$

$$\frac{\partial^2 \psi^{(1)}}{\partial \overline{x}^2} + \frac{\partial^2 \psi^{(1)}}{\partial \overline{y}^2} + 2i \frac{\partial \psi^{(1)}}{\partial \overline{z}} + \mu \psi^{(1)} = -\frac{\partial^2 \psi^{(0)}}{\partial \overline{z}^2} \tag{2.14}$$

It appears that the parabolic equation describes the first term of the outer expansion of ψ in the stretched coordinates (2.9).

2.3 Derivation of the Parabolic Equation by Local Splitting of the Wave Field

The parabolic equation may also be derived as a forward scattering approximation by performing a local wave decomposition (or splitting) of the sound field. This method provides some additional physical insight. Consider again the Helmholtz equation (2.1) and let z designate the direction of preferred propagation. It is convenient to first write (2.1) in matrix form:

$$\frac{\partial}{\partial z}\begin{bmatrix} p \\ \partial p/\partial z \end{bmatrix} = \begin{bmatrix} 0 & 1 \\ (-\nabla_{\perp}^{2} - k^{2}) & 0 \end{bmatrix}\begin{bmatrix} p \\ \partial p/\partial z \end{bmatrix} \tag{2.15}$$

Next define the local waves U and D:

$$\begin{bmatrix} U \\ D \end{bmatrix} = \frac{1}{2}\begin{bmatrix} 1 & i/k_o \\ 1 & -i/k_o \end{bmatrix}\begin{bmatrix} p \\ \partial p/\partial z \end{bmatrix} \tag{2.16}$$

Here D designates a divergent outgoing wave, and U is a convergent ingoing wave (the notations U and D stand for up-going and down-going waves conventionally used in geophysical data processing).

The previous definitions are easily inverted and yield:

$$\begin{aligned} p &= U + D, \\ \frac{\partial p}{\partial z} &= ik_o(D - U) \end{aligned} \tag{2.17}$$

In a homogeneous medium U and D exactly coincide with plane waves propagating in the negative and positive z-direction.

For nonhomogeneous media the splitting of the sound field is not unique and other definitions are also suitable (see McDaniel, 1975b). Substituting (2.17) in (2.15) one obtains:

$$\begin{aligned} \frac{\partial D}{\partial z} &= \frac{i}{2k_o}(k^2 + k_o^2 + \nabla_{\perp}^2)D + \frac{i}{2k_o}(k^2 - k_o^2 + \nabla_{\perp}^2)U \\ \frac{\partial U}{\partial z} &= \frac{i}{2k_o}(k_o^2 - k^2 - \nabla_{\perp}^2)D - \frac{i}{2k_o}(k^2 + k_o^2 + \nabla_{\perp}^2)U \end{aligned} \tag{2.18}$$

This is a coupled set of equations for U and D. Now assume that the backscattered wave U is weak, and may be neglected when compared to the forward travelling wave D. This forward scattering approximation decouples the set (2.18) and yields for D:

$$\frac{\partial D}{\partial z} = \frac{i}{2k_o}(k^2 + k_o^2 + \nabla_{\perp}^2)D \tag{2.19}$$

The outgoing wave D is obtained by solving this equation and it may then be used to calculate U through the last equation (2.18). This process may be iterated, but precise results are usually obtained with a single step. (See, for example, Candel et al. (1980) for numerical calculations based on the forward scattering approximation for one-dimensional wave scattering problems.)

Now, it is more meaningful to cast D in the form of a non-homogeneous plane wave $D = \psi e^{ik_o z}$. The complex amplitude ψ is then the solution of the following equation:

$$2ik_o \frac{\partial \psi}{\partial z} + \nabla_\perp^2 \psi + (k^2 - k_o^2)\, \psi = 0 \tag{2.20}$$

which exactly coincides with the parabolic equation of the previous section.

2.4 Parabolic Equation for a Medium Containing Temperature and Velocity Inhomogeneities

In atmospheric and aeroacoustic wave problems the medium usually contains temperature and velocity fluctuations. For such problems a parabolic equation is conveniently derived from Lighthill's equation:

$$\frac{\partial^2 \rho}{\partial t^2} - c_o^2 \nabla^2 \rho = \frac{\partial^2}{\partial x_i \partial x_j} \rho u_i u_j + \nabla^2 (p - c_o^2 \rho) \tag{2.21}$$

Straightforward calculations (see Candel, 1979) lead to the following parabolic equation:

$$2ik_o \frac{\partial \psi}{\partial z} + \nabla_\perp^2 \psi = 2k_o^2 \psi \left(1 - \frac{i}{k_o} \frac{\partial}{\partial z}\right)\left(\frac{1}{2} \frac{T'}{T_o} + \frac{u_z}{c_o}\right) \tag{2.22}$$

where T' and u_z represent the temperature fluctuation and velocity perturbation respectively in the direction of propagation.

2.5 Numerical Methods of Solution

The parabolic equation was initially solved by Hardin and Tappert (1973) using a split step Fourier algorithm which is first described.

Let us start by formally integrating the parabolic equation (2.8) over a radial distance Δr:

$$\psi(r+\Delta r, z) = \exp[(A+B)\Delta r]\,\psi(r,z) \tag{2.23}$$

where

$$\begin{aligned} A &= \frac{i}{2k_o}\frac{\partial^2}{\partial z^2} \\ B &= \frac{i}{2k_o}\left(k^2 - k_o^2 + \frac{1}{4r^2}\right) \end{aligned} \tag{2.24}$$

Now, for short radial steps it is possible to split the exponential operator as follows:

$$\exp[(A+B)\Delta r] \simeq \exp\left(\frac{B\Delta r}{2}\right)\exp(A\Delta r)\exp\left(\frac{B\Delta r}{2}\right) \tag{2.25}$$

This expression is second order accurate in r. Now $\exp(A\Delta r)$ is most conveniently expressed in Fourier space and expression (2.23) may be written:

$$\begin{aligned} &\psi(r+\Delta r, z) \\ &= \exp\left(\frac{B\Delta r}{2}\right)F^{-1}\left(\exp\left(-i\frac{\Delta r}{2k_o}\alpha^2\right)F\left[\exp\left(\frac{B\Delta r}{2}\right)\psi(r,z)\right]\right) \end{aligned} \tag{2.26}$$

where F designates the Fourier transform operator, and α is the spatial wave number argument of this transform.

This algorithm is quite efficient and has been extensively used in underwater applications. However, it has some drawbacks. Since the fast Fourier transform is used, it is necessary to impose a zero boundary condition at some depth below the ocean floor. It is also necessary to assume that the waves penetrating the sub-bottom are exponentially attenuated. These assumptions provide good results for long range deep-water propagation, but they are inappropriate in other circumstances.

Alternate techniques for solving the parabolic equation are based on the method of lines presented by Lee and Papadakis (1980), or on finite difference methods (for example McDaniel (1975a), Candel

(1978, 1979), and Lee et al. (1981)). Finite difference methods allow general boundary conditions and are well suited to the solution of parabolic problems. Implicit schemes are usually superior to explicit methods because they remain stable for large range step sizes. We shall only describe two common implicit methods.

Let us first write the parabolic equation in the operator form:

$$\frac{\partial \psi}{\partial r} = L\psi \tag{2.27}$$

where $L = a + b\frac{\partial^2}{\partial z^2}$, $a = \frac{ik_o}{2}(N^2 - 1)$, and $b = \frac{i}{2k_o}$. Let Δr and Δz denote the range and depth step sizes and ψ_j^n represents the complex field value at node $(j\Delta z, n\Delta r)$: $\psi_j^n = \psi(j\Delta z, n\Delta r)$. Implicit algorithms may be deduced from the following operator equations:

$$e^{-L\Delta r}\,\psi_j^{n+1} = \psi_j^n \tag{2.28}$$

$$e^{-\frac{L}{2}\Delta r}\,\psi_j^{n+1} = e^{\frac{L}{2}\Delta r}\,\psi_j^n \tag{2.29}$$

Substitution of a second order central difference scheme in the z-direction followed by expansion of the exponential operators leads to the following difference formulas:

$$\left[1 - \Delta r\,(a_j^{n+1} + b_j^{n+1}\delta_z^2)\right]\psi_j^{n+1} = \psi_j^n \tag{2.30}$$

and

$$\left[1 - \frac{\Delta r}{2}\,(a_j^{n+1} + b_j^{n+1}\delta_z^2)\right]\psi_j^{n+1} = \left[1 + \frac{\Delta r}{2}\,(a_j^n + b_j^n\delta_z^2)\right]\psi_j^n \tag{2.31}$$

where

$$\delta_z^2\psi_j = (\psi_{j+1} - 2\psi_j + \psi_{j-1})/(\Delta z)^2$$

The first scheme is fully implicit while the second is of Crank-Nicholson type. Both formulas are unconditionally stable, consistent and convergent. The solution is obtained at each step by solving a system of equations formed by (2.30) and (2.31) with boundary conditions provided in difference form. This system is tridiagonal and its inversion may be performed by classical recursion techniques. Tappert (1977), Lee (1983), and Lee and MacDaniel (1982) give further details on boundary conditions, and Tappert (1977) presents a discussion of the source representation problem.

2.6 Illustrative Calculation

Most calculations described in the literature pertain to underwater sound or seismic applications. Here, we briefly describe the less familiar situation of wave scattering by a viscous core (Rankine) vortex. Additional details may be found in Candel (1979).

A plane wave progressing from left to right impinges on a viscous core vortex (Figure 1). At a radial distance r from the vortex axis the tangential velocity is:

$$u_\theta(r) = \frac{\Gamma}{2\pi r}[1 - \exp(-r^2/R^2)] \tag{2.32}$$

where Γ represents the circulation and R designates a characteristic scale of the core ($R = 4\nu t$ for viscous flow).

The vortex acts as an inhomogeneity which refracts and scatters the incident plane wave. Both mechanisms are well described by the PEM if the maximum tangential velocity remains small, compared to the ambient sound speed. Figures 2, 3, and 4 display some of the numerical results obtained in the analysis of the wave-vortex interaction. The modulus and phase exhibit oscillations typical of scattering phenomena, and the phase lead and lag arise from refraction effects associated with the acceleration or deceleration of the wave fronts by the vortex motion.

3. The Geometrical Approximation

3.1 General Considerations

When the wave motion takes place in a medium whose characteristic scale ℓ, and characteristic time τ, greatly exceed the wavelength λ and period $T = 2\pi/\omega$, the wave field is conveniently described by means of a geometrical theory, or in terms of the more recent theory of waves in slightly dispersive, inhomogeneous and nonlinear media. The geometrical solution of the propagation problem is constructed in three steps:

1. a dispersion relation between the angular frequency ω and the wave vector $\mathbf{k}$ is first obtained,

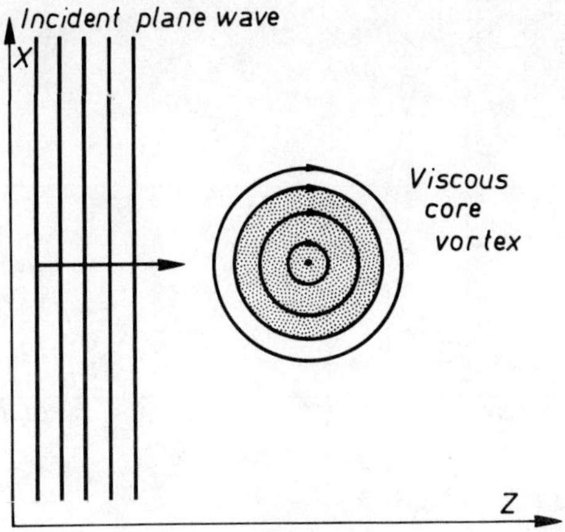

Figure 1. A plane wave propagating from left to right impinges on a viscous core vortex. The vortex is centred at $x_{c^*} = 0.5$, $z_{c^*} = 0.005$, and the tangential velocity reaches its maximum $u_M = 20$ m/s at a distance $R_{M^*} = 0.0389$ of the vortex center. Reduced coordinates for this problem are defined by $x* = x/L$, $z* = z/z_1$, $R_{M^*} = R_M/L$ where L designates the extent of the computational domain in the x-direction and $z_1 = 4\,L^2/\lambda$ is a Fresnel length. The wavelength $\lambda = 4R_M$.

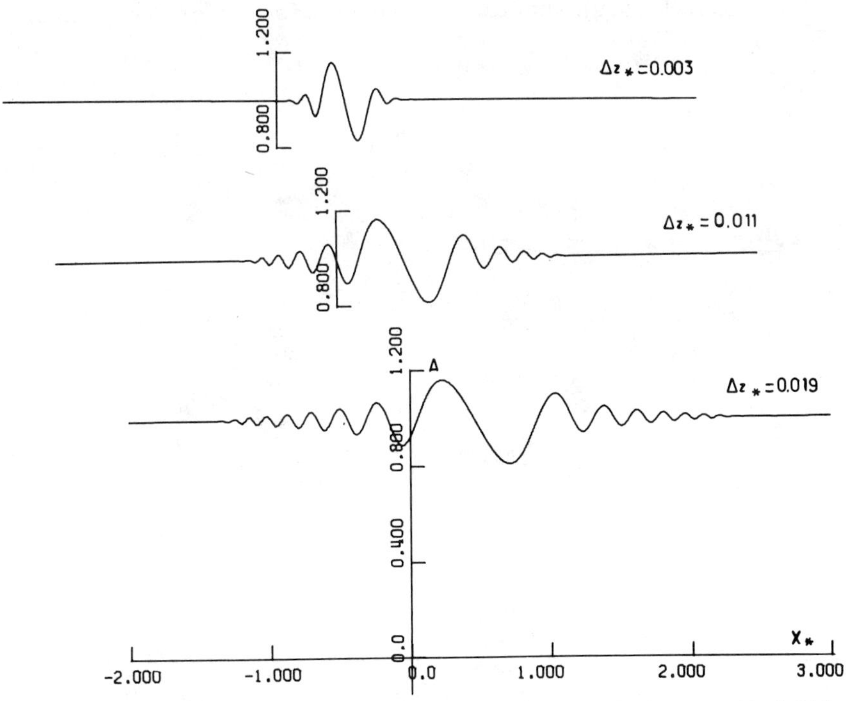

Figure 2. Field modulus in three axial sections, $\Delta z* = 0.003, 0.011,$ and 0.019 for the problem of Figure 1. From Candel (1979). Reprinted with permission of Cambridge University Press.

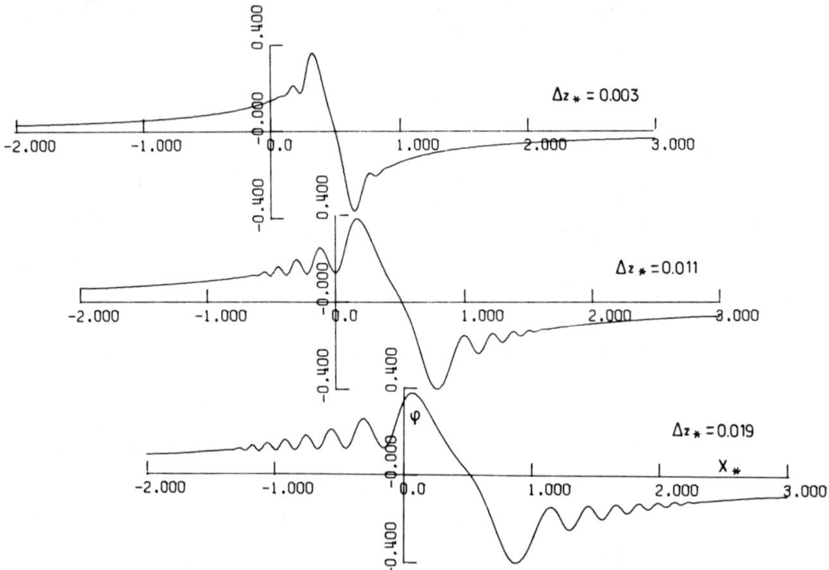

Figure 3. Phase calculated with respect to the incident phase in three axial sections, $\Delta z* = 0.003$, 0.011, and 0.019 for the problem of Figure 1. From Candel (1979). Reprinted with permission of Cambridge University Press.

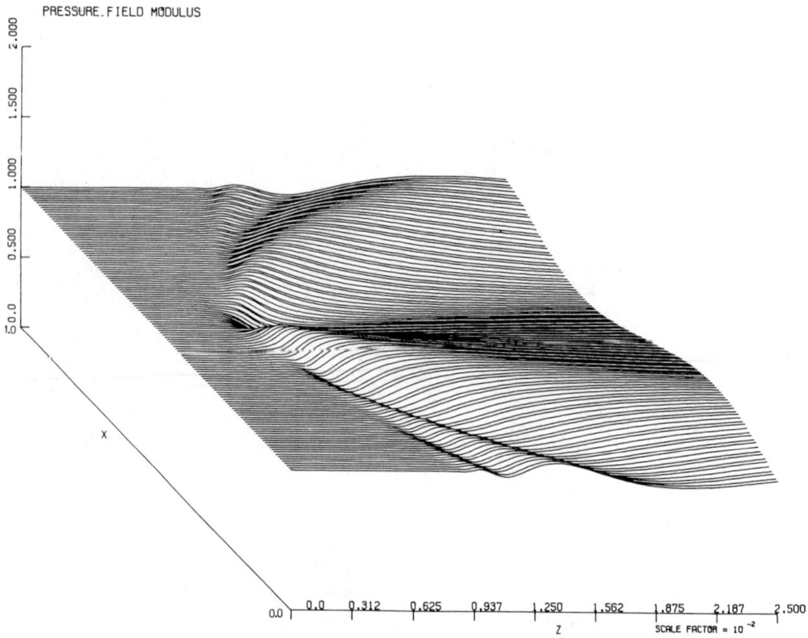

Figure 4. Perspective view of the field modulus for the problem of Figure 1. From Candel (1979). Reprinted with permission of Cambridge University Press.

2. the characteristic lines or rays of the dispersion relation are then determined,

3. the field amplitude is calculated by solving an equation, which generally exhibits a conservation form, and describes the invariance of a field property like the wave action along the characteristic lines.

Basic descriptions of the geometrical approximation may be found in Whitham (1974), Lighthill (1972), Blokhintsev (1946), and Keller (1954).

The rays associated with the dispersion relation are traced in numerous analyses of wave propagation in inhomogeneous media; for instance in the case of electromagnetic waves in the ionosphere, optical beams in graded index fibers, acoustic waves in the atmosphere, or the ocean and seismic waves in the ground. (See for example Quemada (1968), Marcuse (1972), Gossard and Hooke (1975), Urick (1967), and Telford et al. (1976).) The complete determination of the field, comprising the third step of the solution construction, is performed in a more restricted number of studies, and generally on the basis of either particular analytic expressions or farfield approximations.

Another method extensively used in underwater acoustics consists of establishing a relation between the field amplitude and the local ray density. After tracing a large number of trajectories in uniformly distributed directions, it is possible to estimate this density; for instance by calculating the distance between adjacent rays. This procedure is however uneconomical in computation time, and its precision diminishes as the range increases. A more efficient method initiated by Ugincius (1969) consists of calculating the field amplitude by first obtaining the variation of the ray tube cross section along each ray. This may be achieved by integrating a set of ordinary differential equations along the characteristic trajectories. This idea is presented in a more general form in the work of Hayes (1970), and it is specialized to the case of acoustic wave propagation in a stratified atmosphere by Hayes et al. (1970) in calculations of the sonic boom. A complete three-dimensional formulation of the first order geometrical solution is given by Candel (1977a, b). The conservation equation for the wave action is solved by first calculating the variation of the elementary wave-front section along each ray. The ray-tube cross section is then obtained indirectly from the wave-front section. A set of differential equations deduced from the characteristic system allows the determination of this last quantity.

The method is quite general, and may be used to construct first order geometrical solutions to three-dimensional wave problems in nonhomogeneous moving media.

3.2 Basic Formulation of the Geometrical Theory

Detailed presentations of the geometrical approximation may be found in Whitham (1974), Lighthill (1972), Felsen and Marcuvitz (1973), or Candel (1977a, b). Important aspects are outlined in this section.

The problem is to find the field $\mathbf{u}(\mathbf{x}, t)$ solution of a set of equations of the general form:

$$L\left(\frac{\partial}{\partial t}, \nabla; \mathbf{x}, t\right) \mathbf{u}(\mathbf{x}, t) = 0 \tag{3.1}$$

The field is cast in the form of a local plane wave:

$$\mathbf{u}(\mathbf{x}, t) = \mathbf{u}^{\circ}(\mathbf{x}, t)\, e^{i\psi(\mathbf{x}, t)} \tag{3.2}$$

where $\mathbf{u}^{\circ}(\mathbf{x}, t)$ is an amplitude vector and $\psi(\mathbf{x}, t)$ a phase function.

Now let $\mathbf{k} = \nabla\psi$ and $\omega = -\partial\psi/\partial t$ designate the local wave vector and angular frequency.

Substituting (3.2) in the system of equation (3.1), and expanding the amplitude vector in an asymptotic series $\mathbf{u}^{\circ} = \mathbf{u}_0 + \mathbf{u}_1 + \mathbf{u}_2 + \cdots$ one obtains, to the first order, a dispersion relation $D(\omega, \mathbf{k}; \mathbf{x}, t) = 0$ which may be cast in the form:

$$\omega = \Omega(\mathbf{k}, \mathbf{x}, t) \tag{3.3}$$

and to the next orders, equations which take the general form:

$$\frac{\partial A_n}{\partial t} + \nabla \cdot \mathbf{c}_g A_n = f(\mathbf{u}_{n-1}) \tag{3.4}$$

where $\mathbf{c}_g$ designates the local group velocity $\mathbf{c}_g = \partial\Omega/\partial\mathbf{k}$ and A_n is the *wave action* associated with the $\mathbf{u}_n$ term of the geometrical expansion:

$$A_n = A_n(\mathbf{u}_n, \omega, \mathbf{k}; \mathbf{x}, t) \tag{3.5}$$

In many circumstances the first of equations (3.4) describing $\mathbf{u}_0(\mathbf{x}, t)$ takes the conservation form:

$$\frac{\partial}{\partial t}A_0 + \nabla \cdot \mathbf{c}_g A_0 = 0 \tag{3.6}$$

This is the case for acoustic wave propagation in nonhomogeneous moving media. Tables 2 and 3 summarize the main results of the first order geometrical theory for that specific case.

The dispersion relation (3.3) is conveniently solved by the method of characteristics, i.e., by tracing the rays defined by:

$$\begin{aligned} \frac{d^c\mathbf{x}}{dt} &= \frac{\partial\Omega}{\partial\mathbf{k}} \\ \frac{d^c\mathbf{k}}{dt} &= -\nabla\Omega \\ \frac{d^c\omega}{dt} &= \frac{\partial\Omega}{\partial t} \end{aligned} \tag{3.7}$$

where d^c/dt represents differentiation along the characteristic lines:

$$d^c/dt = \partial/\partial t + \mathbf{c}_g \cdot \nabla$$

To obtain the field amplitude $\mathbf{u}_0$ it is then necessary to solve the conservation equation (3.6). A first step in that direction consists of showing that the wave action contained in a wave packet is conserved along the characteristic line (see Whitham (1974), p. 389 or Candel (1977a, b)):

$$A_0 c_g \delta a = \text{constant} \tag{3.8}$$

where δa designates the local elementary ray tube cross section.

The conservation property (3.8) is useful if it is possible to calculate the variation of δa along each ray. It is shown by Candel (1977a, b) that δa may be related to the elementary wave-front cross section $\delta\Sigma$ and that this last quantity may be determined along each ray by integrating a set of ordinary differential equations. For a time-independent medium the construction of the first

$$\pi = \text{pressure}, \quad \underset{\sim}{\varsigma} = \text{velocity}, \quad \sigma = \text{entropy}$$

Field description:

$$\begin{bmatrix} \pi \\ \underset{\sim}{\varsigma} \\ \sigma \end{bmatrix} = \begin{bmatrix} \pi_0 \\ \underset{\sim}{\varsigma}_0 \\ \sigma_0 \end{bmatrix} \qquad \exp[i\psi(x,t)]$$

$$\underset{\sim}{\varsigma}_0 = \frac{\underset{\sim}{\nu}}{\rho c}\pi_0, \quad \sigma_0 = 0$$

Dispersion relation:

$$\omega = kc + \mathbf{k} \cdot \underset{\sim}{\nu} = kc\,(1 + \mathbf{M} \cdot \underset{\sim}{\nu})$$

Group velocity:

$$\mathbf{c}_g = c\underset{\sim}{\nu} + \mathbf{v} = c\,(\underset{\sim}{\nu} + \mathbf{M})$$

Ray equations:

$$\begin{aligned} d^c\mathbf{x}/dt &= c\underset{\sim}{\nu} + \mathbf{v} \\ d^c\mathbf{k}/dt &= -k\nabla c - \mathbf{k} \cdot \nabla \mathbf{v} \\ d^c\omega/dt &= k\partial c/\partial t + \mathbf{k} \cdot \partial \mathbf{v}/\partial t \\ d^c\psi/dt &= 0 \end{aligned}$$

Wave action:

$$A_0 = \frac{\pi_0^2}{\rho c^2}\frac{1}{kc} = \frac{\pi_0^2}{\rho c^2}\,\frac{1 + \mathbf{M} \cdot \underset{\sim}{\nu}}{\omega}$$

Conservation equation:

$$\frac{\partial A_0}{\partial t} + \nabla \cdot \mathbf{c}_g A_0 = 0$$

Table 2. First order geometrical acoustics solution for a time-dependent medium.

Field description:

$$\begin{bmatrix} \pi \\ \underset{\sim}{\zeta} \\ \sigma \end{bmatrix} = \begin{bmatrix} \pi_0 \\ \underset{\sim}{\zeta}_0 \\ \sigma_0 \end{bmatrix} \qquad \exp[ik_0 S(\underset{\sim}{x}) - i\omega t]$$

$$\underset{\sim}{\zeta}_0 = \frac{\underset{\sim}{\nu}}{\rho c}\pi_0, \quad \sigma_0 = 0, \quad k_0 = \omega/c_0$$

Eikonal equation:

$$(\nabla S)^2 = (N - \mathbf{M}\cdot\nabla S)^2$$

$$N = c_0/c, \quad \mathbf{p} = \nabla S, \quad p = N/(1 + \mathbf{M}\cdot\underset{\sim}{\nu})$$

Group velocity:

$$\mathbf{c}_g = c\,(\underset{\sim}{\nu} + \mathbf{M})$$

Ray equation:

$$d^c\mathbf{x}/dS = N^{-1}(\underset{\sim}{\nu} + \mathbf{M})$$
$$d^c\mathbf{p}/dS = N^{-1}[\nabla N - \nabla\mathbf{M}\cdot\mathbf{p}]$$

Energy density:

$$E_0 = \frac{\pi_0^2}{\rho c^2}(1 + \mathbf{M}\cdot\underset{\sim}{\nu}) = A_0\omega$$

Conservation equation:

$$\nabla\cdot\mathbf{c}_g E_0 = 0$$

Table 3. First order geometrical acoustics theory for a time-independent medium.

order geometrical solution reduces to an integration of 18 ordinary differential equations. (Six equations describing the rays, and twelve equations providing $\delta\Sigma$.) A complete derivation of the method, initial conditions, conditions at reflection, and practical details may be found in Candel (1977a, b).

3.3 Illustrative Example

Initial applications of geometrical techniques to aeroacoustic problems may be found in the work of Csanady (1966) and Schubert (1972). These authors use Blokhintsev's principle of conservation of acoustic energy, and their treatment requires use of the geometrical solution in the far field. This approach is questionable because the geometrical solution becomes invalid beyond the diffraction limit.

Complete 3D geometrical solutions are constructed by Candel (1976, 1977a, b) using the general method outlined in the previous section. Applications of geometrical solutions are given by Candel (1975) and Candel et al. (1975). The calculations are performed within the diffraction limit and provide the sound pressure distribution in the geometric near field.

To illustrate this procedure let us consider the problem of sound radiation from a point source embedded in a free jet. The jet is issuing from a convergent circular nozzle at an initial speed of 390 m/s and a static temperature of 870°K. The mean aerodynamic profile of the jet has been determined experimentally, and is conveniently represented by making use of classical similarity properties of free shear flows. Figure 5 shows typical profiles for the index $N = c_0/c$ and Mach number M traced for a few axial sections in the jet. The mean aerodynamic structure is axisymmetric and exhibits strong gradients in the radial direction and slower axial variations.

Now consider a point source located off the jet axis. In this situation, three-dimensional refraction effects occur, and are made evident by tracing rays originating at the source. This is exemplified by Figure 6. The rays launched in a horizontal plane escape from this plane and appear as three-dimensional trajectories. The pressure field amplitude is also strongly affected by the 3D-refraction effects. (See Candel (1977a, b) for a detailed discussion of this aspect.)

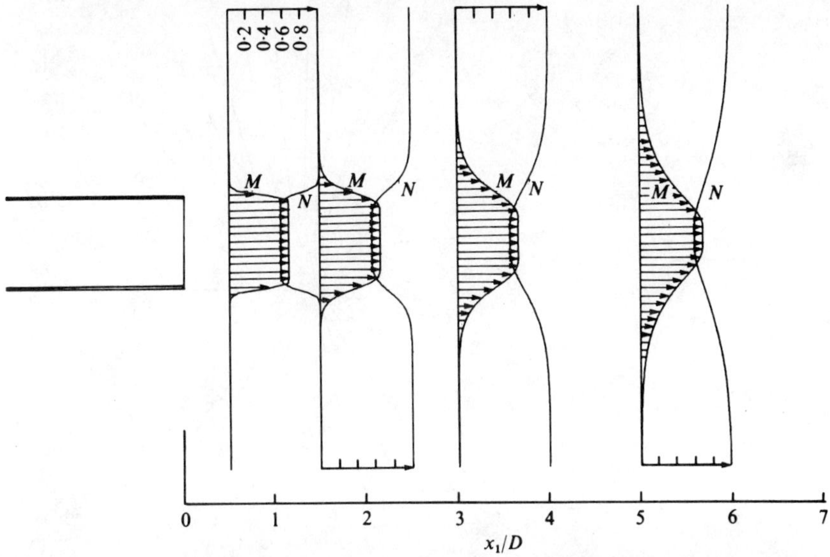

Figure 5. Aerodynamic mean field of a hot subsonic jet. The radial profiles of the mach number M and index N are represented for four axial sections, $x_1 = 0.5D$, $1.5D$, $3D$, and $5D$. Initial jet velocity and static temperature are $U_1^\circ = 390$ m/s and $T_1^\circ = 870$ K. From Candel (1977). Reprinted with permission of Cambridge University Press.

Clearly the geometrical approximation allows the treatment of fairly general 3D propagation problems. An intuitive image of the propagation phenomenon is obtained with this method. However, the geometrical approximation does not account for wave diffraction, and in many instances the rays form envelopes (caustic surfaces) and the calculation of the field amplitude requires special techniques which are not easy to implement numerically.

4. Spectral Methods

4.1 General Considerations

Spectral methods have found extensive applications in fluid dynamics. They are, for example, useful in the direct simulation of turbulence and allow calculations of large scale three-dimensional fluid motion. (See Gottlieb and Orszag (1977) for a general analysis of these methods.) Spectral techniques are also well suited to a

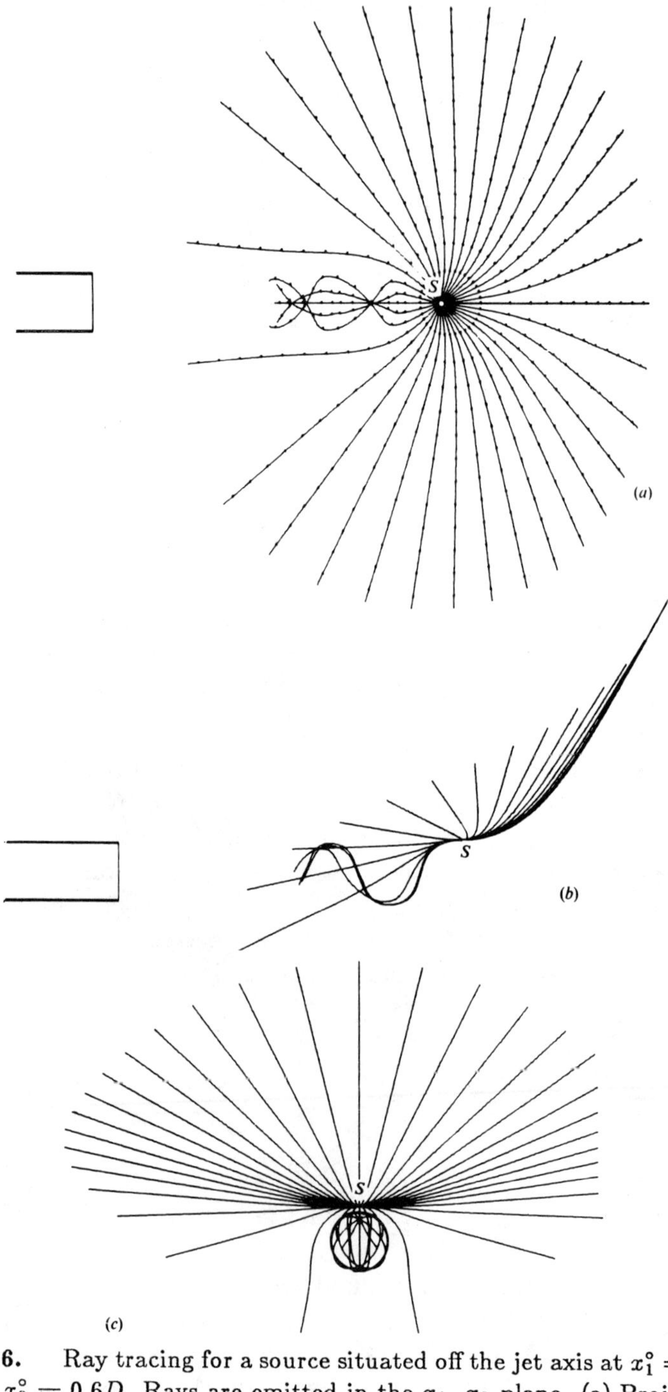

Figure 6. Ray tracing for a source situated off the jet axis at $x_1^\circ = 6D$, $x_2^\circ = 0$, $x_3^\circ = 0.6D$. Rays are emitted in the x_1, x_2 plane. (a) Projection on $x_1 x_2$ plane. (b) Projection on $x_1 x_3$ plane, (c) Projection on $x_2 x_3$ plane.

large class of wave problems. Methods based on Fourier transforms have been proposed for seismic reflectivity migration by Stolt (1978), Claerbout (1976), Gazdag (1978a, b), and Berkhout (1980). The numerical migration relies on the Fast Fourier Transform algorithm (FFT) and some simple Fourier-space filtering operations.

Wave extrapolation and backward projection operations similar to migration have also been developed for acoustic applications. Williams et al. (1980) demonstrate the feasibility of sound source reconstruction by back-projection in the Fourier transform domain. Higgins et al. (1980) report imaging application of field reconstruction methods based on back projection and the FFT algorithm. Stepanishen and Benjamin (1982) and Guyomar (1982) provide further illustration of forward and backward field extrapolation based on FFT methods.

Calculation of near field radiation from planar vibrating structures is performed by Williams and Maynard (1982) with a two-dimensional FFT.

Candel and Crance (1981a, b) and Candel (1983) treat wave problems in layered media by directly synthesizing the propagating field from its Fourier transform spatial spectrum.

Other transforms have been found useful in axisymmetric configurations. In such cases the two-dimensional Fourier transform reduces to a Fourier-Bessel integral. Di Napoli and Deavenport (1979, 1980) explore Green's function calculations in multilayered media based on asymptotic numerical evaluation of the Fourier-Bessel transform. Their method allows fast computations of the sound field radiated by an underwater source.

Frisk, Oppenheim, and Martinez (1980) investigate a method for the determination of the reflection coefficient of the ocean floor based on the Fourier-Bessel/Hankel transform. Candel and Brun (1982) treat point source radiation from a semi-infinite solid towards a light fluid. The field is obtained by making use of numerical Fourier-Bessel transform algorithms developed by Candel (1981a, b, c, d).

Scattering problems in one-, two-, and three-dimensions may also be treated by spectral domain methods as shown by Bojarski (1982).

The general principle of transform domain methods is outlined in Figure 7. The problem initially formulated in the physical domain is analytically or numerically transformed. This yields a propagation problem in the spectral domain which may be solved by analytical methods, algebraic operations, or numerical integration of ordinary differential equations. The spectral domain solution is then inverted by applying a numerical transform. This last operation provides the wave field in physical space.

It is sometimes convenient to modify this basic procedure and develop the solution of the propagation problem by making use of the physical and spectral domains simultaneously as proposed by Bojarski (1982).

We shall first illustrate the basic ideas of spectral methods by treating wave extrapolation (migration) problems. We next consider wave synthesis in layered media for line and point source radiation. General scattering problems in three dimensions are then analyzed with the method of Bojarski.

4.2 Wave Extrapolation in the Wave Number-Frequency Domain

Consider as a first problem field extrapolation of monochromatic waves in a homogeneous medium (Figure 8). The wave motion is described by the Helmholtz equation:

$$\frac{\partial^2 p}{\partial x^2} + \frac{\partial^2 p}{\partial y^2} + \frac{\partial^2 p}{\partial z^2} + k^2 p = 0 \tag{4.1}$$

with $k = \omega/c$.

Now let $\tilde{p}(\alpha, \beta, z)$ designate the 2D-Fourier transform of the pressure field. Then:

$$p(x, y, z) = \int \tilde{p}(\alpha, \beta, z)\, e^{i(\alpha x + \beta y)}\, d\alpha\, d\beta \tag{4.2}$$

Substituting this expression in equation (4.1) one obtains the following transformed problem:

$$\frac{d^2 \tilde{p}}{dz^2} - \gamma^2 \tilde{p} = 0 \tag{4.3}$$

where $\gamma = (\alpha^2 + \beta^2 - k^2)^{1/2}$.

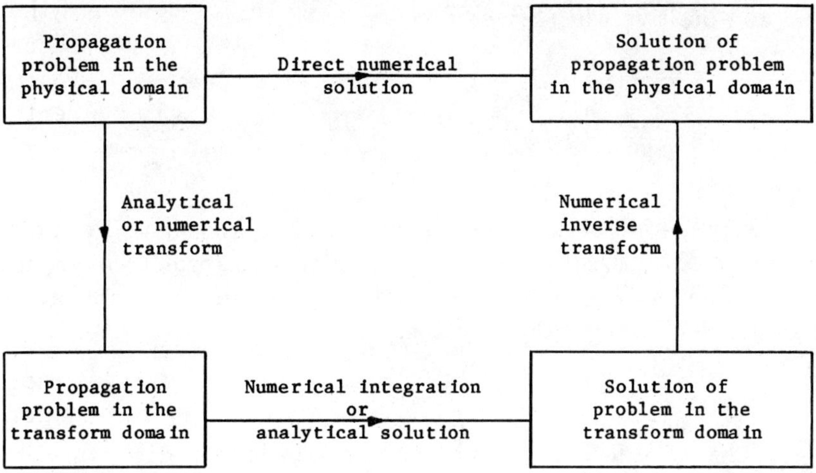

Figure 7. Principle of transform domain methods.

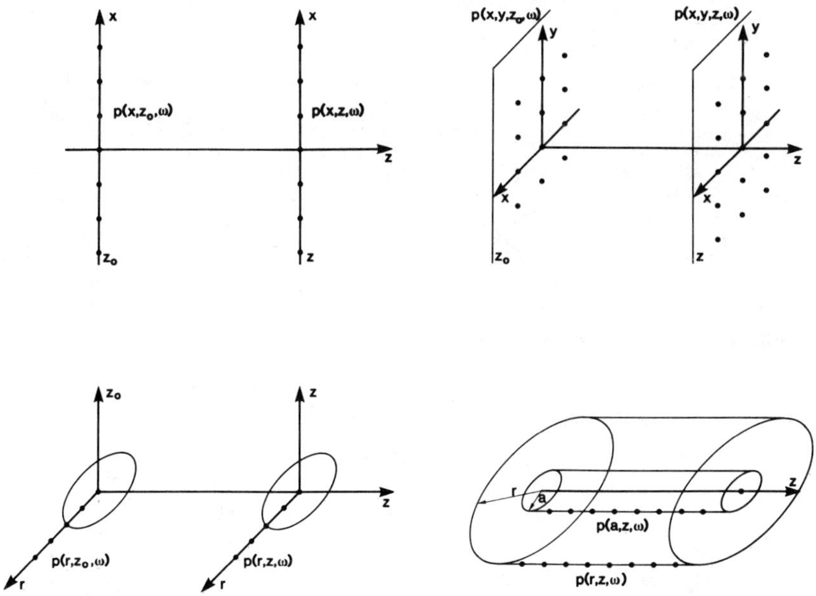

Figure 8. Some field extrapolation geometries.

Among the determinations of γ we select that for which γ is real and positive when $\alpha^2 + \beta^2 > k^2$:

$$\begin{aligned} \gamma &= (\alpha^2 + \beta^2 - k^2)^{1/2} \quad \text{for } \alpha^2 + \beta^2 > k^2 \\ &= -i|k^2 - \alpha^2 - \beta^2|^{1/2} \quad \text{for } \alpha^2 + \beta^2 < k^2 \end{aligned}$$

Now suppose that the pressure field is known in the plane z_0 and let $\tilde{p}(\alpha, \beta, z_0)$ designate its Fourier transform. Making use of equation (4.3), it is a simple matter to obtain the transformed pressure field in any other z-plane:

$$\tilde{p}(\alpha, \beta, z) = \tilde{p}(\alpha, \beta, z_0)\, e^{-\gamma|z-z_0|} \tag{4.4}$$

The wave field in the physical domain is then determined by taking the 2D-Fourier transform of the wave field in the spectral domain.

Field extrapolation thus consists of:

1. taking the 2D-Fourier transform of the known data in the z_0-plane,
2. applying the spatial transfer function or filter, $\exp(-\gamma|z-z_0|)$, to the result of step 1,
3. taking the inverse 2D-Fourier transform of the result of step 2.

This procedure is summarized in block form on Table 4. Similar methods may be used to migrate one-dimensional or axisymmetric fields (Figure 8). For a cylindrically symmetric field, migration along the symmetry z-axis requires a Fourier-Bessel transformation of the data $p(r, z_0)$ available in the z_0 plane. This technique is used by Sheng (1980) to propagate laser beams in unstable cavity resonators.

If the field extrapolation is performed in the radial (transverse) direction, a 1D-Fourier transform is needed, but the spatial frequency transfer function is a little more complex:

$$U(\alpha, r_0, r) = H_0^1(i\gamma r)/H_0^1(i\gamma r_0) \tag{4.5}$$

It turns out that this propagation operator may be generally replaced by its asymptotic estimate:

$$U(\alpha, r_0, r) \simeq (r_0/r)^{1/2}\, e^{-\gamma(r-r_0)} \tag{4.6}$$

Two-dimensional wave field migration:

$p(x, z_0)$	$1D$ Fourier	$\tilde{p}(\alpha, z_0)$	propagation	$\tilde{p}(\alpha, z)$	Inverse $1D$ Fourier	$p(x, z)$
pressure field at z_0	transform		operator $e^{-\gamma\lvert z-z_0\rvert}$ $\gamma = (\alpha^2 - k^2)^{1/2}$		transform	pressure field at z

Three-dimensional wave field migration:

$p(x, y, z_0)$	$2D$ Fourier	$\tilde{p}(\alpha, \beta, z_0)$	propagation	$\tilde{p}(\alpha, \beta, z)$	Inverse $2D$ Fourier	$p(x, y, z)$
pressure field at z_0	transform		operator $e^{-\gamma\lvert z-z_0\rvert}$ $\gamma = (\alpha^2 + \beta^2 - k^2)^{1/2}$		transform	pressure field at z

The wave field is rotationally symmetric about the z-axis. Migration in the z-direction (axial direction).

$p(r, z_0)$	Fourier-Bessel	$\tilde{p}(\varsigma, z_0)$	propagation	$\tilde{p}(\varsigma, z)$	Inverse Fourier-Bessel	$p(r, z)$
pressure field at z_0	transform		operator $e^{-\gamma\lvert z-z_0\rvert}$ $\gamma = (\varsigma^2 - k^2)^{1/2}$		transform	pressure field at z

The wave field is rotationally symmetric about the z-axis. Migration in the radial direction (transverse direction).

$p(r_0, z)$	$1D$ Fourier	$\tilde{p}(r_0, \alpha)$	propagation	$\tilde{p}(r, \alpha)$	Inverse $1D$ Fourier	$p(r, z)$
pressure field at a radius r_0 from the $z-$axis	transform		operator $H_0^1(i\gamma r)/H_0^1(i\gamma r_0)$ $\gamma = (\alpha^2 - k^2)^{1/2}$		transform	pressure field at a radius r from the $x-$axis

Table 4. Field extrapolation procedures.

To illustrate the previous discussion, consider the problem of extrapolating data from the near field to the far field of a dipole source. The geometry of the problem is shown on Figure 9. Now suppose that the wave field is known on the z-axis at $M = 128$ equally spaced sampling points. (These points represent a collection of transducers forming an antenna.) Figure 10 presents the field amplitude extrapolated with procedure 4 of Table 4. The exact field amplitude is shown on Figure 11. The calculations, performed with the FFT algorithm, require special precautions in the transformation of the sampled data. The difficulty is essentially related to the problem of replicated sources corresponding to the spectral aliasing phenomenon. The problem is caused by sampling in the spatial frequency domain, and it is inevitably encountered when the spatial frequency spectrum is not band-limited. Various techniques may be used to reduce this effect. The first consists of extending the computation domain in order to displace the replicated sources further away from the true source. This technique is effective in many cases but other methods may be required. Williams and Maynard (1982) discuss this matter in more detail in relation with the computation of the field radiated by a planar vibrator.

4.3 Direct Synthesis of Waves in Layered Media

Transform methods are well suited to the study of wave motion in layered configurations. Such configurations are of great importance in many areas; like geophysical data processing (Ewing et al. (1957), Claerbout (1976)), underwater sound (Urick (1967), Clay and Medwin (1977), Officer (1958)), optics and electromagnetics (Baños (1966), Wait (1962), Brekhovskikh (1960), Felsen and Marcuvitz (1973)). Layered configurations are often used as models in aeroacoustics. Figure 12 presents a few situations which fall in this category.

Now consider as a first problem, a line source located at $z = -h$ in the presence of a plane interface ($z = 0$), separating medium 1 ($z < 0$) of density ρ_1 and sound speed c_1 from medium 2 ($z > 0$) of density ρ_2 and sound speed c_2. The source radiates monochromatic waves of angular frequency ω and its amplitude is 4π.

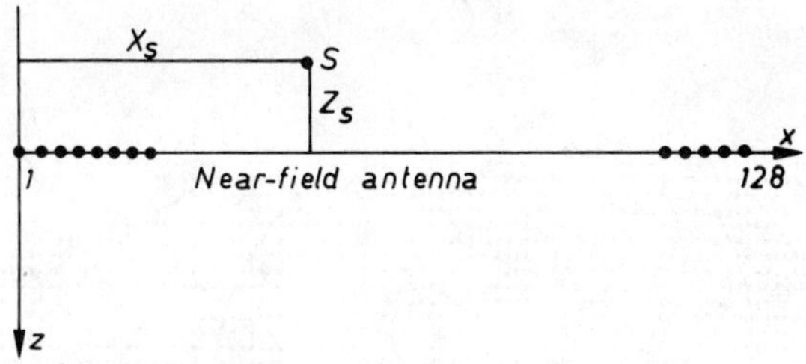

Figure 9. Geometry of the near-field to far-field extrapolation problem. A dipole source is located at $x_{S*} = 50.5\ \Delta x*$, $z_{S*} = -5.5\ \Delta z*$. The reduced coordinates are defined by $x* = x/L$, $z* = z/L$, where L designates the x-spatial extent of the computational domain. The field is sampled at $M = 128$ points with a spatial period $\Delta x* = 1/N$. Here $N = 1024$.

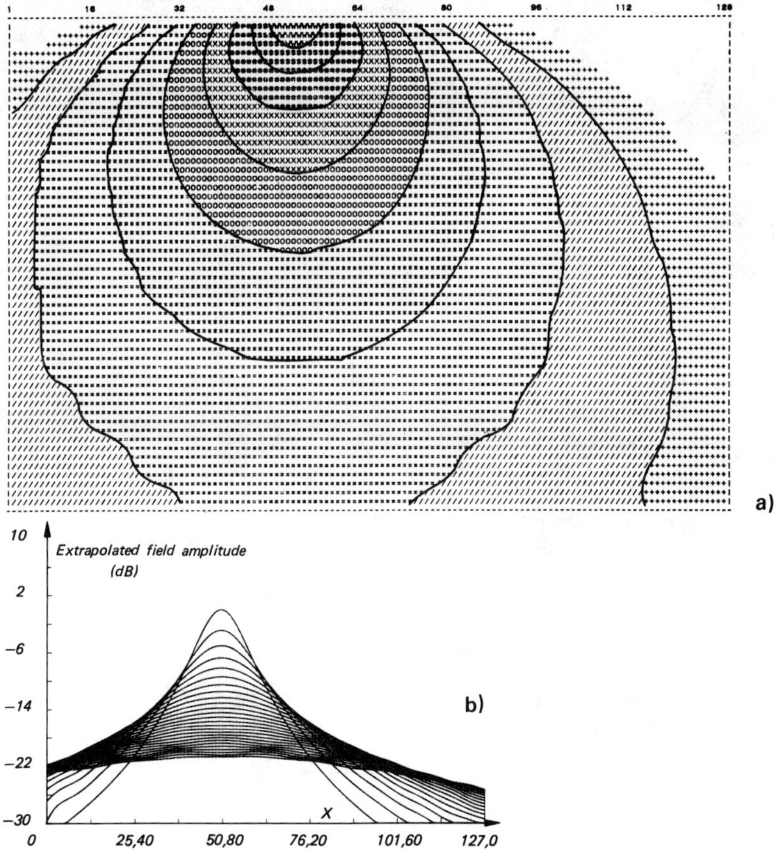

Figure 10. Extrapolated field amplitude for the problem of Figure 9. The reduced wave number is in this case $k* = kL = 80.5(2\pi)$. (a) Constant amplitude contours. (b) Field amplitude in a few radial sections.

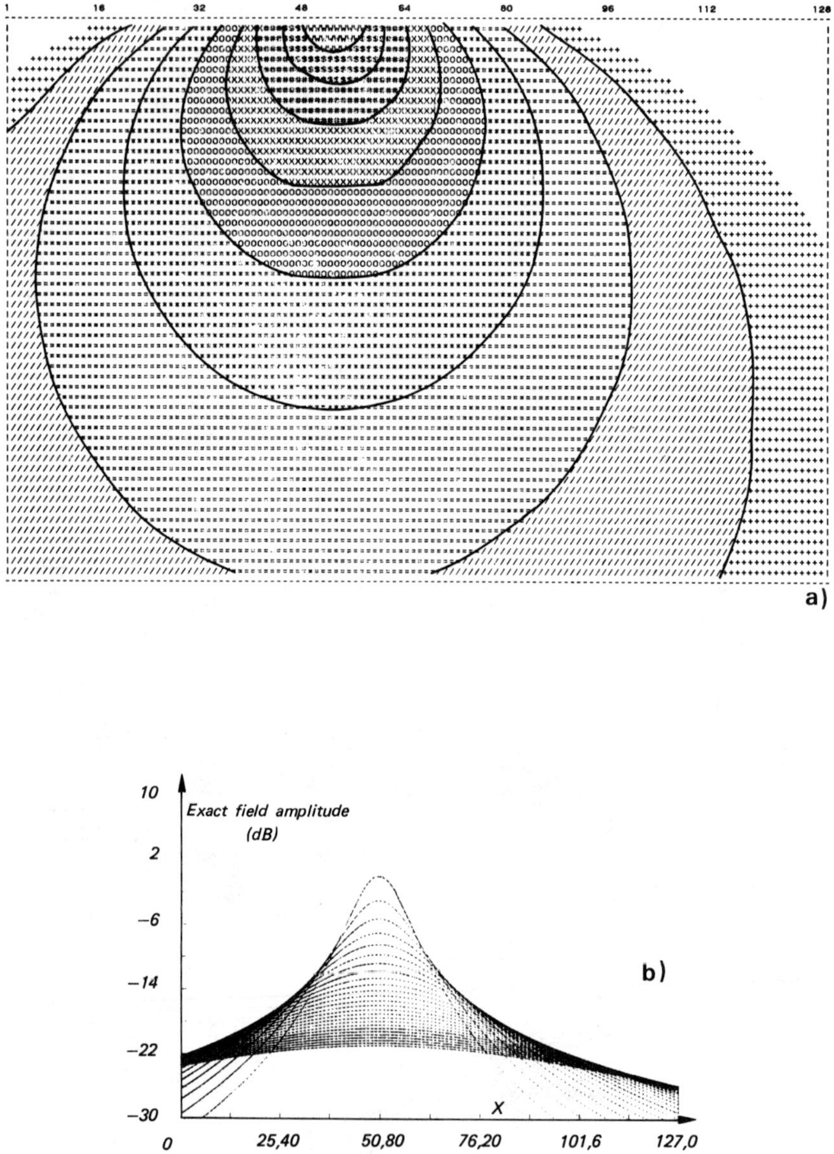

Figure 11. Exact field amplitude for the problem of Fig. 9. The reduced wave number is in this case $k* = kL = 80.5(2\pi)$. (a) Constant amplitude contours. (b) Field amplitude in a few radial sections.

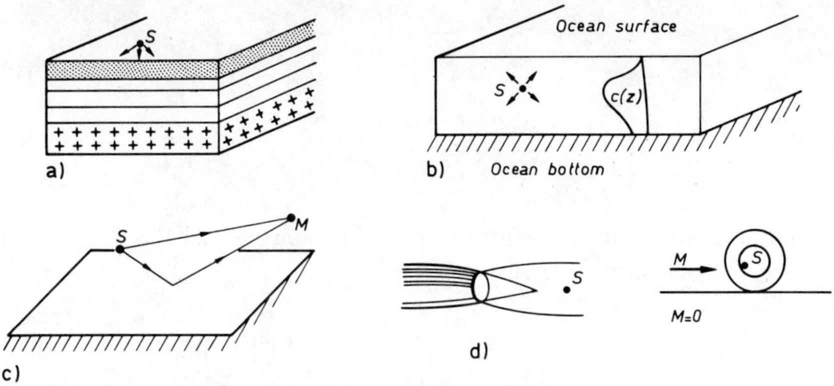

Figure 12. Some typical layered configurations: (a) stratified ground model, (b) layered ocean model for underwater sound applications, (c) a simplified ground model for atmospheric propagation, and (d) a high speed jet flow and the corresponding layered shear-flow model.

The wave field is then the solution of:

$$\begin{aligned} \nabla^2 p_1 + k_1^2 p_1 &= -4\pi\delta(x)\,\delta(z+h) \\ \nabla^2 p_2 + k_2^2 p_2 &= 0 \end{aligned} \tag{4.7}$$

where $k_1 = \omega/c_1$, $k_2 = \omega/c_2$.

Matching conditions at $z = 0$ are obtained by expressing the continuity of pressure and normal displacement η at $z = 0$:

$$p_1(x, 0_-) = p_2(x, 0_+)$$

$$\omega^2 \eta(x, 0_-) = \frac{1}{\rho_1}\frac{\partial p_1}{\partial z}(x, 0_-) = \frac{1}{\rho_2}\frac{\partial p_2}{\partial z}(x, 0_+) = \omega^2 \eta(x, 0_+) \tag{4.8}$$

In addition, the pressure field must satisfy radiation conditions at infinity. The solution is conveniently obtained by taking a Fourier transform in the transverse x-direction (see for example Candel and Crance (1980a, b)).

The final results are:

$$\begin{aligned} p(x, z) = &\int \frac{e^{-\gamma_1|z+h|+i\alpha x}}{xD\gamma_1}\,d\alpha - \int \frac{e^{\gamma_1(z-h)+i\alpha x}}{\gamma_1}\,d\alpha \\ &+ \int e^{\gamma_1(z-h)+i\alpha x}\,\frac{2\rho_2}{\rho_1\gamma_2 + \rho_2\gamma_1}\,d\alpha, \qquad z < 0 \end{aligned} \tag{4.9}$$

$$p(x,z) = \int e^{-\gamma_1 h - \gamma_2 z + i\alpha x} \frac{2\rho_2}{\rho_1\gamma_2 + \rho_2\gamma_1}\, d\alpha, \qquad z > 0 \tag{4.10}$$

The first two terms of expression (4.9) are easily identified as cylindrical waves originating from the line source and the line source image. These terms are respectively:

$$\begin{aligned} G_{1+} &= i\pi H_0^1\{k_1[x^2 + (z+h)^2]^{1/2}\} \\ G_{1-} &= i\pi H_0^1\{k_1[x^2 + (z-h)^2]^{1/2}\} \end{aligned} \tag{4.11}$$

where H_0^1 designates th zero-order Hankel function of the first kind.

The last term of expression (4.9) and the only term of expression (4.10) cannot be evaluated so simply. However they may be estimated numerically by means of the fast Fourier transform algorithm. Practical details are given in Candel and Crance (1981a, b) and will not be repeated here.

It is only worth noting that the size of the discrete transform must be sufficiently large ($N \gtrsim 2048$), and that the singularities located on the integration path, such as branch points, must be treated with special care.

4.4 First Example of Calculation: a Line Source Near an Interface

The computations to be described, were performed with a 2048 point discrete Fourier transform, of which only 128 points are displayed.

Figure 13 corresponds to a case where both regions 1 and 2 have the same density, while the sound speed in region 2 is twice that of region 1.

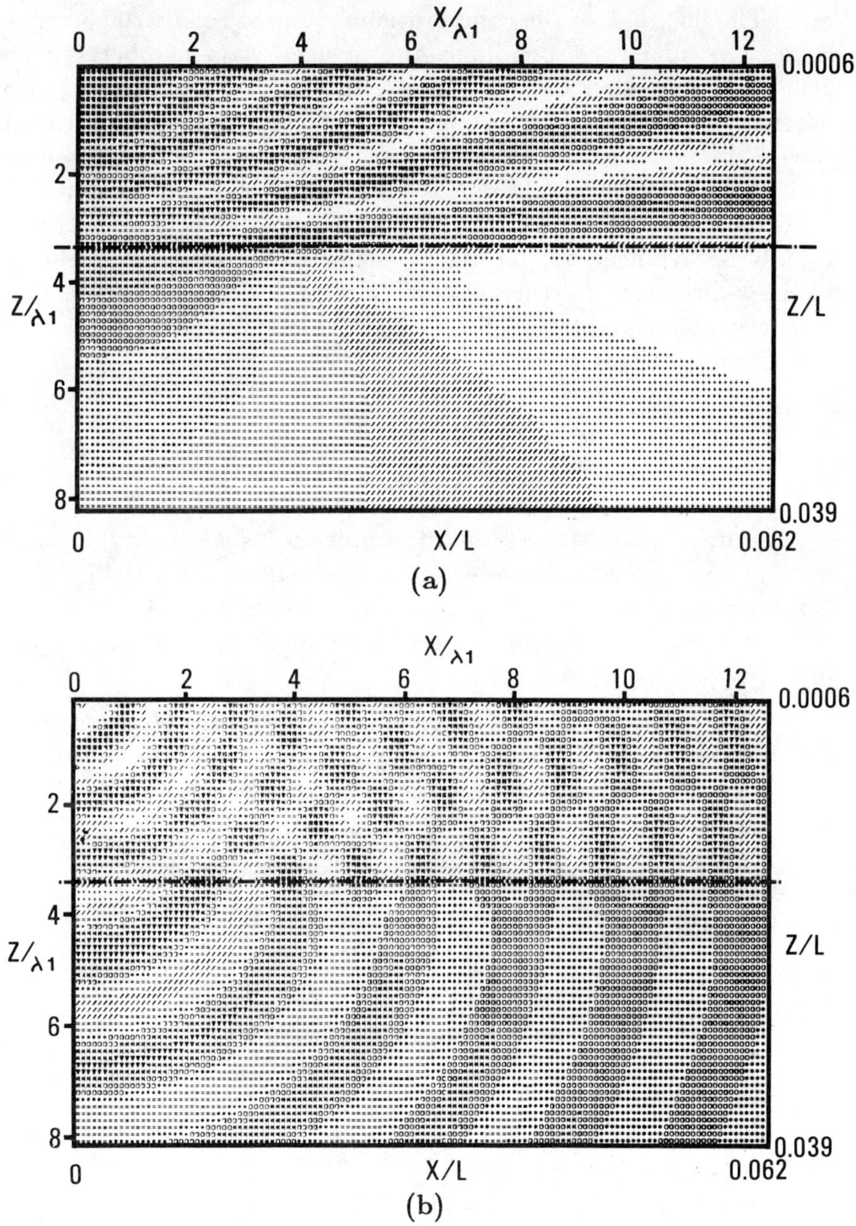

Figure 13. Line source radiation near a plane interface separating two uniform half-spaces. The densities are equal $\rho_1 = \rho_2$, but the sound speed in the lower space ($z > 0$) is twice that of the upper space, $c_2 = 2c_1$. (a) The field amplitude is plotted on a 10-level gray scale spanning the [0, 1] interval. (b) The real part of the field is plotted on a 10-level gray scale spanning the [-1, +1] interval. From Candel and Crance (1981a).

The length L of the computational domain equals 200.5 wavelengths λ_1, and correspondingly the reduced wave number $k_1 L = 200.5(2\pi)$. In region 1 ($z < 0$, above the interface) an interference pattern results from the combination of the direct and reflected waves. In region 2, waves propagate at twice the sound speed of region 1 and their general shape is quasi-cylindrical. The field modulus exhibits extremal values on conjugate hyperbolas (region 1) and rapidly decays near the interfacial plane; it decreases more slowly in the z-direction. Further physical interpretation and comparison with classical asymptotic field estimations may be found in Candel and Crance (1981a).

4.5 Sound Source Radiation in Shear Flow

Fourier synthesis is also applicable to radiation problems in moving layered media (Candel and Crance (1981b), Candel (1983)). Consider once more a line source in a two-dimensional medium in which the flow Mach number and sound speed are only a function of z; $M = M(z)$, $c = c(z)$.

Now, the continuous profiles may be replaced by a staircase function by dividing the shear region in $Q + 1$ plug flows (Figure 14). Each layer is characterized by its index j, Mach number M_j,

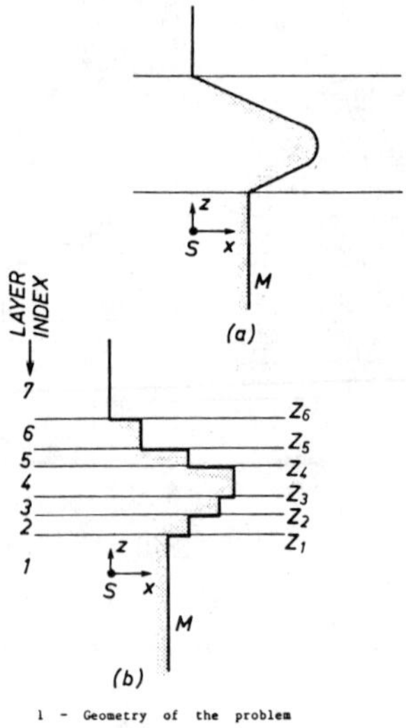

Figure 14. Sound source radiation in two-dimensional shear-flow, (a) continuous shear layer profile, (b) stratified representation.

sound speed c_j, density ρ_j, and transverse dimension d_j. The lower and upper interfaces bounding the jth layer are respectively located at z_j and z_{j+1}. The pressure field in each layer satisfies a convected wave equuation:

$$(1-M_j^2)\frac{\partial^2 p_j}{\partial x^2}+\frac{\partial^2 p_j}{\partial z^2}+2ik_jM_j\frac{\partial p_j}{\partial x}+k_j^2 p_j=-4\pi\delta(x)\,\delta(z) \quad (4.12)$$

where $k_j=\omega/c_j$.

Matching conditions on each interface are obtained by specifying the continuity of pressure and normal displacement:

$$\begin{aligned} p_j\,(x,z_{j+1}) &= p_{j+1}\,(x,z_{j+1}) \\ \eta_j\,(x,z_{j+1}) &= \eta_{j+1}\,(x,z_{j+1}) \end{aligned} \quad (4.13)$$

The local displacement is deduced from the pressure by making use of the linearized momentum equation:

$$\rho_j c_j^2\left(-ik_j+M_j\frac{\partial}{\partial x}\right)^2\eta_j=-\frac{\partial p_j}{\partial z} \quad (4.14)$$

Radiation conditions specified in the first and last layers complete the formulation.

The solution of this problem is obtained by Fourier transformation in the transverse direction (see Candel, 1983). The sound field in the jth layer appears in the form:

$$p_j(x,z)=\int_{-\infty}^{+\infty}\left[A_j(\alpha)\,e^{-\gamma_j(z-z_j)}+B_j(\alpha)\,e^{\gamma_j(z-z_j)}\right]d\alpha \quad (4.15)$$

where

$$\gamma_j=[(1-M_j^2)\alpha^2+2M_jk_j\alpha-k_j^2]^{1/2},$$

and $A_j(\alpha)$, $B_j(\alpha)$ designate the complex amplitudes of outgoing and ingoing components of the plane wave decomposition of the pressure field. The determination of γ_j must be consistent with the radiation condition at infinity. In Fourier space the transfer from layer to layer is described by (2×2) square matrices, and the spectral amplitudes $A_j(\alpha)$, $B_j(\alpha)$ result from simple operations on these matrices.

The solution of the Fourier space problem may also be obtained by direct numerical integration of the transformed equation for each wave number α. Another possibility is to subdivide the medium into layers characterized by linearly varying profiles of sound speed and Mach number. This last method, adopted in some underwater acoustic studies, increases the accuracy at the expense of computer time and programming simplicity.

4.6 Second Example of Calculation: Source Radiation in a Subsonic Shear Flow

Among the many calculations performed we shall only describe the case of line source radiation in a subsonic shear flow.

The computational domain extends over 100.5 wavelength and the FFT has $N = 2048$ points. Only 256 points surrounding the z-axis are displayed. The results shown on Figure 15 correspond to a shear flow represented by five layers. The Mach number decreases monotonically from 0.8 at the source to 0 in the outer region. The Strouhal number based on the thickness δ of the shear layer and the maximum flow velocity is for the present case $\text{St} = f\delta/U_1 = 1.61$.

Convection and refraction effects are clearly important. Wave fronts are convected downstream. A reflected wave forms upstream of the source and interferes with the direct wave. The reflected wave is much weaker downstream. In the outer region the wave fronts become nearly plane downstream and propagate along the shear layer boundary with the phase velocity imposed by the convecting fronts inside the flow. The radiated amplitude reaches its maximum near $\theta \simeq 50°$ corresponding to $\cos^{-1}[1/(1+M_1)]$ and the field amplitude is reduced in the low and high angle directions. These features are all in agreement with high frequency geometrical calculations performed for similar flow structures.

Other calculations corresponding to subsonic and supersonic situations are described in Candel (1983).

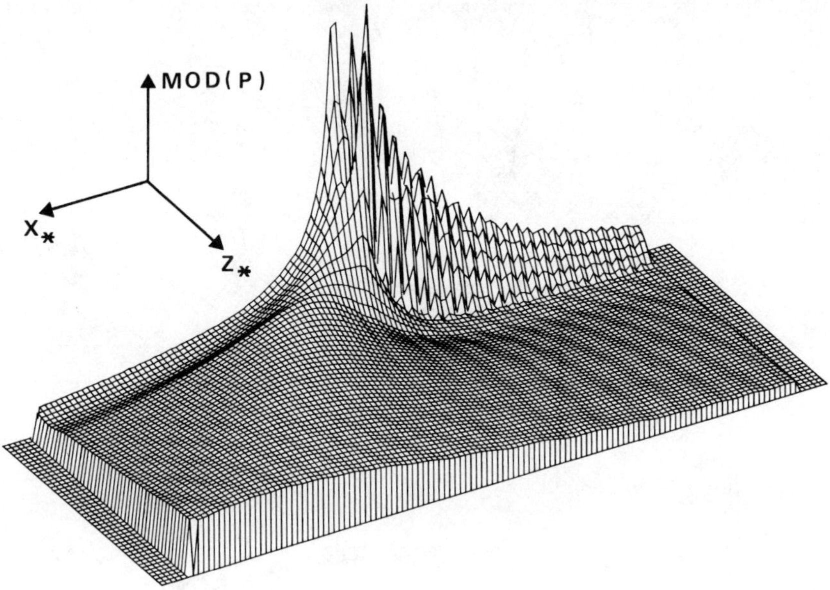

(a)

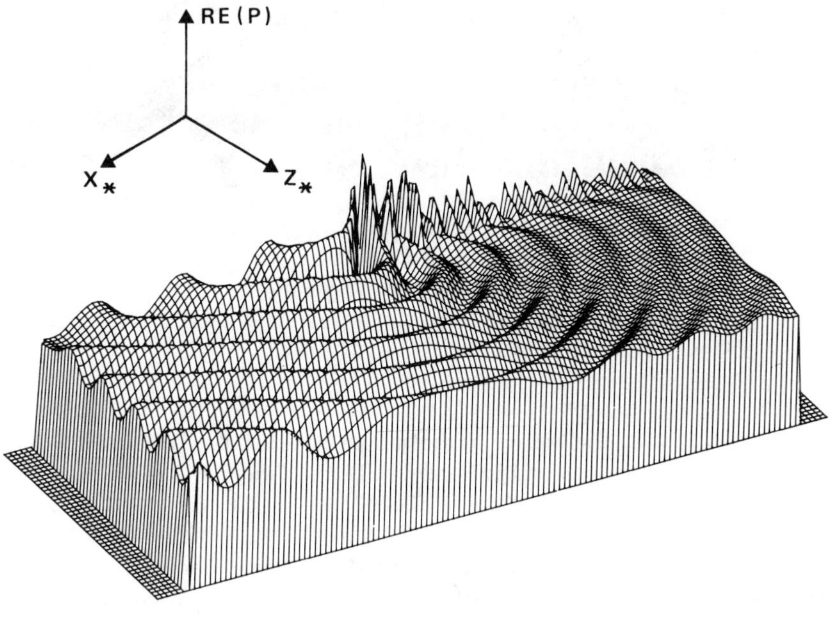

(b)

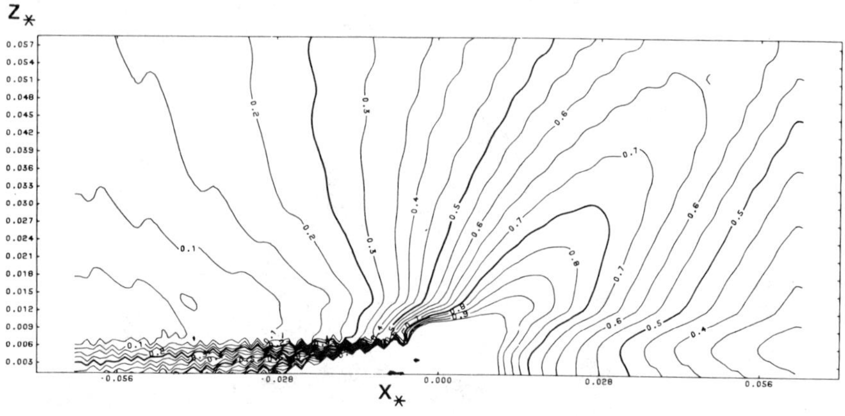

(c)

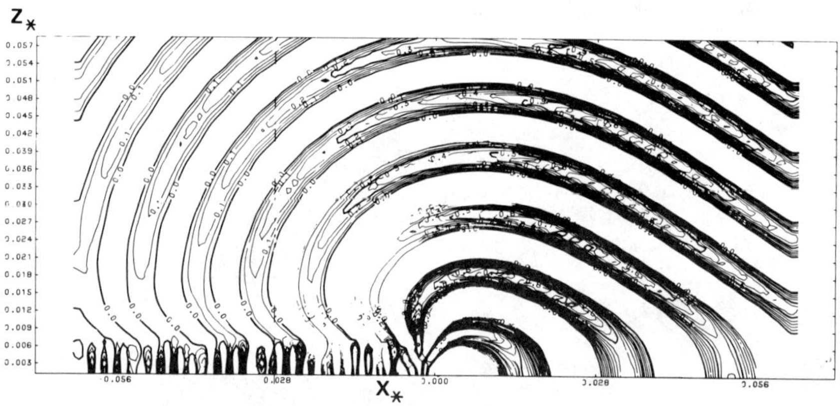

(d)

Figure 15. Sound source radiation in a shear flow. The source is located at $x = 0$, $z = 0$, and its strength is 4π. The flow is represented by five layers $M_1 = 0.8$ for $0 < z* < 0.0067$, $M_2 = 0.6$ for $0.0067 < z* < 0.0087$, $M_3 = 0.4$ for $0.0087 < z* < 0.0107$, $M_4 = 0.2$ for $0.0107 < z* < 0.0128$, $M_5 = 0$ for $0.0128 < z*$ reduced wave number in all layers $k* = kL = 100.5(2\pi)$. (a) and (c) contour plot and transection of field modulus. (b) and (d) contour plot and transection of field real part. From Candel (1983).

4.7 Fourier-Bessel and Hankel Transform Applications to Wave Fields Synthesis

General wave field synthesis requires 2D-Fourier transform calculations. In the special case of rotationally symmetric fields, the 2D-Fourier transform reduces to the zero order Fourier-Bessel transform. This property is used by Di Napoli and Deavenport (1979, 1980) in their *Fast Field Program* calculations of underwater sound source radiation, by Sheng (1980) in laser beam problems, and by Frisk et al. (1980) in the determination of the acoustic reflection coefficient of the ocean bottom.

Progress in the numerical computation of Fourier-Bessel transforms has been recently accomplished (see Candel, 1981a, b, c, d) and allows treatment of a large class of problems. As a first application consider the problem of sound radiation from a source embedded in a solid half-space loaded by a fluid half-space (Figure 16). The source emits monochromatic compression waves. Standard transform methods lead to the following expression for the pressure field in the fluid half-space:

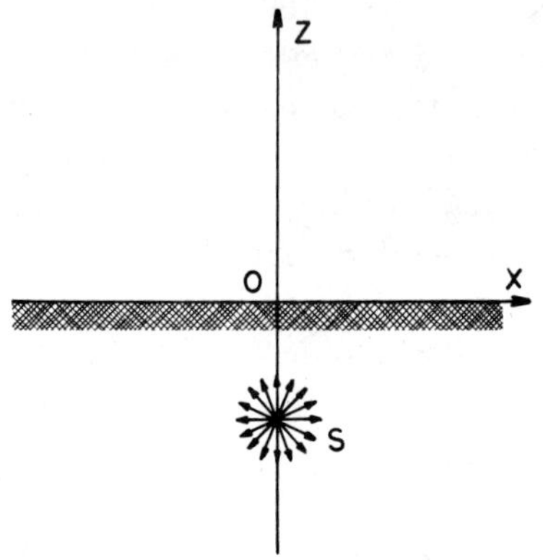

Figure 16. A point source embedded in a solid half-space radiates monochromatic compression waves. The solid is loaded by a very light fluid ($\rho_2/\rho_1 \ll 1$). The source is at $h = -6$ m below the interface.

$$p_2(r,z) = \int_0^\infty G(\varsigma, z)\, J_0(\varsigma, r)\, \varsigma\, d\varsigma \tag{4.16}$$

The pressure field is the Fourier-Bessel transform of the spatial spectrum $G(\varsigma, z)$. This last function is given explicity by Candel and Brun (1982). When the fluid is light (i.e., when the solid is lightly loaded $\rho_2/\rho_1 \ll 1$), the spatial spectrum exhibits two Rayleigh poles located near $\pm k_R$ corresponding to surface waves excited by the source and propagating at a speed c_R. This Rayleigh speed is smaller than the compressional and shear wave speeds in the solid. The main difficulty in treating the integral of equation (4.16) by discrete numerical techniques is associated with the presence of the Rayleigh poles on the integration path. The problem is that of treating a highly singular function by means of a discrete transform.

The difficulty may be overcome by extracting the Rayleigh pole contribution from the spatial spectrum by a partial fraction expansion. Then the discrete Fourier-Bessel transform is applied to the regular part of the spatial spectrum while the Rayleigh pole contribution is evaluated analytically.

Figures 17 and 18 give a typical result of calculation. The source of P-waves is at $h = -12$ m below the fluid/solid interface, the frequency is $f = 19.59$ Hz. P and SV waves propagate in the solid with speeds $c_{\alpha 1} = 3500$ m/s and $c_{\beta 1} = 2020.7$ m/s, and the P-waves in the fluid travel at a speed $c_{\alpha 2} = 333$ m/s. The Rayleigh wave speed is in this case $c_R = 1857.8$ m/s. The Rayleigh pole gives rise to a cylindrical wave with wave fronts locked to the annular Rayleigh wave travelling on the solid surface. The wave field corresponding to the regular part of the spatial spectrum appears on Figure 18. The main lobe is located around the z-axis and a secondary sidelobe is also visible.

4.8 Numerical Transformation of Pole Contributions

As noted above, functions exhibiting poles deserve special care when the poles are near or on the integration path. This situation is encountered in numerous applications of interest. For example, in underwater sound propagation the spatial spectrum exhibits an infinity of poles corresponding to propagating and evanescent modes.

A finite number of these poles are located in the immediate vicinity of the integration path.

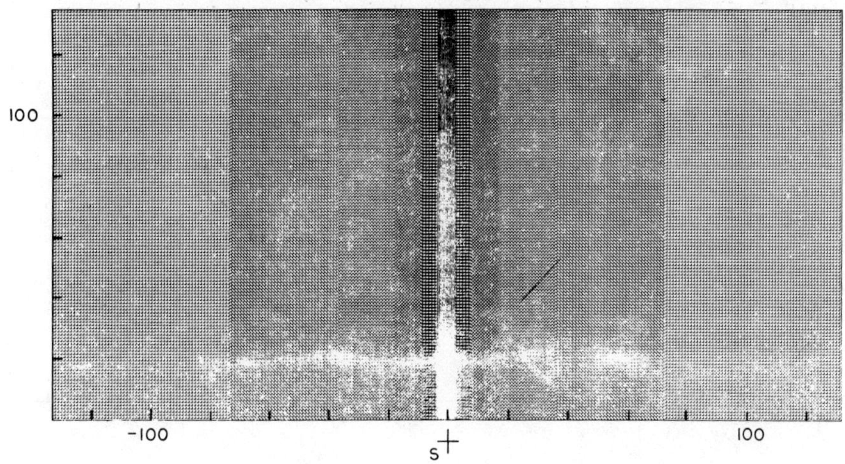

(a)

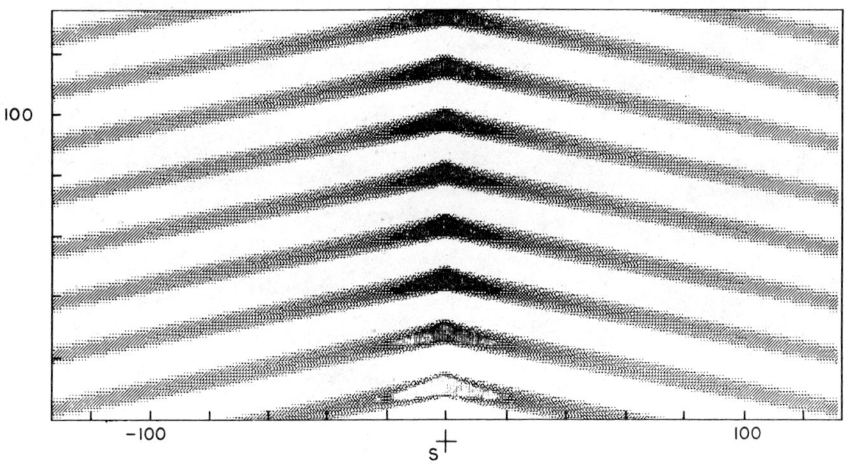

(b)

Figure 17. Structure of the sound field in the fluid medium for the problem of Figure 16. The field modulus is plotted on a logarithmic 10-level gray scale spanning the [-30, 0 dB] interval. The field real part is displayed on a linear 10-level gray scale spanning a $[-p_{\rm ref}, p_{\rm ref}]$ interval. (a) Modulus of the Rayleigh pole contribution, $p_{\rm ref} = 0.01$. (b) Real part of the Rayleigh pole contribution.

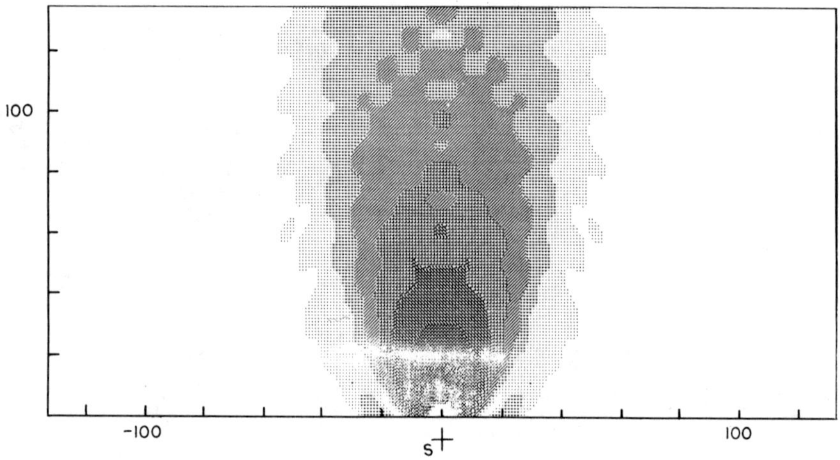

(a)

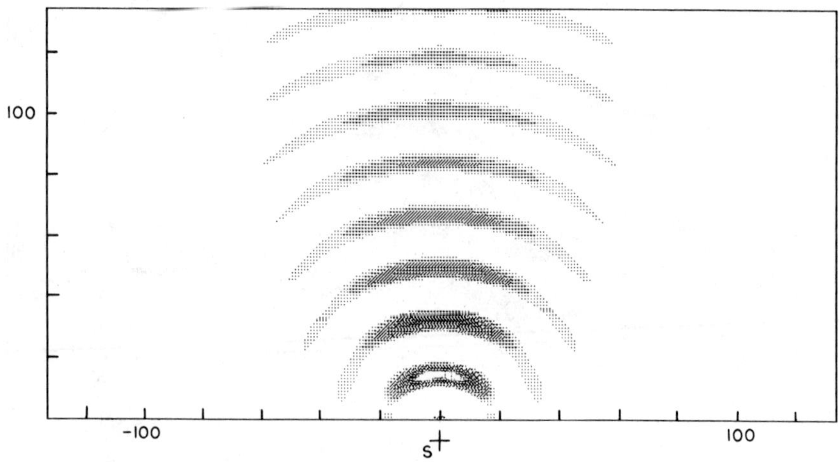

(b)

Figure 18. For the general legend see Figure 17. (a) Modulus of the regular part of the pressure field, $p_{\mathrm{ref}} = 0.16$. (b) Real part of the regular pressure field.

Surprisingly, the difficulty of performing a discrete transform in such circumstances is not mentioned in the papers of Di Napoli and Deavenport (1979, 1980).

A similar problem arises in the numerical evaluation of the Rayleigh integral which describes the radiation of planar vibrators, and it is discussed in a recent paper by Williams and Maynard (1982). These authors present a smoothing procedure which allows correct and accurate calculations of the radiation efficiency of the vibrator.

Alternative procedures for the discrete transformation of functions with poles have been developed by us and are described below.

Consider as a test example the numerical evaluation of:

$$F(x) = \int_{-\infty}^{+\infty} \frac{e^{i\alpha_0 x}}{\alpha - \alpha_0} \, d\alpha \tag{4.17}$$

For $x > 0$ this integral has the following determinations:

$$F(x) = \begin{cases} 2\pi i e^{i\alpha_0 x}, & Im(\alpha_0) > 0 \\ \pi i e^{i\alpha_0 x}, & Im(\alpha_0) = 0 \\ 0, & Im(\alpha_0) < 0 \end{cases} \tag{4.18}$$

Numerical evaluation of $F(x)$ by the FFT algorithm yields good estimates of the exact function when the pole α_0 is sufficiently far from the real axis (Figure 19).

When $Im(\alpha_0) = 0$, the pole lies on the integration path and the result essentially depends on the location of the pole with respect to the sampling points of the FFT (see Figure 20). The correct result is only obtained when the pole is a mid-point of a sampling interval (Figure 21). This is not surprising since the integral should be treated as a Cauchy principle value. Now, in most practical cases the function under transformation has many poles and their locations with respect to the sampling points will be arbitrary. Accurate estimates may be obtained in these circumstances by either: (1) performing local modification of the function under transformation to displace the poles towards sampling interval mid-points, (2) displacing the sampling points located in the neighborhood of each pole to achieve an equidistant distribution around each singularity. Both methods are justifiable approximations and provide acceptable estimates of the pole contributions.

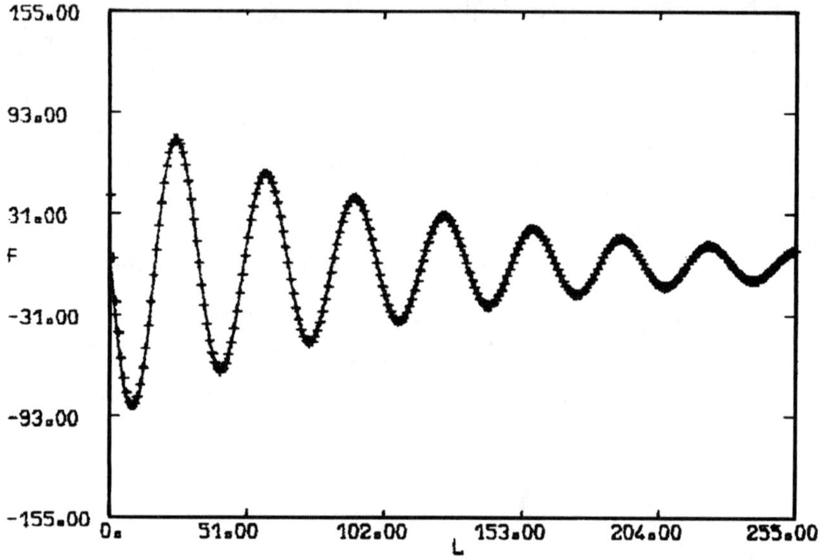

(a)

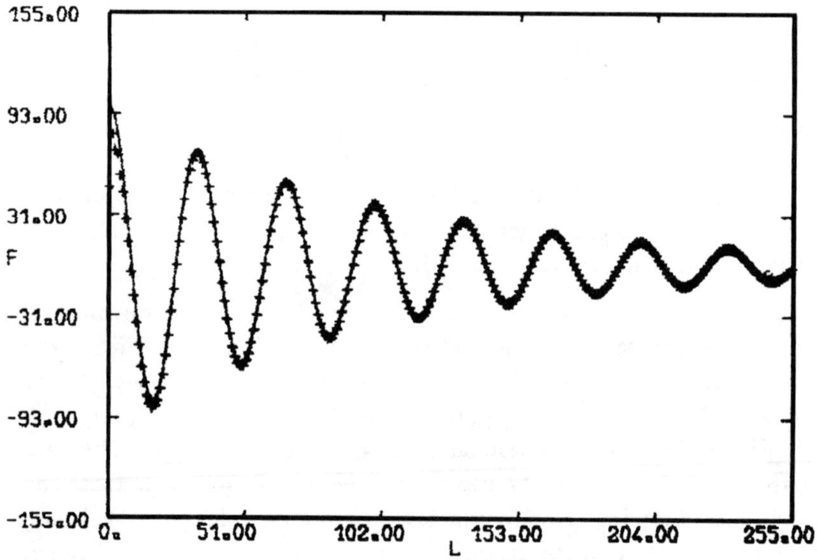

(b)

Figure 19. Fourier transform of a single pole function. The pole is located off the integration path at $\alpha_0 = (15.5 + i0.05)\Delta\alpha$. The symbols correspond to estimates obtained with the FFT algorithm ($N = 512$). The solid lines represent the exact transform. (a) Real part. (b) Imaginary part.

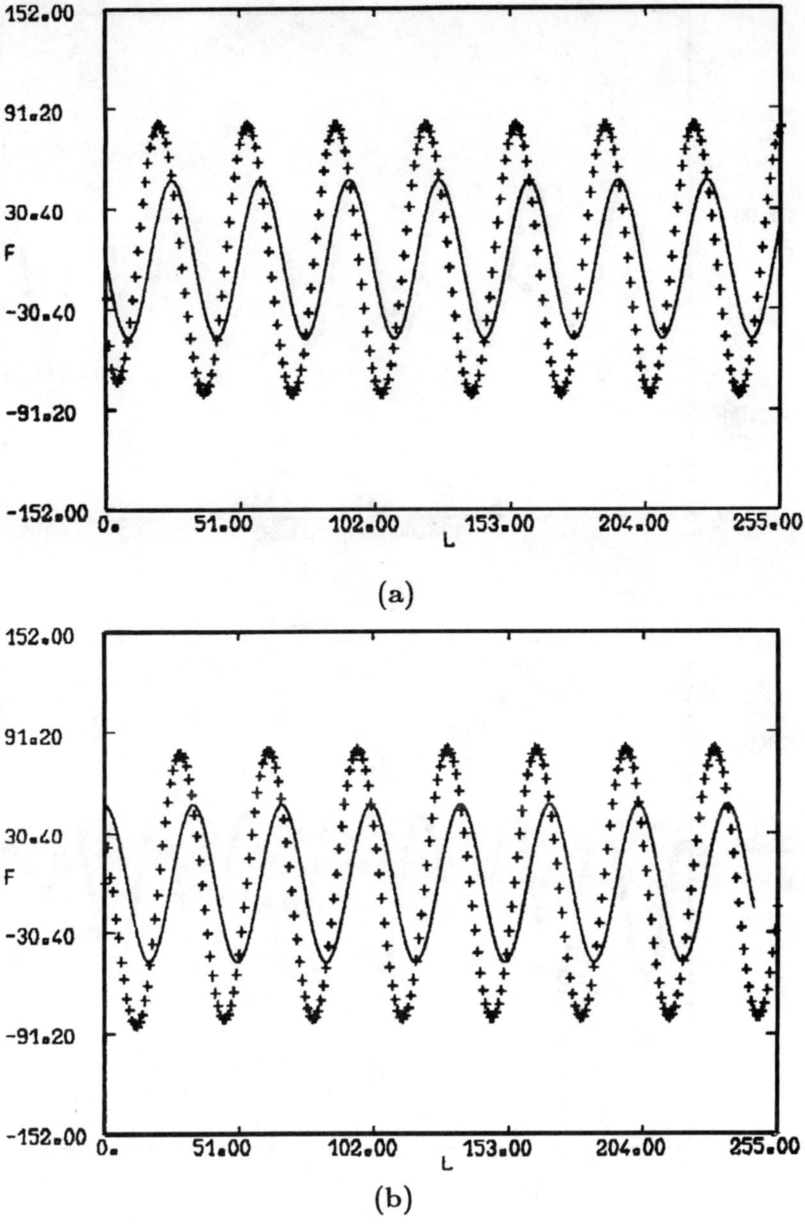

Figure 20. Fourier transform of a simple pole function. The pole is located on the integration path and does not coincide with a sampling interval mid-point: $\alpha_0 = (15.2 + i0.0)\Delta\alpha$. The symbols correspond to estimates obtained with the FFT algorithm ($N = 512$). The solid lines represent the exact transform. (a) Real part. (b) Imaginary part.

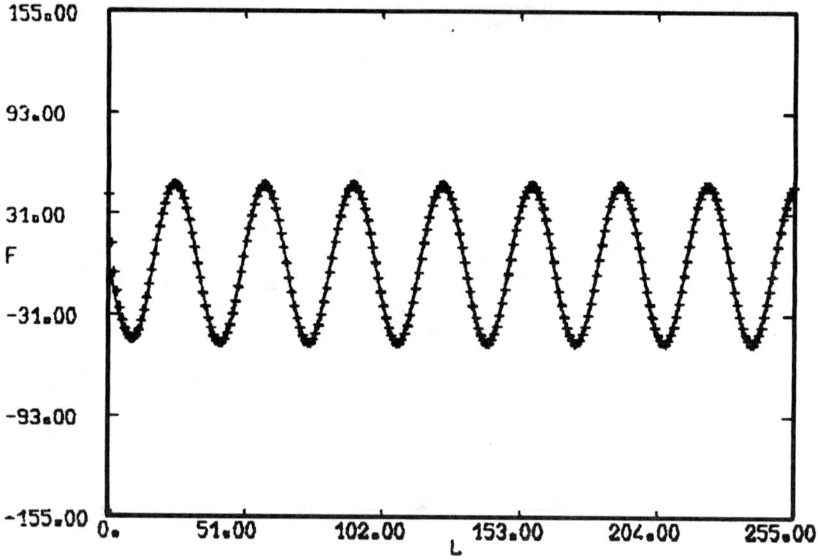

(a)

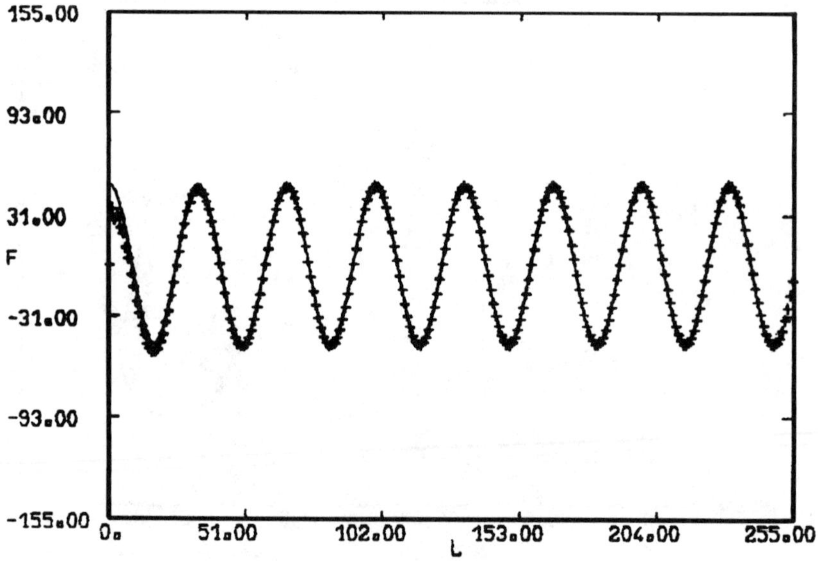

(b)

Figure 21. Fourier transform of a simple pole function. The pole is located on the integration path and coincides with a sampling interval mid-point: $\alpha_0 = (15.5 + i0.0)\Delta\alpha$. The symbols correspond to estimates obtained with the FFT algorithm ($N = 512$). The solid lines represent the exact transform. (a) Real part. (b) Imaginary part.

4.9 Multi-Dimensional Wave Scattering Problems

Spectral methods may be used to solve multi-dimensional wave problems. As a typical example, consider the scattering of a monochromatic wave in a medium containing a bounded inhomogeneous region characterized by a variable sound speed $c(\mathbf{x})$ (Bojarski, 1982). The pressure field satisfies the Helmholtz equation:

$$\nabla^2 p + \frac{\omega^2}{c^2(\mathbf{x})} p = 0 \tag{4.19}$$

Let c_0 designate the sound speed outside the inhomogeneous region. Then (4.19) may be cast in the form:

$$\nabla^2 p + \frac{\omega^2}{c_0^2} p = -f(\mathbf{x}, \omega) \tag{4.20}$$

where

$$f(\mathbf{x}, \omega) = V(\mathbf{x}, \omega)\, p,$$

and

$$V(\mathbf{x}, \omega) = \frac{\omega^2}{c^2} - \frac{\omega^2}{c_0^2}$$

Now let $G(\mathbf{x} \mid \mathbf{x}_0)$ designate the free space Green's function for the Helmholtz equation. Then, the formal solution of equation (4.20) may be written as:

$$\begin{aligned} p(\mathbf{x}) = & \int_v f(\mathbf{x}_0, \omega)\, G(\mathbf{x} \mid \mathbf{x}_0)\, d\mathbf{x}_0 \\ & + \int_s d\mathbf{s}_0 \cdot \left[G(\mathbf{x} \mid \mathbf{x}_0)\, \nabla_0 p - p(\mathbf{x}_0)\, \nabla_0 G(\mathbf{x} \mid \mathbf{x}_0)\right] \end{aligned} \tag{4.21}$$

The surface integral represents the incident fields $p_i(\mathbf{x})$ and the last expression becomes:

$$p(\mathbf{x}) = \int_v f(\mathbf{x}_0, \omega)\, G(\mathbf{x} \mid x_0)\, d\mathbf{x}_0 + p_i(\mathbf{x}) \tag{4.22}$$

This is an integral equation for the pressure field $p(\mathbf{x})$, and it may be solved by the Ritz-Galerkin method. Typically, the solution requires the order of N^3 operations, where N designates the number of discrete cells used in the calculation.

Spectral methods greatly reduce the number of operations required. To see this, it is important to note that the free space Green's function is shift-invariant $G(\mathbf{x} \mid \mathbf{x}_0) = g(\mathbf{x} - \mathbf{x}_0)$, and that as a consequence expression (4.22) takes the form of a convolution integral. Taking the 3D-Fourier transform of this equation then leads to:

$$\tilde{p}(\mathbf{k}) = \tilde{f}(\mathbf{k})\tilde{g}(\mathbf{k}) + \tilde{p}_i(\mathbf{k}) \tag{4.23}$$

This algebraic relation together with

$$f(\mathbf{x}, \omega) = V(\mathbf{x}, \omega)\, p(\mathbf{x}, \omega) \tag{4.24}$$

completely describes the scattering problem.

This set of equations may be solved by Bojarski's iterative procedure

$$\begin{aligned} \tilde{p}_n &= \tilde{g}\tilde{f}_n + \tilde{p}_i \\ p_n &= F^{-1}\tilde{p}_n \\ f_n &= V\, p_n \\ \tilde{f}_{n+1} &= F\, f_n \end{aligned} \tag{4.25}$$

where F designates the 3D-Fourier transform operator. The preceding algorithm requires the order of $3N \log_2 N$ operations, where N is the number of computation cells. The method is quite attractive and some convincing examples of calculation are given by Bojarski (1982).

5. Finite Difference Methods

An enormous amount of literature is concerned with the application of finite difference methods to fluid flow problems. A large number of these studies deal with time-dependent solutions of the Navier-Stokes equations in one, two, or three dimensions. Numerous methods developed in the field of fluid mechanics are applicable to acoustic wave propagation. We shall not attempt to describe all these methods and refer the reader to a review of Turkel (1983) containing more than 300 references to the recent literature.

We shall restrict the present survey to the analysis of papers which are specifically concerned with acoustic wave motion. The number of such studies is surprisingly small compared to those treating fluid flow problems. Indeed, the field of finite difference acoustic wave applications appears to be in its infancy, and most methods used at the present time are of the explicit time marching type. We have found no published report of application to wave problems of the more modern implicit finite difference techniques, which have been so successful in fluid dynamics.

5.1 Direct Solution of the Wave Equation

A classical method (see Richtmeyer and Morton, 1967) for solving the one-dimensional wave equation:

$$\frac{\partial^2 p}{\partial t^2} = c^2 \frac{\partial^2 p}{\partial x^2} \tag{5.1}$$

consists of replacing the second derivatives by second order finite difference expressions:

$$p_j^{n+1} - 2p_j^n + p_j^{n-1} = \left(\frac{c\Delta t}{\Delta x}\right)^2 (p_{j+1}^n - 2p_j^n + p_{j-1}^n) \tag{5.2}$$

where $p_j^n = p(j\Delta x, n\Delta t)$ designates the pressure at point $j\Delta x$ and time $n\Delta t$. Solving equation (5.2) for p_j^{n+1} leads to:

$$p_j^{n+1} = 2p_j^n - p_j^{n-1} + \left(\frac{c\Delta t}{\Delta x}\right)^2 (p_{j+1}^n - 2p_j^n + p_{j-1}^n) \tag{5.3}$$

which permits the calculation of the pressure at time $(n+1)\Delta t$ from the values of the pressure at times $(n-1)\Delta t$ and $n\Delta t$. This explicit scheme is stable if the time increment is not too large and satisfies the Von Neumann stability condition:

$$c\Delta t/\Delta x \leq 1.$$

The scheme (5.3) forms the basis of many studies in acoustics and geophysical wave propagation. Most of these studies are conducted in two dimensions and use the straightforward extension:

$$\begin{aligned} p_{ij}^{n+1} - 2p_{ij}^{n} + p_{ij}^{n-1} &= \left(\frac{c\Delta t}{\Delta x}\right)^2 (p_{i+1j}^{n} - 2p_{ij}^{n} + p_{i-1j}^{n}) \\ &+ \left(\frac{c\Delta t}{\Delta y}\right)^2 (p_{ij+1}^{n} - 2p_{ij}^{n} + p_{ij-1}^{n}) \end{aligned} \tag{5.4}$$

The stability of this 2D-explicit scheme is assured if:

$$\frac{c\Delta t}{(\Delta x^2 + \Delta y^2)^{1/2}} \leq 1.$$

Baumeister (1980, 1981, 1982) uses equation (5.4) to analyze noise propagation in two-dimensional ducts and provides detailed derivations of the boundary conditions. The same scheme is also used by Cabelli (1982). Other applications of explicit finite difference formulas may be found in the work of Alford et al. (1974) in relation with geophysical wave problems.

5.2 Multi-Step Explicit Methods

Multi-step explicit methods have been extensively used to treat multi-dimensional time-dependent problems in fluid mechanics.

The two-step Lax-Wendroff and MacCormack methods are most commonly used. These procedures are second-order accurate in time and space. It has been argued by Maestrello, Bayliss and Turkel (1981) that schemes with higher order accuracy are required for the solution of acoustic radiation problems.

To introduce these multi-step methods we consider wave propagation in a two-dimensional homogeneous medium. The linearized dynamic equations of motion are conveniently written as a first order system:

$$\begin{aligned} \frac{\partial p}{\partial t} + \rho_0 c^2\left(\frac{\partial u}{\partial x} + \frac{\partial v}{\partial y}\right) &= 0 \\ \frac{\partial u}{\partial t} + \frac{1}{\rho_0}\frac{\partial p}{\partial x} &= 0 \\ \frac{\partial v}{\partial t} + \frac{1}{\rho_0}\frac{\partial p}{\partial y} &= 0 \end{aligned} \tag{5.5}$$

or in the more compact form:

$$\frac{\partial U}{\partial t} + \frac{\partial F}{\partial x} + \frac{\partial G}{\partial y} = 0 \tag{5.6}$$

where

$$U = \begin{bmatrix} p \\ u \\ v \end{bmatrix} \quad F = \begin{bmatrix} \rho_0 c^2 u \\ p/\rho_0 \\ 0 \end{bmatrix} \quad G = \begin{bmatrix} \rho_0 c^2 v \\ 0 \\ p/\rho_0 \end{bmatrix}$$

To obtain the solution U at time $t + \Delta t$ from $U(t)$, it is convenient to use the method of splitting, which consists of solving two successive one-dimensional problems (one for each coordinate direction).

Let $L_x(\Delta t)$ and $L_y(\Delta t)$ designate difference operators yielding solutions to the one-dimensional equations:

$$\begin{aligned} \frac{\partial U^{(1)}}{\partial t} + \frac{\partial F}{\partial x} &= 0, \\ \frac{\partial U^{(2)}}{\partial t} + \frac{\partial G}{\partial y} &= 0 \end{aligned} \tag{5.7}$$

The solution at time $t+\Delta t$ is obtained by successively applying L_x and L_y:

$$U_{ij}^{n+1} = L_x(\Delta t)\, L_y(\Delta t)\, U_{ij}^{n} \tag{5.8}$$

This scheme is not second order accurate in time. Second order accuracy is obtained if the one-dimensional operators L_x and L_y are used in a symmetric sequence. The simplest such schemes are:

$$U_{ij}^{n+1} = L_y(\Delta t/2)\, L_x(\Delta t/2)\, L_x(\Delta t/2)\, L_y(\Delta t/2)\, U_{ij}^{n} \tag{5.9}$$

$$U_{ij}^{n+1} = L_y(\Delta t/2)\, L_x(\Delta t)\, L_y(\Delta t/2)\, U_{ij}^{n} \tag{5.10}$$

Repeated application of the one-dimensional operators is also allowed and provides further flexibility to the splitting method.

It is, for instance, possible to write:

$$U_{ij}^{n+1} = \left[L_y\!\left(\frac{\Delta t}{2M}\right)\right]^M L_x(\Delta t) \left[L_y\!\left(\frac{\Delta t}{2M}\right)\right]^M U_{ij}^{n} \tag{5.11}$$

The main advantage of the splitting method is that spatial differencing is only performed in a single dimension at a time. For example, the MacCormack scheme applied in the x-direction is defined by the predictor-corrector expressions:

$$\begin{aligned} \overline{U}_{ij} &= U_{ij}^n - \frac{\Delta t}{\Delta x}(F_{ij}^n - F_{i-1j}^n) \\ U_{ij}^{n+1} &= \frac{1}{2}\left[\overline{U}_{ij} + U_{ij}^n - \frac{\Delta t}{\Delta x}(\overline{F}_{i+1j} - \overline{F}_{ij})\right] \end{aligned} \tag{5.12}$$

In the above equations $\overline{F}$ is evaluated at the intermediate predictor level $\overline{F}_{ij} = F(\overline{U}_{ij})$. The set of expressions given by equation (5.12) form the $L_x(\Delta t)$ operator. Similar formulas may be used to define $L_y(\Delta t)$.

Fourth order accuracy in space is achieved with a scheme of Gottlieb and Turkel (1976). These authors define the x-direction operator $L_x(\Delta t)$ by the following expressions:

$$\begin{aligned} \overline{U}_{ij} &= U_{ij}^n - \frac{\Delta t}{6\Delta x}(-7F_{ij}^n + 8F_{i+1j}^n - F_{i+2j}^n) \\ U_{ij}^{n+1} &= \frac{1}{2}\left[\overline{U}_{ij} + U_{ij}^n - \frac{\Delta t}{6\Delta x}(7\overline{F}_{ij} - 8\overline{F}_{i-1j} + \overline{F}_{i-2j})\right] \end{aligned} \tag{5.13}$$

This scheme is successfully used by Maestrello et al. (1981), Maestrello and Bayliss (1982), and Bayliss and Maestrello (1981) to analyze the interaction of a sound pulse with a shear layer. Numerous practical details are given in these studies, and interesting physical phenomena are described with this method.

5.3 Boundary Conditions

The stability and accuracy of finite difference methods are strongly dependent on a correct specification of boundary conditions. The most critical problem is to set boundary conditions which reduce false reflections at the edges of the computational domain. This important problem is considered by Smith (1974), Reynolds (1978), Bayliss and Turkel (1980, 1982), and some practical schemes are put forward. However, further improvements are still needed in this area.

5.4 Implicit Schemes

Implicit finite difference schemes have not been reported in acoustic wave applications. On the basis of recent numerical experimentation, we feel that these methods have considerable potential. To prove this point, we consider the propagation of acoustic waves in one dimension. In this situation the linearized fluid-dynamic equations take the form:

$$\frac{\partial U}{\partial t} + A\frac{\partial U}{\partial x} = 0 \tag{5.14}$$

where

$$U = \begin{bmatrix} p \\ u \end{bmatrix} \qquad A = \begin{bmatrix} 0 & \rho_0 c^2 \\ 1/\rho_0 & 0 \end{bmatrix}$$

To advance U from t to $t+\Delta t$ we use the formal operator expression:

$$U_j^{n+1} = \exp\left(-A\Delta t\frac{\partial}{\partial x}\right)U_j^n \tag{5.15}$$

An implicit formula is obtained by splitting the exponential operator into two parts:

$$\exp\left(\theta A\Delta t\frac{\partial}{\partial x}\right)U_j^{n+1} = \exp\left[-(1-\theta)A\Delta t\frac{\partial}{\partial x}\right]U_j^n \tag{5.16}$$

where θ is a constant $(0 \le \theta \le 1)$.

Substitution of a central difference operator for the spatial derivative $\partial/\partial x$ and expansion of the exponential operators leads to:

$$(I + \theta A\,\Delta t\,\delta_x)\,U_j^{n+1} = [I - (1-\theta)\,A\,\Delta t\,\delta_x]\,U_j^n \tag{5.17}$$

where

$$\delta_x U_j = (U_{j+1} - U_{j-1})/2\Delta x$$

When $\theta = 0$ the previous scheme is fully explicit, for $\theta = \frac{1}{2}$ it is of Crank-Nicholson type; while for $\theta = 1$, it becomes fully implicit.

It is usually convenient to write expression (5.17) in Δ-form (i.e., solve for $U_j^{n+1} - U_j^n$):

$$(I + \theta A\,\Delta t\,\delta_x)(U_j^{n+1} - U_j^n) = (I - A\,\Delta t\,\delta_x)\,U_j^n \tag{5.18}$$

Expressions (5.17) or (5.18) constitute a block-tridiagonal linear system of equations and they may be solved by fast recursive methods. Schemes (5.17) and (5.18) are particular cases of the more general block implicit methods developed by Briley and McDonald (1980). The extension of the previous method to 2D and 3D problems is treated in these reference.

To illustrate the basic scheme of equation (5.17) we now consider plane wave radiation from a piston in a closed-end duct (Figure 22). The velocity of the piston is given by:

$$\begin{aligned} u(0,t) &= u_0 \sin^2(\pi t/T), \quad 0 \le t \le T \\ &= 0, \quad T < t \end{aligned} \tag{5.19}$$

A pressure pulse is emitted by the piston and it is reflected at the duct end. This process is well described by the implicit formulation (5.17) as demonstrated by the calculation of Figure 23 performed with $\theta = \frac{1}{2}$ scheme and a Courant number $c\Delta t/\Delta x = 10$. The $\theta = \frac{1}{2}$ implicit scheme is unconditionally stable, and large time steps may be used to propagate the wavelets across many computational cells during a single step. This is an advantageous feature of implicit methods and it may bring notable reduction of computation time in 2D or 3D problems.

Figure 22. Geometry of the piston problem.

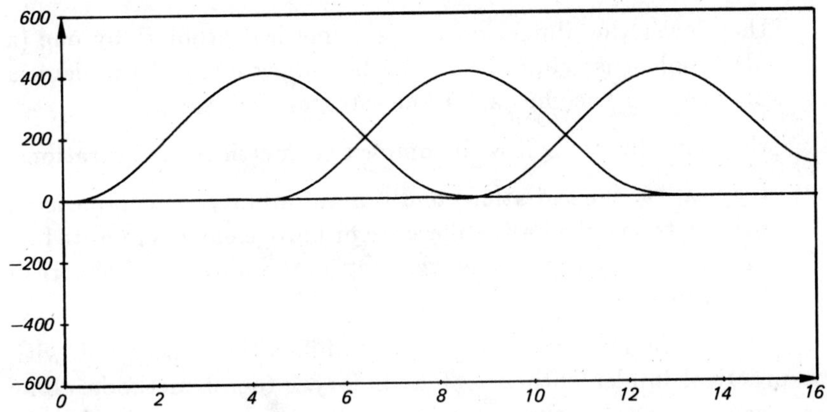

(a)

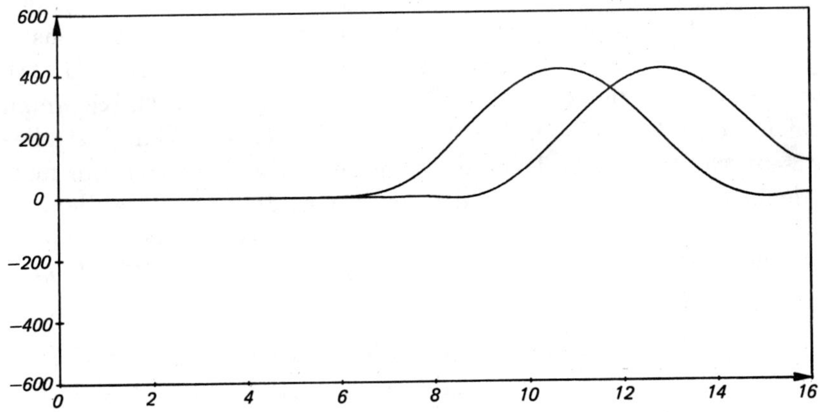

(b)

Figure 23. Pressure pulse propagation for the piston problem of Figure 22. The pulse is displayed at four instants in time, before, and after reflection on the duct end. Computations were performed with a $\theta = \frac{1}{2}$ implicit finite difference scheme for a courant number $c\Delta t/\Delta x = 10$. (Results obtained with the help of M. Kyprianou.)

6. Boundary Integral Methods

Boundary integral methods have been developed in recent years to analyze radiation and scattering problems in two or three dimensions. These methods are useful and attractive because:

1. they lower the dimension of the numerical problem by one (a 3D problem is reduced to a 2D boundary integral problem, a 2D problem becomes a 1D line integral problem),
2. they are able to deal with complex geometrical configurations,
3. they satisfy the radiation conditions at infinity. In comparison, field methods like finite difference or finite elements require far-field boundary conditions which still pose unresolved theoretical problems.

There are however a number of difficulties associated with the practical implementation of boundary integral methods. The first problem arises in the numerical evaluation of integrals of oscillatory and singular functions. Another more essential difficulty is encountered in exterior radiation problems. For certain boundary formulations the solution of the exterior problem is perturbed by the eigensolutions of the internal problem for the same geometry.

Initial applications of boundary integral methods to acoustic problems are due to Chen and Schweikert (1963), Chertok (1964), Copley (1967), Schenck (1968), Burton and Miller (1971), Kleinman and Roach (1974), Jones (1974), and Fillipi (1977). These studies present various integral formulations and discussions of the non-uniqueness problem. Improved formulations and numerical demonstrations are presented by Meyer et al. (1978), Terai (1980), Sayhi et al. (1981), Koopmann and Benner (1982), Hamdi and Ville (1981), and Hamdi (1982).

Most methods developed up to now fall in three general classes: (1) direct formulations, (2) single and double layer potentials formulations, and (3) variational formulations. These three methods are presented after a brief presentation of the classical radiation problem.

6.1 Formulation of the Basic Radiation Problem

For simplicity we only consider the problem of acoustic wave radiation from a closed suurface S vibrating at a radian frequency ω with a normal velocity $\dot{w}(\mathbf{r}_M)$ (Figure 24). We designate by D_E the exterior domain surrounding the vibrating surface S and by D_I the interior of that surface. In accord with most references dealing with boundary integral methods (but in distinction with the usual convention) the normal vector $\mathbf{n}$ is oriented towards the exterior domain D_E.

The steady radiation problem may be formulated in terms of the velocity potential ϕ. The acoustic pressure and velocity are deduced from ϕ through $p = -i\rho_0\omega\phi$ and $\mathbf{v} = -\nabla\phi$ and the governing equations for ϕ are:

$$\nabla^2\phi + k^2\phi = 0 \quad \text{in the exterior domain } D_E, \tag{6.1}$$

$$\frac{\partial\phi}{\partial\mathbf{n}}(\mathbf{r}_M) = -\dot{w}(\mathbf{r}_M) \quad \text{on the vibrating surface, } \mathbf{r}_M \in S \tag{6.2}$$

$$\lim_{R\to\infty} R\left[\frac{\partial\phi}{\partial R} - ik\phi\right] = 0 \quad \text{at infinity} \tag{6.3}$$

The velocity potential satisfies the Helmholtz equation in the exterior domain D_E, the Sommerfeld radiation condition at infinity and the potential normal derivative is specified on the vibrating surface S.

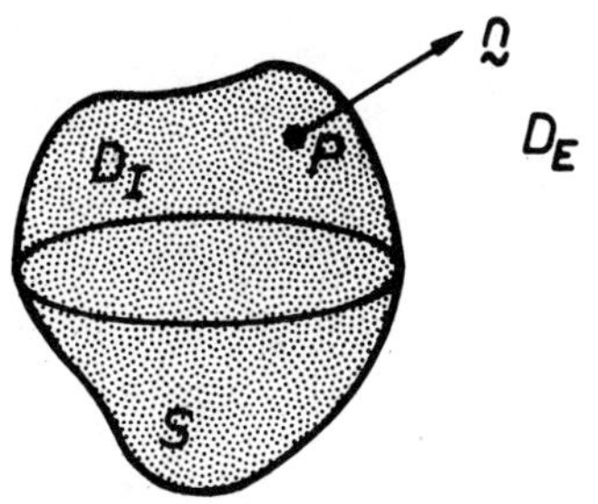

Figure 24. Geometry of the radiation problem.

Integral equations are conveniently deduced from Green's integral relations. Let $G(\mathbf{r}_M \mid \mathbf{r}_p)$ designate the fundamental Green's function of the Helmholtz free space problem:

$$(\nabla^2 + k^2)\, G\,(\mathbf{r}_M \mid \mathbf{r}_p) = -4\pi\delta\,(\mathbf{r}_M - \mathbf{r}_p) \tag{6.4}$$

$$\lim_{R\to\infty} R\left[\frac{\partial G}{\partial R} - ikG\right] = 0 \tag{6.5}$$

This function is well known:

$$G(\mathbf{r}_M \mid \mathbf{r}_p) = \frac{\exp\, ik|\mathbf{r}_M - \mathbf{r}_P|}{|\mathbf{r}_M - \mathbf{r}_P|} \tag{6.6}$$

and it may be used to write the following integral relations:

$$\phi\,(\mathbf{r}_M) = \frac{1}{4\pi}\int_S \left[\phi(\mathbf{r}_p)\frac{\partial G}{\partial \mathbf{n}_P}(\mathbf{r}_M \mid \mathbf{r}_P) - \frac{\partial \phi}{\partial \mathbf{n}_P}(\mathbf{r}_P) G(\mathbf{r}_M \mid \mathbf{r}_P)\right] dS_P, \quad \mathbf{r}_M \in D_E \tag{6.7}$$

$$\phi(\mathbf{r}_M) = \frac{1}{2\pi}⨍_S \left[\phi(\mathbf{r}_p)\frac{\partial G}{\partial \mathbf{n}_P}(\mathbf{r}_M \mid \mathbf{r}_P) - \frac{\partial \phi}{\partial \mathbf{n}_P}(\mathbf{r}_P) G(\mathbf{r}_M \mid \mathbf{r}_P)\right] dS_P, \quad \mathbf{r}_M \in S \tag{6.8}$$

$$0 = \int_S \left[\phi(\mathbf{r}_p)\frac{\partial G}{\partial \mathbf{n}_P}(\mathbf{r}_M \mid \mathbf{r}_P) - \frac{\partial \phi}{\partial \mathbf{n}_P}(\mathbf{r}_P) G(\mathbf{r}_M \mid \mathbf{r}_P)\right] dS_P, \quad \mathbf{r}_M \in D_I \tag{6.9}$$

The sign $⨍$ denotes a principal-value integration which must be used when the point $\mathbf{r}_M$ belongs to the boundary surface S.

6.2 Direct Boundary Integral Formulation

A direct formulation of a boundary integral equation is obtained by making use of Green's expression (6.8) in combination with the boundary condition (6.2):

$$\phi(\mathbf{r}_M) - \frac{1}{2\pi}\int_S \phi(\mathbf{r}_P)\frac{\partial G}{\partial \mathbf{n}_P}(\mathbf{r}_M \mid \mathbf{r}_P)\, dS_P = \frac{1}{2\pi}\int_S G(\mathbf{r}_M \mid \mathbf{r}_P)\, \dot{w}\,(\mathbf{r}_P)\, dS_P, \qquad \mathbf{r}_M \in S \tag{6.10}$$

This is a Fredholm integral equation of the second kind. As suggested by Sayhi et al. (1981) it is possible to avoid the determination of the principal value of the first integral by making use of the Green's function G_0 of the Laplace equation. Expression (6.10) may then be written:

$$\phi(\mathbf{r}_M) - \frac{1}{4\pi}\int_S \left[\phi(\mathbf{r}_P)\frac{\partial}{\partial \mathbf{n}_P}G(\mathbf{r}_M \mid \mathbf{r}_P) - \phi(\mathbf{r}_M)\frac{\partial}{\partial \mathbf{n}_P}G_0(\mathbf{r}_M \mid \mathbf{r}_P)\right] dS_P = \frac{1}{2\pi}\int G(\mathbf{r}_M \mid \mathbf{r}_P)\, \dot{w}\,(\mathbf{r}_P)\, dS_P, \qquad \mathbf{r}_M \in S \tag{6.11}$$

and the integrand appearing in the right hand side is now regular.

Expressions (6.10) or (6.11) are simple and attractive but they give rise to an essential problem. The solution of these equations is not unique for certain values of the wave number k, corresponding to the resonant frequencies of the interior acoustic problem satisfying the same boundary conditions. Numerous techniques have been proposed to overcome this difficulty.

It is for instance possible to associate to the basic integral equation (6.10) or (6.11) a set of additional relations obtained by placing the field point $\mathbf{r}_M$ at various locations in the interior domain D_I. According to the third integral relation (6.9):

$$\int_S \left[\phi(\mathbf{r}_p)\frac{\partial G}{\partial \mathbf{n}_P}(\mathbf{r}_M \mid \mathbf{r}_P) + \dot{w}\,(\mathbf{r}_P)G(\mathbf{r}_M \mid \mathbf{r}_P)\right] dS_P = 0, \qquad \mathbf{r}_M \in D_I \tag{6.12}$$

One or more such equations are then combined to the basic integral equation and it is shown by Koopmann and Benner (1982) that this indeed provides a unique solution for the velocity potential at the surface S.

Another method consists of solving a combination of the basic integral equation (6.10) and an additional integral equation obtained by differentiating (6.10) in the normal direction at the field point $\mathbf{r}_M$. This additional equation takes the form:

$$\frac{1}{2\pi}\int_S \left[\phi(\mathbf{r}_P)\frac{\partial^2 G}{\partial \mathbf{n}_P \partial \mathbf{n}_M}(\mathbf{r}_P \mid \mathbf{r}_M) + \frac{\partial G}{\partial \mathbf{n}_M}(\mathbf{r}_P \mid \mathbf{r}_M)\dot{w}\,(\mathbf{r}_P)\right] dS_P$$
$$= \frac{\partial \phi}{\partial \mathbf{n}_M}(\mathbf{r}_M) = -\dot{w}\,(\mathbf{r}_M) \tag{6.13}$$

The corresponding set of internal resonant frequencies differs from that of the basic integral equation (6.10):

A linear combination of (6.10) and (6.13):

$$\frac{1}{2\pi}\int_S \left[\phi(\mathbf{r}_p)\frac{\partial G}{\partial \mathbf{n}_p}(\mathbf{r}_M \mid \mathbf{r}_P) + \alpha\frac{\partial^2 G}{\partial \mathbf{n}_p \partial \mathbf{n}_M}(\mathbf{r}_M \mid \mathbf{r}_P)\right] dS_p$$
$$+ \frac{1}{2\pi}\int_S \left[G(\mathbf{r}_M \mid \mathbf{r}_P) + \alpha\frac{\partial G}{\partial \mathbf{n}_M}(\mathbf{r}_M \mid \mathbf{r}_P)\,\dot{w}\,(\mathbf{r}_p)\right] dS_p \tag{6.14}$$
$$= \phi(\mathbf{r}_M) - \alpha\dot{w}\,(\mathbf{r}_M), \qquad \mathbf{r}_M \in S$$

then yields a unique solution to the exterior radiation problem. The main difficulty with this last approach comes from the singular behaviour of the $\partial^2 G/\partial \mathbf{n}_p \partial \mathbf{n}_M$ term. Meyer et al. (1978) show that it is possible to deal with this singularity by some special transformations of the integrand. A successful boundary integral method is obtained in this way.

6.3 Indirect Method

The indirect method is well known in the field of hydrodynamics. It consists of writing the potential as a combination of unknown singularities like sources and dipoles:

$$\phi(\mathbf{r}_M) = \int_S \left[\alpha G(\mathbf{r}_M \mid \mathbf{r}_P) + \beta\frac{\partial G}{\partial \mathbf{n}_P}(\mathbf{r}_M \mid \mathbf{r}_P)\right] \mu(\mathbf{r}_P)\, dS_P \tag{6.15}$$

where α and β are constants. The surface distribution $\mu(\mathbf{r}_P)$ may be determined by writing the boundary condition on the surface S:

$$\lim_{\substack{M \to M_0 \\ M_0 \in S}} \frac{\partial}{\partial \mathbf{n}_M} \int \left[\alpha G(\mathbf{r}_M \mid \mathbf{r}_P) + \beta \frac{\partial G}{\partial \mathbf{n}_P}(\mathbf{r}_M \mid \mathbf{r}_P) \right] \mu(\mathbf{r}_P)\, dS_P$$
$$= \frac{\partial \phi}{\partial \mathbf{n}_M}(\mathbf{r}_M) = -\dot{w}\,(\mathbf{r}_M) \tag{6.16}$$

This integral equation has a unique solution $\mu(\mathbf{r}_P)$ and the associated potential $\phi(\mathbf{r}_P)$ is then unique (see Sahyi et al., 1981).

6.4 Variational Formulation

This formulation was recently proposed by Hamdi (1982). It consists of writing the potential in terms of source and dipole distributions:

$$\phi(\mathbf{r}_M) = \int_S \left[\sigma(\mathbf{r}_P) G\,(\mathbf{r}_M \mid \mathbf{r}_P) + \mu\,(\mathbf{r}_P) \frac{\partial G}{\partial \mathbf{n}_P}(\mathbf{r}_M \mid \mathbf{r}_P) \right] dS_P \tag{6.17}$$

It is then shown that $\sigma(\mathbf{r}_P)$ and $\mu(\mathbf{r}_P)$ may be obtained as the stationary solutions of a variational equation. This elegant formulation forms the basis of an efficient numerical algorithm.

6.5 Discretization

To obtain a numerical solution of the boundary integral equations one has to:

(1) discretize the boundary surface S in N panels S_j (the elementary panels are usually plane triangles or quadrilaterals),

(2) approximate the potential or singularity distributions on each panel by standard interpolation methods,

(3) approximate the boundary condition on each panel (the simplest approximation consists of replacing the potential, singularity distribution and normal velocity by a constant average value on each panel),

(4) treat the remaining panel integrals by analytical/numerical methods,

(5) solve a system of linear equations.

To fix the ideas consider equation (6.10). A discrete version of this relation obtained by using the simplest approximation procedure takes the form:

$$\phi_i - A_{ij}\phi_j = B_{ij}\dot{w}_j \tag{6.18}$$

where

$$A_{ij} = \frac{1}{2\pi}\int_{S_j} \frac{\partial}{\partial n_j} G(\mathbf{r}_i \mid \mathbf{r}_j)\, dS \tag{6.19}$$

$$B_{ij} = \frac{1}{2\pi}\int_{S_j} G(\mathbf{r}_i \mid \mathbf{r}_j)\, dS \tag{6.20}$$

It is in general necessary to use more elaborate discretization methods and further practical details may be found in the studies cited in the references.

Acknowledgements

A large part of the author's work on wave synthesis in shear flows and on the development of Fourier-Bessel transform algorithms was performed at the Joint Institute of Aeronautics and Acoustics, Stanford University during a summer visit authorized by ONERA and made possible by a kind invitation issued by Prof. Karamcheti.

References

[1] Alford, R. M., Kelly, K. R., and Boore, D. M. "Accuracy of Finite Difference Modeling of the Acoustic Wave Equation," *Geophysics, 39* (1974), 834–842.

[2] Baer, R. N. "Calculations of Sound Propagation Through an Eddy," *J. Acoust. Soc. Am., 67* (1980), 1180–85.

[3] Baños, A. *Dipole Radiation in the Presence of a Conducting Half Space,* Oxford: Pergamon Press, 1966.

[4] Baumeister, K. J. "Time Dependent Difference Theory for Noise Propagation in a Two-Dimensional Duct," *AIAA Journal, 18* (1980), 1470–76; also *AIAA Paper No. 80-0098.*

[5] ———————. "Numerical Techniques in Linear Duct Acoustics, a Status Report," *Journal of Engineering for Industry, 103* (1981), 270–281.

[6] ———————. "Influence of Exit Impedance on Finite Difference Solution of Transient Acoustic Mode Propagation in Ducts," *Journal of Engineering for Industry," 104* (1982), 113–120.

[7] Bayliss A. and Maestrello, L. "Simulation of Instabilities and Sound Radiation in a Jet," *AIAA Journal, 19* (1981), 835–841; also *AIAA paper No. 80-0960.*

[8] Bayliss, A. and Turkel, E. "Radiation Conditions for Wavelike Equations," *Communications on Pure and Applied Mathematics, 33* (1980), 707.

[9] ———————————. "Far Field Boundary Conditions for Compressible Flows," *J. Computational Physics, 48* (1982), 182–199.

[10] Berkhout, A. J. *Seismic Migration Imaging of Acoustic Energy by Wave Field Extrapolation,* Amsterdam: Elsevier, 1980.

[11] Blokhintsev, D. I. "Acoustics of a Nonhomogeneous Moving Medium," *NACA Tech. Memo. No. 1399,* 1946.

[12] Bojarski, N. N. "The k-Space Formulation of the Scattering Problem in the Time Domain," *J. Acoust. Soc. Am., 72* (1982), 570–584.

[13] Brekhovskikh, L. M. *Waves in Layered Media,* New York: Academic Press, 1960.

[14] Briley, W. R. and McDonald, H. "On the Structure and Use of Linearized Block Implicit Schemes," *J. Computational Physics, 34* (1980), 54–73.

[15] Brock, H. K., Buchal, R. N., and Spofford, C. W. "Modifying the Sound Speed Profile to Improve the Accuracy of the Parabolic Equation Technique," *J. Acoust. Soc. Am., 62* (1977), 543–552.

[16] Burton, A. J. and Miller, G. F. "The Application of Integral Equation Methods to the Numerical Solution of Some Exterior Boundary Problems," *Proc. Roy. Soc., A323* (1971), 201–210.

[17] Cabelli, A. "Duct Acoustics–A Time Dependent Difference Approach for Steady State Solutions," *J. Sound Vib., 85* (1982), 423–434.

[18] Candel, S. M. "Acoustic Conservation Principles and an Application to Plane and Model Propagation in Nozzles and Diffusers," *J. Sound Vib., 41* (1975), 207.

[19] ————. "Application of Geometrical Techniques to Aeroacoustic Problem," *AIAA Paper No. 76-546,* 1976.

[20] ————. "Etudes Théoriques et Expérimentales de la Propagation Acoustique en Milieu Inhomogène et en Mouvement," *thése de Doctorat d'Etat,* Université de Paris VI, *ONERA Publication No. 1977-1,* 1977.

[21] ————. "Numerical Solution of Conservation Equations Arising in Linear Wave Theory: Application to Aeroacoustics," *J. Fluid Mech., 83* (1977), 465–493.

[22] ————. "Résolution Numérique de Problémes de Rayonnement et de Diffusion dans le Cadre de l'approximation Parabolique," *ONERA Publication No. 1978-5,* 1978.

[23] ————. "Numerical Solution of Wave Scattering Problems in the Parabolic Approximation," *J. Fluid Mech., 90* (1979), 465–507.

[24] ————. "Algorithms for the Computation of Even and Odd Fourier-Bessel Transforms," *La Recherche Aérospatiale* (English edition), *1981, No. 5* (1981a), 23–34.

[25] ————. "An Algorithm for the Fourier-Bessel Transform," *Computer Physics Communications, 23* (1981b), 343–353.

[26] ————. "Dual Algorithms for Fast Calculation of the Fourier-Bessel Transform," *IEEE Transactions on Acoustics Speech and Signal Processing, ASSP 29* (1981c), 963–972.

[27] ————. "Simultaneous Calculation of Fourier-Bessel Transforms up to Order N," *J. Computational Physics, 44* (1981d), 243–261.

[28] ————. "Sound Source Radiation in Two-Dimensional Shear Flow," *AIAA Journal, 21* (1983), 221–227; also in *AIAA Paper no. 81-2042.*

[29] Candel, S. M. and Brun, Y. "Rayonnement d'une Source à Proximité d'un Interface Solide-Fluide. Résolution par Transformation de Fourier-Bessel Numérique," *Revue du CETHEDEC, 19, No. 72* (1982), 207–219.

[30] Candel, S. M. and Crance, C. "Direct Fourier Synthesis of Waves in Layered Media and the Method of Stationary Phase," *J. Sound Vib., 74* (1981a), 477–498.

[31] ———. "Direct Fourier Synthesis of Waves: Application to Acoustic f*Source Radiation," *AIAA Journal, 19* (1981b), 290–295; also in *AIAA Paper no. 80-0037.*

[32] Candel, S. M., De Fillipi, F., and Launay, A. "Determination of the Inhomogeneous Structure of a Medium From its Plane Wave Reflection Response, Part I: A Numerical Analysis of the Direct Problem; Part II: A Numerical Approximation," *J. Sound Vib., 68* (1980), 571–595.

[33] Candel, S. M., Guédel A., and Julienne A. "Refraction and Scattering in an Open Wind Tunnel Flow," *Proc. 6th Int. Cong. Instrumentation in Aerospace Simulation Facilities,* Ottawa, 1975, 288.

[34] ———. "Radiation, Refraction, and Scattering of Acoustic Waves in a Free Shear Flow," *AIAA Paper No. 76-544,* 1976.

[35] Chen, L. H. and Schweikert, D. G. "Sound Radiation From an Arbitrary Body," *J. Acoust. Soc. Am., 35* (1963), 1626–32.

[36] Chertock, G. "Sound From Vibrating Surfaces," *J. Acoust. Soc. Am., 36* (1964), 1305–13.

[37] Claerbout, J. F. *Fundamentals of Geophysical Data Processing,* New York: McGraw Hill, 1976.

[38] Clay, C. S. and Medwin, H. *Acoustical Oceanography,* New York: John Wiley, 1977.

[39] Copley, L. G. "Integral Equation Method for Radiation From Vibrating Bodies," *J. Acoust. Soc. Am., 43* (1967), 807–816.

[40] Csanady, G. T. "The Effect of Mean Velocity Variation on Jet Noise," *J. Fluid Mech., 26* (1966), 183.

[41] Ewing, W. M., Jardetzky, W. S., and Press, F. *Elastic Waves in Layered Media,* New York: McGraw-Hill, 1957.

[42] Felsen, F. B. and Marcuvitz, N. *Radiation and Scattering of Waves,* Englewood Cliffs: Prentice Hall, 1973.

[43] Filippi, P. "Layer Potentials and Acoustic Diffraction," *J. Sound Vib., 54* (1977), 473–500.

[44] Fitzgerald, R. M. "Helmholtz Equation as an Initial Value Problem With Application to Acoustic Propagation," *J. Acoust. Soc. Am., 57* (1975), 839–842.

[45] Flatté, S. M. and Tappert, F. D. "Calculation of the Effect of Internal Waves on Oceanic Sound Transmission," *J. Acoust. Soc. Am.,*

58 (1975), 1151–59.

[46] Fock, V. A. "Electromagnetic Diffraction and Propagation Problems," New York: Pergamon Press, 1965.

[47] Frisk, G. V., Oppenheim, A. V., and Martinez, D. R. "A Technique for Measuring the Plane Wave Reflection Coefficient of the Ocean Bottom," *J. Acoust. Soc. Am., 68* (1980), 602–612.

[48] Gazdag, J. "Extrapolation of Seismic Waveforms by Fourier Methods," *IBM J. of Research and Development, 22* (1978), 481–486.

[49] ————. "Wave Equation Migration With the Phase Shift Method," *Geophysics, 43* (1978), 1342–51.

[50] Gossard, E. E. and Hooke, W. H. *Waves in the Atmosphere,* Amsterdam: Elsevier, 1975.

[51] Gottlieb D. and Orszag, S. A. *Numerical Analysis of Spectral Methods: Theory and Applications,* Society for Industrial and Applied Mathematics, Philadelphia, PA., 1977.

[52] Gottlieb, D. and Turkel, E. "Dissipative Two-Four Methods for Time-Dependent Problems," *Mathematics of Computation, 30* (1976), 703–723.

[53] Guyomar, D. "Modes Vibratoires d'un Cristal Piézoélectrique et Méthodes de Reconstruction de l'état de Vibration," *thése de Docteur-Ingénieur,* Université de Technologie de Compiègne, 1982.

[54] Hamdi, M. A. "Formulations Variationnelles par équations Intégrales pour le Calcul de Champs Acoustiques Linéaires Proches et Lointains," *doctoral thesis,* Université de Technologie de Compiègne, 1982.

[55] Hamdi, M. A. and Ville, J. M. "Effect of Finite Length on Sound Radiation From Duct," *AIAA Paper No. 81-1980,* 1981.

[56] Hardin, R. H. and Tappert, F. D. "Application of the Split Step Fourier Method to the Numerical Solution of Nonlinear and Variable Coefficient Wave Equations," *SIAM Rev., 15* (1973), 423.

[57] Hayes, W. P. "Kinematic Wave Theory," *Proc. Roy. Soc., A320* (1970), 209.

[58] Hayes, W. P., Haefli, R. C., and Kulsrud, H. E. "Sonic Boom Propagation in a Stratified Atmosphere With Computer Program," *NASA CR 1299,* 1970.

[59] Higgins, F. P., Norton, S. J., and Linzer, M. "Optical Interferometric Visualization and Computerized Reconstruction of Ultrasonic Fields," *J. Acoust. Soc. Am., 68* (1980), 1169–76.

[60] Hudson, J. A. "A Parabolic Approximation for Elastic Waves," *Wave Motion, 2* (1980), 207–214.

Quarterly Journal of Mechanics and Applied Mathematics, 27 (1974), 129–142.

[62] Keller, J. B. "Geometrical Acoustics, the Theory of Weak Shock Waves," *J. Appl. Phys., 25* (1954), 938.

[63] Kelly, K. R., et al. "Synthetic Seismograms, a Finite Difference Approach," *Geophysics, 41* (1976), 2–42.

[64] Kelley, P. L. "Self Focusing of Optical Beams," *Phys. Rev. Lett., 15* (1965), 1005.

[65] Kleinman, R. E. and Roach, G. F. "Boundary Integral Equations for the Three-Dimensional Helmholtz Equation," *SIAM Review, 16* (1974), 214–236.

[66] Koopmann, G. H. and Benner, H. "Method for Computing the Sound Power of Machines Based on the Helmholtz Integral," *J. Acoust. Soc. Am., 71* (1982), 78–89.

[67] Landers, T. and Claerbout, J. F. "Numerical Calculations of Elastic Waves in Laterally Inhomogeneous Media," *J. Geophys. Res., 77* (1972), 1476–82.

[68] Lee, D. "Effective Methods for Predicting Underwater Acoustic Wave Propagation," *AIAA Paper No. 83-0683,* 1983.

[69] Lee, D. and Papadakis, J. S. "Numerical Solutions of the Parabolic Wave Equation: an Ordinary Differential Equation Approach," *J. Acoust. Soc. Am., 68* (1980), 1482–88.

[70] Lee, D., Botseas, G., and Papadakis, J. S. "Finite Difference Solutions to the Parabolic Wave Equation," *J. Acoust. Soc. Am., 70* (1981), 795–800.

[71] Leontovich, M. A. and Fock, V. A. "Solution of the Problem of Propagation of Electromagnetic Waves Along the Earth's Surface by the Method of Parabolic Equation," *J. Expt. and Theor. Phys., 16, No. 7* (1946), 557 (in Russian); *J. Phys. of the USSR, 10, No. 1* (1946), 136 (in English).

[72] Lighthill, M. J. "The Fourth Annual Fairey Lecture: the Propagation of Sound Through Moving Fluids," *J. Sound Vib., 24* (1972), 471.

[73] Lilly, J. Q. and Miller, T. G. "Target Intensity Enhancement for Repetitively Pulsed Laser Beams," *AIAA Journal, 15* (1977), 434.

[74] McCoy, J. J. "A Parabolic Theory of Stress Wave Propagation Through Inhomogeneous Linearly Elastic Solids," *J. Appl. Mech., 44* (1977), 462–468.

[75] McDaniel, S. T. "Propagation of Normal Mode in the Parabolic Approximation," *J. Acoust. Soc. Am., 57* (1975a), 307–311.

[76] ————————. "Parabolic Approximations for Underwater Sound Propagation," *J. Acoust. Soc. Am., 58* (1975b), 1178–85.

[77] McDaniel, S. T. and Lee, D. "A Finite Difference Treatment of Interface Conditions for the Parabolic Wave Equation: the Horizontal Interface," *J. Acoust. Soc. Am., 71* (1982), 855–858.

[78] Maestrello, L. and Bayliss, A. "Flowfield and Far Field Acoustic Amplification Properties of Heated and Unheated Jets," *AIAA Journal, 20* (1982), 1539–46.

[79] Maestrello, L., Bayliss, A., and Turkel, E. "On the Interaction of a Sound Pulse With the Shear Layer of an Axisymmetric Jet," *J. Sound Vib., 74* (1981), 281–301.

[80] Marcuse, D. *Light Transmission Optics,* New York: Van Nostrand, 1972.

[81] Meyer, W. L., et al. "Boundary Integral Solutions of Three-Dimensional Acoustic Radiation Problems," *J. Sound Vib., 59* (1978), 245–262.

[82] Myers, M. K. and McAninch, G. L. "Parabolic Approximation for Sound Propagation in the Atmosphere," *AIAA Journal, 16* (1978), 836–842.

[83] Di Napoli, F. R. and Deavenport, R. L. "Numerical Models of Underwater Acoustic Propagation," in *Ocean Acoustics*. Ed. J. A. de Santo, Berlin: Springer Verlag, 1979, 79–157.

[84] ————————————————. "Theoretical and Numerical Green's Function Field Solution in a Plane Multilayered Medium," *J. Acoust. Soc. Am., 67* (1980), 92–105.

[85] Officer, C. B. *Introduction to the Theory of Sound Transmission,* New York: McGraw-Hill, 1958.

[86] Palmer, D. R. "Eikonal Approximation and the Parabolic Equation," *J. Acoust. Soc. Am., 60* (1976), 343–354.

[87] Perkins, J. S. and Baer, R. N. "An Approximation to the Three-Dimensional Parabolic Equation Method for Acoustic Propagation," *J. Acoust. Soc. Am., 72* (1982), 515–522.

[88] Quemada, D. *Ondes dans les Plasmas,* Paris: Hermann, 1968.

[89] Reynolds, A. C. "Boundary Conditions for the Numerical Solution of Wave Propagation Problems," *Geophysics, 43* (1978), 1099–1110.

[90] Richtmeyer, R. D. and Morton, K. W. *Difference Methods for Initial-Value Problems,* New York: Interscience, 1967.

[91] De Santo, J. A. "Relation Between the Solution of the Helmholtz and Parabolic Equations for Sound Propagation," *J. Acoust. Soc. Am., 62* (1977), 543–552.

[92] De Santo, J. A., Perkins, J. S., and Baer, R. N. "Corrections to the Parabolic Approximation for Sound Propagation Modelling," *J. Acoust. Soc. Am., 61* (1977), S12(A).

[93] Sayhi, M. N., Ousset, Y., and Verchery, G. "Solution of Radiation Problems by Collocation of Integral Formulations in Terms of Single and Double Layer Potentials," *J. Sound Vib., 74* (1981), 187–204.

[94] Schenck, H. A. "Improved Integral Formulation for Acoustic Radiation Problems," *J. Acoust. Soc. Am., 44* (1968), 41–58.

[95] Schubert, L. K. "Numerical Study of Sound Radiation by a Jet Flow, I. Ray Acoustics," *J. Acoust. Soc. Am., 50* (1972), 960.

[96] Sheng, S. C. "Studies of Laser Resonators and Beam Propagation Using Fast Transform Methods," Gintzon Laboratory, *Report No. 3106,* 1980, Stanford University.

[97] Smith, W. D. "A Nonreflecting Plane Boundary for Wave Propagation Problems," *J. Computational Physics, 15* (1974), 492–503.

[98] Stepanishen, P. R. and Bejamin, K. C. "Forward and Backward Projection of Acoustic Fields Using FFT Methods," *J. Acoust. Soc. Am., 71* (1982), 803–812.

[99] Stolt, R. H. "Migration by Fourier Transform," *Geophysics, 43* (1978), 23–48.

[100] Tappert, F. D. "The Parabolic Approximation Method," in *Wave Propagation and Underwater Acoustics.* Eds. J. B. Keller and J. S. Papadakis, *Lecture Notes in Physics, 70,* Heidelberg: Springer-Verlag, 1977.

[101] Tatarski, V. I. *Wave Propagation in a Turbulent Medium,* New York: Dover, 1961.

[102] Telford, W. M., et al. *Applied Geophysics,* Cambridge: Cambridge University Press, 1976.

[103] Terai, T. "On Calculations of Sound Fields Around Three-Dimensional Objects by Integral Equations Methods," *J. Sound Vib., 69* (1980), 71–100.

[104] Turkel, E. "Progress in Computational Physics," *Computers and Fluids, 11* (1983), 121–144.

[105] Ugincius, P. "Intensity Equations in Ray Acoustics I and II," *J. Acoust. Soc. Am., 45* (1969), 193.

[106] Urick, J. *Principles of Underwater Sound,* New York: McGraw-Hill, 1967.

[107] Wait, J. R. *Electromagnetic Waves in Stratified Media,* Oxford: Pergamon Press, 1962.

[108] Wallace, J. and Lilly, J. Q. "Thermal Blooming of Repetitively

Pulsed Laser Beams," *J. Opt. Soc. Am., 64* (1974), 1651.

[109] Whitham, G. B. *Linear and Nonlinear Waves,* New York: John Wiley, 1974.

[110] Williams, E. G. and Maynard, J. D. "Numerical Evaluation of the Rayleigh Integral for Planar Radiators Using the FFT," *J. Acoust. Soc. Am., 72* (1982), 2020–30.

[111] Williams, E. G., Maynard, J. D., and Skudrzyk, E. "Sound Source Reconstructions Using a Microphone Array," *J. Acoust. Soc. Am., 68* (1980), 340–344.

Nonlinear Acoustic Propagation of Broadband Noise

David G. Crighton
University of Leeds
Leeds, England

Abstract

Broadband jet noise is invariably assumed to propagate according to linear acoustic laws, although there are clear indications from both a limited set of experimental data taken in unusual, though appropriate, circumstances, and from simple weakly nonlinear acoustic theory that nonlinear propagation effects may significantly distort the noise spectrum over ranges of importance in full-scale aircraft flyovers. This paper starts with a review of the experimental situation and discusses the implications for aircraft noise prediction and control that might follow recognition of the importance of nonlinear propagation effects. Three *rational* theoretical approaches to the problem are then examined. The first relies on a simple expansion in amplitude (or, equivalently, frequency or range), and despite its obvious limitations is useful in giving a quantitative indication of the frequency, range and OASPL at which nonlinear spectral distortion can be expected to be significant near the *peak* frequency (and, by implication, even more significant at the higher frequencies which are so important in the determination of Perceived Noise Levels). This method can also incorporate attenuation effects, along with those of spherical spreading, directional nonuniformity, and weak nonlinearity. From it are also derived scaling laws for the variation with jet speed of the nonlinearly induced contribution to the spectrum and its dependence on range.

Next we discuss an exact solution, due to Rudenko and Soluyan (1977), for nonlinear plane wave evolution of a signal which (as in the perturbation expansion) has joint Gaussian statistics. The solution for the evolved pressure spectrum reduces to the perturbation series for low amplitudes and limited ranges in space and in frequency, while at large frequencies the power spectral density decays as ω^{-3}. Previous estimates of the importance of nonlinear effects are confirmed by this solution, though it is not possible to incorporate any realistic form of attenuation into it, nor is it correct at high frequencies because it assumes a one-valued characteristic form of the pressure evolution and excludes contributions to the spectrum from shock waves.

Modifications to the spectrum from the presence of shock waves are discussed in Section 4. For small ranges, in an appropriate sense, shock waves occur only infrequently, and Scott (1982) has shown how their contribution to the spectrum may be completely calculated in terms of two numerically determined functions. This contribution replaces the ω^{-3} decay of the multi-valued waveform spectrum with a universal ω^{-2} decay whose coefficient is given explicitly as a function of range. Restrictions on this solution are discussed, as is the extension to very large ranges where, in the absence of dissipation except on the smallest scales, the waveform is entirely composed of sawtooth ramps, all with slope decreasing as the inverse of range, separated by thin shocks.

Section 5 then describes some observations of non-zero skewness factors in jet noise signals, and of the development of skewness with range. Linear mechanisms cannot account for observations of an increase in the magnitude of the skewness factor with range, neither can nonlinear mechanisms coupled with dissipation. It is shown, however, that incorporation of linear **frequency dispersion** into the weakly nonlinear perturbation scheme of Section 2 does lead to predictions of skewness evolution in general accord with observation.

The paper concludes with a discussion of further work needed, principally on the experimental side, to confirm the theoretical predictions both with regard to jet aircraft noise and with regard to noise from more easily controllable broadband sources; and to establish once and for all the significance of the nonlinear propagation mechanism in the full-scale jet noise problem and in simulations of it on ground-based rigs and models.

1. Introduction

The possibility that jet noise signals are sufficiently intense that their propagation to the very distant field should be calculated with allowance for nonlinear effects (arising to a comparable extent from the nonlinearity of the pressure: density relation and from the convective acceleration term) has been raised from time to time over the thirty years of jet noise research. Locally, of course, nonlinear terms are small compared with linear ones at all points in the acoustic far field; their relative importance is measured by the particle Mach number (ratio of r.m.s. particle velocity in a sound wave to the sound speed) which is never greater than 10^{-2} (at a sound pressure level of 160 dB, above the upper threshold of pain (140 dB)). The structure of the nonlinear terms is such, however, that their effect is cumulative; a sinusoidal plane wave with initial Mach number 10^{-2} would develop a periodic pattern of shock waves after nonlinear evolution over about 10^2 wavelengths. Even weak local nonlinearities might therefore lead to significant waveform distortion (and therefore spectral distortion) in the real jet aircraft noise problem—where the wavelengths are of the order of metres for the peak frequencies and where the propagation ranges of interest are hundreds of metres. This superficial conclusion is weakened by the observation that the real problem involves spherical waves rather than plane, and the r^{-1} amplitude decay of spherical waves greatly offsets the shock-forming tendency of nonlinearity. Against that, however, is the point that for the vitally important matter of aircraft noise certification, correct prediction of levels in the 2–4 kHz range of frequencies is essential to the correct prediction of Perceived Noise Levels, and the nonlinear pumping of even a small amount of energy from the much lower frequencies where most of the energy resides could have an enormous effect on the PNL.

Despite the potential importance of the nonlinear propagation issue it has remained poorly documented and little studied. There was an understandable reluctance on the part of the aeroacoustics community to admit that jet noise levels were high enough for the problem to be serious, and full-scale tests were difficult and expensive, while it was not obvious that the problem could be satisfactorily examined at model scale. Certainly there were difficulties in the correlation of model-scale data with each other and with full-scale data on the basis of linear acoustic theory. Until the rather recent provision of much more accurate values for atmospheric attenuation coefficients at the high frequencies of model-scale jet noise, it was common to find high frequency spectral levels in the far field well above those predicted by the application of linear acoustics (i.e., inverse square law intensity decay and small-signal atmospheric attenuation) to measured data at smaller ranges. Anomalies of this kind have been much reduced with the availability of more accurate values for attenuation, and the *reduced attenuation* problem is at model scale seen more as reflecting inadequacies of liner theory than the possible importance of nonlinear mechanisms.

The same problem arose again, however, in experiments carried out by Gallard and Gower (1978) for British Aerospace and involving full-scale aircraft flyovers. In each test run, the aircraft was flown at constant thrust at a constant climb gradient over a linear array of microphones as shown in Figure 1. Different noise source levels were produced by different thrust settings, but at each thrust condition, far field noise data were obtained at various slant ranges, each making the same angle with the jet exhaust. In this way changes in directivity with changes in thrust were avoided. It again proved impossible to correlate the data on the basis of linear acoustics at frequencies much above 1 kHz; and indeed, the use of the most recent standards for attenuation increased the spectral anomalies. Admittedly, there are many difficulties in the acquisition and processing of data in experiments of this kind, but these were examined by Gallard and Gower, and

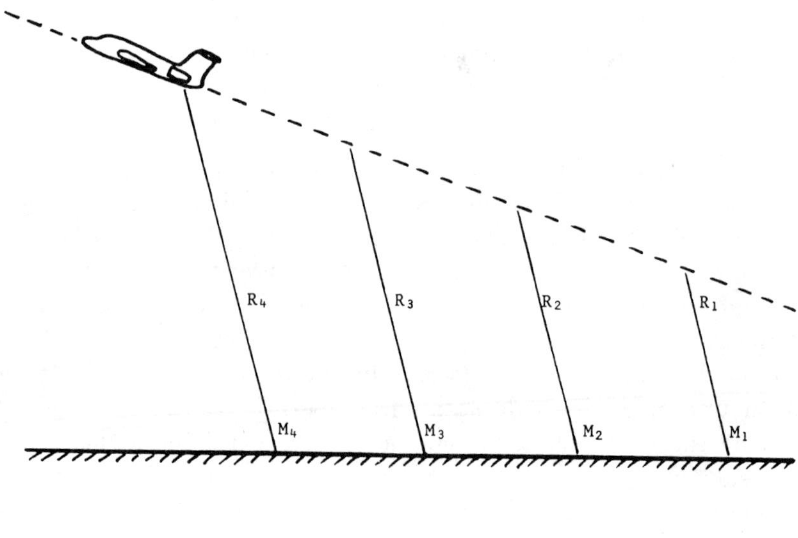

Figure 1. The aircraft flyover procedure used by British Aerospace (Gallard and Gower, 1978). Flights were conducted at constant thrust and constant climb gradient over a linear array of microphones $M_1, M_2, \ldots$, giving noise data at ranges $R_1, R_2, \ldots$ along rays making the same angle with the flight path and the jet exhaust.

again very thoroughly by Howell (1982) and shown to be capable of explaining only a small contribution to the high frequency discrepancies. When that contribution is allowed for, linear acoustics is found to be adequate only at the lowest thrust settings, and there remains a substantial thrust-dependent underprediction of the *effective attenuation* as inferred from these full-scale data. (The reader is referred to Howell (1982) for a full discussion of the interpretation of these experiments.)

Various serious consequences follow, as noted by Crighton and Bashforth (1980), from acceptance of the role of nonlinear waveform distortion in pumping energy to higher frequencies and thus creating a reduced attenuation rate there. Beyond the obvious invalidation of all scaling laws based on linear acoustics these include (i) the possibility that much widely used experimental data may already be contaminated to an unknown degree by nonlinear distortion; (ii) the possibility that tailpipe liners, designed to attenuate high frequency noise generated inside an engine, may be much more effective on ground-based rigs than in the case of real interest, that of aircraft in flight, where nonlinear cascade from low frequencies may, over long ranges, offset the source attenuation provided by the liner; (iii) the possibility that nozzle suppressors of the multi-element kind may affect the nonlinear pumping, and therefore the high frequency levels, in different ways, cases being known (private communication from Mr. L. W. Illston of British Aerospace) in which, of three generally similar chuted or lobed nozzles, two were effective in eliminating spectral distortion, the third much less so; and (iv) the likelihood that the present difficulties in separating out the various source and propagation mechanisms will be greatly increased.

With these points in mind, Crighton and Bashforth (1980) put down the following aims of a theory of nonlinear jet noise propagation:

(1) It should deal with broadband signals directly, rather than with the harmonic waves so widely studied in nonlinear acoustics.

(2) It should lead to criteria with respect to intensity, spectral shape, or whatever, to indicate whether and when nonlinear effects are likely to be important.

(3) It should use only those simple statistical measures of the pressure $q(t)$ at the initial radius r_o that are commonly available in order to predict those quantities for the evolved pressure

$p(r, t)$. Those measures are the power spectral density $S_{qq}(\omega)$ in frequency ω, the overall level q_o^2 and its directivity $q_o^2(\theta, \phi)$, sometimes the directivity of $\frac{1}{3}$-octave bands, and occasionally the skewness factor $\langle q^3 \rangle / q_o^3$.

(4) It should be capable of including attenuation through arbitrary measured or tabulated attenuation coefficients $\alpha(\omega)$, and should not assume special analytic forms, such as the Stokes-Kirchhoff form $\alpha \sim \omega^2$, which do not represent atmospheric attenuation in the audio-frequency range.

(5) It should obviously incorporate effects of **spherical spreading**, but, as the nonuniform directivity of jet noise is an important feature, it should **not** require an assumption of **spherical symmetry**.

(6) It should lead both to scaling laws (analogous to Lighthill's (1952) famous U^8 law for jet noise) which may be useful in diagnosing the presence of spectral distortion, and to an outright prediction of the evolved spectrum $S_{pp}(r, \omega)$ in the very far field from a measured spectrum $S_{qq}(\omega)$ at much shorter ranges (though still in the acoustic far field).

(7) It should be capable of explaining the observations referred to in Section 5 below, of the development with range of positive skewness factors in high speed jet noise pressure fluctuations.

The aim of the present report is to review progress toward a theory capable of meeting this demanding set of requirements. A natural first step involves treating the nonlinear terms as a small perturbation away from linear acoustics, and this is described in Section 2. Although it falls far short of completely meeting the needs outlined above, it has the merit of being simple, capable of development in a number of ways and useful in suggesting circumstances in which nonlinear distortion is likely to be significant and trends by which the presence of that distortion might be diagnosed. In Sections 3 and 4, exact calculations of the spectral evolution are described, the first approach ignoring the presence of shocks in the waveform altogether and the second providing a correction due to shocks in a *small range* approximation. These approaches go much further than the perturbation series, but have other restrictions whose relevance is not yet known. Section 5 then deals with the phenomenon of skewness in high speed jet noise, and shows that the perturbation approach gives results which are qualitatively in accord with all the (limited) experimental observations. The final section summarizes the conclusions of the various theoretical approaches, and emphasizes that the need now is for further experimental work.

All the theoretical studies mentioned are *rational* (though they may have explicit assumptions—for example, that the initial data have a Gaussian probability distribution). The work of Morfey and Howell (1980a) and Howell (1982) does not fall within this category, as it makes a number of assumptions (for example, that the statistics of the evolved spectra is obtained if a numerical constant which would be 2 for a Gaussian distribution is replaced by 1.7, etc.) which cannot be justified on rational grounds, and this work will not be discussed further even thugh it yields a spectrum prediction scheme which closes much of the gap between the British Aerospace data and the predictions of linear theory.

Some mention must, however, be made of the work conducted over several years by Blackstock and his colleagues (Pestorius and Blackstock, 1974; Webster and Blackstock, 1978; and further references cited there). Part of this work dealt with nonlinear propagation of a broadband signal under the dissipative and dispersive mechanisms associated with viscous layers on the walls of a duct. An algorithm was developed, capable of predicting in great detail the evolution of each realization of the pressure as the wave propagated down the duct. The nonlinear distortion of the initial waveform was calculated, over a small range, from the characteristic solution to the equations with all dissipation and dispersion ignored. The evolved waveform was then resolved into its Fourier components, and each of these was corrected for the dispersion and dissipation which it would have experienced in linear propagation over the small increment of range. These modified Fourier components were then synthesized to give a new waveform in real time which was then allowed to propagate over a further increment in range by the same operations. Excellent agreement with experiment was found for the development of individual waveforms. The spectral evolution was then calculated by averaging the results of a large number of realizations calculated in this way.

Another version of this work starts with Fourier components whose amplitudes give a prescribed initial spectrum, and then, in the absence of phase information, assigns to these, phases selected randomly from a uniform distribution over $[-\pi, +\pi]$. A sequence of hypothetical initial signals is thereby created with the required initial spectrum, and each member of the sequence is then allowed to evolve in the manner just described, the evolved spectrum again being calculated by averaging the squares of the evolved Fourier coefficients. Comparisons of the results determined this way with those calculated on the basis of a knowledge of the **actual** initial phase distribution were favorable in the case of signals devoid of skewness; the random phase assignment worked much less well in signals with skewness (hardly surprisingly).

It is highly likely that these methods can be applied to the jet noise spectral evolution problem, though that has not yet been attempted. The disadvantage would, of course, be that although the spectral evolution could probably be calculated reasonably accurately taking into account many of the points (a)–(g) raised above, trends and understanding of them would only be produced from long experience of many different cases.

Another aspect of Blackstock's program was devoted to experiments (Webster and Blackstock, 1978) designed to establish the case for the significance of nonlinear distortion in jet noise signals. A high intensity horn generator was mounted at various heights on a tall tower and noise spectra at a range of heights and intensities measured. The results demonstrated the expected nonlinear energy transfer to higher frequencies; they were also compared with noise data taken from a KC-135A aircraft with high specific thrust turbojet engines. The scalings used for the comparison were standard nonlinear acoustic scalings for single tones, and the conclusion reached that since nonlinear distortion was demonstrably present in the horn experiments it should be expected with confidence in engines of the speed and power of those in the KC-135A. Although Webster and Blackstock (1978) say that *"although the KC-135A is a very noisy aircraft, many other current aircraft produce noise whose spectrum levels are higher than our scaled spectra. One therefore concludes that nonlinear effects are probably common in jet noise,"* this is not a conclusion which is widely accepted. It is hoped that this review will prompt further work on the problem and, if the belief expressed above is found to be widely valid, lead to a reassessment of the mechanisms to be taken into account in the correction of jet noise data to standard ranges and conditions.

2. A Weakly Nonlinear Perturbation Approach

In a situation in which linear propagation theory is generally regarded as satisfactory, and is certainly universally used, it is natural to begin to explore the consequences of nonlinear propagation effects through a weakly nonlinear perturbation scheme. Such a scheme is presented in this section. It has the merits of simplicity and flexibility, and leads to specific predictions which may be useful in diagnosis of the circumstances under which nonlinear spectral distortion is important. It will also be used later, in Section 5, to describe a mechanism by which skewness can be produced, in the course of propagation, in a signal which is initially distributed in a Gaussian manner (and thus devoid of skewness); no alternative mechanism has yet been proposed which is capable of this.

The perturbation approach has the disadvantage that it cannot describe the sepctrum at high frequencies, many octaves up from the peak; the reason for the nonuniformity of the perturbation expansion at high frequencies lies in its obvious incapacity to deal with shock waves, which, though weak, and rare in occurrence, nevertheless dominate the high frequency portion of the spectrum. An approximate method for dealing with this spectral range, due to Scott (1982), is described in Section 4, following an exact analysis, in Section 3, of spectral evolution under nonlinear nondissipative mechanisms. This evolution, first analyzed by Rudenko and Soluyan (1977), (and see related references cited there) ignores the contributions from shocks, and gives our perturbation expansion for small amplitudes and moderate frequencies.

We start (as in Crighton and Bashforth (1980) and Bashforth (1983)) by writing down a well-known model equation for the pressure $p(r,\tau)$ in an outgoing spherically symmetric sound field with weak nonlinearity and negligible dispersion and dissipation; it is (cf. Rudenko and Soluyan (1977), Crighton (1979))

$$\frac{\partial p}{\partial r} + \frac{p}{r} = \beta p \frac{\partial p}{\partial \tau} \tag{2.1}$$

where the variables are dimensional, r is range, τ is retarded time $t-(r-r_o)/a_o$, a_o is the small-signal sound speed, and $\beta = (\gamma + 1)/2\,\rho_o a_o^3$, with ρ_o the mean density and γ the adiabatic exponent. In the derivation of (2.1), the dependent variables are all locally related in the usual linear acoustic fashion, so that $p = \rho_o a_o u$, where u is the radial velocity. It is also assumed that $\omega r/a_o \gg 1$ for all $r \geq r_o$

(the initial radius, at which $p(r,\tau)$ is prescribed as $q(\tau)$) and for all relevant frequencies ω. Equation (2.1) is reduced to plane wave form by writing

$$P = \left(\frac{r}{r_o}\right)p, \qquad x = r_o \ln\left(\frac{r}{r_o}\right), \tag{2.2}$$

which gives

$$\frac{\partial P}{\partial x} = \beta P \frac{\partial P}{\partial \tau}, \qquad P(0,\tau) = q(\tau). \tag{2.3}$$

The exact solution to the problem (2.3) is most naturally written down in characteristic form, and this form will be used in Section 3 to obtain the spectrum of Rudenko and Soluyan (1977). The exact solution can also be written as

$$P(x,\tau) = \sum_{n=0}^{\infty} \frac{(\beta x)^n}{(n+1)!} \frac{\partial^n}{\partial \tau^n}(q^{n+1}) \tag{2.4}$$

and we keep only the first three terms, giving

$$P(x,\tau) = q(\tau) + \frac{\beta x}{2!}\frac{\partial}{\partial \tau}(q^2) + \frac{\beta^2 x^2}{3!}\frac{\partial^2}{\partial \tau^2}(q^3) + 0(\omega^3\beta^3x^3q^4) \tag{2.5}$$

with the frequency ω representing the order of magnitude of $\partial/\partial\tau$. In terms of Fourier transforms defined by

$$f(\tau) = \int_{-\infty}^{+\infty} \tilde{f}(\omega)\exp(i\omega\tau)\,d\omega,$$

equation (2.5) gives

$$\begin{aligned} \tilde{P}(x,\omega) =& \tilde{q}(\omega) + \frac{(i\omega\beta x)}{2!}\int \tilde{q}(\omega')\,\tilde{q}(\omega-\omega')\,d\omega' \\ &+ \frac{(i\omega\beta x)^2}{3!}\int\!\!\int \tilde{q}(\omega-\omega')\,\tilde{q}(\omega'-\omega'')\,\tilde{q}(\omega'')\,d\omega'\,d\omega'' \\ &+ 0\,(\omega^2\beta^3x^3q^4). \end{aligned} \tag{2.6}$$

The first term in each of (2.5) and (2.6) represents the prediction of linear acoustics, while the higher terms represent the production of sum and difference frequency components by quadratic nonlinearity.

Suppose now that $q(\tau)$ is given as a stationary random function of τ, with zero mean and variance $q_o^2 = \langle q^2(\tau)\rangle$. The quantities normally available (from measurements) for prediction of any of the statistical measures of $P(r,\tau)$ are only q_o^2, and the initial power spectral density $S_{qq}(\omega)$ of $\langle q^2\rangle$ or equivalently the autocovariance $R_{qq}(\sigma)$ as a function of time delay σ. In order to use just this information without further—and necessarily quite arbitrary—specification of the higher moments, we assume that q has a joint Gaussian probability distribution at as many points $(\tau_1, \tau_2, \ldots, \tau_n)$ as necessary for our purposes. The entire distribution of q is then specified by the autocovariance, or by the spectrum, and, further, any linear functional of q also has a Gaussian distribution. In particular, the Fourier transform of q and the time derivatives of q have Gaussian distributions. (The reader is referred to Lumley (1970) for an appropriate discussion of the relevant statistical quantities.)

The spectrum and covariance are related by

$$\begin{aligned} R_{qq}(\sigma) &= \langle q(\tau)\, q(\tau+\sigma)\rangle \\ &= \int_{-\infty}^{+\infty} S_{qq}(\omega) \exp(i\omega\sigma)\, d\omega \end{aligned} \tag{2.7}$$

and the spectrum is related to the Fourier coefficients $q(\omega)$ by $\langle \tilde{q}(\omega)\tilde{q}(\omega')\rangle = S_{qq}(\omega)\delta(\omega+\omega')$. If q has joint Gaussian statistics, all odd moments vanish, and fourth moments are related to second through $\langle q_1 q_2 q_3 q_4\rangle = \langle q_1 q_2\rangle\langle q_3 q_4\rangle + \langle q_1 q_3\rangle\langle q_2 q_4\rangle + \langle q_1 q_4\rangle\langle q_2 q_3\rangle$, where q_i may mean $q(\tau_i)$ or $\tilde{q}(\omega_i)$. Application of these relations to (2.5) and (2.6) produces

$$\begin{aligned} R_{PP}(x,\sigma) &= R_{qq}(\sigma) + \beta^2 x^2 q_o^2 \left(\frac{\partial^2 R_{qq}(\sigma)}{\partial\sigma^2}\right) \\ &\quad - \frac{1}{2}\beta^2 x^2 \frac{\partial^2}{\partial\sigma^2}[R_{qq}(\sigma)]^2 + 0(x^4) \end{aligned} \tag{2.8}$$

and

$$\begin{aligned} S_{PP}(x,\omega) &= S_{qq}(\omega)\{1 - \beta^2\omega^2 x^2 q_o^2\} \\ &\quad + \frac{1}{2}\beta^2\omega^2 x^2 \int_{-\infty}^{+\infty} S_{qq}(\omega')\, S_{qq}(\omega-\omega')\, d\omega' + 0(x^4). \end{aligned} \tag{2.9}$$

We emphasize that the Gaussian relation between fourth and second moments is made only for the initial signal $q(\tau)$ and not (as was assumed by Morfey and Howell (1980a) and Howell (1982)) for the evolved signal $P(x,\tau)$. It may be, however, that the Gaussian relation does continue to be satisfied by $P(x,\tau)$ if it is satisfied by $q(\tau)$ and if the relation between P and q is the series (2.4) (leaving aside the question of what that series represents once shocks have formed). We have not been able to prove or disprove this, though particular calculations indicate that a limited form of the Gaussian relation may continue to hold for the evolved signal. Certainly it holds for the one-point distribution, for the differential equation (2.3) shows that

$$\frac{\partial}{\partial x}\langle P^n\rangle = \frac{n\beta}{n+1}\frac{\partial}{\partial \tau}\langle P^{n+1}\rangle = 0, \tag{2.10}$$

and hence that $\langle P^n\rangle(x) = \langle q^n\rangle$. We shall show later, however, that in the presence of dispersion and dissipation the Gaussian assumption **cannot** hold even for the one-point distribution of P if it holds for q.

As implied by the remarks above, the spectral expansion (2.9) is energy-conserving, i.e.,

$$\int_{-\infty}^{+\infty} S_{PP}(x,\omega)\,d\omega = q_o^2 + 0(x^4), \tag{2.11}$$

and hence the term $-\beta^2\omega^2x^2q_o^2S_{qq}(\omega)$ must represent the depletion of the main energy-containing spectral region necessary to produce the cascade to higher frequencies represented by the convolution term in (2.9). However, the convolution term contains difference as well as sum frequencies, and the direction of the cascade of energy away from the initial region depends very much on the low frequency behaviour of the initial spectrum. Rudenko and Soluyan (1977), (p. 252) evaluate the complete series numerically (in the form to be given in Section 3) for a process $q(\tau)$ with **zero integral scale** and show a systematic shift of the spectral peak to lower frequencies as x increases, without significant enhancement of the high frequency components. If, on the other hand, the process has non-zero integral scale, so that $S_{qq}(0) \neq 0$ there **does** seem to be a cascade to higher frequencies. To show this we observe (cf. Section 3) that the convolution terms at **all** orders in x are the terms in the series

$$\Sigma = \sum_{n=1}^{\infty} \frac{(\omega^2\beta^2x^2)^n}{n!}(S_{qq}^*)^n(\omega)$$

where $(S_{qq}^*)^n$ means the nth convolution of S_{qq} with itself. If we take

$$S_{qq}(\omega) = \frac{q_o^2}{\pi^{1/2}\omega_o} \exp\left(-\frac{\omega^2}{\omega_o^2}\right)$$

then

$$(S_{qq}^*)^n(\omega) = \frac{q_o^{2n}}{(n\pi)^{1/2}\omega_o} \exp\left(-\frac{\omega^2}{n\omega_o^2}\right),$$

and the terms in $\sum$ are proportional to $\omega^{2n}\exp(-\omega^2/n\omega_o^2)$, these vanishing increasing rapidly as $\omega \to 0$ and having their maxima at $\omega = \pm n\omega_o$. In this sense the terms in $\sum$ may be regarded as describing successive spectral octaves, and the expansion (2.9) may be regarded as a generalization of an expansion in harmonics (with coefficients proportional to x^n) which would be used if $q(\tau)$ were a sinusoidal signal (see Rudenko and Soluyan (1977) (p. 30)). The expansion in harmonics is nonuniform for large x, reflecting the fact that shocks (or at least steep gradients) will by then have been formed; that is also true of our expansion, but now (since ω is not fixed) the nonuniformity in x is also associated with a nonuniformity in ω at fixed x. The scale on which this nonuniformity emerges will be discussed in Sections 3 and 4. Here we shall use the expansion in its simplest form (2.9) to indicate frequencies, ranges and sound pressure levels at which nonlinear effects are bound to be significant even in the spectral region containing most of the energy (and, presumably, are therefore likely to be even more significant at higher frequencies and larger ranges or sound pressure levels).

Let $\omega_o = 2\pi f_o$ be a typical frequency in $S_{qq}(\omega)$ and consider frequencies $0(\omega_o)$. For the corrections in (2.9) to be small compared with the linear term we require the parameter

$$\begin{aligned}\epsilon_o &= \omega_o \beta r_o q_o \ln(r/r_o) \\ &= \epsilon_* \ln(r/r_o)\end{aligned} \tag{2.12}$$

to be small. Observe that the value of ϵ_* is not sensitive to the choice of initial radius r_o; for provided $\omega r_o/a_o \gg 1$ (as already assumed) and provided r_o is not so large that dissipation effects have already become significant, $r_o q_o$ is (even with nonlinear effects included) a constant as in linear nondissipative acoustics. Take an equivalent *source radius* $r_o = 1$ metre, equivalent source overall sound pressure levels from 120 dB to 180 dB (at $r = r_o$) and frequencies f_o from 250 Hz to 1000 Kz; these are likely to encompass the parameter ranges of current full-scale interest. Then Table 1a gives the values of ϵ_* corresponding to these parameters, and Table 1b the values of r_1 such that $\epsilon_o = \epsilon_* \ln(r_1/r_o) = 1$. These correct those given by Bashforth (1983). Their implication is, for example, that propagation over 1000 m at $f_o = 500$ Hz can only be calculated from linear theory for OASPLs less than 160 dB at 1 metre; and that there will be significant spectral deformation even near the *typical* frequency $f_o = 500$ Hz over a distance as small as 2 metres when the OASPL is as high as 180 dB. The distortions at higher OASPLs, frequencies or ranges will be even more significant.

It might be thought that deviations from spherical symmetry, such as certainly exist in the jet noise field, and linear attenuation mechanisms—which are certainly significant over distances as large as 1000 m even at the lower frequencies—might drastically change these estimates of the importance of nonlinearity. We argue now that this is not the case.

First, if the initial pressure field is directional then, provided again that $\omega r_o/a_o \gg 1$, the evolution equation (2.1) continues to hold along rays specified by a unit vector $\hat{\mathrm{n}} = \hat{\mathrm{n}}(\theta, \phi)$, with the directional dependence appearing only parametrically. Thus,

$$\begin{aligned} S_{pp}(r,\omega,\hat{\mathrm{n}}) =& \left(\frac{r_o}{r}\right)^2 S_{qq}(\omega,\hat{\mathrm{n}}) \left[1 - \omega^2\beta^2 q_o^2(\hat{\mathrm{n}}) r_o^2 \ln^2\left(\frac{r}{r_o}\right)\right] \\ &+ \frac{1}{2}\left(\frac{r_o}{r}\right)^2 \omega^2\beta^2 r_o^2 \ln^2\left(\frac{r}{r_o}\right) \int_{-\infty}^{+\infty} \\ &\times S_{qq}(\omega',\hat{\mathrm{n}})\, S_{qq}(\omega-\omega',\hat{\mathrm{n}})\, d\omega' + \cdots \end{aligned} \tag{2.13}$$

which shows that the estimates of Table 1 remain valid along each $\hat{\mathrm{n}}$, and that spectral distortion will be accompanied by a distortion of the directivity pattern at each frequency.

f_o(Hz) / dB at 1 m	250	500	1000
120	8×10^{-4}	1.6×10^{-3}	3.2×10^{-3}
140	8×10^{-3}	1.6×10^{-2}	3.2×10^{-2}
160	8×10^{-2}	1.6×10^{-1}	3.2×10^{-1}
180	8×10^{-1}	1.6	3.2

Table 1a. Values of $\epsilon_* = \omega_o \beta r_o q_o$ for $r_o = 1m$, $\omega_o = 2\pi f_o$ and various overall sound pressure levels equivalent to q_o.

f_o(Hz) / dB at 1 m	250	500	1000
120	∞	∞	∞
140	2×10^{54}	1.4×10^{27}	3.7×10^{13}
160	2.7×10^{5}	518	23
180	3.5	1.87	1.37

Table 1b. Values of $r_1(m)$ such that $\epsilon_o = \epsilon_* \ln(r_1/r_o) = 1$ ($r_o = 1m$).

As for attenuation effects, we wish to include these in the most general way, since the dominant attenuation mechanisms in atmospheric air in the audio-frequency range cannot be simply represented in thermo-viscosity or bulk viscosity terms. Let $\alpha(\omega)$ denote the usual experimentally determined attenuation coefficient of linear acoustics, a real, even function of ω. (Dispersion should strictly also be included, but is irrelevant here, except in the discussion of skewness to follow in Section 5.) Then the model equation replacing (2.1) is (cf. Crighton (1979) for a discussion of the status of these model equations)

$$\frac{\partial \tilde{p}}{\partial r} + \frac{\tilde{p}}{r} + \alpha(\omega)\tilde{p} = \beta \widetilde{p\frac{\partial p}{\partial \tau}} \tag{2.14}$$

with the tilde indicating Fourier transformation. The physical basis for (2.14) is obvious; it is a composite approximate equation which was the correct limiting forms and incorporates all the effects required in the simplest possible way. Now define

$$\begin{aligned} T(r,\omega) &= \left(\frac{r}{r_o}\right)\exp[\alpha(\omega)(r-r_o)]\tilde{p}(r,\omega), \\ x &= r_o \ln\left(\frac{r}{r_o}\right), \\ \gamma(\omega,\omega') &= \alpha(\omega') + \alpha(\omega-\omega') - \alpha(\omega), \end{aligned} \tag{2.15}$$

and rewrite (2.14) as

$$\begin{aligned} \frac{\partial T}{\partial x} = \frac{1}{2}\beta i\omega \int_{-\infty}^{+\infty} & T(x,\omega')\,T(x,\omega-\omega') \\ & \times \exp\left\{-\gamma r_o\left[\exp\left(\frac{x}{r_o}\right)-1\right]\right\} d\omega' \end{aligned} \tag{2.16}$$

with

$$T(0,\omega) = \tilde{q}(\omega)$$

where, again, $q(\tau) = p(r_o,\tau)$. Differentiate (2.16) with respect to x and determine $\partial T/\partial x$, $\partial^2 T/\partial x^2$ at $x = 0$, and hence the first three terms of an expansion corresponding to (2.5). Then form $T(r,\omega)\,T(r,\omega')$ and average to obtain an expression for the spectrum $S_{TT}(r,\omega)$ in which the $0(x^2)$ term contains a contribution involving the transform of a triple covariance $\langle q(\tau)\,q(\tau+\lambda)\,q(\tau+\mu)\rangle$ together with terms which were present in the absence of dissipation. The triple covariance certainly vanishes if q has joint Gaussian statistics, and even contributes only a vanishing contribution to S_{TT} under the much less restrictive assumption of time-reflexion-symmetry. It follows that the expression for S_{TT} is identical, to $0(x^2)$, to that for S_{PP} in the absence of dissipation, and hence that

$$\begin{aligned} &\left(\frac{r}{r_o}\right)^2 \exp\left[2\alpha_\omega(r-r_o)\right] S_{pp}(r,\omega,\hat{\mathbf{n}}) \\ &= S_{qq}(\omega,\hat{\mathbf{n}})\left[1-\beta^2\omega^2 q_o^2(\hat{\mathbf{n}})\, r_o^2 \ln^2\left(\frac{r}{r_o}\right)\right] \\ &\quad + \frac{1}{2}\omega^2\beta^2 r_o^2 \ln^2\left(\frac{r}{r_o}\right)\int_{-\infty}^{+\infty} S_{qq}(\omega',\hat{\mathbf{n}})\,S_{qq}(\omega-\omega',\hat{\mathbf{n}})\,d\omega' + 0(x^4). \end{aligned} \tag{2.17}$$

Equation (2.17) shows that attenuation affects linear and nonlinear contributions to the spectrum in precisely the same way, and hence that, to $0(x^2)$, *the relative importance of nonlinear terms is independent of attenuation.*

We next give scaling laws for the nonlinearly induced contribution to the spectrum. If the initial spectrum scales on q_o and on a typical frequency ω_o (for the range of ω for which (2.17) is a valid approximation) we can write

$$S_{qq}(\omega) = \omega_o^{-1} q_o^2 \Sigma_{qq}(\Omega) \tag{2.18}$$

with $\Omega = \omega/\omega_o$. If the noise at $r = r_o$ is subsonic jet mixing noise of classical Lighthill (1952) type, then

$$\omega_o \sim U/D, \qquad q_o \sim \rho_o U^2 (D/r_o) M^2, \tag{2.19}$$

where U is the jet exhaust velocity, D the jet diameter, and $M = U/a_o$ the Mach number. The linear terms of (2.17) give

$$S_{TT} \sim \rho_o^2 U^3 M^4 \frac{D^3}{r_o^2} \tag{2.20}$$

which is Lighthill's U^8 law for overall intensity rewritten as a scaling law for the pressure power spectral density. The nonlinear (convolution) term in (2.17) gives

$$S_{TT} \sim \rho_o^2 U^3 M^4 \frac{D^3}{r_o^2} M^{10} \ln^2(r/r_o) \tag{2.21}$$

with a much more rapid variation with M, but with only slow variation with r.

If so-called *excess* or *core* noise (perhaps due to combustion) dominated the field at $r = r_o$, q_o would vary as $U^2 M$, the linear contribution to S_{TT} as $U^3 M^2$ (overall intensity proportional to U^6) and the nonlinear contribution as $U^3 M^2 M^{10} \ln^2(r/r_o)$. The ratio of nonlinear to linear contributions varies in either case as $M^{10} \ln^2(r/r_o)$, whose very rapid variation with M might be used to detect the influence of nonlinear propagation in experimental data.

At supersonic exhaust speeds the noise sources are of Mach wave type, and do not have the Strouhal number frequency scaling.

Instead of (2.19) the appropriate scalings are

$$\omega_o \sim a_o/D, \qquad q_o^2 \sim \rho_o^2 a_o U^3 (D/r_o)^2 \tag{2.22}$$

and lead to

$$S_{TT} \sim \rho_o^2 U^3 \frac{D^3}{r_o^2}$$

and

$$S_{TT} \sim \rho_o^2 U^3 \frac{D^3}{r_o^2} M^3 \ln^2 \left(\frac{r}{r_o}\right) \tag{2.23}$$

for the linear and nonlinear contribution to S_{TT}. Here the growth with M is much slower than in the subsonic case, but still sufficiently rapid to be a useful diagnostic in experiments. One would not, of course, expect to see the scalings like (2.23) in the main energy-containing part of the spectrum, where the usual Lighthill scalings should apply, but rather in the spectral region one or two octaves higher. There the first nonlinear term might be expected to dominate both the original linear contribution and also higher order nonlinear terms which in turn dominate in higher octaves.

Although the scaling laws are insensitive to the shape of the initial spectrum, the coefficients appear to be highly sensitive to shape. Examples given by Crighton and Bashforth (1980) indicate sensitivity to both low and high frequency behaviour of $S_{qq}(\omega)$, and of course neither of these behaviours is unambiguously apparent from the usual kind of full-scale experimental data in 1/3-octave form. These remarks apply again to frequencies somewhat above the 1/3-octave peak frequency, and not to much higher frequencies where, if the assumptions of Sections 3 and 4 hold, the spectrum takes a universal form.

That completes discussion of the perturbation expansion, except for the calculation of skewness to be given in Section 5. The principal value of a perturbation expansion lies in the confidence which it gives to the dominance of the first term, which here is the universally used result of linear acoustics. That term is found here not to dominate, even round the main energy-containing part of the spectrum and even less so at higher frequencies, at quite moderate OASPLs provided the propagation range concerned is that of full-scale flyovers (100 m–1000 m) rather than that (up to 30 m) of ground-based tests or model rigs. The perturbation expansion also indicates that as nonlinearity *begins* to become important, the spectral levels which it affects will vary very rapidly with Mach number compared with the variation subject to linear propagation, and that the linearly generated spectral contributions will be sensitive to the spectral behaviour at much higher and much lower frequencies.

3. Exact Solution for the Spectrum: No Shocks

This section gives a derivation of an exact expression, due to Rudenko and Soluyan (1977) (and references to other Soviet work there), for the spectrum $S_{PP}(x, \omega)$, when P evolves according to (2.3) and $q(\tau)$ has joint Gaussian statistics. The characteristic solution for P is

$$P(x, \tau') = q(\phi), \qquad \tau' = \phi - \beta x q(\phi) \tag{3.1}$$

and we write

$$P(x, \tau) = \frac{1}{2\pi} \int\int P(x, \tau') \, e^{-i\omega(\tau - \tau')} \, d\tau' \, d\omega,$$

insert $P(x, \tau') = q(\phi)$ and transform the τ' integration to one over ϕ, and follow this with an integration by parts. (In order to ignore the integrated contributions at $\phi = \pm\infty$ we should really replace $q(\phi)$ by $\exp(-\epsilon\phi^2)\, q(\phi)$ and let $\epsilon \to 0$ at an appropriate stage; cf. Scott (1982).) This gives

$$P(x, \tau) = \frac{i}{2\pi} \int\int q'(\phi) \exp\left[-i\omega\left(\tau - \phi + \beta x q(\phi)\right)\right] d\phi \frac{d\omega}{\omega}$$

from which the covariance

$$R_{PP}(x, \sigma) \equiv \langle P(x, \tau) \, P(x, \tau + \sigma) \rangle$$

follows as

$$-\frac{1}{4\pi^2} \int\int\int\int e^{-i(\omega_1 + \omega_2)\tau - i\omega_2\sigma + i\omega_1\phi_1 + i\omega_2\phi_2} \times \langle q'(\phi_1) \, q'(\phi_2) e^{-i\omega\beta x q(\phi_1) - i\omega\beta x q(\phi_2)} \rangle \, d\phi_1 \, d\phi_2 \frac{d\omega_1}{\omega_1} \frac{d\omega_2}{\omega_2}.$$

The mean value depends only on $\lambda = \phi_2 - \phi_1$ and if λ is held fixed, the integration over ϕ_1 produces a factor $2\pi\delta(\omega_1 + \omega_2)$, which then enables the ω_2-integration to be performed, with the result

$$R_{PP}(x,\sigma) = \frac{1}{2\pi} \int\int e^{i\omega\sigma - i\omega\lambda} \times \langle q'(\phi_1)\, q'(\phi_2)\, e^{-i\omega\beta x q(\phi_1) + i\omega\beta x q(\phi_2)} \rangle \frac{d\omega}{\omega^2}\, d\lambda.$$

The mean value here can now be written as

$$\frac{1}{\omega^2\beta^2 x^2} \frac{\partial^2}{\partial\phi_1 \partial\phi_2} \langle e^{-i\omega\beta x q(\phi_1) + i\omega\beta x q(\phi_2)} \rangle$$

in which

$$\langle e^{-i\omega\beta x q(\phi_1) + i\omega\beta x q(\phi_2)} \rangle = C_2\,(-\omega\beta x, \omega\beta x)$$

is the two-point characteristic function for the process $q(\phi)$, defined by

$$\begin{aligned} C_2\,(k_1, k_2) &= \langle e^{ik_1 q_1 + ik_2 q_2} \rangle \\ &= \int\int e^{ik_1 q_1 + ik_2 q_2}\, Q\,(q_1, q_2)\, dq_1 dq_2 \end{aligned}$$

where q_1, q_2 are $q(\phi_1)$ and $q(\phi_2)$ and $Q(q_1, q_2)$ is the two-point probability density function (so that $C_2(k_1, k_2)$ is the Fourier transform of the probability density). Now C_2 is a function of λ and not of ϕ_1, ϕ_2 separately, and so $\partial^2/\partial\phi_1\, \partial\phi_2$ may be replaced by $-\partial^2/\partial\lambda^2$. the integral over λ for R_{PP} can then be performed twice by parts; however, in order to ignore the integrated contributions one must recognize that as $\lambda \to \infty$

$$\begin{aligned} C_2 &\to \langle e^{-i\omega\beta x q(\phi_1)} \rangle \langle e^{i\omega\beta x q(\phi_2)} \rangle \\ &= |\langle e^{-i\omega\beta x q(\phi)} \rangle|^2 \\ &= |C_1\,(-\omega\beta x)|^2 \end{aligned}$$

where C_1 is the one-point characteristic function, and must therefore replace C_2 by $C_2 - |C_1|^2$ before integrating by parts. This then gives

$$R_{PP}(x,\sigma) = \frac{1}{2\pi\beta^2 x^2} \int\int e^{i\omega\sigma - i\omega\lambda} \times \left[C_2\,(-\omega\beta x, \omega\beta x) - |C_1\,(-\omega\beta x)|^2\right] \frac{d\omega}{\omega^2}\, d\lambda,$$

from which the expression for the spectrum follows as

$$S_{PP}(x,\omega) = \frac{1}{2\pi\beta^2 x^2 \omega^2} \int e^{-i\omega\lambda} \{C_2 - |C_1|^2\}\, d\lambda. \tag{3.2}$$

This determines the spectrum, given the *entire* two-point statistics of q, for

$$C_2(k_1,k_2) = \sum\sum_{m,n=0}^{\infty} \frac{(ik_1)^m}{m!} \frac{(ik_2)^n}{n!} \langle q_1^m\, q_2^n \rangle$$

and needs specification of *all* the $\langle q_1^m\, q_2^n\rangle$. If, however, q has two-point joint Gaussian statistics, then

$$C_2(k_1,k_2) = \exp\left\{-\frac{1}{2}[q_o^2 k_1^2 + q_o^2 k_2^2 + 2k_1k_2 R_{qq}(\lambda)]\right\} \tag{3.3}$$

and

$$C_1(k) = \exp\left(-\frac{1}{2}q_o^2 k^2\right)$$

and hence

$$\begin{aligned} S_{PP}(x,\omega) =& \frac{1}{2\pi\beta^2 x^2 \omega^2} \exp(-\beta^2 x^2 \omega^2 q_o^2) \\ &\times \int_{-\infty}^{+\infty} e^{-i\omega\lambda} \left[e^{\omega^2\beta^2 x^2 R_{qq}(\lambda)} - 1 \right] d\lambda. \end{aligned} \tag{3.4}$$

Inasmuch as $S_{qq}(\omega)$ is known when $R_{qq}(\lambda)$ is, and vice-versa, this determines the evolved spectrum solely in terms of $S_{qq}(\omega)$. Equation (3.4) corresponds to the result on p. 251 of Rudenko and Soluyan (1977), though their (2.12) clearly has an incorrect sign.

Expanding the exponential gives

$$\begin{aligned} S_{PP}(x,\omega) =& S_{qq}(\omega) + \left(e^{-\omega^2\beta^2x^2q_o^2} - 1\right) S_{qq}(\omega) \\ &+ \sum_{n=2}^{\infty} \frac{(\omega^2\beta^2x^2)^{n-1}}{n!} \frac{1}{2\pi} \int_{-\infty}^{+\infty} R_{qq}^n(\lambda) e^{-i\omega\lambda}\, d\lambda \\ =& S_{qq}(\omega) + \left(e^{-\omega^2\beta^2x^2q_o^2} - 1\right) S_{qq}(\omega) \\ &+ \sum_{n=2}^{\infty} \frac{(\omega^2\beta^2x^2)^{n-1}}{n!} S_{qq}^{*^n}(\omega) \end{aligned} \tag{3.5}$$

where S^{*^n} indicates the n-fold convolution of S with itself. These expressions clearly demonstrate the intricate nonlinear character of the spectral evolution, and show also how the first terms of our perturbation series are recovered and, further, generalize those terms to all orders n. As noted in Section 2, and as remarked by Rudenko and Soluyan (1977) (p. 252), the nature of the spectral distortion depends critically on the low frequency behaviour of S_{qq}. If $S_{qq}(0)$ has a non-zero value, $S_{PP}(x,0)$ also has that value for *all* x, and the very low frequency part of the spectrum is unchanged while the levels at high frequencies are increased. If $S_{qq}(0) = 0$ the low frequency behaviour of S_{PP} is determined by both linear and nonlinear terms and the tendency is to increase S_{PP} for low frequencies and reduce it at higher frequencies, so that the net effect may be a shift of the peak frequency to *lower* values. This trend is illustrated in Figure 2, which is a reproduction of Figure 10.4 of Rudenko and Soluyan. Even here, however, there are indications that the spectral levels *increase* at much higher frequencies. An evaluation of (3.5) for a wide spread of frequencies, ranges and initial spectral shapes would clearly be of much interest and might establish general trends for the direction of nonlinear energy cascade. The low frequency behaviour of jet noise is presumably such that $S_{qq}(0) = 0$; at very low frequencies one presumably has compact quadrupole radiation with a decrease in spectral level as ω^4, though experimental data often show an ω^2 variation which would be consistent with a certain kind of similarity hypothesis. Much more information on the low frequency levels would be necessary before (3.5) could be applied to the nonlinear distortion of a jet noise spectrum with any degree of reliability.

The high frequency limit of equation (3.4) can be found by use of Laplace's method. The covariance is assumed to have an expansion

$$R_{qq}(\lambda) = q_o^2\left(1 - \frac{\lambda^2}{2T_1^2} + \frac{\lambda^4}{4!T_2^4} + \cdots\right) \tag{3.6}$$

where, using turbulence language, T_1 is a Taylor microscale and T_2 a higher order microscale. These are related to the mean squares of derivatives of q by

$$\frac{d^2 R_{qq}}{d\lambda^2}(0) = -\langle\left(\frac{dq}{d\tau}\right)^2\rangle = -\frac{q_o^2}{T_1^2}$$

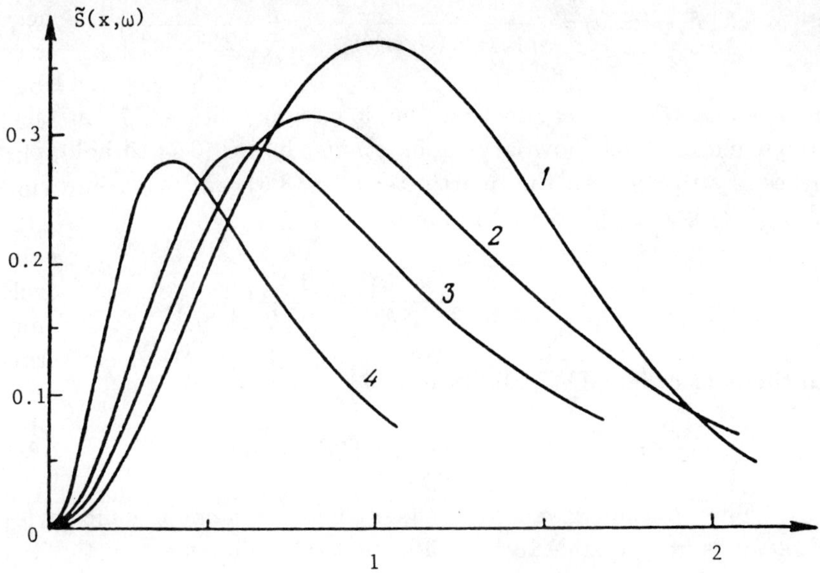

Figure 2. Change in the shape of a broadband spectrum according to (3.5); this is Figure 10.4 of Rudenko and Soluyan (1977), reproduced with permission of the publishers (Plenum Press). The notation has been changed to agree with that of the present paper, the abscissa being ω/ω_o and the ordinate

$$\tilde{S}(x,\omega) = \left(\frac{\pi^{1/2}}{2}\right)\left(\frac{\omega_o}{q_o^2}\right) S_{PP}(x,\omega).$$

Curves 1, 2, 3, 4 show $\tilde{S}(x,\omega)$ as a function of (ω/ω_o) for an initial spectrum

$$S_{qq}(\omega) = \frac{2q_o^2}{\pi^{1/2}\omega_o}\left(\frac{\omega}{\omega_o}\right)^2 e^{-(\omega/\omega_o)^2}$$

(which has maximum value at $\omega = \omega_o$, and makes $\int_{-\infty}^{+\infty} S_{qq}(\omega)d\omega = q_o^2$) and for values 0, $\frac{1}{3}$, $\frac{1}{3^{1/2}}$, $\frac{2}{3^{1/2}}$ of the quantity $\omega_o\beta q_o x$.

and

$$\frac{d^4 R_{qq}}{d\lambda^4}(0) = +\left\langle\left(\frac{d^2 q}{d\tau^2}\right)^2\right\rangle = +\frac{q_o^2}{T_2^4}.$$

If $R_{qq}(\lambda)$ is approximated by its quadratic form near $\lambda = 0$ then it readily follows that

$$S_{PP}(x,\omega) \sim \frac{1}{(2\pi)^{1/2}} \frac{T_1}{(\beta x \omega)^3 q_o} \exp\left(-\frac{T_1^2}{2\beta^2 x^2 q_o^2} \right) \tag{3.7}$$

as $\omega \to \infty$ or $x \to \infty$, so that the spectrum falls as ω^{-3} at large frequencies. To see how large ω or x must be for (3.7) to hold, it is necessary to use also the quartic term in (3.6), and then one finds that (3.7) is multiplied by the factor

$$1 + \frac{1}{\omega^2 \beta^2 x^2} \left[\frac{\omega^2 T_1^2}{2q_0^2} + \frac{1}{8q_o^2} \frac{T_1^4}{T_2^4} \right]$$

to the next order. The validity of (3.7) thus requires

$$\beta x q_o / T_1 \gg 1 \qquad \text{and} \qquad \omega \beta x q_o T_2^2 / T_1^2 \gg 1. \tag{3.8}$$

The solution given here suffers from two main deficiencies. First, it is not possible to generalize it to include any realistic form of attenuation. The dissipative model equation (2.14) can be reduced to the form (2.3) by a transformation of both independent and dependent variables only if the attenuation coefficient $\alpha(\omega) = \alpha_o$, a constant, but this is not representative (except in a phenomenological sense) of any atmospheric attenuation mechanism, even at low frequencies. If $\alpha(\omega) \sim \omega^2$, as for thermo-viscous dissipation or for relaxation mechanisms at low frequencies, then the model equation (2.14) becomes Burgers equation for plane flow, and a generalized Burgers equation for spherical flow. The ordinary Burgers equation has an exact solution for arbitrary initial conditions, but that is so intractable that no expression as simple or complete as (3.4) is in sight. For large times, or large ranges in our problem, the exact solution to Burgers equation takes a sawtooth waveform outside thin shock waves, and the shock wave interactions are equivalent to the collisions of a dynamical system of particles. For these collisions the statistical problem was solved by Kida (1979), with results which will be mentioned only briefly in Section 4 as they apply to waveforms dominated by shock waves and their merging, and these are not thought to be representative of jet noise signals (except possibly within the jet tailpipe). Kida's results may also be applied to the spherical wave problem, for although the generalized Burgers equation does not have an exact general solution, its solutions *do* (after transformation) have the sawtooth waveform (until large ranges at which the mechanism of shock thickening and breakdown is quite different from that for plane waves).

The second main deficiency of the solution (3.4) was recognized by Rudenko and Soluyan (1977) (p. 253) and the necessary correction recently provided by Scott (1982). No account is taken in the derivation of (3.4) of the presence of shock waves. This is clear in the immediate transformation from integration over time τ' to integration over the characteristic variable ϕ, with $-\infty < \phi < +\infty$. Once shocks have formed the $\tau' - \phi$ transformation is not one-one, and at any range x there are sections of the ϕ-axis from which the emanating characteristics have already been swallowed by the shocks and which should be excluded from the integration. How this can be done for small ranges and infrequently occurring shocks will be described in the next section.

4. Spectral Evolution Over Short Ranges; Shocks Included

The solution discussed in Section 3 fails in some frequency intervals at even the shortest ranges, because it takes no account of the formation of shocks. Scott (1982) has shown, however, how this solution may be used and modified so as to deal precisely with the evolution over *small* ranges of the spectrum at *any* frequency. In this section we describe the important steps in Scott's calculation, and give the final results in the notation we have used earlier in this paper.

The evolution of each realization $P(x, \tau)$ is governed by (2.3), namely

$$\begin{gathered}\frac{\partial P}{\partial x} - \beta P \frac{\partial P}{\partial \tau} = 0, \\ P(0, \tau) = q(\tau)\end{gathered} \tag{4.1}$$

where $q(\tau)$ is a Gaussian process which is *assumed* sufficiently smooth that the autocovariance can be expanded in the form (3.6), that is,

$$R_{qq}(\sigma) = q_o^2 \left\{ 1 - \frac{\sigma^2}{2! T_1^2} + \frac{\sigma^4}{4! T_2^4} + \cdots \right\}, \tag{4.2}$$

which implies that the spectrum $S_{qq}(\omega)$ must decay faster than ω^{-5} for large ω. Define

$$P = P_o P_*,$$
$$x = L_o x_*,$$
$$\tau = T_o \tau_*,$$
$$\sigma = T_o \sigma_*,$$

with

$$P_o = q_o \left(\frac{T_2}{T_1}\right)^2,$$
$$L_o = \frac{T_1}{\beta q_o},$$
$$T_o = \frac{T_2^2}{T_1}. \tag{4.3}$$

Then

$$\frac{\partial P_*}{\partial x_*} - P_* \frac{\partial P_*}{\partial \tau_*} = 0,$$

$$P_*(0, \tau_*) = \frac{1}{P_o} q(\tau_*)$$

and

$$\langle P_*(0, \tau_*)\, P_*(0, \tau_* + \sigma_*) \rangle = \frac{q_o^2}{P_o^2} - \frac{\sigma_*^2}{2!} + \frac{\sigma_*^4}{4!} + \cdots, \tag{4.4}$$

so that, dropping the asterisk, we can deal with $P(x, \tau)$ normalized so that

$$\frac{\partial P}{\partial x} - P \frac{\partial P}{\partial \tau} = 0,$$

$$P(0, \tau) = u_o(\tau)$$

with

$$\langle u_o(\tau)\, u_o(\tau + \sigma) \rangle = \left(\frac{T_1}{T_2}\right)^4 - \frac{\sigma^2}{2!} + \frac{\sigma^4}{4!} + \cdots, \tag{4.5}$$

which implies that

$$\langle \left(\frac{du_o}{d\tau}\right)^2 \rangle = \langle \left(\frac{d^2 u_o}{d\tau^2}\right)^2 \rangle = 1.$$

This normalization (although not always with quite the same notation) is that used by Scott.

The first point is now that the solution of Rudenko and Soluyan applies, when shocks have formed, *not to the shocked waveform* $P(x,\tau)$ developing from the initial $u_o(\tau)$, *but instead to the shock-free waveform* $\hat{P}(x,\tau)$ developing from an appropriate $\hat{u}_o(\tau)$—where by suitable choice of $\hat{u}_o(\tau)$ (a different choice for each x considered) $\hat{P}(x,\tau)$ is shock-free for all ranges x_1 less than the range x considered, and $\hat{P}(x,\tau) = P(x,\tau)$. Thus $\hat{P}(x_1,\tau)$ is shock-free for $0 \leq x_1 < x$, and develops shocks at $x_1 = x$ of precisely the strength and location of those in the real signal $P(x,\tau)$, while the shock-free portions of $\hat{P}(x,\tau)$ and $P(x,\tau)$ coincide also, so that $\hat{P}(x,\tau)$ is precisely the required signal. To achieve this, consider the values ϕ, $-\infty < \phi < \infty$, taken by the characteristic variable, and for any fixed range x, examine whether or not the characteristic emanating from $\tau = \phi$ at $x_1 = 0$ has been absorbed by a shock at ranges $x_1 \leq x$. For ϕ such that the characteristic *has not* been absorbed define $\hat{u}_o(\phi) = u_o(\phi)$. For any interval $\phi_1 \leq \phi \leq \phi_2$ whose characteristics *have* run into a shock at or before range x (ϕ_1 labelling the bottom of a shock at range x, ϕ_2 the top of the same shock), define $\hat{u}_o(\phi)$ as a *linear function* over the interval. For each such linear portion we have $d\hat{u}_o/d\phi = x^{-1}$, because top and bottom of the same shock have the same τ, and $\tau = \phi_1 - xu_o(\phi_1) = \phi_2 - xu_o(\phi_2)$.

Given $\hat{u}_o(\phi)$ defined in this way (see Figure 3), let $\hat{P}(x,\tau)$ be the waveform evolving from $\hat{u}_o(\tau)$ at $x = 0$. Then from the characteristic solution

$$\hat{P}(x,\tau) = \hat{u}_o(\phi), \qquad \tau = \phi - x\hat{u}_o(\phi) \tag{4.6}$$

we see that the linear segments of $\hat{u}_o(\phi)$ corresponding to the intervals $\{Z; \phi_1 \leq \phi \leq \phi_2\}$ transform at range x into shocks between the values $u_o(\phi_2)$ and $u_o(\phi_1)$, while the remaining portions of $\hat{u}_o(\phi)$ transform at range x into the lossless portions $P(x,\tau)$ between the shocks. Further, the linear portions have only finite gradient at ranges less than x, while by construction the remainder of $\hat{u}_o(\phi)$ transforms into a shock-free waveform at such ranges. Therefore $\hat{P}(x,\tau)$ is precisely the required single-valued, but shocked, waveform at range x. Equation (3.2) now gives the spectrum of P at range x in terms of the one and two point characteristic functions, C_1 and C_2, of the distribution of $\hat{u}_o$, rather than of the actual initial distribution q. Inasmuch as $\hat{u}_o(\phi)$ is in principle completely specified for

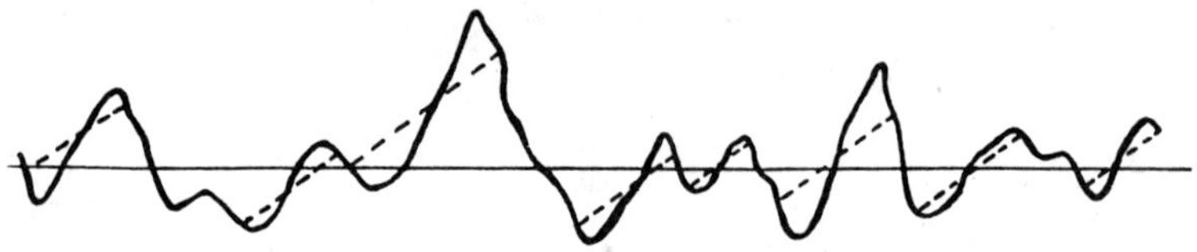

Figure 3a. A typical waveform $u_o(\phi)$. The dotted lines all have gradient x^{-1} and indicate the portions of the waveform which have transformed into shocks at or before x. The dotted lines cut off lobes of equal area (see Whitham, 1974).

Figure 3b. The waveform $\hat{u}_o(\phi)$ defined by the dotted lines of Figure 3a together with the continuous arcs of $u_o(\phi)$ which join them.

each range x, this formally determines the spectral evolution with precise allowance for the presence of shocks (within the framework of the lossless model problem defined by (2.3)).

The next step is to show that for ranges $x \ll 1$ the intervals Z of ϕ can be simply characterized, and in each of these intervals the behaviour of $u_o(\phi)$ can be approximated in a simple way. Now expression (3.2) for the spectrum (with $\beta = 1$ and with $\hat{u}_o(\phi)$ replacing $q(\phi)$) is

$$S_{PP}(x,\omega) = \frac{1}{2\pi x^2\omega^2}\int e^{-i\omega\lambda} \left[\langle e^{-i\omega x\hat{u}_o(\phi_1)+i\omega x\hat{u}_o(\phi_2)}\rangle - |\langle e^{-i\omega x\hat{u}_o(\phi)}\rangle|^2\right] d\lambda$$

with $\lambda = \phi_2 - \phi_1$. Write this as

$$\begin{aligned} S_{PP}(x,\omega) &= \frac{1}{2\pi x^2\omega^2}\int e^{-i\omega\lambda} f(-\lambda)\, d\lambda \\ &= \frac{1}{2\pi x^2\omega^2}\lim_{A\to\infty}\frac{1}{2A}\int_{-A}^{+A}\int_{-A}^{+A} f(\phi_1-\phi_2)\, e^{i\omega(\phi_1-\phi_2)} d\phi_1\, d\phi_2 \\ &= \frac{1}{2\pi x^2\omega^2}\lim_{A\to\infty}\frac{1}{2A}\{\langle |I|^2\rangle - |\langle I\rangle|^2\} \end{aligned} \tag{4.7}$$

with

$$I = \int_{-A}^{+A} \exp\left[i\omega\left(\phi - x\hat{u}_o(\phi)\right)\right] d\phi \tag{4.8}$$

$$= \int_{-A}^{+A} \exp\left[i\omega\left(\phi - xu_o(\phi)\right)\right] d\phi$$

$$+ \int_Z e^{i\omega\phi}\left\{e^{-i\omega x\hat{u}_o(\phi)} - e^{-i\omega xu_o(\phi)}\right\} d\phi \tag{4.9}$$

where the last integral runs only over the intervals Z whose characteristics have run into shocks before range x. Here the first term alone would give rise (on squaring) to the Rudenko and Soluyan spectrum, while the integral over Z gives the correction due to shocks. In each interval of Z, $\hat{u}_o(\phi)$ is simply a linear function, and the problem is then, as stated earlier, to characterize the limits of these intervals and the approximate behaviour of $u_o(\phi)$ in them.

To this end introduce the Gaussian process

$$\alpha(\phi) = x^{-2}\left(x\frac{du_o}{d\phi} - 1\right) \tag{4.10}$$

whose probability density function is

$$F(\alpha) = \frac{x}{(2\pi)^{1/2}} e^{-\alpha} \exp\left[-\frac{1}{2}(\alpha^2 x^2 + x^{-2})\right].$$

The characteristic through ϕ at $\tau = 0$ will have run into a shock before range x if $\alpha(\phi) > 0$, and if x is small the form of $F(\alpha)$ indicates that shocks form only rarely, and that whenever a ϕ is found for which $\alpha(\phi) > 0$, that value of $\alpha(\phi)$ is only at most $0(1)$. Thus in the intervals which form Z, the typical estimate of $du_o/d\phi$ is

$$\frac{du_o}{d\phi} = x^{-1} + 0(x).$$

The first derivative of u_o is thus large when x is small, and since higher order odd derivatives of u_o are correlated with the first, this means that a naive Taylor expansion of $u_o(\phi)$ in Z cannot be made. Consider, however, the process

$$W(\phi) = u_o(\phi) + R_o'(\phi - \phi_o)\frac{du_o}{d\phi}\bigg|_{\phi=\phi_o},$$

a Gaussian process of zero mean for any ϕ_o (R_o being the autocovariance of the process $u_o(\phi)$). Since

$$\langle W(\phi) \frac{du_o}{d\phi}\bigg|_{\phi=\phi_o} \rangle = 0,$$

$W(\phi)$ is uncorrelated with the value of $\alpha(\phi_o)$ and *may* therefore be expanded in a Taylor series about ϕ_o. If we choose ϕ_o such that

$$\frac{d^2u_o}{d\phi^2} = 0 \qquad \text{at } \phi = \phi_o$$

and use now α for $\alpha(\phi_o)$, we find that

$$W(\phi) = u_o(\phi_o) + o(\phi - \phi_o)^2$$

and hence that

$$u_o(\phi) = u_o(\phi_o) + (1 + \alpha x^2)\psi - \frac{1}{6}x^2\psi^3 + o(x^2) \qquad (4.11)$$

where we take $\phi - \phi_o$ to be $0(x)$ and write $\phi - \phi_o = x\psi$. (The naive expansion would lack the term $-\frac{1}{6}x^2\psi^3$.) With the aid of the expansion (4.11) the approximate form of the characteristic relation $\tau = \phi - xu_o(\phi)$ is

$$\frac{\tau - \tau_o}{x^3} = \frac{1}{6}\psi^3 - \alpha\psi \qquad (4.12)$$

where $\tau_o = \phi_o - xu_o(\phi_o)$, which shows that $\psi(\tau)$ is a triple valued function of τ when τ lies in the range

$$-\frac{1}{3}(2\alpha)^{3/2} < \frac{\tau - \tau_o}{x^3} < \frac{1}{3}(2\alpha)^{3/2}.$$

In order that the equal areas rule for shock fitting be satisfied, the shock must be located at $\tau = \tau_o$, and then the interval (a member of Z) of values of ϕ whose characteristics have run into a shock before range x is given by

$$-(6\alpha)^{1/2}x \leq \phi - \phi_o \leq +(6\alpha)^{1/2}x. \qquad (4.13)$$

The description of the shock wave contribution to the spectrum now follows from the ensuing identification of the shocks. Find all intervals Z in which $\alpha(\phi) > 0$ and in each find ϕ_o such that $d^2u_o/d\phi^2 = 0$ there. Then for x small the region around each ϕ_o defined as Z is given by the inequalities (4.13), and in that region the initial function $u_o(\phi)$ may be represented as in (4.11), while $\hat{u}_o(\phi)$ is a linear function over the same range. Given this specification of the functions occurring in I (equation (4.9)) and the probability density function $F(\alpha)$, the rest of Scott's calculation of S_{PP} is essentially a matter of manipulation.

With this description, the contribution to I from interval Z can be written as

$$\begin{aligned} &x \exp(i\omega\tau_o) \int_{-(6\alpha)^{1/2}}^{+(6\alpha)^{1/2}} \left\{1 - \exp[i\Omega(\frac{1}{6}\psi^3 - \alpha\psi)]\right\} d\psi \\ &\quad = x\,(6/\Omega)^{1/3}\, G\,(6^{1/3}\Omega^{2/3}\alpha)\, \exp(i\omega\tau_o) \end{aligned} \tag{4.14}$$

in terms of a dimensionless frequency $\Omega = \omega x^3$ and the function

$$G\,(y) = \int_{-y^{1/2}}^{+y^{1/2}} \{1 - \exp[i(\psi^3 - y\psi)]\}\, d\psi; \tag{4.15}$$

such contributions are to be summed over all intervals Z. When the resulting sum is added to the first term in I (in (4.9)) and the total is squared as in (4.7) to get the spectrum we have

$$S_{PP} = S_1 + S_2 + S_3 \tag{4.16}$$

with S_1 the Rudenko and Soluyan expression, S_3 coming from the square of the shock contribution alone and S_2 coming from the cross terms. The upshot of a rather involved calculation is that

$$S_3\,(x, \omega) = \frac{3\exp(-1/2x^2)}{2\pi^2\omega^2} \int_o^\infty (6\Omega^2)^{-\frac{1}{3}} |G\,(6^{\frac{1}{3}}\Omega^{\frac{2}{3}}\alpha)|^2\, e^{-\alpha}\, d\alpha, \tag{4.17}$$

while $S_2(x, \omega)$ is given by a more complicated expression which, for $\Omega = \omega x^3 = 0(1)$, reduces to

$$\begin{aligned} S_2\,(x, \omega) = &\frac{3\exp(-1/2x^2)}{2\pi^2\omega^2} \times \int_{-\infty}^{+\infty} d\psi \int_o^\infty d\alpha \left(\frac{6}{\Omega}\right)^{1/3} \\ &\times G\,(6^{1/3}\Omega^{2/3}\alpha) \exp\left[i\Omega\left(\frac{1}{6}\psi^3 - \alpha\psi\right) - \alpha\right], \end{aligned} \tag{4.18}$$

this result actually holding for all $\omega \gg 1$. For $\omega = 0(1)$, on the other hand, it is found that

$$S_2(x,\omega) = \left(\frac{3}{\pi}\right)^{3/2} \frac{x^5}{2^{1/2}} \exp(-1/2x^2) \int_{-\infty}^{+\infty} e^{i\omega\phi}\left\{e^{i\omega R_o'(\phi)} - 1\right\} d\phi \tag{4.19}$$

and the limit of this as $\omega \to \infty$ can be shown to equal to the limit of (4.18) as $\Omega \to 0$, so that (4.18) and (4.19) cover all frequencies between them.

For $\omega = 0(1)$ the dominant term in the sum (4.16) is the term S_1 of Rudenko and Soluyan, while for $\omega \gg 1$ the expressions for S_2 and S_3 can be combined in the form

$$S_2 + S_3 \sim \omega^{-2} \exp(-1/2x^2)[\Phi(\Omega) + \Psi(\Omega)] \tag{4.20}$$

with

$$\begin{aligned}\Phi(\Omega) = & \frac{1}{2\pi^2} \int_{-\infty}^{+\infty} d\psi \int_o^{\infty} d\alpha \left(\frac{6}{\Omega}\right)^{1/3} \\ & \times G\,(6^{1/3}\Omega^{2/3}\alpha) \exp\left[i\Omega\left(\frac{1}{6}\psi^3 - \alpha\psi\right) - \alpha\right], \end{aligned} \tag{4.21}$$

$$\Psi(\Omega) = \frac{3}{2\pi^2} \int_0^{\infty} (6\Omega^2)^{-\frac{1}{3}} |G\,(6^{1/3}\Omega^{2/3}\alpha)|^2\, e^{-\alpha}\, d\alpha. \tag{4.22}$$

These functions have been computed by Scott; they can be estimated for large values of Ω, with the result that $S_2/S_3 \to 0$ and

$$S_3 \sim \frac{6}{\pi^2\omega^2} \exp\left(-\frac{1}{2x^2}\right) \tag{4.23}$$

whose ω^{-2} decay must dominate the ω^{-3} decay of the Rudenko and Soluyan spectrum S_1 at sufficiently large ω. Scott states that the ω^{-3} decay holds for all $\omega \gg x^{-2}$ if the covariance $R_o(\phi)$ is analytic at $\phi = 0$, though our estimate in (3.8) indicates that $\omega \gg x^{-1}$ is sufficient. In any event, when $\omega = 0(x^{-3})$ we then have $S_1 + S_2 + S_3$ in an asymptotic universal form, namely

$$S\,(x,\omega) \sim \omega^{-2} \exp\left(-\frac{1}{2x^2}\right)\left[\frac{1}{(2\pi)^{1/2}\Omega} + \Phi(\Omega) + \Psi(\Omega)\right] \tag{4.24}$$

which can be, and indeed was calculated numerically once and for all by Scott (1982, Figure 3). For $\Omega \gg 1$ we have $\Phi(\Omega) \to 6/\pi^2$ while the other terms vanish, and we recover the universal ω^{-2} $\exp(-1/2x^2)$ decay law (4.23).

These results are all, of course, for the spectrum of the dimensionless signal originally written as $P_*(x_*, \tau_*)$ with respect to dimensionless frequency. They may be easily converted to the dimensional variables used elsewhere in this paper; in particular, the result (4.23) may prove to be particularly important and, in the notation of Sections 2 and 3 can be written

$$S_{PP}(x,\omega) \sim q_o^2 \frac{T_2^2}{T_1^3} \frac{6}{\pi^2\omega^2} \exp\left(-\frac{T_1^2}{2\beta^2 x^2 q_o^2}\right). \tag{4.25}$$

Scott emphasizes that these results, relying on the existence of at least the fourth derivative of the covariance, impose severe smoothness restrictions on the *source* q; he says in particular that the theory cannot be applied to the jet noise problem because, in the frequency range $\Omega \gg 1$ (to which (4.23) applies), jet noise spectra do not appear to decay faster than ω^{-5}, and in fact appear to decay only as ω^{-2}. Those remarks apply, however, only to high speed jet noise in the distant field, where the ω^{-2} may represent a *confirmation* of (4.23) rather than a contradiction of the requirements underlying (4.23). We have found no data for power spectral densities at short ranges which allow one to decide whether or not these requirements are met, and there are absolutely no data from which estimates of the scales T_1 and T_2 could be obtained. Consequently, while Scott has provided a theory which completely and rationally allows for the effects of shock waves (except at the very highest frequencies, comparable with the inverse shock passage time, where diffusive effects must be taken into detailed account) and which gives remarkably simple, explicit and universal expressions for the spectral decay with x and with ω for *small* ranges, we have no data from which any test of that theory can yet be made.

The spectral evolution problem for the lossless Burgers equation (2.3) has also been solved by Burgers (1974) and by Kida (1979) in the large range limit. For sufficiently large ranges (but not so large that attenuation has become significant over the whole waveform), the waveform $P(x, \tau)$ in each realization of an ensemble has a simple form; it consists of sawtooth ramps, each with the same slope (a multiple of $(-x)^{-1}$), separated by thin shocks which may be treated as discontinuities propagating in accordance with the rules of *weak shock theory* (see Whitham (1974) for a description of this and its relation to the full Burgers equation with thermoviscous attenuation included). Burgers and Kida rephrase this description of the waveform in terms of a geometrical construction involving the tangency of a parabolic arc with the curve defined by the integral of the initial signal distribution $q(\tau)$, and obtain the probability density functions for the location and advance velocity of one and two shocks. From the two-shock density the pressure covariance $R_{PP}(x,\sigma)$ and spectrum $S_{PP}(x,\omega)$ are calculated as $x \to \infty$ with results which depend (as do the entire dynamics) critically on whether the integral

$$J(x) = \int_o^\infty < P(x,\tau)\, P(x,\tau+\sigma) > d\sigma$$

vanishes or not; it follows from Burgers equation that $dJ(x)/dx = 0$ and hence that the vanishing or otherwise of J remains significant for all x. For high frequencies it is found in either case that $S_{PP}(x,\omega) \sim \omega^{-2}$ (this corresponding simply to the sawtooth waveform), with a coefficient whose value and x-dependence depend on J. When $J = 0$, $S_{PP}(x,\omega) \sim \omega^{-2}x^{-\frac{3}{2}}$, with a coefficient that can be given explicitly in terms of the parameters defining the initial distribution $q(\tau)$. If $J = 0$, the low-frequency spectral asymptotes are $S_{PP}(x,\omega) \sim \omega^2$, while when $J \neq 0$, $S_{PP}(x,\omega) \sim J + A\omega^2$. The spectral problem in this large range limit is essentially fully solved by Burgers (1974), who dealt with $J \neq 0$, and by Kida (1979), who deal analytically with $J = 0$ and numerically with both cases, calculating numerically various statistical measures which were found in all cases to agree remarkably well with the asymptotic estimates. We give no further details of these studies here, because we have no evidence that jet noise waveforms propagating in the atmosphere over large ranges ever develop fully into the sawtooth ramps with repeated shock wave mergings whose statistics are described by the work of Burgers and Kida. It is possible, however, that this description may be needed for sound propagation at high intensities in the jet engine tailpipe, where the overall sound pressure levels may be very high and where the alleviating effects of spherical spreading are absent.

5. Generation of Skewness by Coupling of Nonlinear Distortion and Linear Dispersion

Statistical asymmetry in a random pressure field p can be at least partly characterized by the value of the skewness factor

$$Q_3 = \langle p^3 \rangle / \langle p^2 \rangle^{3/2}. \tag{5.1}$$

This factor vanishes whenever the one-point probability density function is an even function (so that positive and negative values of p with the same magnitude occur with equal frequency), and it certainly vanishes for p with a Gaussian distribution. If, then, the initial pressure $q(t)$ is Gaussian and the propagation over range x, say, is governed by *linear* mechanisms, it follows that $p(x,t)$ remains Gaussian and, in particular, $Q_3(x)$ remains zero. It is not easy to see how this situation would change if the propagation were nonlinear, for the nonlinear effects do not change the values of $p(x,t)$ which actually occur in each realization of an ensemble; the waveform is distorted, but contains just the initial values $q(t)$ and nothing more. We shall show, however, that if the nonlinear distortion is coupled with linear frequency dispersion, the evolved signal $p(x,t)$ **can** indeed contain values other than those present in $q(t)$, and so, that $Q_3(x)$ can be non-zero even when $Q_3(0)$ is zero. Moreover, for the most common mechanism producing dispersion in atmospheric air, it turns out that the skewness produced is always positive, so that the coupling of nonlinear distortion and linear dispersion tends on average to increase the magnitude of positive pressure peaks and to decrease the magnitudes of negative pressure excursions. Before demonstrating this, however, we discuss the only two reports (which to our knowledge have been written) which describe the occurrence of non-zero skewness in jet noise signals.

Non-zero skewness was observed by Ffowcs Williams, Simson, and Virchis (1975) to be present in jet noise signals which exhibited a quality which they called "crackle". This quality was readily identifiable subjectively, but was not discernible in the spectra; no way was found to distinguish a "crackling" from a "non-crackling" jet by examination of the noise spectra alone. The crackling jet did, however, have a distinctive pressure waveform, consisting of large, steeply rising, narrow pressure peaks joined by negative pressure regions of smaller amplitude, slower rates of change and greater width. No such typical asymmetry between positive and negative pressure parts of the waveform was observed in jets which were not judged to exhibit crackle. Corresponding to these features, non-crackling jets were found to have small skewness factors Q_3; if $Q_3 < 0.3$ the average listener would not detect the quality of crackle, whereas if $Q_3 > 0.7$ the crackle would be noticeable to the listener. The asymmetric waveforms, the crackle quality, and the positive skewness factors were found both in high speed full-scale tests with a Concorde Olympus 593 engine and in laboratory model jets run at similar exhaust conditions.

Ffowcs Williams, Simson, and Virchis suggested that *nonlinear effects in the near field* were responsible for the crackle; the rapid increases to large positive pressure values were postulated to arise from the violent and abrupt formation of large eddies through a snaking type of deflexion of the whole jet column. However, they were not able to quantify this idea in any way, or even to provide a specific explanation of the mechanism involved (most mechanisms would, of course, leave some trace of their existence in the spectra).

We are more inclined to think that the skewness arises, at least in part, *in the course of propagation* rather than in the near field, provided the near field levels are high enough to make nonlinear distortion significant over the ranges concerned. In this we receive some support from the interesting results obtained by Szewczyk (1978). He was not primarily interested in the nonlinear propagation issue or in the crackle phenomenon, but rather, in the differences between the noise fields (within the downstream *cone of silence*) generated by subsonic and supersonic jets. In examining these differences he noted that the waveforms from subsonic jets were *regular and not skewed*, while those from supersonic jets were *extremely spiky and obviously skewed*. Corresponding to the latter, skewness factors between 0.782 and 0.804 were computed, the small differences here arising from different data sampling rates and data record lengths. Non-zero skewness was not found in data from jets with exhaust speeds less than 1400 f.p.s. These features are consistent with those noticed by Ffowcs Williams, Simson, and Virchis.

Szewczyk's new and important discovery was that Q_3 *increased with range*, at least initially. At 5 nozzle diameters away, the waveform was regular and not skewed, but by 60 diameters had attained its spiky skewed shape with an increase of skewness factor from 0 to about 0.8. Examination of data taken from a full-scale Rolls-Royce Viper turbojet also revealed an increase of skewness with range, though here (Szewczyk, private communication) the increase in Q_3 over a given distance varied widely from one day to another (under nominally identical exhaust conditions). Szewczyk (1978) further conducted experiments with a Hartmann generator as the noise source, confirming the increase of skewness with range but showing also that the rate of increase of skewness is dependent upon the source intensity.

There is a clear indication in all this that skewness develops with range because of coupling between nonlinear distortion of the waveform and its linear distortion and decay under dissipative and dispersive mechanisms. The coupling is necessary because nonlinear effects alone cannot produce skewness (or cannot change a given initial skewness factor), as is shown in (2.10), while purely linear mechanisms would leave a Gaussian distribution Gaussian and would tend to *produce* a Gaussian distribution from one initially non-Gaussian. Further, the dominant (relaxation) mechanisms producing dispersion and dissipation in sound waves in atmospheric air at frequencies of interest in full-scale jet noise are known to vary very widely with changes in relative humidity (cf. Monk (1969), where changes involving factors of as much as 1000 are reported for the relaxation times in atmospheric air as the relative humidity changes from 20% to 70%). The importance of these mechanisms would be consistent with the wide variation in skewness factor noted by Szewczyk in full-scale experiments conducted in the open air.

To calculate the initial development of skewness we use the expansion of Section 2 for the quantity $T(x,\omega)$ of (2.15); this quantity is simply the pressure at range r and frequency ω compensated for the effects of spherical divergence and atmospheric attenuation and dispersion. To incorporate the latter we write

$$\alpha(\omega) = \hat{\alpha}(\omega) + i\hat{\delta}(\omega) \tag{5.2}$$

where $\hat{\alpha}(\omega)$ is even in ω and represents attenuation while $\hat{\delta}(\omega)$ is odd in ω and represents dispersion; these quantities are not independent, but are essentially Hilbert transforms of each other, in accordance with the Kramers-Krönig relations (cf. Morfey and Howell (1980b), where $\hat{\delta}(\omega)$ is deduced in this way from measurements of $\hat{\alpha}(\omega)$). Now the expansion of $T(x,\omega_1)$ can be written symbolically as

$$T_{\omega_1} = q_1 + xC_{q_1q_1} + \frac{x^2}{2}(D_{q_1q_1q_1} + \Gamma_{q_1q_1}) + 0(x^3) \tag{5.3}$$

where C and D represent quadratic and cubic convolution terms independent of α, while the term $\Gamma_{q_1q_1}$ is

$$-\frac{1}{2}\beta i\omega_1\int_{-\infty}^{+\infty}\gamma(\omega_1,\omega_1')\,\tilde{q}(\omega_1')\,\tilde{q}(\omega_1-\omega_1')\,d\omega_1',$$

where, as in Section 2,

$$\gamma(\omega,\Omega)=\alpha(\Omega)+\alpha(\omega-\Omega)-\alpha(\Omega).$$

From (5.3) we form

$$\langle T_{\omega_1}T_{\omega_2}T_{\omega_3}\rangle = \left(\frac{r}{r_o}\right)^3 \exp\{[\alpha_{\omega_1}+\alpha_{\omega_2}+\alpha_{\omega_3}](r-r_o)\}\langle\tilde{P}_{\omega_1}\tilde{P}_{\omega_2}\tilde{P}_{\omega_3}\rangle$$

and from

$$S_{PPP}(\omega_1,\omega_2)=\frac{\langle P_{\omega_1}P_{\omega_2}P_{\omega_3}\rangle}{\delta\,(\omega_1+\omega_2+\omega_3)}$$

we can calculate

$$\langle p^3\rangle=\int_{-\infty}^{+\infty}\int_{-\infty}^{+\infty}S_{PPP}(\omega_1,\omega_2)\,d\omega_1\,d\omega_2.$$

In the calculation of $\langle\tilde{P}_{\omega_1}\tilde{P}_{\omega_2}\tilde{P}_{\omega_3}\rangle$, fourth moments like $\langle\tilde{q}(\omega_1)\tilde{q}(\omega_2)\tilde{q}(\omega_3')\tilde{q}(\omega_3-\omega_3')\rangle$ occur with coefficients linear in γ, as well as terms independent of γ which make zero contribution to $\langle p^3\rangle$. The fourth moments are related to second moments because $\tilde{q}(\omega)$ has joint Gaussian statistics, and after reduction one comes to the relation

$$\langle p^3\rangle=\left(\frac{r_o}{r}\right)^3\left(\frac{3\beta x^2}{4}\right)\mathrm{Re}\int_{-\infty}^{+\infty}\int_{-\infty}^{+\infty}I(\omega_1,\omega_2)\,d\omega_1\,d\omega_2,$$

with

$$I(\omega_1,\omega_2)=i(\omega_1+\omega_2)\,S_{qq}(\omega_1)\,S_{qq}(\omega_2)\times[\overline{\alpha}(\omega_1)+\overline{\alpha}(\omega_2)-\overline{\alpha}(\omega_1+\omega_2)] \times e^{-[\alpha(\omega_1)+\alpha(\omega_2)+\alpha(-\omega_1-\omega_2)](r-r_o)}$$

where $\overline{\alpha} = \hat{\alpha} - i\hat{\delta}$ is the complex conjugate of α. This gives

$$\frac{\partial^2 \langle p^3 \rangle}{\partial x^2} = \left(\frac{r_o}{r}\right)^3 \left(\frac{3\beta}{2}\right) \int_{-\infty}^{+\infty} \int_{-\infty}^{+\infty} S_{qq}(\omega_1)\, S_{qq}(\omega_2) \times (\omega_1 + \omega_2) \Big[\hat{\delta}(\omega_1) + \hat{\delta}(\omega_2) - \hat{\delta}(\omega_1 + \omega_2) \Big] \, d\omega_1 \, d\omega_2, \tag{5.4}$$

a result which was given by Bashforth (1983). It shows that the magnitude of $\langle p^3 \rangle$ initially grows because of dispersion effects coupled with nonlinear effects, and that dissipation itself plays no role at first.

The mean square pressure, compensated for spherical spreading, can only *decrease* with x, so that a minimum estimate of the development of skewness is provided by

$$\left. \frac{d^2 Q_3}{dx^2} \right|_{x=0} = \frac{3\beta}{2} \int_{-\infty}^{+\infty} \int_{-\infty}^{+\infty} S_{qq}(\omega_1)\, S_{qq}(\omega_2) (\omega_1 + \omega_2) \times \Big[\hat{\delta}(\omega_1) + \hat{\delta}(\omega_2) - \hat{\delta}(\omega_1 + \omega_2) \Big] \, d\omega_1 \, d\omega_2. \tag{5.5}$$

This clearly shows that skewness is amplitude—dependent, and, inasmuch as $\hat{\delta}(\omega)$ varies over a large range with relative humidity, is highly sensitive to atmospheric conditions. There seems in principle no reason why negative rather than positive skewness should not be produced by the coupling of nonlinearity and dispersion. Suppose, however, that the dispersion is produced by the relaxation of a single degree of freedom, as is claimed to be the case for atmospheric air under typical conditions, the relaxing mode then being the vibrational mode of N_2 (Hodgson, 1973). The dispersion relation for a wave exp $(ikx - i\omega t)$ then follows from the model equation

$$\left(1 + T\frac{\partial}{\partial t}\right)\left(\frac{\partial u}{\partial t} + a_\infty \frac{\partial u}{\partial x}\right) = (a_\infty - a_o)\frac{\partial u}{\partial x} \tag{5.6}$$

(cf. Lighthill, 1956; Rudenko and Soluyan, 1977) as

$$k = \frac{\omega\,(1 - i\omega T)}{a_o - a_\infty i\omega T},$$

so that

$$\hat{\alpha}(\omega) = \frac{(a_\infty - a_o)\,\omega^2 T}{(a_o^2 + a_\infty^2 \omega^2 T^2)}, \tag{5.7}$$

$$\hat{\delta}(\omega) = \frac{\omega\,(a_o + a_\infty \omega^2 T^2)}{(a_o^2 + a_\infty^2 \omega^2 T^2)}. \tag{5.8}$$

Here T is the relaxation time, a_o and a_∞ the equilibrium and frozen sound speeds, respectively. If (5.8) is used in (5.5) one finds, after some manipulation, that

$$\begin{aligned} \left.\frac{d^2 Q_3}{dx^2}\right|_{x=0} &= 3\beta a_o a_\infty (a_\infty - a_o)\,T^2 \\ &\times \int_0^\infty \int_0^\infty d\omega_1\, d\omega_2\, S_{qq}(\omega_1)\, S_{qq}(\omega_2) \\ &\times (a_o^2 + a_\infty^2 \omega_1^2 T^2)^{-1} \\ &\times (a_o^2 + a_\infty^2 \omega_2^2 T^2)^{-1} \big(a_o^2 + a_\infty^2 (\omega_1 + \omega_2)^2 T^2\big)^{-1} \\ &\times \big\{12 a_o^4 \omega_1 \omega_2 + 6 a_o^2 a_\infty^2 \omega_1 \omega_2 (\omega_1^2 + \omega_2^2)\,T^2 \\ &+ 2 a_\infty^4 \omega_1 \omega_2 \big[(\omega_1^2 - \omega_2^2)^2 + \omega_1^2 \omega_2^2\big]\,T^4\big\}, \end{aligned} \tag{5.9}$$

which is always positive (since $a_\infty > a_o$).

A simple form of this result applies if $\Omega T \ll (a_o/a_\infty)$ for all Ω for which $S_{qq}(\Omega)$ is significantly different from zero. It is

$$\begin{aligned} \left.\frac{d^2 Q_3}{dx^2}\right|_{x=0} &= \frac{9(\gamma+1)}{2}\,\frac{a_\infty (a_\infty - a_o)\,T^2}{\rho_o a_o^6} \\ &\times \left(\int_\infty^\infty \omega^2 S_{qq}(\omega)\, d\omega\right)^2 \end{aligned} \tag{5.10}$$

$$= \frac{9(\gamma+1)}{2}\,\frac{a_\infty (a_\infty - a_o)\,T^2}{\rho_o a_o^6} \langle \dot{q}^2 \rangle^2 \tag{5.11}$$

$$= \frac{9(\gamma+1)}{2}\,\frac{a_\infty (a_\infty - a_o)\,T^2}{\rho_o a_o^6}\,\frac{q_o^4}{T_1^4} \tag{5.12}$$

in terms of the Taylor microscale T_1 introduced in Section 3.

There are no published data with which one could compare the initial growth of Q_3 as predicted by (5.12). However, the mechanism proposed here does have all the properties which have so far been qualitatively observed, and expressions like (5.12) may indicate the parameters that need to be measured in future work.

6. Concluding Remarks

A number of striking predictions have been made in the work reported in this survey, and these may be summarized as follows:

(1) Significant spectral distortion due to nonlinear propagation effects is to be expected in broadband jet noise signals which have levels equivalent to 160 dB or more at $r_o = 1$ m (i.e., to 120 dB or more at 100 m). At such levels the distortion will be significant even around the spectral peak, and more significant at higher frequencies.

(2) When the distortion is slight enough for it to be treated by perturbation methods, it is found that the relative importance of nonlinear terms is a weak function ($\ln^2 r/r_o$) of range and a very strong function of exhaust Mach number, and is independent of atmospheric attenuation.

(3) The magnitude of the spectral distortion is in general sensitive to details of the initial spectral shape, and it is not possible to give a simple general criterion according to which nonlinear distortion will be significant or not.

(4) In an appropriate short range and high frequency approximation the spectrum takes a universal ω^{-2} form corresponding to the presence of weak rare shocks in the waveform (provided an appropriate smoothness assumption is made with regard to the initial signal $q(\tau)$, this, in particular, precluding the presence of shocks in $q(\tau)$).

(5) Skewness may be produced in initially non-skewed signals by the coupling between nonlinear distortion and linear dispersion in the atmosphere. Relaxation mechanisms produce only positive skewness, as observed in jet noise signals.

At this stage there is an abundance of theoretical predictions and almost no experimental data with which to compare them. The necessary experiments do not have to involve full scale flight exercises; laboratory experiments are important because of the much wider range of parameters which can be covered. The parameter ranges in which nonlinear propagation effects would be expected to be important can be found from the scalings given in Section 2 from the perturbation approach. Note that the very weak ($\ln^2 r/r_o$) dependence on range means that large distances of the full-scale kind are not needed in the spherical wave problem, and that the same effects should be detectable on laboratory scales. Similar effects have indeed long been seen in laboratory jet noise; if far field model scale data are corrected by linear theory to smaller ranges, it is common to find high frequency levels which steadily increase and eventually, at sufficiently short ranges, exceed the levels at the spectral *peak*. Those inferred high frequency levels then greatly exceed the levels measured directly at small ranges. It has generally been assumed that the high frequency discrepancies which arise in this way are caused by inadequate knowledge of linear attenuation mechanisms at the higher frequencies (perhaps up to 50 kHz) of importance in model scale tests. Nonlinear distortion could provide an alternative explanation, at model scale or full scale.

With regard to (4), it is common to observe extensive regions of ω^{-2} decay in the spectral densities of high speed jets. Such regions correspond to the presence of shocks in the waveforms. What has received no discussion is the question of whether the ω^{-2} spectra are found at all ranges, and thus presumably associated with the shocks generated in the near field in the manner postulated by Ffowcs Williams, Simson, and Virchis (1975), or whether they evolve with range from spectra with a much more rapid decay. In the latter case the shocks are presumably generated by nonlinear propagation distortion in the manner analyzed by Scott (1982), and then the spectral scalings (with range and other parameters) will be quite different. Experiments involving a noise source of the Hartmann generator kind would seem to be useful here in separating out the shock production mechanisms.

Extensions of the studies of Ffowcs Williams, Simson, and Virchis (1975) and Szewczyk (1978) are clearly also called for, preferably again with a noise source of the Hartmann type and, since the phenomenon seems to be highly sensitive to atmospheric conditions, over a range of carefully controlled humidities. Such experiments could be valuable in confirming the general understanding of nonlinear propagation of broadband noise which has been reviewed here.

We end with a note of caution in relation to the aircraft noise problem. Noise levels certainly have been greatly reduced by the introduction of high bypass ratio turbofan engines, and this trend has been followed, albeit with less spectacular reductions, in the latest series of engines. It is not certain, however, that these reduced levels have made the nonlinear propagation issue irrelevant, and of concern only in connection with very high speed exhaust flows. Spectral distortion seems to be sensitive to spectral shape in a way as yet not understood, and there is no guarantee that it will not be significant in, for example, high thrust developments of current *quiet* engines.

References

[1] Bashforth, S. *Ph. D. thesis,* University of Leeds, Leeds, England, 1983.

[2] Burgers, J. M. "The Nonlinear Diffusion Equation," Amsterdam: D. Reidel, 1974.

[3] Crighton, D. G. "Model Equations of Nonlinear Acoustics," *Ann. Rev. Fluid Mech., 11* (1979), 11–33.

[4] Crighton, D. G. and Bashforth, S. "Nonlinear Propagation of Broadband Jet Noise," *AIAA-80-1039,* 1980.

[5] Ffowcs Williams, J. E., Simson, J., and Virchis, V. J. "'Crackle': An Annoying Component of Jet Noise," *J. Fluid Mech., 71* (1975), 251–71.

[6] Gallard, J. and Gower, C. "Study of the Attenuation of Sound in the Atmosphere Using Existing In-Flight Noise Data," *British Aerospace Report 554,* April 1978.

[7] Hodgson, J. P. "Vibrational Relaxation Effects in Weak Shock Waves in Air and the Structure of Sonic Bangs," *J. Fluid Mech., 58* (1973), 187–96.

[8] Howell, G. P. *Ph. D. thesis,* Univ. of Southampton, Southampton, England, 1982.

[9] Kida, S. "Asymptotic Properties of Burgers Turbulence," *J. Fluid Mech., 93* (1979), 337–77.

[10] Lighthill, M. J. "On Sound Generated Aerodynamically. Part I, General Theory," *Proc. R. Soc. Lond., A211* (1952), 564–87.

[11] ———. "Viscosity Effects in Sound Waves of Finite Amplitude," in *Surveys in Mechanics.* Eds. G. K. Batchelor and R. M. Davies, Cambridge: Cambridge University Press, 1956, 250–351.

[12] Lumley, J. L. *Stochastic Tools in Turbulence,* New York: Academic Press, 1970.

[13] Monk, R. G. "Thermal Relaxation in Humid Air," *J. Acoust. Soc. Am., 46* (1969), 580–86.

[14] Morfey, C. L. and Howell, G. P. "Nonlinear Propagation of Aircraft Noise in the Atmosphere," *AIAA-80-1041,* 1980a.

[15] ———————. "Speed of Sound in Air as a Function of Frequency and Humidity," *J. Acoust. Soc. Am., 68* (1980b), 1525–27.

[16] Pestorius, F. M. and Blackstock, D. T. "Propagation of Finite-Amplitude Noise," in *Finite-Amplitude Effects in Fluids,* Proc. 5th International Symposium on Nonlinear Acoustics, Copenhagen, 1973, ed. L. Bjørnø, Guildford, England: IPC Press, 1974.

[17] Rudenko, O. V. and Soluyan, S. I. *Theoretical Foundations of Nonlinear Acoustics.* Trans. R. T. Beyer, New York: Consultants Bureau (Plenum), 1977.

[18] Scott, J. F. "The Nonlinear Propagation of Acoustic Noise," *Proc. R. Soc. Lond., A383* (1982), 55–70.

[19] Szewczyk, V. M. *Ph.D. thesis,* University of Southampton, Southampton, England, 1978.

[20] Webster, D. A. and Blackstock, D. T. "Experimental Investigation of Outdoor Propagation of Finite-Amplitude Noise," *NASA CR-2992,* 1978.

[21] Whitham, G. B. *Linear and Nonlinear Waves,* New York: Wiley-Interscience, 1974.

Acoustic Propagation in Partially Choked Ducts

Ali H. Nayfeh
Virginia Polytechnic Institute and State University
Blacksburg, VA 24061
and
Yarmouk University
Irbid, Jordan

Abstract

Sound propagation in partially choked hard- or soft-walled ducts is discussed. Techniques for its theoretical analysis are still in the development stage, because linear theory is inadequate to describe acoustic propagation in regions of near-sonic mean flows and the ducts have varying cross sections and carry mean flows with strong axial and transverse gradients. The linear three-dimensional problem has been successfully treated using the wave envelope technique, whereas the quasi-one-dimensional continuous nonlinear problem has been successfully treated by using spectral methods or the method of matched asymptotic expansions. Efforts are underway to combine the wave envelope technique with either spectral methods or the method of matched asymptotic expansions to treat the nonlinear three-dimensional problem. The wave envelope technique is based on solving for the envelopes of the quasi-parallel acoustic modes that exist in the duct instead of solving for the actual wave. It is applicable to hard-walled as well as lined ducts, strong as well as weak transverse gradients, and strong as well as weak axial gradients. The numerical results show that converging ducts produce substantial refractions toward the duct's center for axisymmetric waves propagating against near choked flows. Spinning modes get refracted toward an annulus away from the duct's center. The degree of refraction and intensification at the throat of a partially choked duct decreases with increasing spinning mode number and/or real part of wall admittance. The acoustic signals become arbitrarily large as the throat Mach number approaches sonic conditions because the linear acoustic equations are singular there. The nonlinearity generates higher harmonics and an acoustic streaming, resulting in a decrease in the energy carried by the fundamental. The strength of the nonlinearity increases with increasing source strength, frequency, or mean flow Mach number. Beyond a given nonlinear strength, acoustic shocks form.

1. Introduction

The trend toward the use of high bypass turbojet engines has decreased jet noise and resulted in noise emissions from the inlet nacelles being responsible for an increasing proportion of community annoyance. Consequently, effective control measures are required to reduce the inlet noise to acceptable levels. One promising approach to the reduction of inlet noise is the use of a high subsonic Mach number inlet, or partially choked inlet, in conjunction with an acoustic duct linear (Greatrex, 1961; Sobel and Welliver, 1961; Lumsdaine, 1972; Klujber et al., 1973; Nayfeh, Kaiser, and Shaker, 1979).

The concept of sound reduction by choked or partially choked inlets has been investigated experimentally at great length. Several experimental investigators tested actual jet engines with various shapes of centerbodies as well as experimental ducts to choke the flow (Sobel and Welliver, 1961; Greatrex, 1961; Sawhill, 1966; Anderson, 1966; Cawthorne et al.,1967; Chestnutt and Stewart,1967; Chestnutt, 1968; Chestnutt and Clark, 1971; Anderson et al., 1972; Hawking and Lawson, 1972; Benzakein et al., 1973). The amount of noise reduction depends on how the choking is achieved; in the above reported experiments, the amount of noise reduction is as high as 33 dBs and as low as 3 dBs. Moreover, these experiments show that the choking may be accompanied by a loss in the compressor efficiency. Thus, the optimum choking configuration is the one accompanied by no loss or a minor loss in the compressor efficiency. Hence, many experimental investigators (Higgins et al., 1968; Miller and Abbott, 1972; Klujber, 1973; Groth, 1974; Savkar and Kazin, 1974; Abbott, 1974; Koch et al., 1974; Miller, 1975) studied the possibility of attaining a significant noise reduction with a minor loss in compressor efficiency. The results of these experiments show that the loss in the compressor efficiency can be minimized by carefully designing the centerbody. Testing a J-85 turbojet engine and using a translating centerbody inlet with a radial vane, Groth (1974) measured a 40 dB reduction in a fully-choked inlet while maintaining a recovery pressure of 92.1%.

In summary, most, but not all of the reported experimental investigations have noted significant reductions of the noise levels when the inlet is choked or partially choked. The geometry of the inlet, the geometry of the centerbody, and the operating conditions seem to have an effect on the acoustic as well as on the aerodynamic performance. Most of the potential noise reduction is achieved by operating in the partially choked state (mean Mach numbers in the throat of 0.8–0.9). Furthermore, although the experimental studies have demonstrated that the choked inlet is a viable noise-reduction technique, they have not provided insight into the physical mechanisms that are responsible for the noise reduction or that explain the differences among the several experimental studies. Moreover, although the problem of sound propagation in uniform ducts (two-dimensional, rectangular, circular, annular), with or without mean flows, for hard as well lined ducts, has been studied extensively (Nayfeh, Kaiser, and Telionis, 1975a; Vaidya and Dean, 1977), techniques for the theoretical analysis of sound propagation through regions of near-sonic mean flow in converging-diverging ducts are still in the development stage. Any successful technique must take into account the variations in the cross sections of the ducts, the effects of mean flows with strong axial and transverse gradients, and the effects of nonlinearities because linear acoustic equations are inadequate to describe acoustic propagation in regions of near-sonic mean flows. So far, the available techniques treat either the general linear problem or the quasi-one-dimensional nonlinear problem. These problems are discussed, respectively, in the following two sections.

2. Linear Problem

Several analytical as well as numerical techniques have been developed for the analysis of linear wave propagation in nonuniform ducts carrying mean flows. Surveys of these techniques were made by Nayfeh, Kaiser, and Telionis (1975a), Nayfeh (1976), Vaidya and Dean (1977), and Nayfeh, Kaiser, and Shaker (1979). A survey is presented here for the problem of waves propagating in variable-area ducts carrying high subsonic mean flows. The approaches that have been developed so far include quasi-one-dimensional approximations, multiple-scale solutions, finite-element methods, and the wave-envelope technique.

The earlier studies of this problem treated the quasi-one-dimensional case. Eisenberg and Kao (1971) analyzed plane waves in a specially-chosen variable-area duct that yields an equation with constant coefficients. Numerical results of their analytical solution exhibit a singularity as the mean Mach number approaches unity. Davis and Johnson (1974) obtained numerical solutions of the acoustic equations for certain acoustic boundary conditions, while King and Karamcheti (1975) obtained similar numerical results using the method of characteristics. Myers and Callegari (1977) analyzed the behavior of the linearized acoustic equations at the throat of the duct to exhibit explicitly the form of the singularity for two special types of area variation. The quasi-one-dimensional approximation neglects or averages the transverse mean-flow gradients, which can be very significant, especially if the flow approaches enough to sonic conditions. Moreover, with this approach, one cannot account for large wall admittances and/or the propagation of modes other than the lowest mode.

Nayfeh and co-workers studied extensively the propagation of various acoustic modes in ducts having slowly-varying cross sections and carrying general mean flows. Nayfeh, Telionis, and Lekoudis (1975) discussed the acoustic propagation in lined plane ducts with varying cross sections and sheared mean flow. This work was extended to annular ducts by Nayfeh, Kaiser, and Telionis (1975b). Nayfeh and Kaiser (1975) extended this approach to analyze wave propagation in a variable-area duct carrying compressible sheared mean flows. Nayfeh et al. (1980) carried out an experimental study of sound propagating in variable-area ducts with and without mean flows. The axial variations were small. The experimental data are in reasonable agreement with the multiple-scales solution. Although the method of multiple scales can determine the transmission and attenuation of all modes including the effects of transverse as well as axial gradients and large wall admittances, it is limited to slow axial variations of the duct cross section. Moreover, the expansion needs to be carried out to higher orders in order to determine reflection and intermodal coupling of the acoustic signal.

Several finite-element models have been developed for the analysis of acoustic propagation in nonuniform ducts with compressible mean flows. Sigman, Majjigi, and Zinn (1978) applied a finite-element approach to the governing equations which were expressed in terms of a velocity-potential; thus their analysis is limited to irrotational mean flows. Majjigi, Sigman, and Zinn (1979) expanded their method so that soft-wall ducts can be analyzed. Since the introduction of these finite-element models, several finite-element approaches were employed which consider rotational mean flows. A general compressible mean flow is considered in the finite-element model of Abrahamson (1977) but the results presented are for an incompressible case. Abrahamson (1979) studied the possibility of reducing the computational effort associated with the finite-element analysis. A comparison of the transmission and reflection coefficients evaluated using a finite-element scheme with those from the method of weighted residuals was performed by Eversman, Astley, and Thanh (1977) in which good agreement was observed. Astley, Walkington, and Eversman (1979) extended their original finite-element procedure. Also, Astley, Walkington, and Eversman (1980) applied a finite-element analysis to ducts with a peripherally varying liner. In these studies, the mean flow model is rotational but it appears that the refractive effect of a finite boundary-layer thickness at the duct walls was ignored. A large amount of computation will be required with these purely numerical techniques at high frequencies because a large number of elements are needed to resolve the smallest wavelength and the highest mode. Thus, the computational time increases rapidly with increasing frequency and duct length. No numerical results have been presented to date for cases of high subsonic mean flows.

Nayfeh and co-workers (Nayfeh, Kaiser, and Shaker, 1979; Nayfeh, Shaker, and Kaiser, 1980) developed the wave-envelope technique for the analysis of sound transmission and attenuation in an infinite, hard-walled or lined circular duct carrying a compressible, sheared mean flow, and having a variable-area cross section. The acoustic disturbance is expressed as a superposition of the quasi-parallel duct modes. An explicit description of the fast axial variation of the acoustic disturbance is given and only the slower variations of the mode amplitudes and phases are calculated. The method is valid for large as well as small axial variations as long as the mean flow does not separate. Using the wave-envelope technique, Nayfeh (1980) and Nayfeh, Kelly, and Watson (1982) analyzed the propagation of symmetric and spinning acoustic modes in partially choked converging circular lined and hard-walled ducts. Kelly, Nayfeh, and Watson (1982) used the wave-envelope technique to analyze the propagation of symmetric and spinning acoustic modes in partially choked converging-diverging ducts. Uenishi and Myers (1983) applied the wave-envelope technique to the problem of wave propagation in partially choked two-dimensional converging-diverging ducts. In this section, the wave-envelope technique is described and applied to circular and two-dimensional ducts. Moreover, some of the available numerical results are discussed.

Velocities, lengths, and time are made dimensionless by using the reference speed of sound c_a (value in mean flow at the center of the duct entrance), the radius R_0 of the circular duct or the half-width d_0 of the two-dimensional duct in the uniform region, and R_0/c_a or d_0/c_a, respectively. The pressure p is made dimensionless by using $\rho_a c_a^2$, the density ρ and temperature T are made dimensionless by using their corresponding reference values, while the viscosity μ and the thermal conductivity κ are made dimensionless by using their wall values in the uniform section. In terms of these dimensionless variables, the equations which describe the unsteady viscous flow in a duct are (see for example, Schlichting, 1968):

Conservation of Mass

$$\frac{\partial \rho}{\partial t} + \nabla \cdot (\rho \vec{v}) = 0 \tag{2.1}$$

Conservation of Momentum

$$\rho\left(\frac{\partial \vec{v}}{\partial t} + \vec{v} \cdot \nabla \vec{v}\right) = -\nabla p + \frac{1}{\text{Re}} \nabla \cdot \underline{\tau} \tag{2.2}$$

Conservation of Energy

$$\begin{aligned} &\rho\left(\frac{\partial T}{\partial t} + \vec{v} \cdot \nabla T\right) - (\gamma - 1)\left(\frac{\partial p}{\partial t} + \vec{v} \cdot \nabla p\right) \\ &\quad = \frac{1}{\text{Re}}\left[\frac{1}{Pr} \nabla \cdot (\kappa \nabla T) + (\gamma - 1)\Phi\right] \end{aligned} \tag{2.3}$$

Equation of State

For a perfect gas,

$$\gamma p = \rho T \tag{2.4}$$

For a Newtonian fluid, the dimensionless viscous-stress tensor $\underline{\tau}$ and the dimensionless dissipation function Φ are related to $\vec{v}$ by

$$\underline{\tau} = \mu[\nabla\vec{v} + (\nabla\vec{v})^T] + \lambda\nabla\cdot\vec{v}$$
$$\Phi = \underline{\tau} : \nabla\vec{v}$$
$$= \sum_{j=1}^{3}\sum_{i=1}^{3} \tau_{ij}\partial v_i/\partial x_j$$

where $(\nabla\vec{v})^T$ denotes the transpose of $\nabla\vec{v}$.

In general, the ducts carry a high subsonic, steady, sheared mean flow that satisfies equations (2.1) through (2.4). The presence of sound in the ducts results in the perturbation of the flow quantities so that

$$q(\vec{r},t) = q_0(\vec{r}) + q_1(\vec{r},t) \tag{2.5}$$

where q stands for any flow quantity, q_0 is the mean-flow part, and q_1 is the acoustic part. Substituting equation (2.5) into equations (2.1) through (2.4) and eliminating the mean-flow quantities, one obtains the following acoustic equations:

$$\frac{\partial\rho_1}{\partial t} + \nabla\cdot(\rho_0\vec{v}_1 + \rho_1\vec{v}_0) = \text{NL} \tag{2.6}$$

$$\begin{aligned}\rho_0\left(\frac{\partial\vec{v}_1}{\partial t} + \vec{v}_0\cdot\nabla\vec{v}_1 + \vec{v}_1\cdot\nabla\vec{v}_0\right) + \rho_1\vec{v}_0\cdot\nabla\vec{v}_0 \\ = -\nabla p_1 + \frac{1}{\text{Re}}\nabla\cdot\underline{\tau}_1 + \text{NL}\end{aligned} \tag{2.7}$$

$$\begin{aligned}\rho_0\left(\frac{\partial T}{\partial t} + \vec{v}_0\cdot\nabla T_1 + v_1\cdot\nabla T_0\right) + \rho_1\vec{v}_0\cdot\nabla T_0 \\ -(\gamma-1)\left(\frac{\partial p_1}{\partial t} + \vec{v}_0\cdot\nabla p_1 + \vec{v}_1\cdot\nabla p_0\right) \\ = \frac{1}{\text{Re}}\left[\frac{1}{Pr}\nabla\cdot(\kappa_0\nabla T_1 + \kappa_1\nabla T_0) + (\gamma-1)\,\Phi_1\right] + \text{NL}\end{aligned} \tag{2.8}$$

$$\frac{p_1}{p_0} = \frac{\rho_1}{\rho_0} + \frac{T_1}{T_0} \tag{2.9}$$

where $\underline{\tau}_1$ and Φ_1 are linear in the acoustic quantities and NL stands for the nonlinear terms in the acoustic quantities.

Solution to equations (2.6) through (2.9) subject to general initial and boundary conditions is not available yet. To determine solutions for the propagation of sound in ducts, researchers have used simplifying assumptions. Here, the nonlinear and viscous terms in the acoustic equations are neglected, and the mean flow is taken to be a function of the axial and transverse coordinates only. Further, we neglect swirling mean flows. The assumption of linearization is not valid for high sound-pressure levels. The effects of the nonlinear acoustic properties of the lining material become significant when the sound-pressure level exceeds about 130 dB (re 0.0002 dyne/cm^2), while the effects of the gas nonlinearity become significant when the sound pressure level exceeds about 160 dB. In particular, the nonlinearity of the gas must be included when the mean flow is transonic (i.e., near the throat) as discussed in the next section. Nayfeh (1973a) showed that the viscous terms in the acoustic equations produce an effective admittance at the wall that leads to small dispersion and attenuation. For lined ducts, this admittance produced by the acoustic boundary layer may be neglected, but it cannot be neglected for hard-walled ducts, as demonstrated analytically and experimentally by Pestorius and Blackstock (1973).

2.1 Cylindrical Ducts

In terms of the cylindrical coordinate system (r, θ, x), each acoustic quantity $q_1(r, x, \theta, t)$ is expressed for sinusoidal time variations as

$$q_1(r, x, \theta, t) = \sum_{m=0}^{\infty} q_{1_m}(r, x) \exp[-i(\omega t + m\theta)] \tag{2.10}$$

where ω is the dimensionless frequency. Then, the acoustic equations (2.6)–(2.9) in cylindrical coordinates become

$$\begin{aligned} &-i\omega\rho_1 + \frac{\partial}{\partial x}(\rho_0 u_1 + u_0 \rho_1) + \frac{i\rho_0 m}{r} w_1 \\ &+ \frac{1}{r}\frac{\partial}{\partial r}(r\rho_0 v_1 + r v_0 \rho_1) = 0 \end{aligned} \tag{2.11}$$

$$\rho_0\left[-i\omega u_1 + \frac{\partial}{\partial x}(u_0 u_1) + v_0\frac{\partial u_1}{\partial r} + v_1\frac{\partial u_0}{\partial r}\right] + \rho_1\left[u_0\frac{\partial u_0}{\partial x} + v_0\frac{\partial u_0}{\partial r}\right] = -\frac{\partial p_1}{\partial x} \tag{2.12}$$

$$\rho_0\left[-i\omega v_1 + \frac{\partial}{\partial r}(v_0 v_1) + u_0\frac{\partial v_1}{\partial x} + u_1\frac{\partial v_0}{\partial x}\right] + \rho_1\left[v_0\frac{\partial v_0}{\partial r} + u_0\frac{\partial v_0}{\partial x}\right] = -\frac{\partial p_1}{\partial r} \tag{2.13}$$

$$\rho_0\left[-i\omega w_1 + v_0\frac{\partial w_1}{\partial r} + \frac{v_0 w_1}{r} + u_0\frac{\partial w_1}{\partial x}\right] = -\frac{\mathrm{im}}{r}p_1 \tag{2.14}$$

$$\begin{aligned}&\rho_0\left[-i\omega T_1 + v_0\frac{\partial T_1}{\partial r} + u_0\frac{\partial T_1}{\partial x} + v_1\frac{\partial T_0}{\partial r} + u_1\frac{\partial T_0}{\partial x}\right]\\ &+ \rho_1\left[v_0\frac{\partial T_0}{\partial r} + u_0\frac{\partial T_0}{\partial x}\right]\\ &-(\gamma-1)\left[-i\omega p_1 + u_0\frac{\partial p_1}{\partial x} + v_0\frac{\partial p_1}{\partial r} + u_1\frac{\partial p_0}{\partial x} + v_1\frac{\partial p_0}{\partial r}\right]\\ &= 0\end{aligned} \tag{2.15}$$

$$\frac{p_1}{p_0} = \frac{\rho_1}{\rho_0} + \frac{T_1}{T_0} \tag{2.16}$$

where u, v, and w are the velocities in the axial, radial, and aximuthal directions, respectively, and the subscript m has been suppressed.

The boundary conditions are based on the assumption that the duct wall is lined with a point-reacting acoustic material whose specific acoustic admittance β may vary along the duct. For no-slip mean flows, a requirement of continuity of the particle displacement gives

$$v_1 - R'u_1 = \frac{\beta}{\rho_w c_w} p_1\sqrt{1+R'^2} \quad \text{at } r = R \tag{2.17}$$

where R' is the slope of the wall and the subscript w refers to values at the wall.

An approximate solution to equations (2.11)–(2.17) is sought in the form

$$\begin{aligned} p_1 &\approx \sum_{n=1}^{N} \Big\{ A_n(x)\, \psi_n^p(r,x) \exp\big(i \int k_n(x)\, dx\big) \\ &\quad + \tilde{A}_n(x)\, \tilde{\psi}_n^p(r,x) \exp\big(i \int \tilde{k}_n(x)\, dx\big) \Big\} \\ u_1 &\approx \sum_{n=1}^{N} \Big\{ A_n(x)\, \psi_n^u(r,x) \exp\big(i \int k_n(x)\, dx\big) \\ &\quad + \tilde{A}_n(x)\, \tilde{\psi}_n^u(r,x) \exp\big(i \int \tilde{k}_n(x)\, dx\big) \Big\} \end{aligned} \tag{2.19}$$

with similar expressions for v_1, w_1, T_1, and ρ_1, where the tilde refers to upstream propagation, the $\psi_n(r,x)$ are the quasi-parallel mode shapes corresponding to the quasi-parallel propagation coefficients $k_n(x)$, and the $A_n(x)$ are complex functions whose moduli and arguments represent, in some sense, the amplitudes and the phases of the (m,n) modes. The circumferential mode number m is assumed to be specified and the corresponding subscript on A, ψ, and k is not explicitly stated; each variable is expressed as a summation over a finite number of radial modes n, with $n = 1$ denoting the fundamental radial mode rather than the conventional $n = 0$. Since k_n is complex, the exponential factor contains an estimate of the attenuation of the (m,n) mode as well as the axial oscillations of the acoustic modes.

Since the ψ_n are the quasi-parallel mode shapes, they are the solutions of the following problem:

$$-i\hat{\omega}\psi^\rho + ik\rho_0\psi^u + \frac{i\rho_0 m}{r}\psi^w + \frac{1}{r}\frac{\partial}{\partial r}(r\rho_0\psi^v) = 0 \tag{2.20}$$

$$-i\rho_0\hat{\omega}\psi^u + \rho_0\frac{\partial u_0}{\partial r}\psi^v + ik\psi^p = 0 \tag{2.21}$$

$$-i\rho_0\hat{\omega}\psi^v + \frac{\partial \psi^p}{\partial r} = 0 \tag{2.22}$$

$$-i\rho_0\hat{\omega}\psi^w + \frac{im}{r}\psi^p = 0 \tag{2.23}$$

$$-i\rho_0\hat{\omega}\psi^T + \rho_0\frac{\partial T_0}{\partial r}\psi^v + i(\gamma-1)\hat{\omega}\psi^p = 0 \tag{2.24}$$

$$\frac{\psi^p}{p_0} = \frac{\psi^\rho}{\rho_0} + \frac{\psi^T}{T_0} \tag{2.25}$$

$$\psi^v - \frac{\beta}{\rho_w c_w}\psi^p = 0 \quad \text{at } r = R \tag{2.26}$$

where

$$\hat{\omega} = \omega - ku_0 \tag{2.27}$$

Equations (2.20)–(2.27) can be combined to yield the well-known problem for parallel-duct eigenfunctions (Nayfeh et al., 1975a):

$$\frac{\partial^2\psi^p}{\partial r^2} + \left[\frac{1}{r} + \frac{T_0'}{T_0} + \frac{2ku_0'}{\hat{\omega}}\right]\frac{\partial\psi^p}{\partial r} + \left[\frac{\hat{\omega}^2}{T_0} - k^2 - \frac{m^2}{r^2}\right]\psi^p = 0 \tag{2.28}$$

$$\frac{\partial\psi^p}{\partial r} - i\frac{\omega\beta}{T_w^{1/2}}\psi^p = 0 \quad \text{at } r = R \tag{2.29}$$

At each axial location, the solution of equations (2.28) and (2.29) yields $\psi_n^p(r,x)$ and its corresponding propagation coefficient $k_n(x)$. The basic functions are normalized so that

$$\int_0^R r\left[\psi_n^p(r,x)\right]^2 dr = 1 \tag{2.30}$$

Then, equations (2.20)–(2.25) are used to express the mode shapes of the other flow variables in terms of ψ_n^p and k_n.

Since the transverse dependence in the assumed solution is chosen a priori, it cannot satisfy equations (2.11)–(2.17) exactly. Thus, the assumed solution is subjected to constraints. The deviations from the quasi-parallel solution are required to be orthogonal to every solution of the adjoint quasi-parallel problem (Nayfeh, 1973b, 1981). These constraints yield the following $2N$ equations for the A's:

$$\sum_{n=1}^{2N} f_{mn}\frac{dA_n}{dx} = \sum_{n=1}^{2N} g_{mn}A_n \tag{2.31}$$

where

$$m = 1, 2, 3, \cdots, 2N,$$

$$\begin{aligned} f_{mn} = \Bigg(\int_0^R & [\phi_{1m}(\rho_0\psi_n^u + u_0\psi^\rho) + \phi_{2m}(\rho_0 u_0\psi_n^u + \psi_n^p) \\ & + \phi_{3m}(\rho_0 u_0\psi_n^v) + \phi_{4m}\rho_0 u_0\psi_n^w \\ & + \phi_{5m}(\rho_0 u_0\psi_n^T - (\gamma-1)u_0\psi_n^p)]\, dr\Bigg) e^{i\int k_n dx} \end{aligned} \tag{2.32}$$

$$\begin{aligned} g_{mn} = -\Bigg\{\int_0^R & \Bigg(\phi_{1m}\frac{\partial}{\partial x}(\rho_0\psi_n^u + u_0\psi_n^\rho) - rv_0\psi_n^\rho\frac{\partial}{\partial r}\left(\frac{\phi_{1m}}{r}\right) \\ & + \phi_{2m}\left[\rho_0\frac{\partial}{\partial x}(u_0\psi_n^u) + \psi_n^\rho\left(u_0\frac{\partial u_0}{\partial x} + v_0\frac{\partial u_0}{\partial r}\right) + \frac{\partial\psi_n^p}{\partial x}\right] \\ & - \psi_n^u\frac{\partial}{\partial r}(\rho_0 v_0\phi_{2m}) \\ & + \phi_{3m}\left[\rho_0 u_0\frac{\partial\psi_n^v}{\partial x} + \rho_0\psi_n^u\frac{\partial v_0}{\partial x} + \psi_n^\rho\left(v_0\frac{\partial v_0}{\partial r} + u_0\frac{\partial v_0}{\partial x}\right)\right] \\ & - v_0\psi_n^v\frac{\partial}{\partial r}(\rho_0\phi_{3m}) + \phi_{4m}\left[\frac{\rho_0 v_0\psi_n^w}{r} + \rho_0 u_0\frac{\partial\psi_n^w}{\partial x}\right] \\ & - \psi_n^w\frac{\partial}{\partial r}(\rho_0 v_0\phi_{4m}) + \phi_{5m}\Bigg[\rho_0 u_0\frac{\partial\psi_n^T}{\partial x} + \rho_0\psi_n^u\frac{\partial T_0}{\partial x} \\ & + \psi_n^\rho(v_0\frac{\partial T_0}{\partial r} + u_0\frac{\partial T_0}{\partial x}) \\ & -(\gamma-1)\left(u_0\frac{\partial\psi_n^p}{\partial x} + \psi_n^u\frac{\partial p_0}{\partial x} + \psi_n^v\frac{\partial p_0}{\partial r}\right)\Bigg] \\ & - \psi_n^T\frac{\partial}{\partial r}(\rho_0 v_0\phi_{5m}) + (\gamma-1)\psi_n^p\frac{\partial}{\partial r}(v_0\phi_{5m}) \\ & + i\phi_{1m}(k_n - k_m)(\rho_0\psi_n^u + u_0\psi_n^\rho) \\ & + i\phi_{2m}(k_n - k_m)(\psi_n^p + \rho_0 u_0\psi_n^u) \\ & + i\phi_{3m}(k_n - k_m)\rho_0 u_0\psi_n^v + i\phi_{4m}(k_n - k_m)\rho_0 u_0\psi_n^w \\ & + i\phi_{5m}(k_n - k_m)\, u_0\left[\rho_0\psi_n^T - (\gamma-1)\,\psi_n^p\right]\Bigg)\, dr \\ & + \rho_0\phi_{1m}\left[R'\psi_n^u + \frac{\beta}{\rho_w c_w}\psi_n^p(\sqrt{1+R'^2} - 1)\right]_{r=R}\Bigg\} e^{i\int k_n dx} \end{aligned} \tag{2.33}$$

For convenience the upstream modes are now denoted by A_n, $n = N+1, \ldots, 2N$, that is

$$A_{N+n} = \tilde{A}_n, \qquad n = 1, 2, 3, \ldots, N$$

Here, the ϕ_{nm} are solutions of the adjoint problem; they are related to ψ_n^p by

$$\begin{aligned} \phi_{1n} &= \frac{rT_0\psi_n^p}{\hat{\omega}}, \\ \phi_{2n} &= \frac{k}{\hat{\omega}}\phi_{1n}, \\ \phi_{3n} &= \frac{irT_0}{\hat{\omega}}\frac{\partial}{\partial r}\left(\frac{\hat{\omega}\phi_{1n}}{rT_0}\right), \end{aligned} \tag{2.34}$$

$$\begin{aligned} \phi_{4n} &= \frac{m\phi_{1n}}{r\hat{\omega}}, \\ \phi_{5n} &= \frac{\phi_{1n}}{T_0}, \\ \phi_{6n} &= -i\rho_0\hat{\omega}\phi_{1n} \end{aligned} \tag{2.35}$$

Writing equation (2.31) in matrix form, $FdA/dx = GA$, and solving for dA/dx, one obtains

$$dA/dx = F^{-1}GA \tag{2.36}$$

where A is a column vector whose elements are the A_n. A Runge-Kutta forward-integration technique is used to solve equation (2.36) for the function A at each axial station. Since the problem is linear, one can determine the solution for any problem subject to general boundary conditions at the two ends of the duct by a linear combination of $2N$ linearly independent solutions.

The linearly independent solutions are obtained by setting all mode amplitudes except one equal to zero at $x = 0$ and integrating equation (2.36) to $x = L$. One such integration for each of the $2N$ modes allows one to obtain the transfer matrices TR_1, TR_2, TR_3, and TR_4 which are defined by

$$\begin{aligned} B^+(L) &= TR_1B^+(0) + TR_2B^-(0) \\ B^-(L) &= TR_3B^+(0) + TR_4B^-(0) \end{aligned} \tag{2.37}$$

where $B^+(x)$ is a column vector of the amplitudes $A_n e^{i\int k_n dx}$ of the right–running modes and $B^-(x)$ is a column vector of the amplitudes $\tilde{A}_n e^{i\int \tilde{k}_n dx}$ of the left-running modes.

To evaluate the coefficients f_{mn} and g_{mn} of the set of ordinary-differential equations for the mode amplitudes A_n, one must specify all mean-flow variables u_0, ρ_0, T_0, p_0, v_0, and their first partial derivatives with respect to both x and r. The form of the g_{mn} as given in equation (2.33) would also require some second derivatives. However, those terms involving second derivatives have been integrated by parts and the resulting expression for the g_{mn}, which requires only first derivatives of the mean flow, has been used in the numerical development.

Although the program is general and can accept any mean flow, a simplified model of the mean flow has been employed in the numerical results presented here because the effort involved in calculating the actual mean profiles is substantial. This model uses one-dimensional gas-dynamics theory to describe the mean-flow variables in the inviscid core; thus u_0, p_0, T_0, and ρ_0 are constant across the duct cross section except in the region of the wall boundary layer. The radial velocity v_0 is calculated to be a linear function of r consistent with the mean-continuity equation and the flow tangency condition at the wall. The wall is assumed to be adiabatic and the temperature profile is assumed to be related to the velocity profile by

$$\frac{T_0}{T_c} = 1 + \frac{1}{2} r_1 (\gamma - 1) M_c^2 \left[1 - \left(\frac{u_0}{u_c} \right)^2 \right] \tag{2.38}$$

where the subscript c refers to values in the inviscid core, $r_1 = 0.8$ is the recovery factor, and $\gamma = 1.4$ is the ratio of the gas specific heats. The initial dimensionless displacement thickness of the boundary layer is assumed to be equal to 0.02. Specification of the mean Mach number and displacement thickness at $x = 0$ permits their subsequent axial variations to be calculated from mass-flow considerations. The one-dimensional gas dynamics theory provides the axial variations of T_c, ρ_c, u_c, etc. and then the boundary-layer profiles are computed. In the case of lined ducts, the liner is taken to be uniform and point-reacting so that it can be represented by a dimensionless admittance β that is independent of x in the form

$$\beta = \left[R_e \left(1 - \frac{i\omega}{\omega_0} \right) + i \cot \left(\frac{\omega d_c}{\sqrt{T_w}} \right) \right]^{-1} \tag{2.39}$$

where R_e is the resistance of the facing sheet, ω_0 is the characteristic frequency of the facing sheet, and d_c is the depth of the backing cavity of the liner. Here,

$$T_w = T_{ad} = T_c \left[1 + \frac{1}{2}(\gamma - 1) r_1 M_c^2 \right]. \tag{2.40}$$

As mentioned earlier, since the effort involved in calculating the actual mean profiles in the boundary layer is substantial, all investigations of the effect of shear layers on sound propagation in ducts have used various approximations of the mean flow (Nayfeh, Kaiser, and Telionis 1975a); such as linear, quadratic, quarter-sine, Pohlhausen, and $1/N$ power-law profiles. Nayfeh, Kaiser, and Shaker (1974) evaluated the influence of these approximations on the attenuation rates for the case of an incompressible mean flow. Their results show that the attenuation rates from the various profiles can be correlated by using the shape factor (ratio of displacement thickness to momentum thickness) and the displacement thickness as scaling parameters. Thus, the quarter-sine, quadratic, Pohlhausen, and Blasius profiles, which have the shape factors 2.66, 2.50, 2.55, and 2.55, approximately produce the same attenuation rates for a given displacement thickness. Nayfeh (1980) showed that these conclusions apply in the case of high Mach number mean flows. Therefore, the numerical results presented here are for the case of a quadratic velocity profile.

Figure 1 shows the axial variation of the absolute value of the pressure at the duct centerline when the first axisymmetric right-running mode is input at $x = 0$ in a hard-walled converging duct as shown in Figure 2; its radius is assumed to vary with axial distance according to

$$R = 1 + .15 \left(-1 + \cos \frac{\pi x}{L} \right). \tag{2.41}$$

It can be seen that as the throat Mach number increases toward unity, the amplitude becomes very large. This is to be expected because the wave numbers of the right-running modes become very large. This can be seen more easily from an examination of the wave

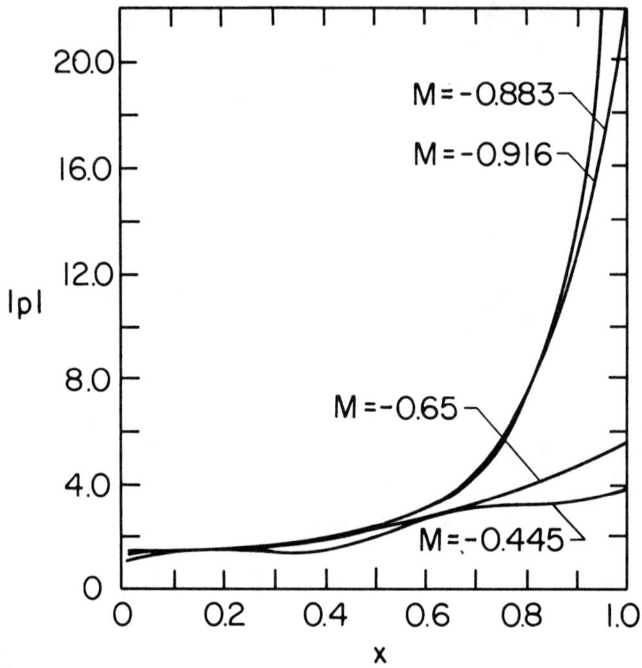

Figure 1. Effect of mean Mach number on the variation of the absolute value of the pressure at the centerline with axial distance in a hard walled duct when the first upstream mode is input at $x = 0$, $\omega = 9$.

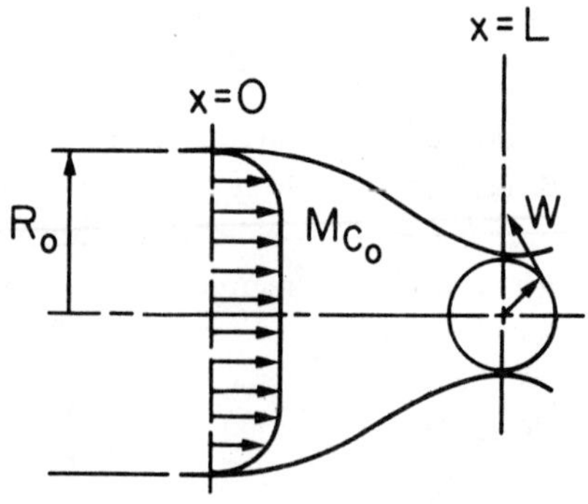

Figure 2. Flow configuration for converging duct.

numbers for the case of uniform flow; that is

$$k = \left\{ -M_c\omega \mp \left[\omega^2 - (1 - M_c^2)\,\kappa^2\right]^{1/2} \right\} (1 - M_c^2)^{-1} \tag{2.42}$$

where M_c is the Mach number and κ is the eigenvalue. As $M_c \to 1$, one of the values of k approaches infinity, while the other remains bounded. Thus, the upstream propagating portion of the acoustic wave travels at the local sound speed relative to the flow and thus tends to *pile-up* near the throat if the flow approaches sonic conditions there. This unusual behavior of the solution is due to linearization of the acoustic equations, as demonstrated numerically by Eisenberg and Kao (1971) and Hersh and Liu (1976) and analytically by Myers and Callegari (1977) for the one-dimensional case. Thus, the inclusion of the nonlinear terms becomes necessary at high Mach numbers. Moreover, although the numerical results may qualitatively show the correct behavior for cases of M_c greater than about 0.80 for the case of hard-walled ducts, they may be quantitatively in error because the linearized theory may not be valid at these high Mach numbers. However, for lined ducts, the linear theory may be applicable at high mean Mach numbers, depending on the admittance of the liner. Figure 3 shows the effect of varying the real part of the admittance β on the axial variation of the centerline pressure for $M_t = -.883$ and $\omega = 9$ when the lowest right-running mode is incident at $x = 0$. It can be seen that the intensification of the sound at the centerline decreases significantly as the real part of β increases.

Figure 4 shows that there is no intensification for waves propagating with the flow. The axial variations of the centerline and wall pressures are shown for the case in which the first left-running mode is incident at $x = L$ in a hard-walled duct when $M_t = -.916$ and $\omega = 9$. Adding more left-running modes to the input at $x = L$ does not produce any significant intensification, suggesting that the linear theory is applicable to cases of waves propagating with the flow even if the mean Mach number is high. It is not clear why the centerline pressure fluctuates rapidly whereas the wall pressure is reasonably smooth.

Figure 5 shows the pressure profiles at the throat of the duct for the case of the lowest axisymmetric right-running mode being incident at $x = 0$ for $\omega = 9$. The waves are propagating against the flow, and hence the boundary-layer is expected to refract the waves

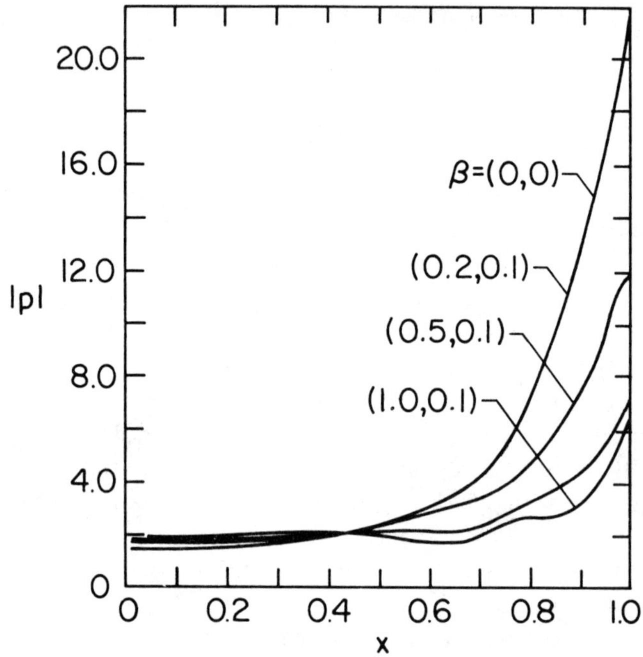

Figure 3. Effect of admittance on the variation of the absolute value of the pressure at the duct centerline with axial distance when the first upstream mode is input at $x = 0$, $\omega = 9$, and $M_t = -0.883$.

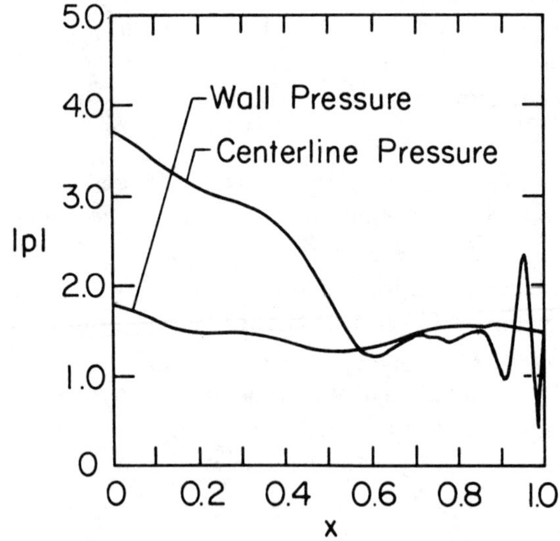

Figure 4. Variation of the wall and centerline pressures with axial distance when the first downstream mode is input at $x = L$, $\omega = 9$, and $M_t = -0.916$.

toward the center (Nayfeh, Kaiser, and Telionis, 1975a). However, the boundary-layer is assumed to be artificially thin so that its refractive effects are small. This is shown in Figure 5 for the cases $M_t = -0.445$ and -0.65. It can be seen that, when M_t increases toward unity, the axial variations tend to refract strongly the waves toward the centerline. At $M_t = -0.883$, the centerline pressure is about 54 dB above the wall value. Other numerical results show that the centerline pressure is about 72.94 dB above the wall value when $M_t = -0.916$.

Figures 6 and 7 compare the acoustic pressure profiles for an axisymmetric mode and three spinning modes ($m = 1, 2, 3$) in a hard-wall duct. The input for these cases is the same except for the circumferential (spinning) mode number. The profiles at the exit are presented. The throat mean Mach number is $-.883$, the inlet displacement thickness is .001, the frequency is 9, and seven modes are considered. At $x = 0$ the lowest right-running mode is incident and it has the same value in all cases so that the acoustic signal propagates upstream. As in Figure 5, the maximum pressure amplitude in the axisymmetric case occurs at the centerline of the throat. With increasing spinning mode number one sees that the maximum amplitude decreases and shifts toward the duct wall at the throat. This indicates that, whereas the refractive effect of the axial gradients is still present, the focusing and intensification is weakened by the asymmetries.

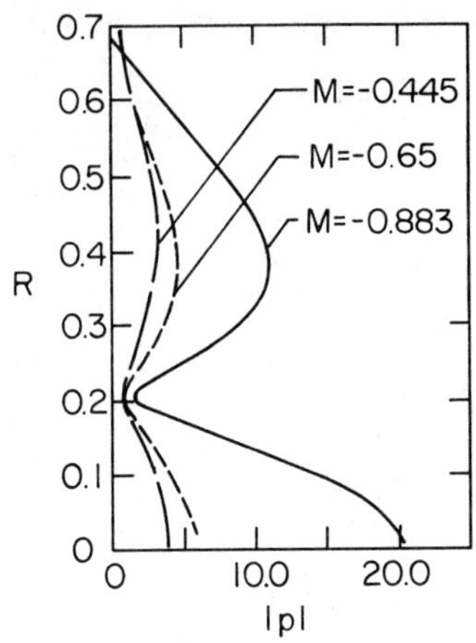

Figure 5. Effect of mean Mach number on the acoustic pressure at the throat of a hard walled duct when the first upstream mode is input at $x = 0$, $\omega = 9$.

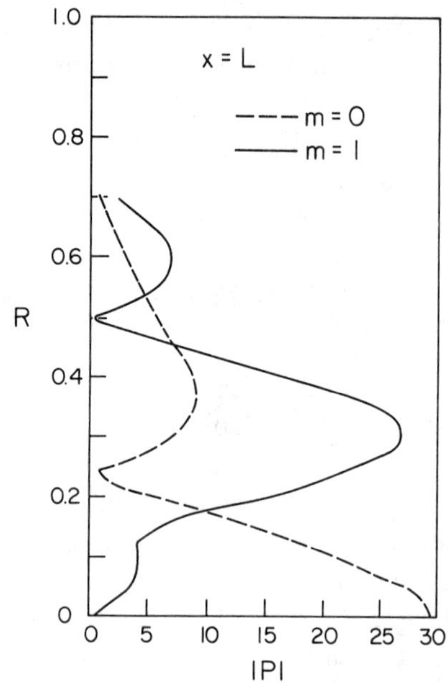

Figure 6. Comparison of the acoustic pressure profiles at the throat of a hard-walled duct for $m = 0$ and $m = 1$ when the first mode is input at $x = 0$, $\omega = 9$, and $M_t = -0.883$.

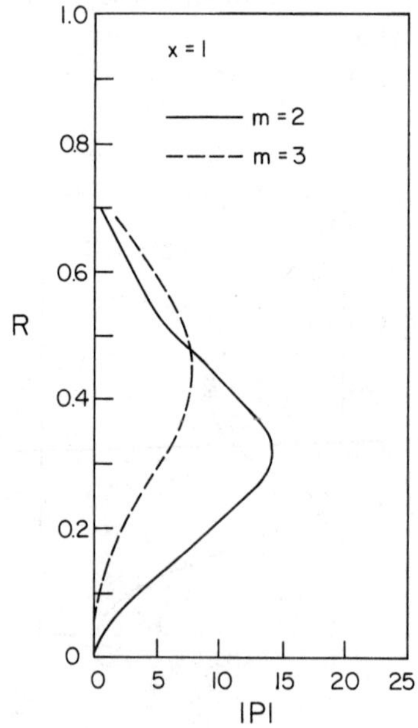

Figure 7. Comparison of the acoustic pressure profiles at the throat of a hard-walled duct for $m = 2$ and $m = 3$ when the first mode is incident at $x = 0$, $\omega = 9$, and $M_t = -0.883$.

Figure 8 shows the variation of the amplitude of the pressure with both the streamwise and radial distances for the case of the lowest three modes being incident at $x = 0$ in a hard-walled duct when $M_t = -0.916$ and $\omega = 9$. It can be easily seen that little refraction takes place except in a small interval near the throat, where the upstream mean Mach number exceeds about 0.8, in agreement with the results of Figure 5.

Figure 9 shows that the magnitude of the refraction of the sound at the duct throat is a function of the frequency. It shows the variation of the magnitude of the pressure with radial distance at the throat of a hard-walled duct when the lowest mode is incident at $x = 0$ for the case $M_t = -0.883$. Although the refraction of the different modes is substantial and increases with increasing frequency, the refraction of the pressure profiles does not lend itself to this simple interpretation because the wave pattern is one of interference among the different modes. It can be easily seen from Figure 9 that little refraction takes place when $\omega = 1$. Thus, the one-dimensional approaches while containing the basic physics of the problem, require higher mean Mach numbers and or incident sound amplitudes to show substantial nonlinear behavior. However, the refraction for the case $\omega = 5$ is much more than that for the case $\omega = 9$ and slightly more than for the case $\omega = 15$.

In the next two figures, we present numerical results for the converging-diverging duct shown in Figure 10 and given by

$$R = 1 + .15\left(-1 + \cos\frac{2\pi x}{L}\right) \tag{2.43}$$

Figure 11 shows the acoustic pressure profiles across the duct radius at the three axial locations $x = 0, \frac{1}{2}L, L$ for the case of $M_t = -0.883$, $\delta^* = 0.02$ and $\omega = 9$ when the lowest axisymmetric mode is incident at $x = 0$. The peak acoustic pressure is greatly reduced at $x = L$, as compared to the profile at the throat, and this maximum value is shifted from the duct centerline. The continuous mode reflections, that occur in the diverging section of the duct, reduce the exit acoustic signal and refract the sound to an annulus at about half the radius of the duct. Figure 12 shows the effect of varying the mean flow Mach number on the centerline acoustic pressure. The input data is the same as that used to produce the preceding figure. For a near sonic flow ($M_t = -.883$), the growth in the amplitude of the

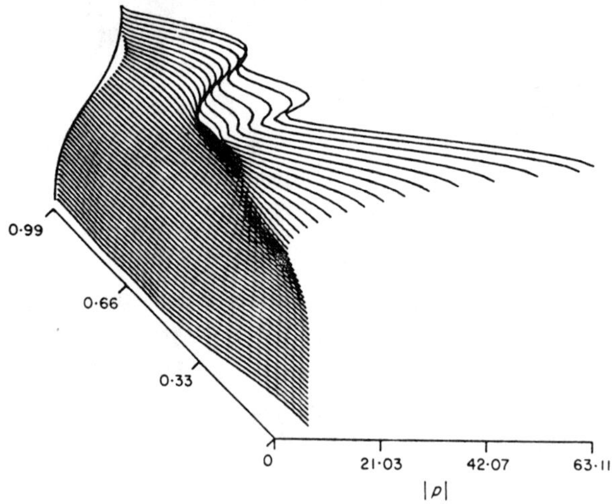

Figure 8. Variation of the acoustic signal with both axial and radial distances in a hard-walled converging duct when the lowest axisymmetric mode is input at $x = 0$.

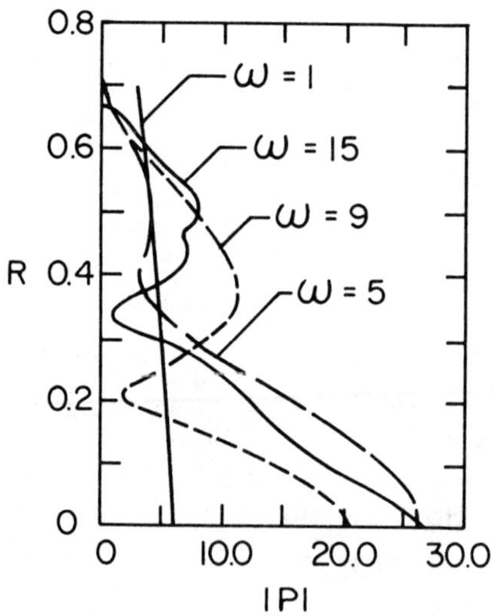

Figure 9. Effect of frequency on the acoustic pressure at the throat of a hard-walled duct when the lowest upstream mode is input at $x = 0$, $M_t = -0.883$.

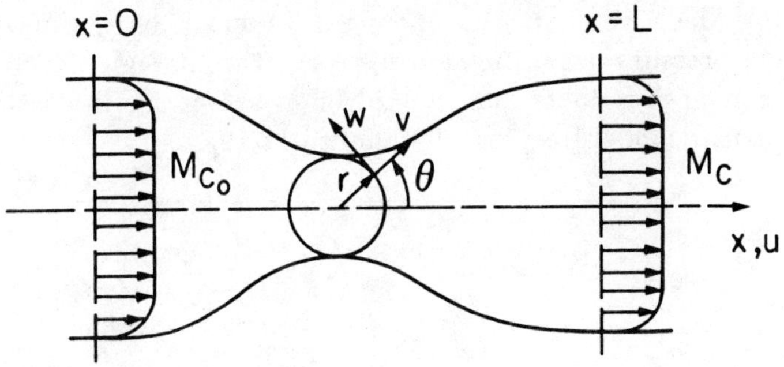

Figure 10. Flow configuration of converging-diverging duct.

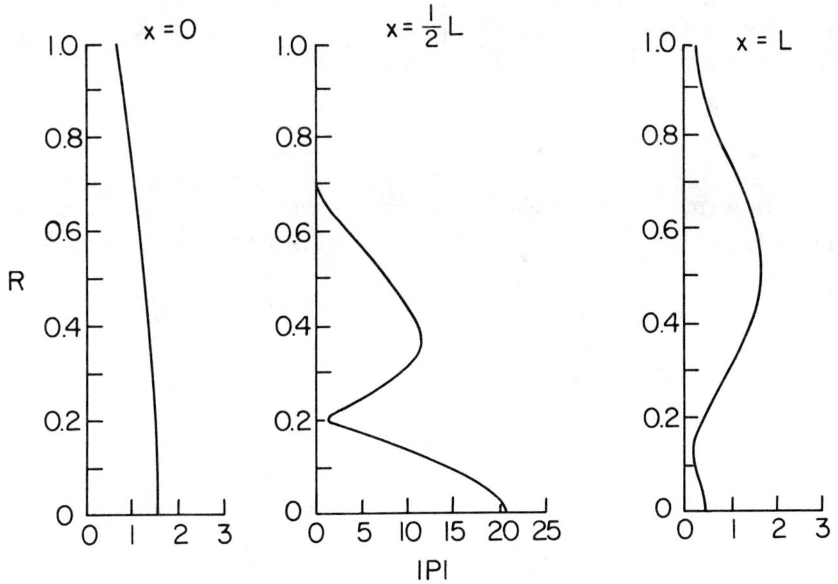

Figure 11. Acoustic pressure profiles at the entrance, throat, and exit of a hard-walled converging-diverging duct when the lowest axisymmetric mode is input for $\omega = 9$ and $M_t = -0.883$.

pressure at the throat contrasts sharply to those of the two lower speed flows. Also, at $x = L$ the centerline pressure decreases with increasing Mach number. For $M_t = -.449$ and $-.65$ the maximum centerline pressure occurs upstream of the throat region. As in the case of converging ducts, one should be aware of the limitations of using a linear model for partially-choked flows.

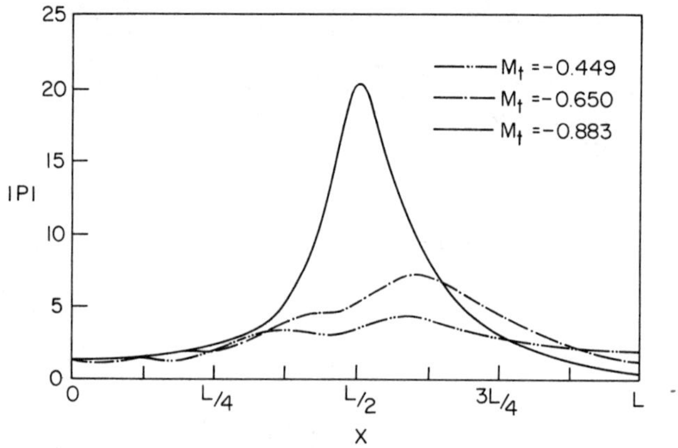

Figure 12. Axial variation of the centerline acoustic pressure in a hard-walled converging-diverging duct when the lowest axisymmetric mode is input at $x = 0$, $\omega = 9$, and $M_t = -0.883$.

2.2 Two-Dimensional Ducts

In terms of the Cartesian coordinate system (x, y) each acoustic quantity $q_1(x, y, t)$ is expressed for sinusoidal time variations as

$$q_1(x, y, t) = q_1{}^*(x, y)\, e^{-i\omega t} \tag{2.44}$$

Then the acoustic equations (2.6)–(2.9) become

$$-i\omega\rho_1 + \frac{\partial}{\partial x}(\rho_0 u_1 + u_0\rho_1) + \frac{\partial}{\partial y}(\rho_0 v_1 + v_0\rho_1) = 0 \tag{2.45}$$

$$\begin{aligned}\rho_0\left[-i\omega u_1 + \frac{\partial}{\partial x}(u_0 u_1) + v_0\frac{\partial u_1}{\partial y} + v_1\frac{\partial u_0}{\partial y}\right]& \\ + \rho_1\left[u_0\frac{\partial u_0}{\partial x} + v_0\frac{\partial u_0}{\partial y}\right] &= -\frac{\partial p_1}{\partial x}\end{aligned} \tag{2.46}$$

$$\begin{aligned}\rho_0\left[-i\omega v_1 + \frac{\partial}{\partial y}(v_0 v_1) + u_0\frac{\partial v_1}{\partial x} + u_1\frac{\partial v_0}{\partial x}\right] \\ + \rho_1\left[v_0\frac{\partial v_0}{\partial y} + u_0\frac{\partial v_0}{\partial x}\right] = -\frac{\partial p_1}{\partial y}\end{aligned} \tag{2.47}$$

$$\begin{aligned}\rho_0\left[-i\omega T_1 + v_0\frac{\partial T_1}{\partial y} + u_0\frac{\partial T_1}{\partial x} + v_1\frac{\partial T_0}{\partial y} + u_1\frac{\partial T_0}{\partial x}\right] \\ + \rho_1\left[v_0\frac{\partial T_0}{\partial y} + u_0\frac{\partial T_0}{\partial x}\right] - (\gamma-1) \\ \times\left[-i\omega p_1 + u_0\frac{\partial p_1}{\partial x} + v_0\frac{\partial p_1}{\partial y} + u_1\frac{\partial p_0}{\partial x} + v_1\frac{\partial p_0}{\partial y}\right] = 0\end{aligned} \tag{2.48}$$

$$\frac{p_1}{p_0} = \frac{\rho_1}{\rho_0} + \frac{T_1}{T_0} \tag{2.49}$$

where u and v are the velocities in the axial and transverse directions and the star has been suppressed. In terms of Cartesian coordinates, the boundary condition (2.17) is replaced with

$$v_1 \mp d' u_1 = \frac{\beta}{\rho_w c_w} p_1 \sqrt{1 + d'^2} \qquad \text{at } y = \pm d \tag{2.50}$$

An approximate solution of equations (2.45)–(2.50) is sought in the form

$$\begin{aligned}p_1 \approx \sum_{n=1}^{N}\Big(A_n(x)\,\psi_n^p(y,x)\,\exp(i\textstyle\int k_n(x)\,dx) \\ + \tilde{A}_n(x)\tilde{\psi}_n^p(y,x)\,\exp(i\textstyle\int \tilde{k}_n(x)\,dx)\Big)\end{aligned} \tag{2.51}$$

$$\begin{aligned}u_1 \approx \sum_{n=1}^{N}\Big(A_n(x)\,\psi_n^u(y,x)\,\exp(i\textstyle\int k_n(x)\,dx) \\ + \tilde{A}_n(x)\tilde{\psi}_n^u(y,x)\,\exp(i\textstyle\int \tilde{k}_n(x)\,dx)\Big)\end{aligned} \tag{2.52}$$

with similar expressions for v_1, T_1, and ρ_1, where, as before, the tilde refers to upstream propagation, the $\psi_n(y,x)$ are the quasi-parallel mode shapes corresponding to the quasi-parallel propagation coefficients $k_n(x)$, and the $A_n(x)$ are complex functions whose moduli and arguments represent, in some sense, the amplitudes and the phases of the nth mode. Since the ψ_n are the quasi-parallel mode shapes, they are the solutions of the following problem:

$$-i\hat{\omega}\psi^\rho + ik\rho_0\psi^u + \frac{\partial}{\partial y}(\rho_0\psi^v) = 0 \tag{2.53}$$

$$-i\rho_0\hat{\omega}\psi^u + \rho_0\frac{\partial u_0}{\partial y}\psi^v + ik\psi^p = 0 \tag{2.54}$$

$$-i\rho_0\hat{\omega}\psi^v + \frac{\partial \psi^p}{\partial y} = 0 \tag{2.55}$$

$$-i\rho_0\hat{\omega}\psi^T + \rho_0\frac{\partial T_0}{\partial y}\psi^v + i(\gamma-1)\hat{\omega}\psi^p = 0 \tag{2.56}$$

$$\frac{\psi^p}{p_0} = \frac{\psi^\rho}{\rho_0} + \frac{\psi^T}{T_0} \tag{2.57}$$

$$\psi^v - \frac{\beta}{\rho_w c_w}\psi^p = 0 \qquad \text{at } y = \pm d \tag{2.58}$$

where $\hat{\omega}$ is defined by equation (2.27). Equations (2.53)–(2.58) can be combined to yield the well-known problem for parallel eigenfunctions in two-dimensional ducts (Nayfeh, Kaiser, and Telionis, 1975a):

$$\frac{\partial^2\psi^p}{\partial y^2} + \left[\frac{T_0'}{T_0} + \frac{2ku_0'}{\hat{\omega}}\right]\frac{\partial\psi^p}{\partial y} + \left(\frac{\hat{\omega}^2}{T_0} - k^2\right)\psi^p = 0 \tag{2.59}$$

$$\frac{\partial\psi^p}{\partial y} - \frac{i\omega\beta}{T_w^{1/2}}\psi^p = 0 \qquad \text{at } y = \pm d \tag{2.60}$$

The eigenfunctions are normalized so that

$$\int_{-d}^{d}[\psi^p(y,x)]^2\,dy = 1 \tag{2.61}$$

Since the transverse dependence in the assumed solution is chosen a priori, it cannot satisfy equations (2.45)–(2.50) exactly. Thus, as in the cylindrical case, the deviations from the quasi-parallel solution are required to be orthogonal to every solution of the adjoint quasi-parallel problem. These constraints yield the $2N$ equations (2.31) for the evolution of the A's with x, where

$$f_{mn} = \left(\int_{-d}^{d} [\phi_{1m}(\rho_0 \psi_n^u + u_0 \psi_n^\rho) + \phi_{2m}(\rho_0 u_0 \psi_n^u + \psi_n^p) + \phi_{3m}\rho_0 u_0 \psi_n^v + \phi_{4m}(\rho_0 u_0 \psi_n^T - (\gamma - 1)\, u_0 \psi_n^p)]\, dy \right) e^{i \int k_n dx} \tag{2.62}$$

$$\begin{aligned} g_{mn} = -\Bigg\{ \int_{-d}^{d} \Bigg(& \phi_{1m} \frac{\partial}{\partial x} (\rho_0 \psi_n^u + u_0 \psi_n^\rho) - v_0 \psi_n^\rho \frac{\partial \phi_{1m}}{\partial y} \\ & + \rho_0 \phi_{2m} \frac{\partial}{\partial x} (u_0 \psi_n^u) + \phi_{2m} \psi_n^\rho \left(u_0 \frac{\partial u_0}{\partial x} + v_0 \frac{\partial u_0}{\partial y} \right) \\ & + \phi_{2m} \frac{\partial \psi_n^p}{\partial x} - \psi_n^u \frac{\partial}{\partial y} (\rho_0 v_0 \phi_{2m}) + \rho_0 \phi_{3m} \left(u_0 \frac{\partial \psi_n^v}{\partial x} + \phi_n^u \frac{\partial v_0}{\partial x} \right) \\ & + \phi_{3m} \psi_n^\rho \left(v_0 \frac{\partial v_0}{\partial y} + u_0 \frac{\partial v_0}{\partial x} \right) - v_0 \psi_n^v \frac{\partial}{\partial y} (\rho_0 \phi_{3m}) \\ & + \rho_0 \phi_{4m} \left(u_0 \frac{\partial \psi_n^T}{\partial x} + \psi_n^u \frac{\partial T_0}{\partial x} \right) + \phi_{4m} \psi_n^\rho \left(v_0 \frac{\partial T_0}{\partial y} + u_0 \frac{\partial T_0}{\partial x} \right) \\ & - (\gamma - 1)\, \phi_{4m} \left(u_0 \frac{\partial \psi_n^p}{\partial x} + \psi_n^u \frac{\partial p_0}{\partial x} + \psi_n^v \frac{\partial p_0}{\partial y} \right) \\ & - \psi_n^T \frac{\partial}{\partial y} (\rho_0 v_0 \phi_{4m}) + (\gamma - 1) \psi_n^p \frac{\partial}{\partial y} (v_0 \phi_{4m}) \\ & + i \phi_{1m} (k_n - k_m)(\rho_0 \psi_n^u + u_0 \psi_n^\rho) \\ & + i \phi_{2m} (k_n - k_m)(\psi_n^p + \rho_0 u_0 \psi_n^u) + i \phi_{3m} (k_n - k_m) \rho_0 u_0 \psi_n^v \\ & + i \phi_{4m} (k_n - k_m)\, u_0 (\rho_0 \psi_n^T - (\gamma - 1)\, \psi_n^p) \Bigg) dy \\ & + \left[\rho_0 \phi_{1m} d' \psi_n^u + \frac{\rho_0 \beta}{\rho_w c_w} \phi_{1m} \psi_n^p (\sqrt{1 + d'^2} - 1) \right]_{y=d} \\ & + \left[-\rho_0 \phi_{1m} d' \psi_n^u + \frac{\rho_0 \beta}{\rho_w c_w} \phi_{1m} \psi_n^p (\sqrt{1 + d'^2} - 1) \right]_{y=-d} \Bigg\} e^{i \int k_n dx} \end{aligned} \tag{2.63}$$

In equations (2.62) and (2.63), the ϕ_{nm} are the eigenfunctions of the adjoint quasi-parallel problem, equations (2.53)–(2.58). They are determined by multiplying equations (2.53)–(2.57) by the functions ϕ_1, ϕ_2, ϕ_3, ϕ_4, and ϕ_5, respectively, adding the resulting equations, integrating the result by parts from $y = -d$ to $y = +d$ to transfer the derivatives from the ψ's to the ϕ's, and obtaining

$$\begin{aligned}&\int_{-d}^{d} \psi^p[-i\hat{\omega}\phi_1 - p_0 T_0 \phi_5]\, dy + \int_{-d}^{d} i\rho_0 \psi^u[-\hat{\omega}\phi_2 + k\phi_1]\, dy \\ &+ \int_{-d}^{d} \rho_0 \psi^v\left[-i\hat{\omega}\phi_3 - \frac{\partial \phi_1}{\partial y} + \frac{\partial u_0}{\partial y}\phi_2 + \frac{\partial T_0}{\partial y}\phi_4\right] dy \\ &+ \int_{-d}^{d} \psi^p\left[ik\phi_2 - \frac{\partial \phi_3}{\partial y} + i(\gamma-1)\,\hat{\omega}\phi_4 + \rho_0 T_0 \phi_5\right] dy \\ &+ \int_{-d}^{d} \rho_0 \psi^T[-i\hat{\omega}\phi_4 - p_0\phi_5]\, dy + [\rho_0\phi_1\psi^v + \phi_3\psi^p]_{-d}^{d} = 0\end{aligned} \tag{2.64}$$

Then, the adjoint equations are obtained by setting each of the brackets in the integrands of equation (2.64) equal to zero; that is

$$i\hat{\omega}\phi_1 + p_0 T_0 \phi_5 = 0 \tag{2.65}$$

$$-\hat{\omega}\phi_2 + k\phi_1 = 0 \tag{2.66}$$

$$-i\hat{\omega}\phi_3 - \frac{\partial \phi_1}{\partial y} + \frac{\partial u_0}{\partial y}\phi_2 + \frac{\partial T_0}{\partial y}\phi_4 = 0 \tag{2.67}$$

$$ik\phi_2 - \frac{\partial \phi_3}{\partial y} + i(\gamma - 1)\,\hat{\omega}\phi_4 + \rho_0 T_0 \phi_5 = 0 \tag{2.68}$$

$$i\hat{\omega}\phi_4 + p_0\phi_5 = 0 \tag{2.69}$$

Substituting for ψ^v from equation (2.58) into the remaining term in equation (2.64), we have

$$\left[\psi^p\left(\phi_3 + \frac{\rho_0\beta\phi_1}{\rho_w c_w}\right)\right]_{y=d} - \left[\psi^p\left(\phi_3 + \frac{\rho_0\beta\phi_1}{\rho_w c_w}\right)\right]_{y=-d} = 0 \tag{2.70}$$

The adjoint boundary condition are obtained by setting each of the coefficients of $\psi^p(d, x)$ and $\psi^p(-d, x)$ in equation (2.70) equal to zero.

The result is

$$\phi_3 + \frac{\rho_0 \beta}{\rho_w c_w} \phi_1 = 0 \qquad \text{at } y = \pm d \tag{2.71}$$

Solving equations (2.65)–(2.67) and (2.69) for the ϕ_n for $n \geq 2$ in terms of ϕ_1 yields

$$\phi_2 = \frac{k}{\hat{\omega}} \phi_1 = \frac{k T_0}{\hat{\omega}^2} \eta \tag{2.72}$$

$$\phi_3 = \frac{i T_0}{\hat{\omega}} \frac{\partial}{\partial y} \left(\frac{\hat{\omega} \phi_1}{T_0} \right) = \frac{i T_0}{\hat{\omega}} \frac{\partial \eta}{\partial y} \tag{2.73}$$

$$\phi_4 = \frac{\phi_1}{T_0} = \frac{\eta}{\hat{\omega}} \tag{2.74}$$

$$\phi_5 = -\frac{i \hat{\omega}}{p_0 T_0} \phi_1 = -\frac{i \eta}{p_0} \tag{2.75}$$

where

$$\eta = \frac{\hat{\omega} \phi_1}{T_0} \tag{2.76}$$

Substituting equations (2.72)–(2.76) into equations (2.68) and (2.71) yields

$$\frac{\partial}{\partial y} \left[\frac{T_0}{\hat{\omega}^2} \frac{\partial \eta}{\partial y} \right] + \left[1 - \frac{k^2 T_0}{\hat{\omega}^2} \right] \eta = 0 \tag{2.77}$$

$$\frac{\partial \eta}{\partial y} - \frac{i \omega \beta}{T_w^{1/2}} \eta = 0 \qquad \text{at } y = \pm d \tag{2.78}$$

One can easily show by multiplying equation (2.59) with $T_0/\hat{\omega}^2$ that equation (2.77) is its self-adjoint. Hence, the solutions of (2.77) and (2.78) are the same as the solutions of equations (2.59) and (2.60); that is, $\eta \propto \psi^p$. Consequently, we do not need to solve the adjoint equations to solve for the ϕ_n. Instead, we calculate them from equations (2.72)–(2.76) once ψ^p is calculated.

Uenishi and Myers (1983) presented numerical results for waves propagating in a hard-walled converging-diverging duct by employing a quasi-one-dimensional mean flow model. As in the case of axisymmetric waves propagating in a cylindrical converging-diverging duct, they found (Figure 13) that the axial gradients tend to refract upstream propagating waves toward the centerline of the duct at the throat. The strength of the refraction increases with increasing throat Mach number and the present linear analysis breaks down as $M_t \to -1$.

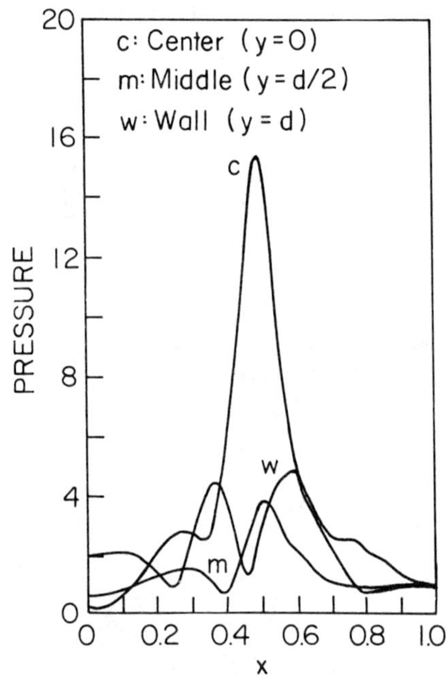

Figure 13. Variation of the acoustic pressure in a hard-walled two-dimensional duct when the lowest mode is input at $x = 0$, $\omega = 7.5$, and $M_t = -0.85$.

3. Nonlinear Problem

The numerical results presented in the preceding section as well as the other results of Nayfeh (1980), Nayfeh, Kelly, and Watson (1982), Kelly, Nayfeh, and Watson (1982), and Uenishi and Myers (1983) show that converging or converging-diverging ducts produce substantial refractions toward the duct center for axisymmetric or two-dimensional waves propagating against near choked flows. In fact, the amplitudes of such waves at the throat of the duct become arbitrarily large as the mean Mach number of the flow approaches unity. As pointed out in the preceding section, this unusual behavior is due to the linearization of the acoustic equations. Myers and Callegari (1977) studied the propagation of quasi-one-dimensional sound waves in a converging-diverging duct containing a quasi-one-dimensional near choked flow. They confirmed the singular behavior of the linear theory and explicitly determined the form of the singularity for two special types of area variation. This linear theory is discussed in Section 3.2.

The singular behavior at the throat of a choked converging or converging-diverging duct implies that linear acoustic theory cannot be used to describe the upstream propagation of waves in choked or partially choked ducts. Consequently, nonlinear effects must be included in any theory that describes the upstream propagation of sound through regions of high subsonic Mach numbers. However, owing to the difficulty of the nonlinear three-dimensional problem, two approaches that incorporate nonlinear effects in quasi-one-dimensional models have been developed. The first approach (Callegari and Myers, 1977, 1979; Myers and Callegari, 1978) uses the method of matched asymptotic expansions and it is discussed in Section 3.3. The second approach (Nayfeh, Shaker, and Kaiser, 1978; Nayfeh, Kelly, and Watson, 1981) uses spectral methods and it is discussed in Section 3.4.

The problem considered is that of a quasi-one-dimensional, inviscid, nonlinear flow in the hard-walled duct shown in Figure 10. In terms of the dimensionless quantities introduced in Section 2.1, the governing equations are

Conservation of Mass

$$\frac{1}{\rho}\frac{\partial \rho}{\partial t} + \frac{u}{\rho}\frac{\partial \rho}{\partial x} + \frac{\partial u}{\partial x} + u\frac{A'(x)}{A(x)} = 0 \tag{3.1}$$

Conservation of Momentum

$$\frac{\partial u}{\partial t} + u\frac{\partial u}{\partial x} + \frac{1}{\rho}\frac{\partial p}{\partial x} = 0 \tag{3.2}$$

Conservation of Energy

$$\rho\left(\frac{\partial T}{\partial t} + u\frac{\partial T}{\partial x}\right) - (\gamma - 1)\left(\frac{\partial p}{\partial t} + u\frac{\partial p}{\partial x}\right) = 0 \tag{3.3}$$

Equation of State

$$\gamma p = \rho T \tag{3.4}$$

where $A(x)$ is the cross-sectional area of the duct. The energy and state equations (3.3) and (3.4) can be replaced with the isentropic state equation

$$\gamma p = \rho^\gamma \tag{3.5}$$

It should be noted that the above quasi-one-dimensional equations are valid only when the acoustic wavelength is sufficiently large and the area variations are sufficiently slow. Since the quasi-one-dimensional model neglects or averages over the cross section, it cannot account for multi-dimensional effects which may be significant in the high Mach number regions. However, to gain insight into the mechanisms that operate in near sonic regions and to develop the mathematical techniques needed to analyze these mechanisms, researchers treated the one-dimensional problem.

3.1 The Basic Steady Flow

Before discussing the acoustic problem we need the mean flow or the basic steady flow. Denoting the mean flow by the subscript $_0$, we find from equations (3.1), (3.2), and (3.5) that it is governed by

$$\frac{u_0}{\rho_0}\rho_0' + u_0' + u_0\frac{A'}{A} = 0 \tag{3.6}$$

$$u_0 u_0' + \frac{1}{\rho_0}p_0' = 0 \tag{3.7}$$

$$\gamma p_0 = \rho_0^\gamma \tag{3.8}$$

Eliminating ρ_0 and p_0 from equations (3.6)–(3.8) yields

$$\frac{(1 - M^2)\,M'}{M\left[1 + \frac{1}{2}(\gamma - 1)\,M^2\right]} = -\frac{A'}{A} \tag{3.9}$$

where

$$M = \frac{u_0}{c_0}$$

and

$$
\begin{aligned}
c_0^2 &= \frac{\partial p_0}{\partial \rho_0} = \rho_0^{\gamma-1} \\
&= \left[1 + \frac{1}{2}(\gamma - 1)\, M^2\right]^{-1}
\end{aligned} \tag{3.10}
$$

Integrating equation (3.9) gives

$$
\alpha^s M^s(x)\left[1 + \frac{1}{2}(\gamma - 1)M_t^2\right] = M_t^s\left[1 + \frac{1}{2}(\gamma - 1)M^2\,(x)\right] \tag{3.11}
$$

where M_t is the throat Mach number,

$$
\alpha = \frac{A(x)}{A(\frac{1}{2}L)} \qquad \text{and} \qquad s = \frac{2\,(\gamma - 1)}{\gamma + 1} \tag{3.12}
$$

For a given value of M_t and a given area variation $\alpha(x)$, equation (3.11) can be solved numerically to determine the distribution of M with axial distance x. Since we are interested in choked or partially choked flows, we let $M_t = 1 - \epsilon$, where ϵ is a small positive parameter. In this case, an approximate solution to equation (3.11) can be obtained in powers of ϵ. The result is

$$
M(x) = M_0 - \frac{2\epsilon^2 M_0[1 + \frac{1}{2}(\gamma - 1)\, M_0^2]}{(\gamma + 1)(1 - M_0^2)} + \cdots \tag{3.13}
$$

where M_0 is given by

$$
\frac{1}{2}(\gamma + 1)\, \alpha^s M_0^s = 1 + \frac{1}{2}\,(\gamma - 1)\, M_0^2 \tag{3.14}
$$

It is clear that the straightforward expansion (3.13) breaks down near the throat. In fact, the second term becomes $0(\epsilon)$ if $1 - M_0 = 0(\epsilon)$.

For a given area variation $\alpha(x)$, one can numerically solve equation (3.14) to determine $M_0(x)$. However, in the neighborhood of the throat, an approximate solution can be obtained as follows. We let $\alpha = 1 + f(x)$, where $f(x)$ is positive everywhere except at the throat where it is zero. Then, we rewrite equation (3.14) as

$$
\frac{2[1 + \frac{1}{2}(\gamma - 1)\, M_0^2]}{(\gamma + 1)\, M_0^s} - 1 = [1 + f(x)]^s - 1
$$

Hence,

$$\frac{1}{2}s(2-s)(M_0-1)^2 \approx sf(x)$$

or

$$M_0 \approx 1-\sqrt{\frac{1}{2}(\gamma+1)\,f(x)} \tag{3.15}$$

For the special area variation $f(x) = a(x-\frac{1}{2}L)^2$, considered by Callegari and Myers,

$$M_0 \approx 1-\sqrt{\frac{1}{2}a(\gamma+1)}\;|x-\frac{1}{2}L| \tag{3.16}$$

Thus, $M_0 - 1 = 0(x-\frac{1}{2}L)$ as $x \to \frac{1}{2}L$, and the expansion (3.13)) breaks down when $x-\frac{1}{2}L = 0(\epsilon)$.

To determine an approximation to the solution of equation (3.11) when $M_t = 1-\epsilon$ that is valid near the throat, we rewrite it as

$$\frac{[1+f(x)]^s[1+\frac{1}{2}(\gamma-1)(1-\epsilon)^2]}{\frac{1}{2}(\gamma+1)(1-\epsilon)^s}-1=\frac{2[1+\frac{1}{2}(\gamma-1)\,M^2\,(x)]}{(\gamma+1)\,M^s\,(x)}-1$$

Hence,

$$\frac{1}{2}s(2-s)(M-1)^2 \approx sf(x)+\frac{1}{2}s(2-s)\,\epsilon^2$$

or

$$M \approx 1-\sqrt{\epsilon^2+\frac{1}{2}(\gamma+1)\,f(x)} \tag{3.17}$$

For the special area variation $f(x) = a(x-\frac{1}{2}L)^2$, (3.17) becomes

$$M \approx 1-\sqrt{\epsilon^2+\frac{1}{2}a(\gamma+1)(x-\frac{1}{2}L)^2} \tag{3.18}$$

It is clear that any expansion of (3.18) for small ϵ keeping x fixed is not valid when $x-\frac{1}{2}L = 0(\epsilon)$.

3.2 The Linear Acoustic Problem

Before treating the acoustic problem, we change the dependent variables in the governing equations from ρ, p, and u into the Riemann invariants. To accomplish this, we add $\pm c(c^2 = \partial p/\partial \rho)$ times equation (3.1) to equation (3.2) and obtain

$$\frac{\partial J_1}{\partial t} + (u+c)\frac{\partial J_1}{\partial x} + \frac{A'}{A}cu = 0 \tag{3.19}$$

$$\frac{\partial J_2}{\partial t} + (u-c)\frac{\partial J_2}{\partial x} - \frac{A'}{A}cu = 0 \tag{3.20}$$

where the Riemann invariants J_1 and J_2 are given by

$$J_1 = u + \int \frac{c}{\rho} d\rho = u + \frac{2c}{\gamma - 1} \tag{3.21}$$

$$J_2 = u - \int \frac{c}{\rho} d\rho = u - \frac{2c}{\gamma - 1} \tag{3.22}$$

The presence of sound in the ducts results in the perturbation of the flow quantities so that

$$u = u_0(x) + u_1(x,t) \quad \text{and} \quad c = c_0(x) + c_1(x,t) \tag{3.23}$$

where u_1 and c_1 are the acoustic parts. Then, (3.21) and (3.22) can be rewritten as

$$J_1 = u_0 + \frac{2c_0}{\gamma - 1} + P(x,t) \quad \text{and} \quad J_2 = u_0 - \frac{2c_0}{\gamma - 1} + Q(x,t) \tag{3.24}$$

where

$$P = u_1 + \frac{2c_1}{\gamma - 1} \quad \text{and} \quad Q = u_1 - \frac{2c_1}{\gamma - 1} \tag{3.25}$$

or

$$u_1 = \frac{1}{2}(P+Q) \quad \text{and} \quad c_1 = \frac{1}{4}(\gamma - 1)(P - Q) \tag{3.26}$$

Substituting equations (3.23)–(3.26) into equations (3.19) and (3.20) and subtracting the basic steady parts, we obtain the nonlinear acoustic equations

$$\begin{aligned}\frac{\partial P}{\partial t} + \left[u_0 + c_0 + \frac{1}{4}(\gamma+1)P + \frac{1}{4}(3-\gamma)\,Q\right]\frac{\partial P}{\partial x}\\ + B_{11}P + B_{12}Q + \frac{(\gamma-1)A'}{8A}(P^2 - Q^2) = 0\end{aligned} \tag{3.27}$$

$$\begin{aligned}\frac{\partial Q}{\partial t} + \left[u_0 - c_0 + \frac{1}{4}(3-\gamma)P + \frac{1}{4}(\gamma+1)\,Q\right]\frac{\partial Q}{\partial x}\\ + B_{21}P + B_{22}Q - \frac{(\gamma-1)A'}{8A}(P^2 - Q^2) = 0\end{aligned} \tag{3.28}$$

where

$$B_{11} = \frac{1}{4}(\gamma+1)\left[u_0' + \frac{2c_0'}{\gamma-1}\right] + \frac{A'}{4A}[2c_0 + (\gamma-1)u_0] \tag{3.29}$$

$$B_{12} = \frac{1}{4}(3-\gamma)\left[u_0' + \frac{2c_0'}{\gamma-1}\right] + \frac{A'}{4A}[2c_0 - (\gamma-1)u_0] \tag{3.30}$$

$$B_{21} = \frac{1}{4}(3-\gamma)\left[u_0' - \frac{2c_0'}{\gamma-1}\right] - \frac{A'}{4A}[2c_0 + (\gamma-1)u_0] \tag{3.31}$$

$$B_{22} = \frac{1}{4}(\gamma+1)\left[u_0' - \frac{2c_0'}{\gamma-1}\right] - \frac{A'}{4A}[2c_0 - (\gamma-1)u_0] \tag{3.32}$$

Neglecting the nonlinear terms in equations (3.27) and (3.28) yields the linear acoustic equations

$$\frac{\partial P}{\partial t} + (u_0 + c_0)\frac{\partial P}{\partial x} + B_{11}P + B_{12}Q = 0 \tag{3.33}$$

$$\frac{\partial Q}{\partial t} + (u_0 - c_0)\frac{\partial Q}{\partial x} + B_{21}P + B_{22}Q = 0 \tag{3.34}$$

For sinusoidal time variations, we let

$$P = \psi_1(x)\,e^{-i\omega t} \qquad \text{and} \qquad Q = \psi_2(x)\,e^{-i\omega t} \tag{3.35}$$

in equations (3.33) and (3.34) and obtain

$$(u_0 + c_0)\,\psi_1' + (B_{11} - i\omega)\,\psi_1 + B_{12}\psi_2 = 0 \tag{3.36}$$
$$(u_0 - c_0)\,\psi_2' + B_{21}\psi_1 + (B_{22} - i\omega)\,\psi_2 = 0 \tag{3.37}$$

In general, the coupled equations (3.36) and (3.37) cannot be solved analytically for an arbitrary $M(x)$ and hence an arbitrary area variation $\alpha(x)$. Moreover, since equation (3.37) is singular at the location where $u_0 = c_0$ or $M = 1$, any numerical procedure will fail there and hence it must be implemented by an analytical resolution of the singular behavior. Furthermore, the linearized acoustic equations do not approximate uniformly the sound signal near the throat of the duct as the mean flow Mach number approaches unity. In this case, u, ρ, and p become arbitrarily large near the throat and the nonlinear terms must be taken into account. Nonlinear effects are accounted for by using the method of matched asymptotic expansions in the following section and by using spectral methods in Section 3.4.

An analytical solution to equations (3.36) and (3.37) was obtained by Tsien (e.g., Crocco and Cheng, 1956) for the Mach number distribution

$$M_0(x) = \left(1 - k|x - \frac{1}{2}L|\right)\left[\frac{1}{2}(\gamma+1) - \frac{1}{2}(\gamma-1)\left(1 - k|x - \frac{1}{2}L|\right)^2\right]^{1/2} \tag{3.38}$$

where $|x - \frac{1}{2}L| < k^{-1}$. Expanding equation (3.38) for small $|x - \frac{1}{2}L|$ and comparing the result with equation (3.16), we find that they are asymptotically equivalent as $|x - \frac{1}{2}L| \to 0$ if $k^2 = 2a/(\gamma+1)$. For the Mach number distribution (3.38), Tsien expressed the solutions of equations (3.36) and (3.37) in terms of hypergeometric functions of complex argument. Near the throat, M_0 is given by equation (3.16), and the solutions of equations (3.36) and (3.37) have the form

$$\psi_1 = B^{\pm} + \cdots \qquad \text{and} \qquad \psi_2 = \frac{A^{\pm}}{|x - \frac{1}{2}L|}\, e^{-i\omega\theta} + \cdots \tag{3.39a}$$

$$\theta = a^{-1/2}\, sgn\,(x - \frac{1}{2}L)\, \ln|x - \frac{1}{2}L| \tag{3.39b}$$

Therefore, P remains finite, whereas Q becomes infinite at a sonic throat.

3.3 The Method of Matched Asymptotic Expansions

As pointed out, the linear solution breaks down in the neighborhood of the throat and hence it cannot approximate uniformly the sound wave. To determine an expansion that is uniformly valid, Callegari and Myers (1977, 1979) and Myers and Callegari (1978) took the linear solution as the first term in the outer expansion, obtained an inner expansion that is valid near the throat, and matched both expansions. The outer expansion is taken as a power series expansion in the parameter δ, the amplitude of the wave. Moreover, the terms in the expansion are expanded further in powers of ϵ, the deviation of the throat mean flow Mach number from unity, and the outer expansion is expressed as a Fourier series in the input frequency ω; that is,

$$P^0 = \delta \sum_{n=0}^{\infty} \psi_{1n}(x)\, e^{-in\omega t} + 0(\delta\epsilon) \tag{3.40a}$$

$$Q^0 = \delta \sum_{n=0}^{\infty} \psi_{2n}(x)\, e^{-in\omega t} + 0(\delta\epsilon) \tag{3.40b}$$

where ψ_{1n} and ψ_{2n} are given by equations (3.36) and (3.37) with ω being replaced with $n\omega$.

The singular behavior of the mean flow and the linearized solutions led Callegari and Myers to introduce the stretching transformation

$$X = \sqrt{\frac{\gamma+1}{2}}\, \frac{x - \frac{1}{2}L}{\epsilon} \tag{3.41}$$

Then, they sought an inner expansion in powers of ϵ in the form

$$P^i = \epsilon P_1(X,t) + \epsilon^2 P_2(X,t) + \cdots \tag{3.42}$$

$$Q^i = \epsilon Q_1(X,t) + \epsilon^2 Q_2(X,t) + \cdots \tag{3.43}$$

Substituting equations (3.42) and (3.43) into equations (3.27) and (3.28), using the approximation (3.18) to the Mach number near the throat, and equating the coefficient of each power of ϵ to zero, we obtain

$$\frac{\partial P_1}{\partial X} = 0 \tag{3.44}$$

$$\frac{\partial Q_1}{\partial t} + \left[\frac{1}{4}(3-\gamma)P_1 + \frac{1}{4}(\gamma+1)Q_1 - M_1(X)\right]\frac{\partial Q_1}{\partial X} - \frac{M_1'}{\gamma+1}[(3-\gamma)P_1 + (\gamma+1)Q_1] = 0 \tag{3.45}$$

$$\frac{\partial P_2}{\partial X} = -\frac{1}{2}\frac{\partial P_1}{\partial t} - \frac{\gamma-1}{\gamma+1}M_1'P_1 \tag{3.46}$$

It follows from equation (3.44) that $P_1 = P_1(t)$. Then, matching the one-term inner expansion ϵP_1 with the one-term outer expansion given by equation (3.40a) demands that $\delta = 0(\epsilon)$. But matching Q^i and Q^0 demands that $\delta = 0(\epsilon^2)$. Hence, $P_1 \equiv 0$ and equation (3.45) becomes

$$\frac{\partial Q_1}{\partial t} + \left[\frac{1}{4}(\gamma+1)Q_1 - M_1\right]\frac{\partial Q_1}{\partial X} - M_1'Q_1 = 0 \tag{3.47}$$

Letting

$$\eta = \frac{1}{4}(\gamma+1)Q_1 - M_1 \tag{3.48}$$

we rewrite equation (3.47) as

$$\frac{\partial \eta}{\partial t} + \eta\frac{\partial \eta}{\partial X} = aX \tag{3.49}$$

Using the method of characteristics, one replaces equation (3.49) with

$$\frac{d\eta}{d\tau} = aX \tag{3.50}$$

$$\frac{dt}{d\tau} = 1 \tag{3.51}$$

$$\frac{dX}{d\tau} = \eta \tag{3.52}$$

whose solutions were expressed by Callegari and Myers as

$$\eta = -\sqrt{aX^2 + \phi(\tau)} \tag{3.53}$$

where

$$t = -\frac{1}{\sqrt{a}} \ln|X + \sqrt{X^2 + \phi/a}| + \frac{1}{\sqrt{a}} \ln \frac{2}{\epsilon}\sqrt{\frac{\gamma+1}{2}} + \tau \tag{3.54}$$

with τ being the characteristic parameter. Combining equations (3.53) and (3.48) gives

$$Q^i = \frac{4\epsilon}{\gamma+1}\left[\sqrt{aX^2+1} - \sqrt{aX^2+\phi(\tau)}\right] + \cdots \tag{3.55}$$

With P_1 being zero, equation (3.46) yields $P_2 = P_2(t)$ and hence

$$P^i = \epsilon^2 P_2(t) + \cdots \tag{3.56}$$

To determine the coefficients $A_n^{\pm}$ and $B_n^{\pm}$ and the function $\phi(\tau)$, we need to apply the initial conditions and match the inner and outer expansions. Using Van Dyke's matching principle to match (3.40a) and (3.56), Callegari and Myers (1979) obtained

$$P^i = \epsilon^2\left[\frac{1}{4}(\gamma+1)\,A_0^{\pm} + \frac{1}{2}(\gamma+1)\,B_0^{\pm} + \sum_{n=1}^{\infty} B_n^{\pm}(1+d_n^{\pm})\,e^{-in\omega t}\right] + \cdots \tag{3.57}$$

where

$$d_n^{\pm} = \frac{in\omega(\gamma+1) + 2(\gamma-1)\sqrt{a}\;sgn(x-\frac{1}{2}L)}{in\omega(\gamma+1) + 4\sqrt{a}\;sgn(x-\frac{1}{2}L)}$$

Since P^i is independent of X, one can equate the two expressions in equation (3.57) to relate the B_n^{+} to the B_n^{-}.

To match (3.40b) and (3.55), we determine the one-term inner expansion of Q^0 and obtain

$$(Q^0)^i = -\frac{2\epsilon^2}{|x-\frac{1}{2}L|} \times\left[\frac{1}{2}a_0^{\pm} + \sum_{n=1}^{\infty} a_n^{\pm}\cos n\omega\,(t+\theta) - \sum_{n=1}^{\infty} b_n^{\pm}\sin n\omega\,(t+\theta)\right] \tag{3.58}$$

where $A_n^\pm = a_n^\pm - ib_n^\pm$ for $n \geq 1$ and $A_0^\pm = \frac{1}{2}a_0^\pm$. Then, we determine the one-term outer expansion of Q^i and obtain (Callegari and Myers, 1979)

$$(Q^i)^0 = \left(\frac{3}{\gamma+1}\right)^{3/2} \frac{\epsilon^2}{\sqrt{a}|x - \frac{1}{2}L|}[1 - \phi(\tau)] \tag{3.59}$$

In the outer region $x > \frac{1}{2}L$, the outer limit of equation (3.54) is

$$t + \frac{1}{\sqrt{a}} \ln\left(x - \frac{1}{2}L\right) = \tau \tag{3.60}$$

Equating equation (3.58) when $x > \frac{1}{2}L$ and equation (3.59) and using equation (3.60) yields the Fourier-series expansion

$$\phi(\tau) = 1 + 2\left(\frac{\gamma+1}{2}\right)^{3/2} \sqrt{a}\left[\frac{1}{2}a_0^+ + \sum_{n=1}^{\infty}(a_n^+ \cos n\omega\tau - b_n^+ \sin n\omega\tau)\right] \tag{3.61}$$

For $x < \frac{1}{2}L$, the outer limit of equation (3.54) is

$$t - \frac{1}{\sqrt{a}} \ln\left(\frac{1}{2}L - x\right) = \tau + \frac{1}{\sqrt{a}} \ln \frac{2a(\gamma+1)}{\epsilon^2} - \frac{1}{\sqrt{a}} \ln \phi = D \tag{3.62}$$

Equating equation (3.58) when $x < \frac{1}{2}L$ and equation (3.59) and using equation (3.62) yields the Fourier-series expansion

$$\phi(\tau) = 1 + 2\left(\frac{\gamma+1}{2}\right)^{3/2} \sqrt{a}\left[\frac{1}{2}a_0^- + \sum_{n=1}^{\infty}(a_n^- \cos n\omega D - b_n^- \sin n\omega D)\right] \tag{3.63}$$

Eliminating $\phi(\tau)$ from equations (3.61) and (3.63) yields

$$\begin{aligned} &\frac{1}{2}a_0^+ + \sum_{n=1}^{\infty}(a_n^+ \cos n\omega\tau - b_n^+ \sin n\omega\tau) \\ &\quad = \frac{1}{2}a_0^- + \sum_{n=1}^{\infty}(a_n^- \cos n\omega D - b_n^- \sin n\omega D) \end{aligned} \tag{3.64}$$

where D is related to τ and ϕ by equation (3.62). For a given value of τ, equation (3.64) connects the coefficients of the wave harmonics downstream of the throat with those upstream of the throat. Instead of eliminating the parameter τ and obtaining a denumerable set of equations relating the upstream and downstream quantities, Callegari and Myers (1979) used the boundary conditions to reduce equation (3.64) to a denumerable set of equations for the a_n^+ and b_n^+.

For the case of a harmonic acoustic velocity source located downstream of the throat at $x = x_s > \frac{1}{2}L$ and an anechoic termination upstream of the throat at $x = x_e < \frac{1}{2}L$, Callegari and Myers (1979) solved the resulting equations for the a_n^+ and b_n^+ and presented numerical results for the Tsien's Mach number distribution. They calculated the variation of the transmitted power with the source strength and the deviation of the throat Mach number from unity. Figure 14 shows one of their results for the variation of the upstream transmitted total power with the deviation of the throat Mach number from unity. In this case, a shock wave develops for $\epsilon = 1 - M_t < 0.06$. We note that the present analysis is not valid in regions of shock formation. Figure 14 shows that the linear theory nearly predicts the correct total transmitted power for $\epsilon > 0.1$. Below this value, the predictions of the linear theory exceed those of the nonlinear theory. However, we should keep in mind that the linear theory is not valid as $\epsilon \to 0$ because it is singular.

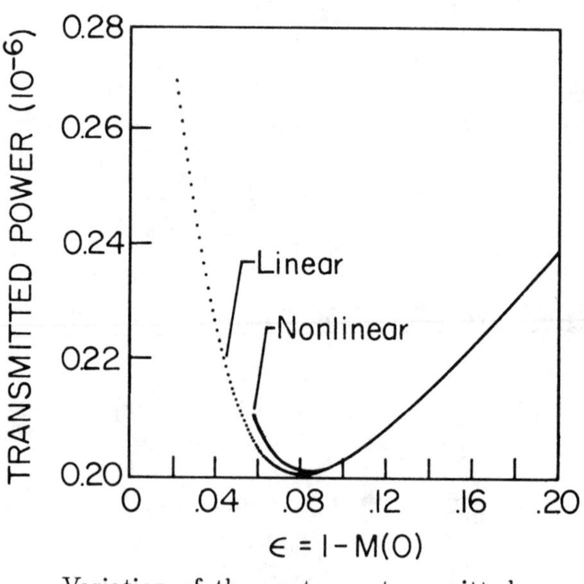

Figure 14. Variation of the upstream transmitted power with the deviation of the throat Mach number from unity.

The present analytical approach is for a specific mean flow Mach number distribution and hence a specific area distribution. The analysis needs to be modified for example if $f(x) = bx^4$ instead of ax^2 or if $f(x) = a^+x^2$ upstream of the throat and bx^4 downstream of the throat. Moreover, the linear solution needs to be calculated numerically for a general area variation and its inner limit must be determined to accomplish the matching. Furthermore, it is not clear yet how to extend such an analysis to the practical problem (i.e., three-dimensional case). As an alternate technique, Nayfeh and co-workers developed a spectral method for the treatment of this problem. The method is not restricted by area variation as long as shocks do not develop in the duct.

3.4 A Spectral Method

Nayfeh and co-workers represented the solution of equations (3.1)–(3.4) as the sum of a steady term plus higher harmonics due to the acoustic disturbance; that is

$$u(x,t) = u_{10}(x) + \sum_{n=1}^{N} \left[u_{1n}(x)\, e^{-in\omega t} + \overline{u}_{1n}(x)\, e^{in\omega t}\right] \tag{3.65}$$

$$\rho(x,t) = \rho_{10}(x) + \sum_{n=1}^{N} \left[\rho_{1n}(x)\, e^{-in\omega t} + \overline{\rho}_{1n}(x)\, e^{in\omega t}\right] \tag{3.66}$$

$$T(x,t) = T_{10}(x) + \sum_{n=1}^{N} \left[T_{1n}(x)\, e^{-in\omega t} + \overline{T}_{1n}(x)\, e^{in\omega t}\right] \tag{3.67}$$

with the pressure being eliminated using equation (3.4). The steady terms are not solutions of the steady form of equations (3.1)–(3.4); they include both the steady mean flow and the acoustic streaming. Nayfeh and co-workers did not use the isentropic equation (3.5) to avoid the difficulties involved in separating the effects of the various harmonics. Substituting equations (3.65)–(3.67) into (3.1)–(3.4), eliminating p, and equating the coefficients of equal powers of $\exp(in\omega t)$ to zero, we obtain the following coupled set of first-order nonlinear ordinary-differential equations:

$$A(x,y)\,\frac{dy}{dx} = B(x,y) \tag{3.68}$$

where y is a column vector of the unknowns. The coefficient matrix A and the inhomogeneous column vector B are defined in the Appendix.

To solve equations (3.68), we first transform them into real form by rewriting them as

$$(A_r + iA_i)\left(\frac{dy_r}{dx} + i\frac{dy_i}{dx}\right) = B_r + iB_i \tag{3.69}$$

which upon separation of the real and imaginary parts yields

$$A_r\frac{dy_r}{dx} - A_i\frac{dy_i}{dx} = B_r \tag{3.70}$$

$$A_i\frac{dy_r}{dx} + A_r\frac{dy_i}{dx} = B_i \tag{3.71}$$

Since the imaginary parts of the steady terms are zero, their imaginary parts in equations (3.70) and (3.71) are discarded. Equations (3.70) and (3.71) can be rewritten as

$$A^*(x, y^*)\,\frac{dy^*}{dx} = B^*(x, y^*) \tag{3.72}$$

where A^* is a $3(2N+1) \times 3(2N+1)$ real matrix and y^* and B^* are $3(2N+1)$ column vectors. Specifying impedance conditions at both ends of the duct, one can numerically solve the $3(2N+1)$ coupled system of first-order nonlinear ordinary-differential equations.

Equation (3.72) and the impedance boundary conditions constitute a nonlinear two-point boundary-value problem, which Nayfeh and co-workers solved numerically for the case of a fundamental wave entering a duct at its left end when its right end is infinite in length. As the wave propagates towards the throat, higher harmonics as well as a steady-stream component are produced by the nonlinear effects. Owing to the axial gradients of the mean flow and the duct cross section, part of the incoming wave and the components produced by the nonlinearity is transmitted and the other part is reflected. Since the duct's entrance (i.e., $x = 0$) and exit (i.e., $x = L$) are away from the throat, linear theory is expected to hold there. By matching the linear and nonlinear expressions at $x = 0$ and $x = L$, Nayfeh, Shaker, and Kaiser (1978) determined the boundary conditions for the nonlinear problem in terms of the right- and left-running waves. The input signal consists of a right-running fundamental density wave whose strength is ρ_I. As a result of the reflections, a left-running fundamental wave having the strength $c_{r1}\rho_I$ exists at $x = 0$. Here, c_{r1} is an unknown reflection coefficient of the fundamental wave. Therefore, the boundary conditions for the fundamental wave at the entrance can be expressed as

$$\begin{aligned} u_{11}(0) &= (1 - c_{r1})\rho_I, \\ \rho_{11}(0) &= (1 + c_{r1})\rho_I, \\ T_{11}(0) &= (\gamma - 1)(1 + c_{r1})\rho_I \end{aligned} \tag{3.73}$$

For the higher harmonics, only left-running waves caused by the reflections exist at the duct entrance. Thus, the boundary conditions at $x = 0$ are expressed as the product of the strength ρ_I of the input disturbance and the unknown reflection coefficients c_{rn} according to

$$\begin{aligned} u_{1n}(0) &= -c_{rn}\rho_I, \\ \rho_{1n}(0) &= c_{rn}\rho_I, \\ T_{1n}(0) &= (\gamma - 1)\, c_{rn}\rho_I \end{aligned} \tag{3.74}$$

for $n \geq 2$. The infinite uniform termination is equivalent to having only right-running waves in the uniform section of the duct for $x > L$. This condition can be represented mathematically as

$$u_{1n} = \frac{\sqrt{T_{10}}}{\rho_{10}}\rho_{1n} \qquad \text{at } x = L \tag{3.75}$$

for $n \geq 1$, where T_{10} and ρ_{10} are determined from the mean flow values. Separating equation (3.75) into real and imaginary parts yields $2N$ boundary conditions.

Nayfeh, Shaker, and Kaiser developed a computer code to solve equations (3.72)–(3.75) using a fourth-order Runge-Kutta integration scheme. The magnitude of the input signal ρ_I and the reflection coefficients c_{rn} are specified and the program integrates through the duct to determine the conditions at $x = L$. A Newton-Raphson iteration scheme is used to increment the c_{rn} to satisfy the boundary conditions at $x = L$. Such an iteration on the c_{rn} is essential in the nonlinear case; further, these transmission and reflection characteristics are functions of the strength of the input signal. Using a central-difference scheme to solve the present problem would allow a direct imposition of the appropriate boundary conditions but would require an iteration on the nonlinear terms of equation (3.72). The central-difference method has not been implemented to date on this problem.

Nayfeh, Shaker, and Kaiser (1978) found that when the strength of the input signal and/or the throat Mach number is increased, the numerical procedure eventually produces either a strong oscillation or abrupt jumps in the acoustic signal near the duct throat. They found that refinement of the numerical step size does not produce any qualitative change in the instabilities. Moreover, an increase in the number of harmonics does not cure the problem. In fact, they calculated cases in which a doubling of the number of harmonics causes a discontinuity or instability to develop. Checking the determinant $|A^*|$ of the coefficient matrix A^*, they found that in every instability case $|A^*|$ approaches zero at an axial position that depends on the mean flow Mach number, the strength and frequency of the input signal, and the number of harmonics. If the mesh size is sufficiently coarse, the calculations may proceed through the duct without developing an instability. However, refining the step size would show the instability. This observation indicates that the difficulty results from a combination of the basic properties of the coefficient matrix and the numerical procedure. Since the linear problem is singular as $M_t \to 1$, the determinant of A^* is nonzero only as a result of deviations of M_t from unity and the correct nonlinear terms, which are small. If the correct reflection coefficients are used, then the numerical procedure can account for the nonlinear terms and hence, although $|A^*|$ is small, it is not zero. Deviations from the correct reflection coefficients might produce incorrect nonlinear effects that make $|A^*|$ zero or close to zero, and hence, instabilities would develop.

Nayfeh, Kelly, and Watson (1981) modified the computer code of Nayfeh, Shaker, and Kaiser (1978) by replacing (a) the Runge-Kutta procedure with the more accurate Adams-PECE variable step-variable order method, (b) the Newton-Raphson iteration procedure by the Levenberg-Marquardt algorithm, and (c) the matrix inversion of A^* with the singular value decomposition of A^* by which one can easily monitor the development of singularities. Thus, at each axial step, A^* is decomposed as

$$A^* = USV^T \tag{3.76}$$

where U and V are orthogonal matrices and S is a diagonal matrix that shows if A^* is singular. If one or more elements of S is zero, then A^* is singular. Substituting equation (3.76) into equation (3.72) yields

$$USV^T \frac{dy^*}{dx} = B^* \tag{3.77}$$

which implies that

$$SV^T \frac{dy^*}{dx} = U^T B^* = G \tag{3.78}$$

Since S is a diagonal matrix, equation (3.78) can be rewritten as

$$V^T \frac{dy^*}{dx} = \begin{bmatrix} \frac{G_1}{\sigma_1} \\ \frac{G_2}{\sigma_2} \\ \vdots \\ \frac{G_n}{\sigma_n} \end{bmatrix} = H \tag{3.79}$$

where the σ_n are the elements of S. The vector H determines whether any of the derivatives is singular. There is no need to solve for dy^*/dx to realize the singular behavior because V is orthogonal. Premultiplying H by V might shift the singularity to another component but it will not eliminate the singularity from equation (3.72). Consequently, one needs to determine the c_{rn} in such a way that none of the elements of H is singular and the boundary conditions at $x = L$ are satisfied. This numerical procedure was applied by Nayfeh, Kelly, and Watson (1981) to examine acoustic propagation through a converging-diverging duct. Its radius is given by $R = 1 - 0.1(1 - \cos \frac{2\pi x}{L})$ and it connects to straight sections at $x = 0$ and $x = L$.

As a first step in the numerical procedure, one needs to determine the range of reflection coefficients that keep the elements of H and hence equation (3.72) regular. Nayfeh, Kelly, and Watson (1981) chose to classify the problem as singular if any of the elements of H exceeded 100. One can use other tolerances. To begin with, they considered only the reflection coefficient c_{r1} and assumed all other c_{rn} to be zero. By varying c_{r1}, they determined the region of admissible reflection coefficients shown in Figure 15. Increasing the number of c_{rn} indicates that the admissible region did not shrink appreciably. Figure 16 shows the development of the higher harmonics and the rapid increase in the intensity of the disturbance near the throat. The higher harmonics do not become significant until the throat is approached. There is a reduction in the amplitude of the fundamental at the exit; this is the result of the transfer of energy from the fundamental to the higher harmonics caused by the nonlinearity. Figure 17 illustrates the intensification of the fundamental at the throat as the source strength ρ_I or mean flow Mach number increases. Beyond the last point in each figure, the program failed to converge. Figure 18 shows the reduction of the intensity of the fundamental at the exit as the source strength or mean flow Mach number increases.

The present numerical procedure is limited to the case of continuous solutions. For moderate values of the source strength ρ_I and moderate values of $\epsilon = 1 - M_t$, Nayfeh, Kelly, and Watson were able to determine reflection coefficients that simultaneously avoid the singularity at the throat and satisfy the exit boundary conditions. However, as ρ_I is increased and/or ϵ is decreased, their computer program could not avoid the singularity indicating that the problem does not possess a continuous solution and a shock needs to be introduced. Figure 19 shows the variation of the Mach number with time. These curves are periodic in time with period 2π. Each

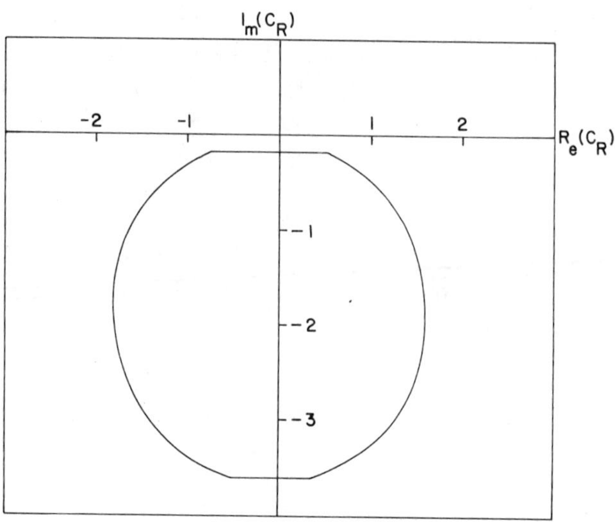

Figure 15. Region of admissible reflection coefficients using two harmonics and one reflection coefficient: $\omega = 1$, $\rho_I = .005$, $M_0 = -.4$.

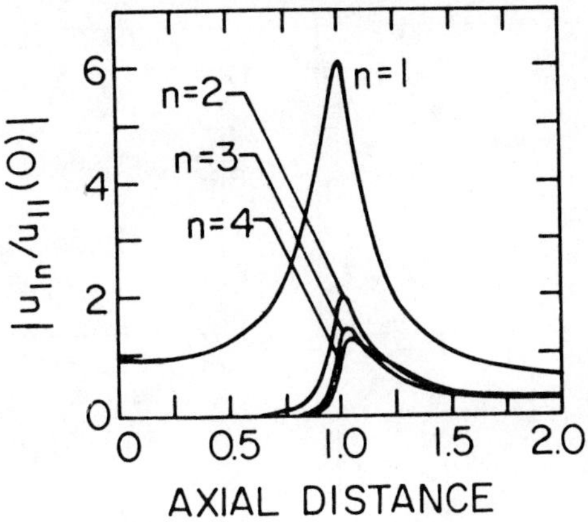

Figure 16. Development of harmonic amplitudes: $\omega = 1$, $M_0 = -0.4$, $\rho_I = 0.0045$, $C_{r1} = 0.057 - 0.155i$, $C_{r2} = 0.007 - 0.012i$, $C_{r3} = 0.003$, $C_{r4} = 0.001 + 0.002i$.

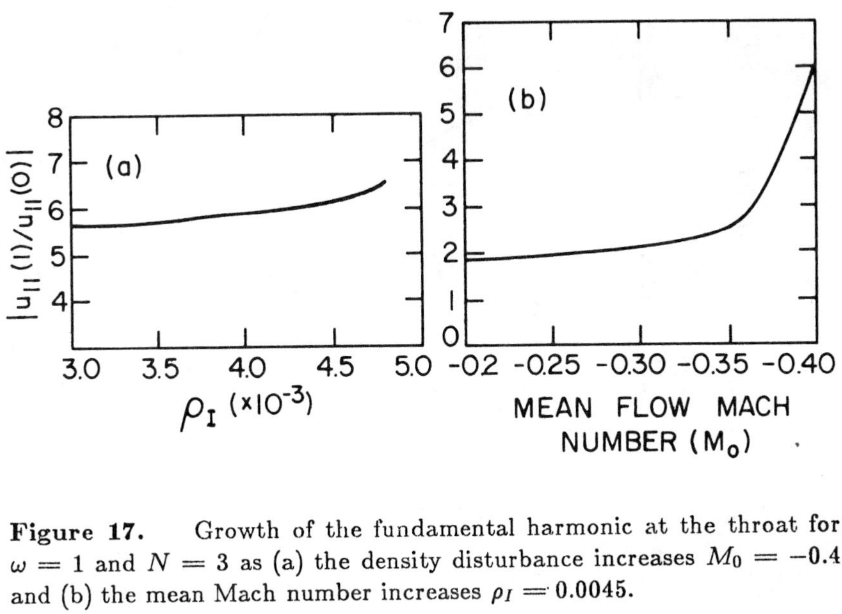

Figure 17. Growth of the fundamental harmonic at the throat for $\omega = 1$ and $N = 3$ as (a) the density disturbance increases $M_0 = -0.4$ and (b) the mean Mach number increases $\rho_I = 0.0045$.

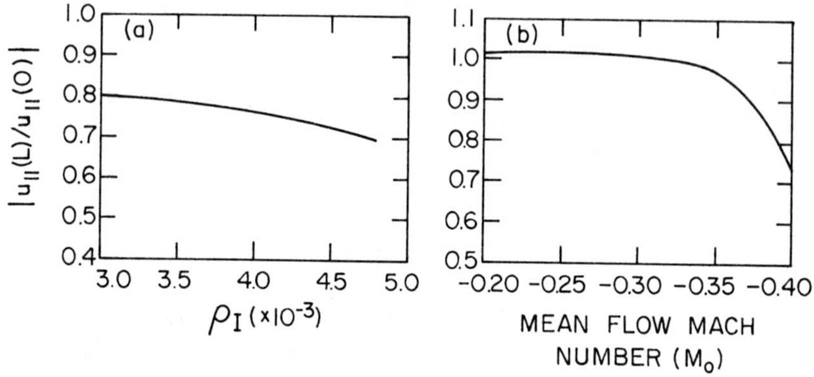

Figure 18. Reduction of the fundamental harmonic at the exit for $\omega = 1$ and $N = 3$ as (a) the density disturbance increases $M_0 = -0.4$ and (b) the mean flow Mach number increases $\rho_I = 0.0045$.

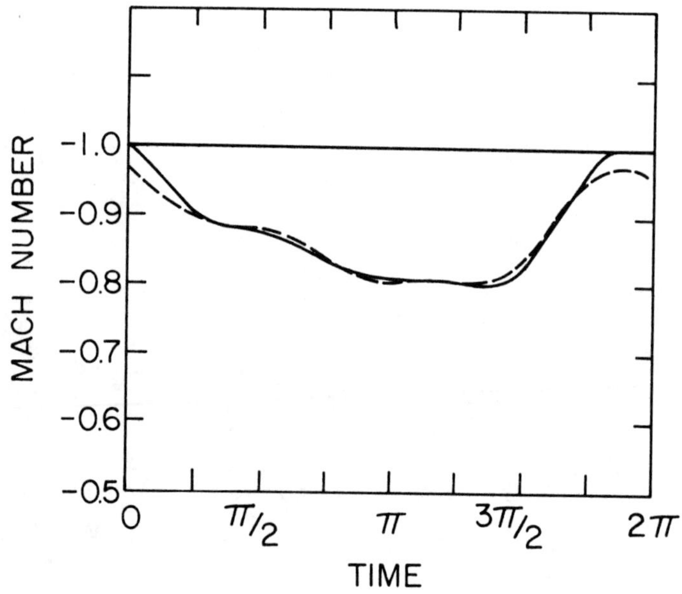

Figure 19. Comparison of Mach number variation between a singular case and a non-singular case, $\omega = 1$, $N = 3$, $M_0 = -0.4$, ——, $\rho_I = 0.0049$; - - - -, $\rho_I = 0.0045$.

curve is calculated by using three harmonics, the acoustic streaming, and the mean flow. For the case, $\rho_I = 0.0049$, a singularity appears just downstream of the throat and this is where the Mach number is calculated. As is seen, the Mach number exceeds unity at parts of the cycle, which is a necessary condition for a shock to appear. In the case $\rho_I = 0.0045$, no singularity is encountered and the iteration scheme converges. The time history of the Mach number is computed at the throat; throughout its time variation it is subsonic. Thus, when $\rho_I \geq 0.0049$, a shock must be introduced and fitted to continuous solutions that satisfy the end boundary conditions. The shock fitting procedure is still in the development stage.

4. Concluding Remarks

The linear three-dimensional acoustic problem is successfully treated by using the wave envelope technique. It is applicable to hard-walled as well as lined ducts, large as well as small axial gradients, and large as well as small transverse gradients. Nonlinear effects are successfully treated in the case of quasi-one-dimensional flows by using either spectral methods or the method of matched asymptotic expansions as long as shock waves do not form. A shock fitting technique needs to be developed to treat cases of strong source strengths and high mean flow Mach numbers. The wave envelope technique needs to be combined with either spectral methods or the method of matched asymptotic expansions to treat the three-dimensional nonlinear problem.*

Appendix

If y is the column vector of the unknowns $\overline{u}_{1n}, \ldots, u_{1n}, \overline{\rho}_{1n}, \ldots,$

$\rho_{1n}, \overline{T}_{1n}, \ldots, T_{1n}$, the coefficients of equation (3.68) take the following forms. The matrix $A(x,y)$ is

$$A(x,y) = \begin{bmatrix} \alpha_{11} & \alpha_{12} & \alpha_{13} \\ \alpha_{21} & \alpha_{22} & \alpha_{23} \\ \alpha_{31} & \alpha_{32} & \alpha_{33} \end{bmatrix},$$

*This work was supported by the Fluid Dynamics Program of the Office of Naval Research.

where the α_{ij} are $(2N+1)\times(2N+1)$ submatrices defined by

$$\alpha_{nm} = \begin{bmatrix} E_{nm}(0) & \overline{E}_{nm}(1) & \overline{E}_{nm}(2) & \cdots \\ E_{nm}(1) & E_{nm}(0) & \overline{E}_{nm}(1) & \cdots \\ E_{nm}(2) & E_{nm}(1) & E_{nm}(0) & \cdots \\ \vdots & \vdots & \vdots & \end{bmatrix},$$

in which

$$E_{11}(0) = \rho_{10}, \qquad E_{12}(0) = u_{10}, \qquad E_{13}(0) = 0,$$
$$E_{11}(n) = \rho_{1n}, \qquad E_{12}(n) = u_{1n}, \qquad E_{13}(n) = 0,$$

$$E_{21}(0) = F_1(0), \qquad E_{22}(0) = T_{10}/\gamma, \qquad E_{23}(0) = \rho_{10}/\gamma,$$
$$E_{21}(n) = F_1(n), \qquad E_{22}(n) = T_{1n}/\gamma, \qquad E_{23}(n) = \rho_{1n}/\gamma,$$

$$E_{31}(0) = 0, \qquad E_{32}(0) = (1-\gamma)\,F_2(0), \qquad E_{33}(0) = E_{21}(0),$$
$$E_{31}(n) = 0, \qquad E_{32}(n) = (1-\gamma)\,F_2(n), \qquad E_{33}(n) = E_{21}(n),$$

$$F_1(0) = \sum_{j=-N}^{N} u_{1j}\bar{\rho}_{1j}, \quad \text{where } (\ )_{1,-n} \equiv (^{-})_{1,n},$$
$$F_1(1) = \sum_{j=-N}^{N-1} \bar{\rho}_{1j} u_{1j+1},$$
$$F_1(2) = \rho_{11}u_{11} + \sum_{j=-N}^{N-2} \bar{\rho}_{1j} u_{1j+2},$$

$F_2(n)$ has the same form as $F_1(n)$ but with ρ replaced by T.

The vector $B(x, y)$ is

$$B(x,y) = \begin{bmatrix} B_1(n) \\ B_2(n) \\ B_3(n) \end{bmatrix}, \qquad n = -N, \ldots, 0, \ldots, N,$$

where

$$B_1(n) = -[in\omega\rho_{1n} + F_1(n)\,A_s],$$
$$B_2(n) = -F_5(n),$$
$$B_3(n) = -[F_3(n) + (1-\gamma)\,F_4(n)],$$

in which

$$A_s = \frac{1}{A}\frac{dA}{dx},$$

$$F_3(0) = \sum_{j=-N}^{N} \bar{\rho}_{1j} i\omega j T_{1j},$$

$$F_3(1) = \sum_{j=-N}^{N-1} \bar{\rho}_{1j} i\omega (j+1)\, T_{1j+1}, \qquad \text{etc.}$$

$F_4(n)$ has the same form as $F_3(n)$ but with ρ, T being replaced by T, ρ, respectively, $F_5(n)$ has the same form as $F_3(n)$ with T being replaced by u.

References

[1] Abbott, J. M. "Aeroacoustic Performance of Scale Model Sonic Inlets—Takeoff/Air Approach Noise Reduction," *AIAA Paper No. 74-202,* 1974.

[2] Abrahamson, A. L. "A Finite Element Model of Sound Propagation in a Nonuniform Circular Duct Containing Compressible Flow," *AIAA Paper 77-1301,* 1977.

[3] ————————. "Acoustic Duct Liner Optimization Using Finite Elements," *AIAA Paper 79-0662,* 1979.

[4] Anderson, A. O. "An Acoustic Evaluation on the Effect of Choking Using a Modal Supersonic Inlet," *Boeing Document D6A-10378-1,* 1966.

[5] Anderson, R., et al. "Boeing/Aeritalia Sonic Inlet Feasibility Study," *Boeing Document D6-40208,* 1972.

[6] Astley, R. J., Walkington, N. J., and Eversman, W. "The Application of Finite Element Techniques to Acoustic Transmission in Lined Ducts With Flow," *AIAA Paper 79-0660,* 1979.

[7] ——————. "Transmission in Flow Ducts With Peripherally Varying Linings," *AIAA Paper 80-1015,* 1980.

[8] Benzakein, M. J., Kazin, S. B., and Savell, C. T. "Multiple Pure-Tone Noise Generation and Control," *AIAA Paper 73-1021,* 1973.

[9] Callegari, A. J. and Myers, M. K. "Nonlinear Effects on Sound in Nearly Sonic Duct Flows," *AIAA Paper 77-1296,* 1977.

[10] ——————. "Sound Transmission in Ducts Containing Nearly Choked Flows," *AIAA Paper 79-0623,* 1979.

[11] Cawthorne, J. M., Morris, G. J., and Hayes, C. "Measurement of Performance, Inlet Flow Characteristics, and Radiated Noise for a Turbojet Engine Having Choked Inlet Flow," *NASA TN D-3929,* 1967.

[12] Chestnutt, D. "Noise Reduction by Means of Inlet-Guide-Vane Choking in an Axial-Flow Compressor," *NASA TN D-4682,* 1968.

[13] Chestnutt, D. and Clark, L. R. "Noise Reduction by Means of Variable-Geometry Inlet Guide Vanes in a Cascade Apparatus," *NASA TM X-2392,* 1971.

[14] Chestnutt, D. and Stewart, N. D. "Axial Flow Compressor Noise Reduction by Means of Inlet Guide Vane Choking," *NASA Langley Working Paper, LWP-473,* 1967.

[15] Crocco, L. and Cheng, S. I. *Theory of Combustion Instability in Liquid Propellant Rocket Motors,* London: Butterworths, Appendix B, 1956.

[16] Davis, S. S. and Johnson, M. L. "Propagation of Plane Waves in a Variable Area Duct Carrying a Compressible Subsonic Flow," presented at the 87th Meeting of the Acoustical Society of America, New York, 1974.

[17] Eisenberg, N. A. and Kao, T. W. "Propagation of Sound Through a Variable-Area Duct With a Steady Compressible Flow," *Journal of the Acoustical Society of America, 49* (1971), 169–75.

[18] Eversman, W., Astley, R. J., and Thanh, V. P. "Transmission in Nonuniform Ducts—A Comparative Evaluation of Finite Element and Weighted Residuals Computation Schemes," *AIAA Paper 77-1299,* 1977.

[19] Greatrex, F. B. "By-Pass Engine Noise," *Trans. SAE, 69* (1961), 312–24.

[20] Groth, H. W. "Sonic Inlet Noise Attenuation and Performance With a J-85 Turbojet Engine as a Noise Source," *AIAA Paper 74-91,* 1974.

[21] Hawking, D. L. and Lawson, M. V. "Theoretical Investigation of

Supersonic Rotor Noise," *Loughborough University Report TT7213,* 1972.

[22] Hersh, A. S. and Liu, C. Y. "Sound Propagation in Choked Ducts," *NASA CR-135123,* 1976.

[23] Higgins, C. C., Smith, J. N., and Wise, W. H. "Sonic Throat Inlets," *NASA SP-189* (1968), 197–215.

[24] Kelly, J. J., Nayfeh, A. H., and Watson, L. T. "Acoustic Propagation in Partially Choked Converging-Diverging Ducts," *Journal of Sound and Vibration, 81* (1982), 519–34.

[25] King, L. S. and Karamcheti, K. "Propagation of Plane Waves in the Flow Through a Variable Area Duct," *Progress in Astronautics and Aeronautics, Vol. 37, Aeroacoustics,* ed. H. T. Nagamatsu, Cambridge: MIT Press, 1975, 403–18.

[26] Klujber, F. "Results of an Experimental Program for the Development of Sonic Inlets for Turbofan Engines," *AIAA Paper 73-222,* 1973.

[27] Klujber, F., et al. "Investigation of Noise Suppression by Sonic Inlets for Turbofan Engines," *Boeing Document D6-40855, NASA CR-121126, CR-121127,* 1973.

[28] Koch, R. L., Ciskowski, T. M., and Garzon, J. R. "Turbofan Noise Reduction Using a New Sonic Inlet," *AIAA Paper 74-1098,* 1974.

[29] Lumsdaine, E. "Development of a Sonic Inlet for Jet Aircraft," *Internoise 1972 Proceedings* (1972), 501–06.

[30] Majjigi, R. K., Sigman, R. K., and Zinn, B. T. "Wave Propagation in Ducts Using the Finite Element Method," *AIAA Paper 79-0659,* 1979.

[31] Miller, B. A. "Experimentally Determined Aeroacoustic Performance and Control of Several Sonic Inlets," *AIAA Paper 75-1184,* 1975.

[32] Miller, B. A. and Abbott, J. M. "Aerodynamic and Acoustic Performance of Two Choked Flow Inlets Under Static Condition," *NASA-TM-X-2629-E7008,* 1972.

[33] Myers, M. K. and Callegari, A. J. "On the Singular Behavior of Linear Acoustic Theory in Near-Sonic Duct Flows," *Journal of Sound and Vibration, 51* (1977), 517–31.

[34] ———. "Transmission of Sound Through High-Subsonic Flows in Non-Uniform Ducts," *AIAA Paper 78-1155,* 1978.

[35] Nayfeh, A. H. "Effect of the Acoustic Boundary Layer on the Wave Propagation in Ducts," *Journal of the Acoustical Society of America,* (1973a), 1737–42.

[36] ———. *Perturbation Methods,* New York: Wiley-Interscience, Chapter 6, 1973b.

[37] ———. "Sound Propagation Through Nonuniform Ducts, *Proceedings of the Society of Engineering Science at NASA Langley Research Center,* 1976.

[38] ———. "Intensification and Refraction of Acoustical Signals in Partially-Choked Converging Ducts," *Journal of Sound and Vibration, 73* (1980), 519–31.

[39] ———. *Introduction to Perturbation Techniques,* New York: Wiley-Interscience, Chapter 15, 1981.

[40] Nayfeh, A. H. and Kaiser, J. E. "Effect of a Compressible Sheared Mean Flow on Sound Transmission Through Variable-Area Plane Ducts," *AIAA Paper 75-128,* 1975.

[41] Nayfeh, A. H., et al. "A Comparison of Experiment and Theory for Sound Propagation in Variable-Area Ducts," *Journal of Sound and Vibration, 71* (1980), 241–59.

[42] Nayfeh, A. H., Kaiser, J. E., and Shaker, B. S. "Effect of Mean-Velocity Profile Shapes on Sound Transmission Through Two-Dimensional Ducts," *Journal of Sound and Vibration, 34* (1974), 413–23.

[43] ———. "A Wave-Envelope Analysis of Sound Propagation in Nonuniform Circular Ducts With Compressible Mean Flows," *NASA Contractor Report 3109,* March 1979.

[44] Nayfeh, A. H., Kaiser, J. E., and Telionis, D. P. "Acoustics of Aircraft Engine-Duct Systems, *AIAA Journal, 13* (1975a), 130–153.

[45] ———. "Transmission of Sound Through Annular Ducts of Varying Cross Sections," *AIAA Journal, 13* (1975b), 60–65.

[46] Nayfeh, A. H., Kelly, J. J., and Watson, L. T. "Nonlinear Propagation in Near Sonic Flows," *Journal of Sound and Vibration, 75* (1981), 359–70.

[47] ———. "Propagation of Spinning Acoustic Modes in Partially Choked Converging Ducts," *Journal of the Acoustical Society of America, 71* (1982), 796–802.

[48] Nayfeh, A. H., Shaker, B. S., and Kaiser, J. E. "Computation of Nonlinear One-Dimensional Waves in Near-Sonic Flows," *AIAA Journal, 16* (1978), 1154–59.

[49] ———. "Transmission of Sound Through Nonuniform Circular Ducts With Compressible Mean Flows," *AIAA Journal, 18* (1980), 515–25.

[50] Nayfeh, A. H., Telionis, D. P., and Lekoudis, S. G. "Acoustic Propagation in Ducts With Varying Cross Sections and Sheared Mean Flow," *Progress in Astronautics and Aeronautics, Vol. 37, Aeroacoustics,* ed. H. T. Nagamatsu, Cambridge: MIT Press, 1975, 333–51.

[51] Pestorius, F. M. and Blackstock, D. T. "Nonlinear Distortion in the Propagation of Intense Acoustic Noise," *Interagency Symposium on University Research in Transportation Noise Proceedings, Vol. II* (1973), 565–77.

[52] Savkar, S. D. and Kazin, S. B. "Some Aspects of Fan Noise Suppression Using High Mach Number Inlets," *AIAA Paper 74-554,* 1974.

[53] Sawhill, R. H. "Investigation of Inlet Airflow Choking as a Mean of Supressing Compressor Noise—5' Inlet Model," *Boeing Documents D6A-10155-1,* 1966.

[54] Schlichting, H. *Boundary-Layer Theory,* New York: McGraw-Hill, 1968, pp. 66 and 254.

[55] Sigman, R. K., Majjigi, R. K., and Zinn, B. T. "Determination of Turbofan Inlet Acoustics Using Finite Elements," *AIAA Journal, 16* (1978), 1139–45.

[56] Sobel, J. A. and Welliver, A. D. "Sonic Block Silencing for Axial and Screw-Type Compressor," *Noise Control, 7* (1961), 9–11.

[57] Uenishi, K. and Myers, M. K. "An Analysis of the Two-Dimensional Acoustic Field in a Nonuniform Duct Carrying Compressible Flow," *AIAA Paper 83-0669,* 1983.

[58] Vaidya, P. G. and Dean, P. D. "State of the Art of Duct Acoustics," *AIAA Paper 77-1279,* 1977.